ENCYCLOPEDIA OF PHYSICS

EDITED BY

S. FLÜGGE

VOLUME XLIX/1

GEOPHYSICS III

PART I

GROUP EDITOR

J. BARTELS †

WITH 246 FIGURES

SPRINGER-VERLAG
BERLIN · HEIDELBERG · NEW YORK
1966

HANDBUCH DER PHYSIK

HERAUSGEGEBEN VON

S. FLÜGGE

BAND XLIX/1

GEOPHYSIK III

TEIL I

REDAKTION

J. BARTELS †

MIT 246 FIGUREN

SPRINGER-VERLAG
BERLIN · HEIDELBERG · NEW YORK
1966

ISBN-13: 978-3-642-46040-1 e-ISBN-13: 978-3-642-46038-8
DOI: 10.1007/978-3-642-46038-8

Softcover reprint of the hardcover 1st edition 1966

Library of Congress Catalog Card Number A 56-2942

Titel Nr. 5792

JULIUS BARTELS 1899 bis 1964

In memoriam Julius Bartels.

Herausgeber und Verlag trauern um JULIUS BARTELS, der uns am 6. März 1964 durch einen unerwarteten Tod entrissen wurde. Von den ersten Anfängen dieses Handbuches an hat er die Gruppe der geophysikalischen Bände redigiert, deren erste beiden, die Nummern 47 und 48 der gesamten Reihe, schon 1956 und 1957 erscheinen konnten. Das Erscheinen des dritten und letzten Geophysikbandes, Nummer 49 des Handbuches, dessen Vorbereitung damals bereits weit vorgeschritten war, mußte sich unbestimmt verzögern infolge des unvorhersehbaren Aufschwunges, den die mit dem Internationalen Geophysikalischen Jahr und den Anfängen der Raumforschung einsetzende stürmische Entwicklung den für diesen Band reservierten Teilen der Geophysik brachte. So ist es BARTELS selbst, der noch mit großer Aktivität an dieser Entwicklung teilgenommen hat, nicht mehr vergönnt gewesen, dies Werk zu vollenden; in rastloser Tätigkeit bis zuletzt hat er sich frühzeitig aufgerieben, den Überlebenden den Abschluß des Werkes als Vermächtnis hinterlassend.

BARTELS selbst hatte bereits vorgesehen, in Anbetracht der gegenüber der ursprünglichen Planung so völlig veränderten Lage den fehlenden Band 49 in mindestens zwei Teilbände aufzuspalten. Der Wunsch, in einem sich so schnell entwickelnden Gebiet die Beiträge möglichst bald dem Leser vorlegen zu können, hat uns bewogen, diese Aufteilung noch etwas weiterzuführen, als es ursprünglich beabsichtigt war, und bei der Aufteilung dem baldigen Erscheinen mehr Vorrang vor der systematischen Anordnung der Beiträge zu geben, als es in diesem Handbuch sonst üblich ist. Demgemäß legen wir hier einen ersten, relativ schmalen Teilband vor, der ebenso wie der ihm bald folgende zweite Teilband, der der Ionosphäre gewidmet sein wird, noch weitgehend von BARTELS selbst vorbereitet worden ist. Es ist beabsichtigt den weiten Stoffbereich, zu dem BARTELS in diesem Rahmen nicht mehr entscheidend beitragen konnte, in den Händen des neuen Bandherausgebers, Professor Dr. KARL RAWER, in zwei weiteren Teilbänden völlig neu aufzubauen.

S. FLÜGGE

Contents.

The Aurora.

By

S.-I. AKASOFU, S. CHAPMAN and A. B. MEINEL.

With 114 Figures.

1. Introduction. The aurora is a distinct luminosity, most often visible only from places in high latitudes. It is emitted light proceeding from definite regions in the upper atmosphere. The auroral light is *initiated* by the passage, through the upper atmosphere, of energetic charged particles coming from outside. The particles include both electrons and protons. Most of the visible auroral light is due to the ionization or excitation of the upper atmospheric constituents by incoming electrons and also by secondary electrons produced thereby. Protons, after having a number of ionizing and charge-exchange collisions, are finally slowed down to form neutral hydrogen atoms; when these are in an excited state during their passage down, they emit some of the hydrogen lines such as H_α, H_β (Sect. 52 and 64).

The aurora is a complex phenomenon whose description involves many characteristics, some of them intrinsic and others relative to other phenomena or circumstances. Some of its chief intrinsic characteristics are geometrical — the form and location and space distribution of the region of emission of the light.

The aurora is also a changing phenomenon. The time element in the description may be partly supplied by specifying the changes and movement of form of the auroral emitting volume, and of the variation of luminosity within it. A further important (relative) element in the description of an aurora is its close association with various magnetic activities and also with some ionospheric phenomena. The term "aurora" is now extended to a volume of subvisual light and also to atmospheric light artificially produced by nuclear explosion processes.

The varied spatial and temporal characteristics of auroras, together with some closely related geophysical phenomena, constitute a body of data that may as a whole be comprised under the name *auroral morphology*.

The aurora has much to tell us about the physical and chemical properties of the upper atmosphere. The aurora presents problems relating to its immediate production in the upper atmosphere. At present, the energy spectrum of the incoming particles above the auroral ionosphere is known only incompletely, so that the relation between the various optical processes and the incoming flux has not yet been fully explored in a quantitative way.

Still more difficult problems are concerned with the more distant causes. The sun is regarded as the ultimate source of the energy, whatever its nature may be; it may be solar particles themselves, of different types. But how such energy is fed into the magnetospheric region to produce an enhanced injection of energetic particles into the auroral ionosphere, is still not clear . The answers to this question will be closely related to the properties of the solar plasma impinging upon the magnetosphere and to the interaction between them. The nature of the radiation belts that encircle the earth at a few radii from the earth and beyond, is undoubtedly important in understanding the origin of the aurora.

I. The geographical distribution of the aurora.

The aurora occurs commonly only in high latitudes, both northern and southern. The northern aurora is called the *aurora borealis* or *northern lights*, and the southern one is called the *aurora australis* or *southern lights*; together they are the *aurora polaris*. But at times the aurora becomes visible from places far beyond its usual limits. It is seen from middle and low latitudes, sometimes even from the tropics; such exceptional ones may be called *aurora tropicalis*.

2. The auroral division of the globe. The frequency with which auroras are visible is roughly linked with geomagnetic latitude[1]. In connection with the frequency of auroral appearance, it is convenient to divide the earth's surface into five parts (CHAPMAN 1953 [*14*]). The regions of the earth within the circles of geomagnetic latitude 60° north or south are called the *auroral regions*. Each includes 6.7 per cent of the earth's surface. In each auroral region (N and S) the frequency is greatest along the *auroral zone*. This zone is a narrow band, about 3° wide, lying about equally on either side of an oval line that coincides approximately with the circle of 67° geomagnetic latitude. The region enclosed by the zone is called the *auroral cap* (N or S).

The parts between 45° and 60° geomagnetic latitude (N and S) are called the *subauroral zones*. Each of these belts includes about 7.9 per cent of the earth's surface. The remaining region of the earth lying between the circles of geomagnetic latitude 45° north and south is called the *minauroral belt*. This region includes 70.7 per cent of the earth's surface. The name minauroral is given to this region because it is the part of the globe where the frequency of auroral occurrence is minimal.

3. The auroral zones. *a) The northern auroral zone.* As one goes poleward from temperature latitudes, the aurora is seen more often. But LOOMIS (1860, [*15*]), after an extensive collection of records of its appearance, found that this northward increase of frequency of visibility does not continue to the pole. His data enabled him in 1860 to envisage an oval band in which the aurora is most frequently visible. His diagram is reproduced here as Fig. 1. Further, he noticed that the center of the oval is not at the geographic pole, but at a point about 10° away towards NW Greenland.

FRITZ (1873, 1881 [*2*]), using much more data covering the period from 503 B.C. to A.D. 1872, confirmed LOOMIS' conclusions and constructed his well-known map of *isochasms*, the lines of equal average annual frequency of auroral visibility expressed by M nights per year. The maximum frequency of auroral visibility was found to lie approximately along LOOMIS' oval band. The isochasms are approximately parallel to this band, and therefore have the same center. This is neither the geographical nor the magnetic dip pole; it is at the *geomagnetic pole*, which at the epoch 1945 (VESTINE, LANGE, LAPORTE and SCOTT 1947 [*16*]) had the geographic coordinates 78°.6 N, 70°.1 W. Geomagnetic coordinates (latitude, longitude) are conventionally referred to a pole at 78°.5 N, 69° W (see footnote to Sect. 2).

VESTINE (1944 [*17*]) revised FRITZ's isochasmic diagram, with the aid of additional data covering more than a century, including the two International

[1] Geomagnetic latitude (abbrev. gm. lat.) is used in analogy to geographical latitude and is associated with the geomagnetic axis, conventionally defined by one of its poles, at 78°.5 N, 69° W. One of the authors (S. CHAPMAN: 1963 [*14a*]) has proposed to replace the term "geomagnetic coordinates" by the more concise expression "dipole coordinates". See the map of the northern hemisphere with geomagnetic (or dipole) coordinates (latitude and longitude) in Fig. 4.

Polar Years. His changes of their form and location were slight, but he added more isochasms within the auroral cap (Fig. 2). He also improved the numeration of the isochasms, expressing the frequency by the percentage of nights of visible aurora among all possible nights (that is, excluding nights on which the sky is fully obscured by clouds, and periods as in summer — in high latitudes — when the sky is never dark).

The isochasmic diagrams represent a *relative* characteristic of the aurora, since they refer to its visibility to *observers*, perhaps near the horizon, far away. A more

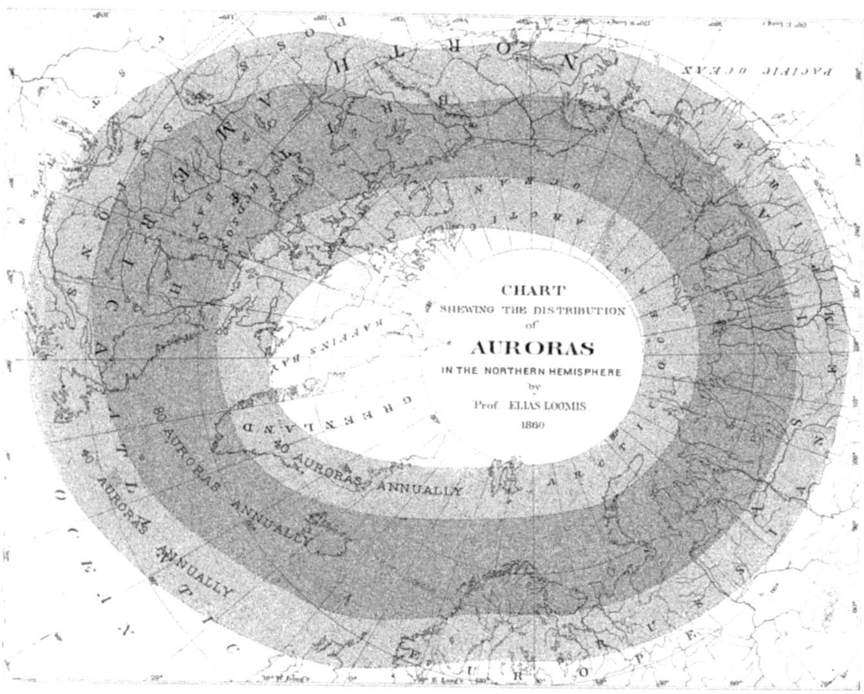

Fig. 1. The geographic distribution of the frequency of auroras in the northern hemisphere. (Loomis, E., 1860 [*15*].)

intrinsic diagram of an associated kind, showing the frequency of *overhead* location of the aurora at each place, is called *isoauroral*, the lines being *isoaurores* (Chapman 1953 [*17*]). Feldstein and Solomatina (1961 [*18*]), using the IGY (1957—1958) photographic auroral observations at 42 stations in the northern hemisphere, obtained the appearance probability of the overhead aurora (Fig. 3). Although their data were collected in a much shorter period than Vestine's, their curves of the maximum frequency agree fairly well and are substantially identical, except over northern Canada, particularly over the region of Hudson's Bay; see Fig. 4. Vestine's curve is located about 4° north of that given by Feldstein and Solomatina. Hultqvist (1958, 1959, 1961 [*19*] to [*21*]) and Vestine and Sibley (1959 [*22*]) support theoretically Feldstein's modification. This is further discussed in detail in Sect. 4.

Davis (1962 [*23*]) defined the term "*incidence*" signifying the total number of auroral borders crossing every 0.5° meridional segment. His analysis was based on data from five Alaskan all-sky camera stations between gm latitude 60° and

70° N (Fig. 5). According to this definition, the normalized curve of the incidence indicates that the center line of the auroral zone is around gm lat. 66°∼67° N over Alaska, with the half-width of 2°.

b) The southern auroral zone. Following earlier attempts by Davies (1933 [*24*]), and by White and Geddes (1939 [*25*]), an isochasmic diagram was constructed

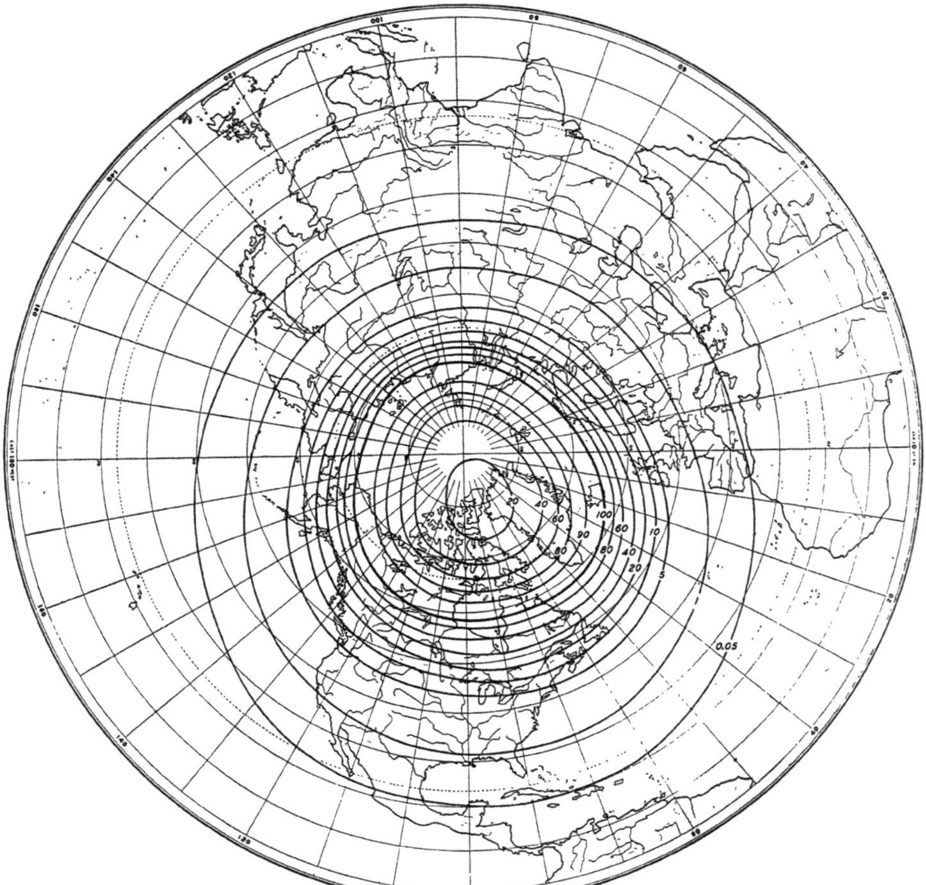

Fig. 2. The geographic distribution of the frequency of auroras in the northern hemisphere. (Vestine, E. H., 1944 [*17*].)

for the southern hemisphere by Vestine and Snyder (1945 [*26*]). It is based on far less material than that for the northern hemisphere, and contains fewer isochasms.

The IGY data, both photographic and visual, have been extensively analyzed to improve the frequency curves in the southern hemisphere by Gartlein, Nack and Sprague (1960 [*27*]), Feldstein (1960 [*28*]), and Bond and Jacka (1960 [*29*]). The curves of the maximum isoaurores obtained by them differ greatly, however, from each other; this is mainly because of the lack of data in the southern hemisphere. Fig. 6 shows the maximum isoaurores given by White and Geddes, Vestine and Snyder, Bond and Jacka, and Feldstein. The discrepancy between FELDSTEIN'S curve and that of BOND and JACKA is particularly serious around Macquarie Island, where FELDSTEIN'S maximum isoaurore is located about 6° south of that given by BOND and JACKA. Further extensive observation is necessary to determine the southern maximum isoaurore with the same accuracy as the northern one (see also SCHNEIDER (1961 [30])).

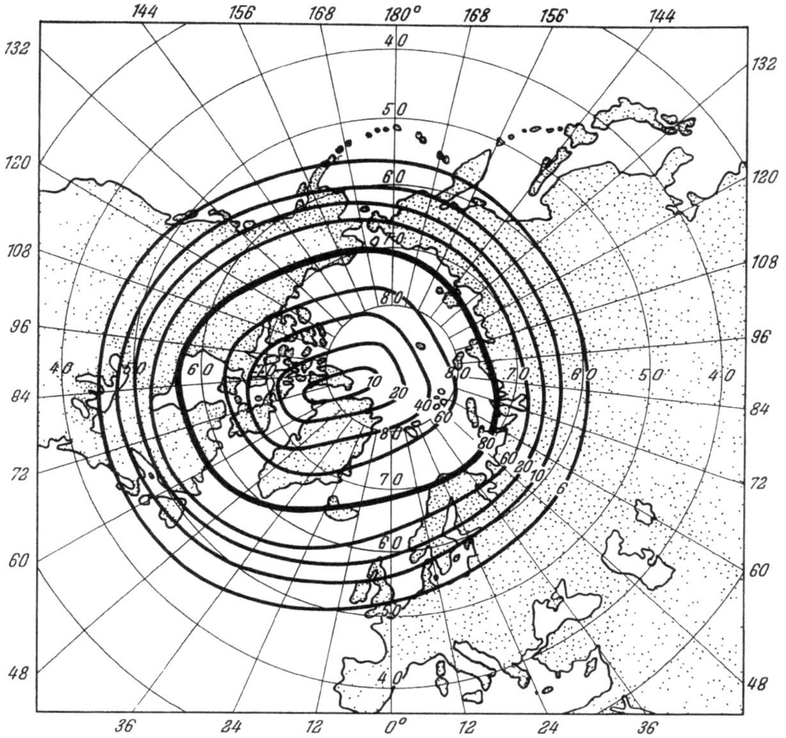

Fig. 3. Isoaurores in the northern hemisphere. (FELDSTEIN, Y. L., and E. K. SOLOMATINA, 1961 [18].)

4. The theoretical auroral zones. QUENBY and WEBBER (1959 [31]) and HULTQVIST (1959 [20], 1961 [21]) have shown *calculated* auroral zones, based on the assumption that the aurora is produced by energetic charged particles impinging on the auroral ionosphere, guided by the earth's magnetic field lines, from a circular source in the equatorial plane. Because at a distance of several earth radii the dipolar component of the earth's field is most important, so that circular symmetry could be reasonably assumed, HULTQVIST projected, along the geomagnetic field lines, a circle of radius 5.6 earth radii in the equatorial plane to the earth's surface.

In the northern hemisphere, the projection is oval, with the longest diameter approximately in the plane containing the 170° and 350° geomagnetic meridians. The curve agrees well with VESTINE'S maximum isochasm, except over the region of Hudson's Bay, Canada. It is found, however, that FELDSTEIN'S revised curve (Fig. 4) agrees with HULTQVIST'S curve in that region. Therefore, in the northern hemisphere the agreement between the observed and calculated zones is well established.

On the other hand, in the southern hemisphere, none of the observed isoaurores agrees with Hultqvist's with the accuracy shown in the northern hemisphere.

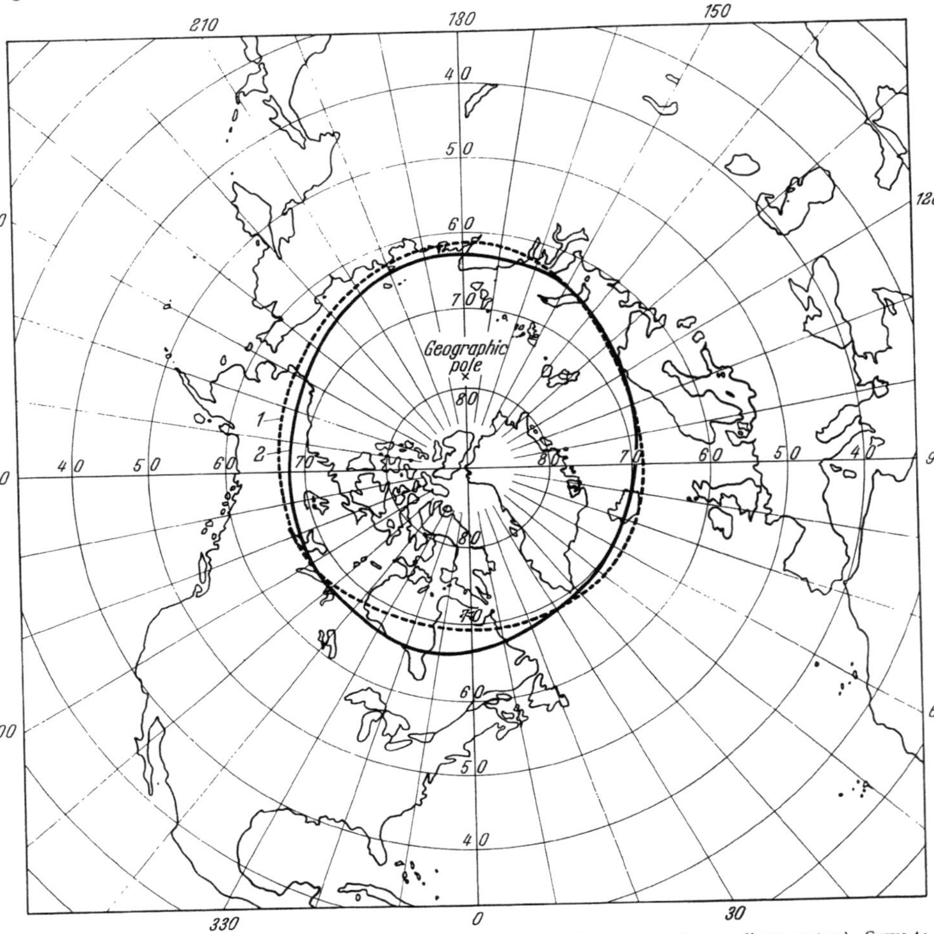

Fig. 4. Northern auroral zones determined from observations (on a map with geomagnetic co-ordinate system). Curve 1: Vestine (1944); Curve 2: Feldstein and Solomatina (1961). (Hultqvist, B., 1961 [21].)

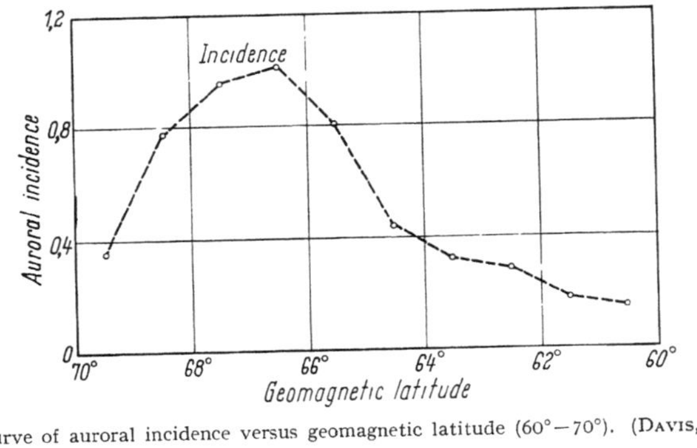

Fig. 5. Average curve of auroral incidence versus geomagnetic latitude (60°–70°). (Davis, T. N., 1962 [23].)

The harmonic coefficients of the geomagnetic field for epoch 1945, obtained by VESTINE, LANGE, LAPORTE and SCOTT [*16*], on which HULTQVIST's calculation was based, were obtained from much fewer data points there than those in the northern polar regions. However, HULTQVIST concluded that although this fact

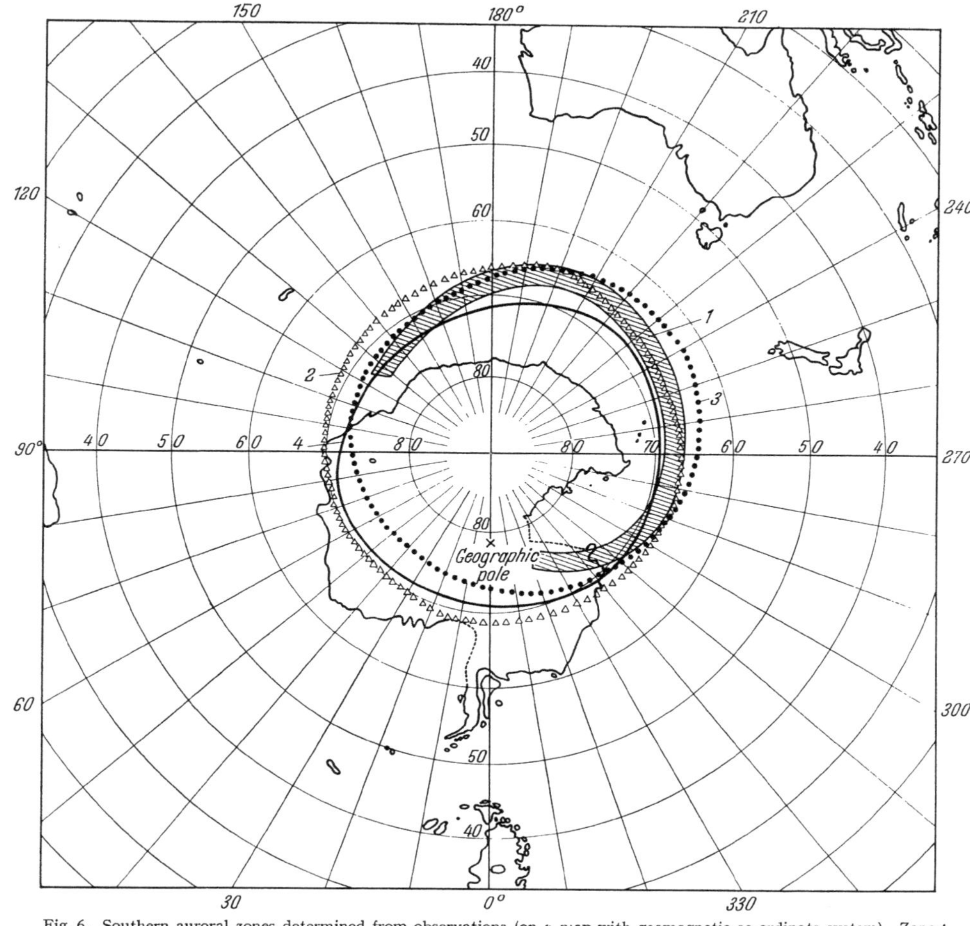

Fig. 6. Southern auroral zones determined from observations (on a map with geomagnetic co-ordinate system). Zone 1 is according to WHITE and GEDDES (7), zone 2 to VESTINE and SNYDER (8), zone 3 to BOND and JACKA (15), and zone 4 to FELDSTEIN (4). (HULTQVIST, B., 1961 [*21*].)

may partially account for the discrepancy, it seems unlikely that a difference as large as 6° in the regions to the south of New Zealand is primarily attributable to errors in the harmonic analysis of the geomagnetic field.

VESTINE and SIBLEY (1960 [*32*]) produced a calculated auroral zone on a somewhat different basis. It is known that for a spiralling charged particle in the earth's geomagnetic field the integral

$$I = \int (w_s/w)\, ds = \int \left(1 - \frac{B}{B_m}\right)^{\frac{1}{2}} ds \qquad (4.1)$$

is an invariant (for details, see Sect. 47). Here w is the speed of a charged particle, w_s its component along a geomagnetic field line, and ds is the field line-element;

B denotes the field intensity of the earth's field at a point on the field line and B_m the field intensity at the mirror point on the same field line.

If a group of energetic charged particles are released at a point in space (for example, by a nuclear explosion), they oscillate between the two hemispheres along the field lines and drift westwards or eastwards depending on the sign (positive or negative) of the charge (Sect. 47). They will eventually form a surface around the

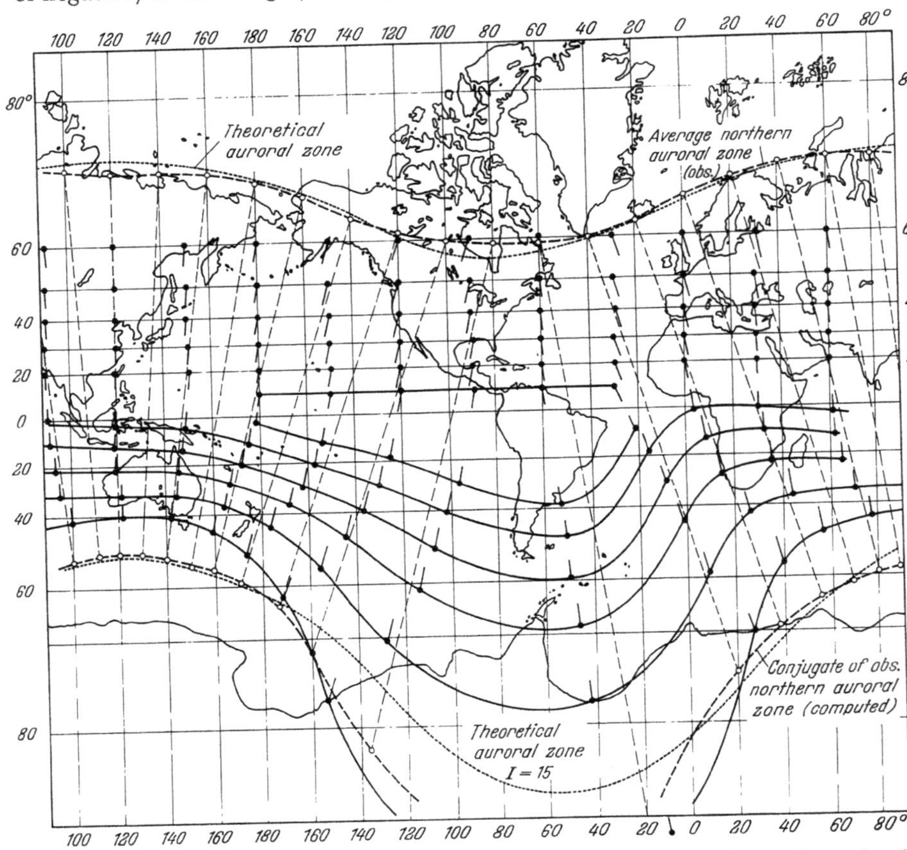

Fig. 7. Approximate intersections of lines of force of the geomagnetic field with the earth's surface, northern and southern hemispheres, based on first 48 Gauss coefficients. (VESTINE, E. H., and W. L. SIBLEY, 1959 [22].)

earth, and the surface thus defined is called a *magnetic surface*. The surface is made up by segments of the field lines with the same value of I; the field intensity B_m determines the "edge" of the surface at which they mirror.

VESTINE and SIBLEY found that choosing $I=15.7$ (in units of the earth radius) and $B_m=0.5$ gauss they can reproduce fairly well VESTINE's maximum isochasm in the northern hemisphere (Fig. 7). They confirmed HULTQVIST's conclusion that over Hudson's Bay VESTINE's isochasm is about 4° north of the calculated one. The agreement in the southern hemisphere is not so good as that in the northern hemisphere. In their calculation, VESTINE and SIBLEY used FINCH and LEATON's first 48 coefficients for the epoch of 1955.

The studies by HULTQVIST and by VESTINE and SIBLEY have made it possible to determine pairs of points, one in each hemisphere, which are linked by the lines of force. These points are called *conjugate points* (Fig. 7). There are several causes which are known to distort seriously the outer geomagnetic field; as yet their effects at the earth's surface are within the uncertainty of the spherical harmonic analysis. Therefore the actual location of the field lines in space is likely to

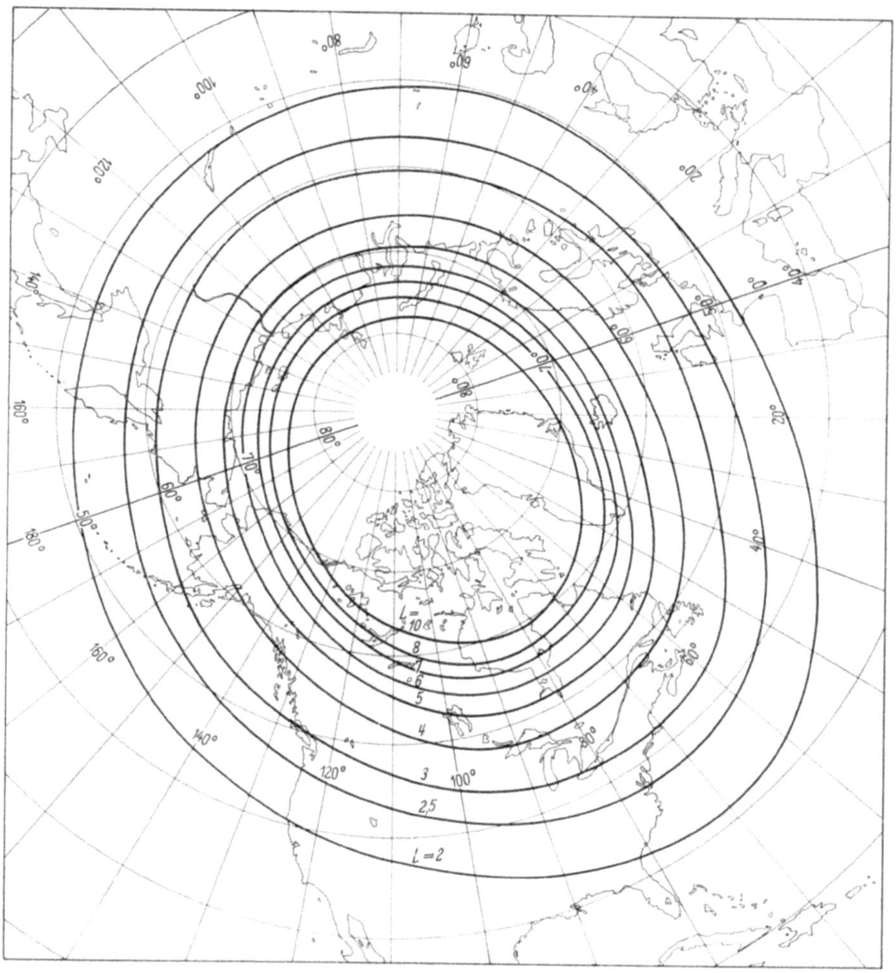

Fig. 8. McIlwain's L lines. (McIlwain 1961, private communication.)

differ from that calculated from the present spherical harmonic analysis. Nevertheless, a number of geophysical phenomena associated with energetic charged particles indicate a striking similarity in an area around conjugate points, suggesting that the areas are linked by the same bundle of field lines (see Sect. 31).

McIlwain (1961 [33]) proposed a new coordinate system for mapping the distribution of magnetically trapped particles. McIlwain showed that a magnetic surface can be specified with a reasonable accuracy, by a single parameter called L, which is expressed in units of the earth radius. If the earth's field were a pure dipole, the magnetic surfaces would be surfaces of revolution generated by the rotation of the lines of force around the dipole axis. In this special case the parameter L signifies the geocentric distance of the point at which the field line crosses the equator. The lines of intersection of magnetic surfaces with the earth's surface may be called L lines (Fig. 8). The L lines appear to be closely related to the isochasms or isoaurores. The isoaurore of maximum frequency agrees fairly well with the line $L=6.0$. (See also Stone (1963 [34]).)

The close resemblance between the maximum isochasms (or the maximum isoaurore) and the theoretical ones supports the conception now widely held, that auroral particles move on the magnetic surfaces.

5. The distribution of the aurora in the polar cap. *a) General.* Auroral observation, particularly for the auroras in the auroral cap, has been greatly advanced by the past three international efforts, the *International Polar Years I* (1882/83) and *II* (1932/33), and particularly by the *International Geophysical Year*. The large amount of auroral data collected during the IGY, photographic and visual, has greatly increased our knowledge of the aurora in such little inhabited regions. The coming IQSY (1964/65)[1] will add further information on the aurora during a period of low solar activity. Because studies during the IGY seem to confirm the existence and nature of the auroras that appear in the auroral cap only during magnetically quiet periods, the IQSY will be the most suitable period to investigate the nature of such auroras.

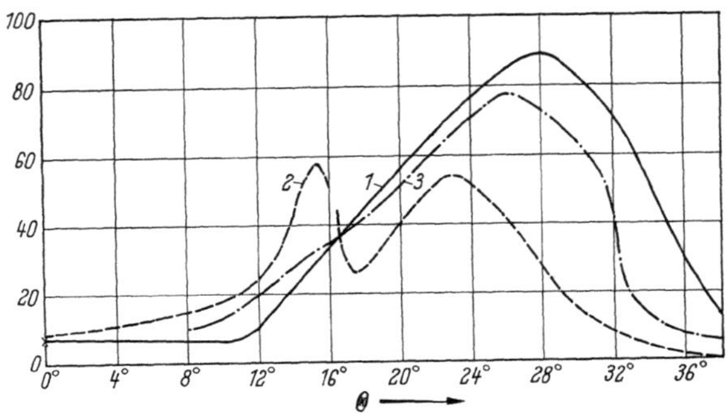

Fig. 9. Latitudinal distribution of aurora appearance frequency in the zenith in the northern hemisphere. *1* magneticallydisturbed days. *2* magnetically-quiet days. *3* all days. (Feldstein, Y. I., 1962 [*35*].)

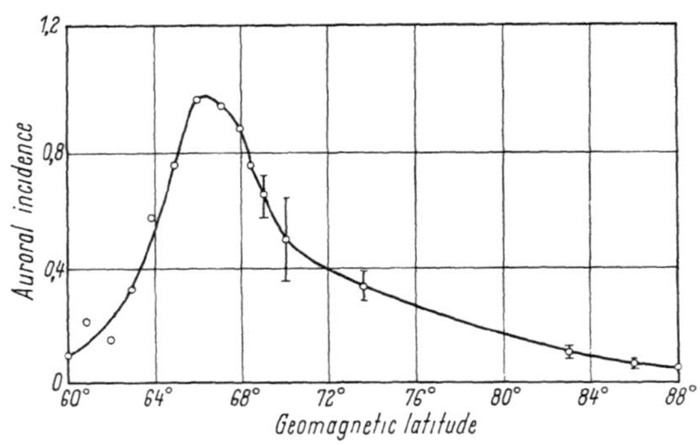

Fig. 10. The incidence of auroral forms at geomagnetic latitudes 60°–90°. (Davis, T. N., 1962 [*36*].)

Fig. 9 shows, for the northern hemisphere, the latitudinal distribution of the appearance frequency of overhead aurora for magnetically quiet days, disturbed days and for all days during the IGY (Feldstein 1962 [35]). For both magnetically disturbed days and all days, the appearance frequency seems to decrease monotonically towards the geomagnetic pole. Davis (1962 [36]) also made a similar study, in terms of the incidence, using the data collected from Thule (gm lat. 88.0° N), Alert (gm lat. 85.7° N), Resolute Bay (gm lat. 83.1° N), Baker Lake (gm lat. 73.8° N) and Churchill (gm lat. 68.8° N), together with five Alaskan stations (Fig. 10). His curve seems to show a much steeper decrease from the peak, as one moves inward from the auroral zone towards the pole, than Feldstein's.

[1] IQSY = International Years of the Quiet Sun.

b) Inner auroral zone. The diurnal occurrence frequency curve of the aurora, for stations well inside the auroral cap, shows two maxima, one around midnight and one in the early morning (for details see Sect. 19). LASSEN (1959 [*37*]), using his visual observations at Godhavn (gm lat. 79.9° N), together with some other information, inferred that there is an inner auroral zone located approximately 10° away from the geomagnetic pole. At Godhavn, the midnight maximum in the diurnal occurrence frequency curve is produced by brilliant auroras approaching the zenith from the south-east, whereas the second maximum is produced by diffuse bands with faint rays, appearing sporadically at a high elevation, around 6 hr. local time.

He showed that the places at which the morning auroras are observed near the zenith, when plotted on a polar map, lie along one of HULTQVIST's oval curves, a circle in the equatorial plane projected along the field line on the earth's surface. He suggested thus that such an oval curve indicates the existence of an inner auroral zone. FELDSTEIN and SOLOMATINA (1961 [*18*]) later obtained isoauroras for such morning auroras (Fig. 11). NIKOLSKY (1961 [*38*]) states that the second auroral zone inferred by LASSEN agrees with the position of this zone determined earlier by him from the analysis of magnetic records.

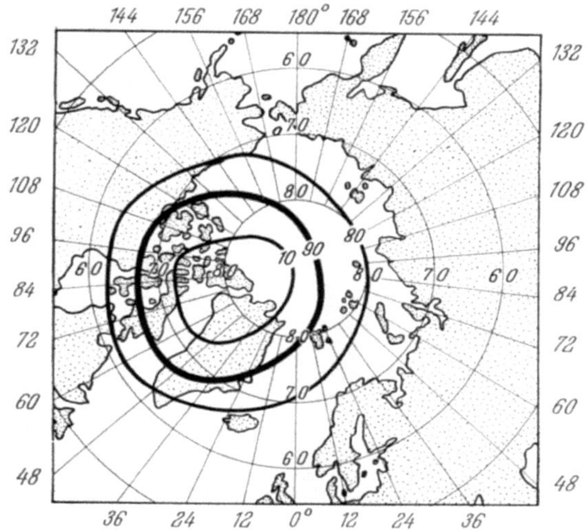

Fig. 11. Isolines of frequency of aurora appearances in the zenith at the hours of the morning maximum. (FELDSTEIN, Y.I., and E. K. SOLOMATINA, 1961 [*18*].)

As we discuss later (Sect. 19), the nature and characteristics of the auroras in the auroral cap seem to differ in several ways from those in the auroral zone. Fig. 9 shows that such a second "zone" appears during quiet periods, but not during disturbed periods. The alignment of the auroral arcs or bands there does not seem to be parallel to the curve suggested by LASSEN or FELDSTEIN (cf. DAVIS 1962 [*36*]; DENHOLM 1961 [*39*]: for details see Sect. 14). Further, a detailed analysis of auroral data in the Antarctic regions does not seem to indicate any definite sign of the existence of an inner auroral *zone* (HATHERTON 1960 [*40*]; HATHERTON and MIDWINTER 1960 [*41*]; DENHOLM 1961 [*39*]). Thus, any suggested relations to theories such as that of ALFVÉN (1955 [*43*]) remain obscure. However, the investigation of the characteristics of such polar cap auroras will greatly increase our knowledge of the auroral morphology. (See also KHOROSHEVA (1963 [*42*]).)

6. The distribution of the aurora in the subauroral zone and in the minauroral zone. As Figs. 2, 3 and 10 indicate, the frequency of appearance of auroras outside the auroral zone decreases more steeply towards the equator than towards the pole. MCINNES and ROBERTSON (1960 [*44*]) showed that over the British Isles (along gm long. 80°), in 1957, the number of nights with overhead auroras was 55 at gm lat. 64° N, 10 at gm lat. 57° N, and only 2 at gm lat. 54° N. The auroras in the subauroral zone appear during medium and great magnetic disturbances, when the center of the auroral activity shifts equatorwards.

The number M for the outlying isochasms is still rather uncertain. The outermost isochasm yet constructed is the one numbered 0.1 by FRITZ and VESTINE, implying that at places along this line the aurora can be seen, on the average over a long period, only once in ten years. This isochasm passes over the Mediterranean Sea; thus the Greeks and Romans had some, though few, opportunities of seeing the aurora, and it is mentioned in their writings.

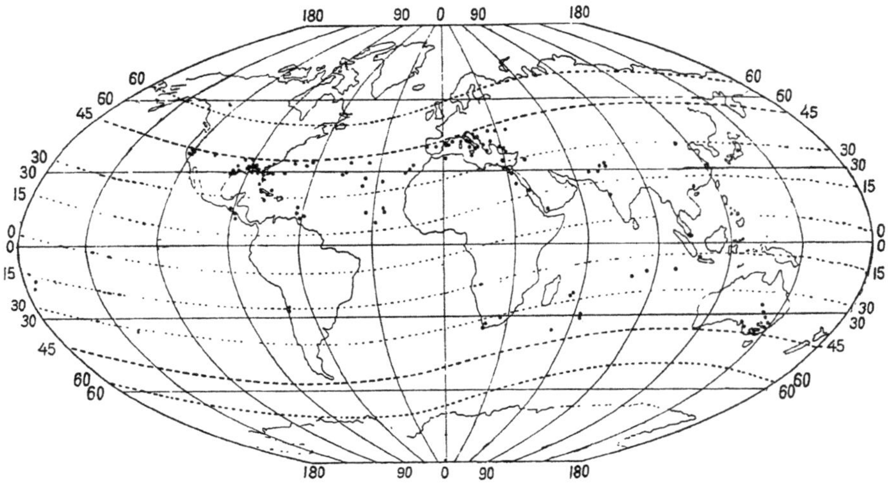

Fig. 12. Part of the globe, showing 0°, 30°, and 60° circles of geographic (gg) and geomagnetic (gm) latitude, also 45° (gm) Dots indicate places in the minauroral belt (extending between the N and S circles of gm latitude 45°) for which the observation of an aurora has been recorded. (KIMBALL, D. S., Yale Observatory 1956, reproduced from [11].)

The outermost isoaurore made so far is numbered 6 by FELDSTEIN, implying that at places along this line overhead auroras were seen during 6% of the IGY period. This isoaurore passes over England, the border line between Canada and U.S.A., the southern tip of continental Alaska, Kamchatka Pen., middle Siberia, and the southern edge of the Baltic Sea.

Table 1.

Date, Year	Observatory, and gm dipole latitude
September 1, 1859 . .	Honolulu (21° N)
February 4, 1872 . .	Bombay (10° N)
September 25, 1909 .	Singapore (10° S)
May 13, 1921	Samoa (16° S)
August 19/20, 1950 . .	Photograph from Greece (estimated 36° N)
September 13, 1957 .	Mexico (30° N)
September 23, 1957 .	Mexico
February 11, 1958 . .	Mexico

The aurora has been seen sometimes from places well to the south of the isochasm numbered 0.1, even down to less than gm latitude 10°, as at Aden and Bombay on February 4, 1872, perhaps the most outstanding aurora on record. Auroras seen from the minauroral belt usually lie in the northern half of the observer's sky, and so do many of the auroras seen from the subauroral zone.

KIMBALL (cf. CHAPMAN 1957 [11]) made an extensive study of some of the historic auroral displays over the entire globe, partly based on LOOMIS' memoirs. In Fig. 12, dots indicate places in the minauroral belt for which the observation of an aurora has been recorded. Some of the historic auroral displays are listed in Table 1.

We add here ABBOTT'S [45] photographic estimate of the aurora, taken in Greece, on August 19/20, 1950; the aurora seemed to descend to gm lat. 36° N. We have added also in the above Table three great auroral displays observed during the IGY. AKASOFU and CHAPMAN (1962 [46]) made a detailed analysis of the distribution of the aurora of February 11, 1958 (Sect. 35): the aurora was recorded by an all-sky camera located even at Memanbetsu (gm lat. 34.0° N), Japan. SCHLOBOHM, LEADABRAND, DYCE, DOLPHIN and BERG (1959 [47]) observed a radar reflection (106.1 Mc) from a point about 900 km north of Stanford (gm lat. 43° N) at an altitude of 300 km on that day. Some photometric data on that day are discussed in Sect. 17.

7. Time variation of the average auroral distribution. *a) General.* The auroral zone is a statistical conception, and the zone of the maximum isochasmic number is not necessarily the main belt of auroral activity at all epochs. The distribution

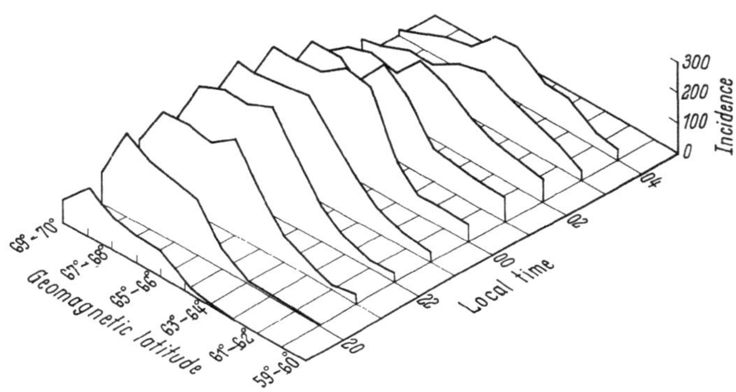

Fig. 13. Isometric plot of auroral incidence vs. geomagnetic latitude and local time data from 59 nights during the 1957-1958 observing season. Observing stations are located at 61-1/2°, 64-1/2°, 65-1/2°, 66-1/2°, and 68-1/2°. (DAVIS, T. N., 1962 [23].)

during a limited short period of order 30 minutes can greatly differ from this average. An extreme example, a drastic change of the distribution, occurred during the great magnetic storm of February 11, 1958. At about the maximum epoch of the main phase, the aurora was completely absent from the region of the auroral zone, and the region where overhead auroras were seen descended to gm lat. 50° N in the North American Continent (AKASOFU and CHAPMAN 1962 [46], see also Sect. 35). Such a phenomenon is often referred to as the "equatorward shift" of the auroral zone. As mentioned already, the determination of the isochasms was originally based on long-term statistical studies, spanning many decades, including a number of sunspot cycles. The auroral zones thus defined should be regarded as a limiting case, and such a statistical result should not be confused with any instantaneous distribution. (See also AKASOFU and CHAPMAN (1963 [48]).)

The distribution of the aurora has been discussed statistically for different local times or absolute hours (universal time or geomagnetic time, etc.), for different seasons, or epochs in the sunspot cycles, or different degrees of world-wide magnetic activity.

b) Daily variation. In terms of the incidence, DAVIS (1962 [23]) made a detailed study of the daily variation of the auroral distribution over the Alaskan sky; his analysis is based on data from 59 nights during the 1957/58 observing season. The result is shown here in an isometric form as Fig. 13.

In the early evening, the aurora, if any, is at latitudes around 70°. As the evening progresses, auroras tend to occur at lower latitudes, producing a rather broad peak between 65° and 70°. At about midnight, the peak in the distribution occurs approximately at 66°. After midnight, the maximum becomes less pronounced and the distribution becomes more uniform over a wider range of latitude.

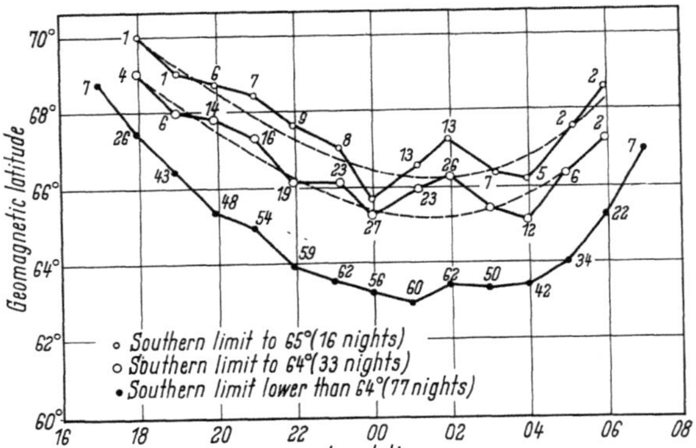

Fig. 14. Average southern extent for each of the three groups of displays extending to geomagnetic latitude 65°, 64°, and beyond 64°. The number beside each point gives the number of displays contributing to that particular meanposition. (Davis, T. N., 1962 [23].)

Davis and Kimball (1960 [49]) also showed the average southern extent of three groups of auroral displays, occurring during the 1957—1958 season. The auroral displays are grouped together according to their ranges of the minimum latitudes attained by southernmost auroras for each display, namely 65°, 64° and lower than 65° (Fig. 14). Again, the southward progression in the early evening is clearly seen for each group. Then at about 2 hr. local time, the curves seem to show a northward retreat of the auroras. Davis (1962 [23]) indicated, however, that this does not seem to be due to actual northward motion of individual auroral forms, but rather due to the disappearance of individual forms, which is apparently more pronounced at lower latitudes. His conclusion is based on the fact that at all times at every point in the auroral zone the number of auroral forms crossing the zenith from north to south is larger than that from south to north (Sect. 21).

c) Seasonal variation. The seasonal variation of the auroral distribution over the entire auroral cap has not been studied, although it is known that at a station to the south of the (northern) auroral zone the appearance frequency increases in equinoctial months (cf. Meinel, Negaard and Chamberlain 1954 [50]).

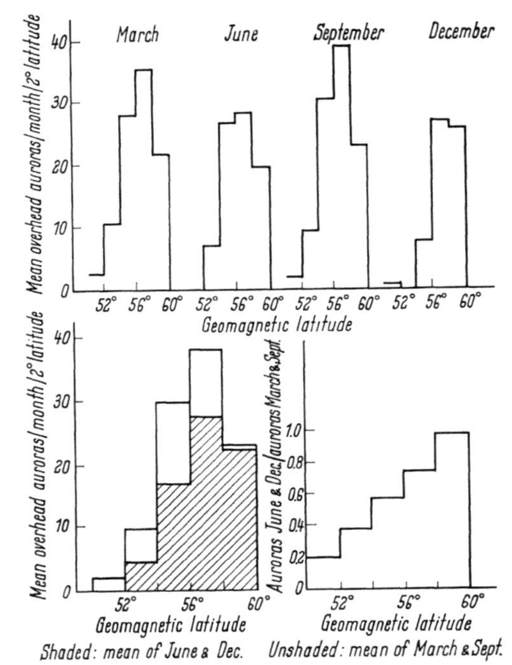

Fig. 15. Comparison of the southern extension of the auroral zone in equinoctial months with that in solstital months. (Gartlein, C. W., and R. K. Moore, 1951 [51].)

Gartlein and Moore (1951 [51]), using visual data collected between 1939 and 1950, studied the seasonal variation of auroral distribution over the northern U.S. Fig. 15 shows the southern extension of the auroral belt in each season. The upper diagram shows the average number of auroras seen overhead throughout the period of the study in the indicated months, as a function of the geomagnetic latitude. The apparent decrease from the 56°—58° zone to the 58°—60° zone is due to the lack of data in that region.

The lower diagrams seem to indicate that the number of auroras seen in the northermost region (58°—60°) varies relatively little throughout the year, but at lower latitudes during equinoctial periods the belt in which overhead auroras are seen becomes, on the average, broad. Such a study must be supplemented in the future, particularly in the regions to the north of gm lat. 58°, by all-sky camera data.

d) Sunspot cycle. It has long been suggested that the main belt of auroras contracts during periods of low sunspot number; and conversely, at times of large sunspot number, the radius of the main belt is enlarged, moving towards lower gm latitudes. Observing auroras at a station to the south of the (northern) auroral zone, the appearance frequency of auroras increases with increase of the sunspot number (cf. MEINEL, NEGAARD and CHAMBERLAIN 1954 [*50*]; see also Sect. 32a and Fig. 51). However, no detailed study of the sunspot cycle variation of the auroral distribution has yet been made in any systematic way. It is expected that iso-auroral diagrams will be constructed from data obtained during the IQSY, which will enable us to compare them with the IGY isoauroral diagrams, such as that produced by FELDSTEIN and SOLOMATINA (1960 [*18*]).

So far, the information available on this subject is fragmentary. DAVIES (1948 [*52*]) summarized visual auroral observations made, between 1943—1947, at Clyde (gm lat. 81.5° N), Chesterfield (gm lat. 73.3° N), Churchill (gm lat. 68.8° N), Ottawa (gm lat. 56.5° N), St. Johns (gm lat. 58.7° N), Prince Rupert (gm lat. 58.6° N), and Saskatoon (gm lat. 60.9° N). He showed that to the north of the auroral zone, the number of auroras, and the number of nights of auroras decreased progressively from minimum (1944) to maximum (1947), whereas to the south, auroras were much more frequent at sunspot maximum than at sunspot minimum. He further pointed out that during the period of 1932/33, at sunspot minimum, at Point Barrow, auroras were seen to the northeast of the station, suggesting a contraction of the main belt. PATON and MCINNES (1954 [*53*]) indicated also that in the winter of 1952/53 auroras observed from North Greenland (gm lat. 80°) were overhead or to the north and east.

It is not certain that the main belt of auroras contracts in such a drastic manner to the circle of gm latitude 80°. As we discuss in Sects. 5b and 19, there are apparently two types of aurora in the auroral cap. One of them seems to appear well inside the auroral cap, and tends to occur more frequently during periods of magnetic calmness. It may well be that during sunspot minimum such an aurora appears relatively more frequently than during sunspot maximum, and that at the minimum the auroral zone aurora remains there with much less activity. In fact, ELVEY, LEINBACH, HESSLER and NOXON (1955 [*54*]) suggested that during 1951—1953 there was a type of auroras around gm latitudes 67° and 69° which are independent of magnetic activity. EVANS and THOMAS (1959 [*55*]), using visual observations from Halley Bay (gm lat. 74.7° S) and elsewhere, found also a narrow zone, centered around gm lat. 71.7°, in which quiet arcs appear irrespective of magnetic activity. SHERET and THOMAS (1961 [*56*]) reported also that the zone stayed approximately at the same place in 1959.

II. The auroral forms.

8. Introduction. The aurora shows great variety in its form, structure, activity, motion, brightness and color. In this and the next chapter, our description is limited in the sense that it is based on visual and photographic records which have certain patterns of the color sensitivity characteristics for different wave lengths. In fact, human eyes are rather insensitive in the wave length regions of λ 3900 and of λ 6300, where some of the most characteristic auroral spectral bands or lines exist. It is not surprising, therefore, if photometric equipments observe an entirely different form or structure of the aurora. Human eyes happen to be most sensitive around one of the major lines, λ 5577 [OI] from atomic oxygen, so that visual observations are mainly based on the luminosity profile of that line.

There are two basic forms of the aurora, namely *bands* and *surfaces*. Their forms are discussed in detail in this chapter, their dynamical features in part B of this Chapter, and their spectral features in Chap. VII.

A. The bands.

9. The classification and definition of bands. A band is a "ribbon-like" form, that appears to extend upwards from a nearly constant height above the earth's surface; its horizontal extent much exceeds its extent in angular elevation. The simplest type of band appears as a circular arch with nearly homogenous brightness along its horizontal extent; it is called a *homogeneous arc*. Any features additional to this simplest form can be regarded as an indication of certain types of activity (Table 2).

Table 2.

	Additional feature	Name
Homogeneous arc plus	rays	Rayed arc
	waves or folds	Homogeneous band
	rays, waves or folds	Rayed band
	pulsations	Pulsating arc or band

When apparent vertical striations (the rays) are added to a homogeneous arc, the arc is called a *rayed arc* (for details, see Sect. 13). When the arc becomes wavy or folded, but without vertical striations, it is called a *homogeneous band*; if it has ray structure, it is called a *rayed band*. The scale of the folds has a wide variety, perhaps 10$\sim$300 km. A rayed band with a well-developed fold (of order 100 km) used to be called *drapery* or horseshoe type aurora. It is shown later (Sect. 13) that some of the rays are also small scale folds.

If an arc or band changes its brightness within a period of a few seconds to a few minutes, it is said to be *pulsating*. Most of the arcs or bands seem to change their brightness, particularly in the evening when auroras to the east are active. But some of the homogeneous and diffuse arcs, which are a little brighter than the Milky Way, undergo little change in their brightness.

Sometimes an arc or band is so diffuse that only some bright portions (most likely due to its folds) are conspicuous; they are referred to as *isolated rays*. It is not yet known whether there exist really isolated rays in the sense of "columns" or "shafts" of light.

Arcs and bands can only be recognized definitely when their lower border can be clearly seen. For some observers, particularly in the subauroral zone, the lower border is located below the horizon, and the upper part of the band can only be seen as a *glow*.

Further, when a rayed arc or band is seen at the magnetic zenith, the rays or subsidiary folds are seen to converge towards a small region, resulting in an apparent fan-shape form. Such a form is often called a *corona*. We notice, therefore, that the same aurora can be seen as a rayed band, a corona or a glow, depending on the observer's location relative to the aurora. In Fig. 15a and b, all-sky camera photographs of homogeneous arcs and a rayed band are shown. The photograph of Fig. 15a was taken from Fort Yukon (gm lat. 67.0° N) at 2337 AST (Alaska Standard Time) on February 15, 1958. At least five separate

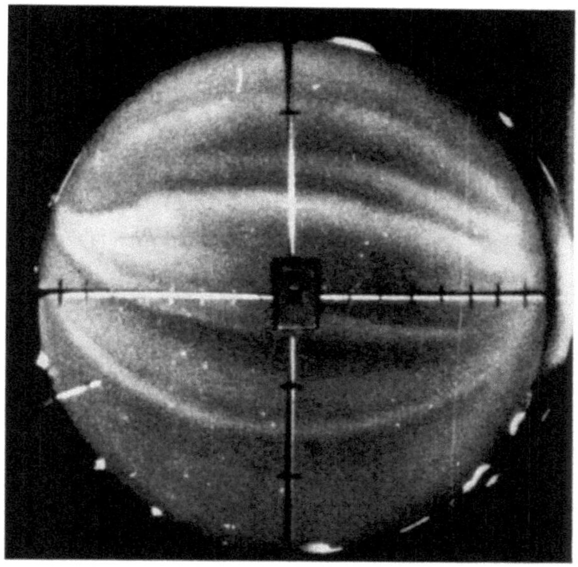

Fig. 15a. All-sky camera photograph taken from Fort Yukon, Alaska (geomag. lat. 60.7° N) at 0937 G.M.T. on February 16, 1958; the local time was 2337, 15 February. The photograph shows at least five separate arcs. They extend in the geomagnetic east-west direction.

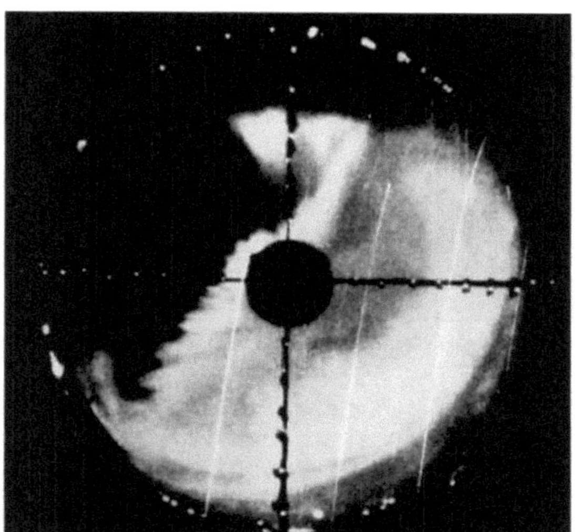

Fig. 15b. All-sky camera photograph taken at Kotzebue, Alaska (geomag. lat. 63.6° N) at 1119 G.M.T., 23 September 1957 (0019 local time). It shows an exceptionally fine example of a rayed arc. A large polar magnetic disturbance of order 1500 γ was observed at the same time at College, Alaska.

arcs are seen. Fig. 15b was from Kotzebue (gm lat. 63.6° N) at 0119 AST on September 23, 1957. It shows exceptionally fine examples of a rayed arc. Figs. 16a and b show color photographs of a rayed band and a corona, taken from College.

10. The east-west length of a band. A most striking characteristic of auroras is their very thin ribbon-like structure. The east-west extent is particularly important in connection with the origin of auroral particles.

a

b

Fig. 16a and b. Color photographs of the aurora. (a) A rayed band; (b) a corona (AKASOFU, S.-I.).

Under the best observing conditions, an observer may see auroras over an area of radius about 1000 km. It is common to observe arcs extending from horizon to horizon (see Fig. 15a), a distance of over 2000 km (ELVEY 1957 [57]). Some pre-IGY evidence suggested that arcs or bands may extend over at least 5000 km, but there was no definite pre-IGY observation to show that *individual* arcs do indeed extend over such a great distance. It has become possible through the IGY all-sky camera network to see such an example, although even the best network was not always close enough to study the subject in great detail.

Figs. 17a and b show an example of a long arc extending at least 5000 km in the east-west direction. The arc moved from the auroral zone into the auroral cap during a moderate break-up (Fig. 17a) and then slowly returned to the auroral zone with a certain increase of brightness (Fig. 17b) (AKASOFU 1963 [*58*]).

There is also some other evidence, though less certain: at a certain epoch of the great auroral display of February 11, 1958, a group of auroras seemed to extend from the north of Sakhalin (Siberia) to as far as the south of Newfoundland (Canada), suggesting an extension of about 10000 km (Sect. 35b). Some visual observations conducted over the northern part of the United States indicate that some arcs extend from coast to coast (at least 4500 km) approximately along the border line between Canada and the USA, (GARTLEIN 1958 [*59*]).

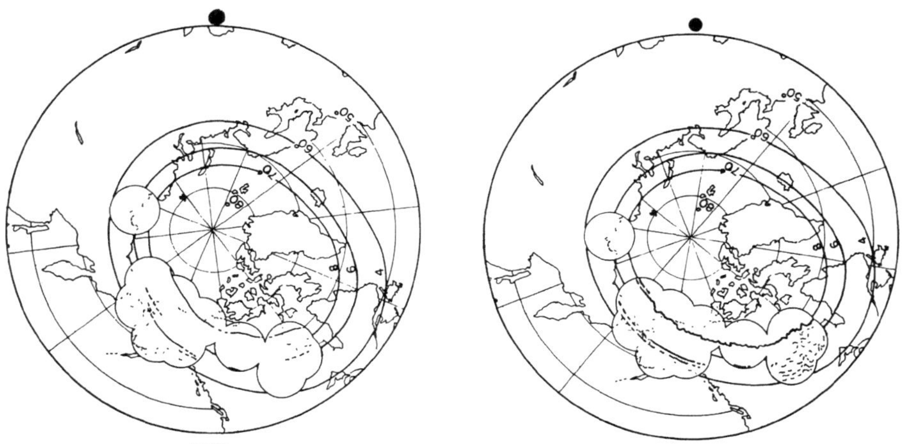

0830
a

0930
b

Fig. 17a and b. The distribution of the auroras on March 24, 1958. (AKASOFU, S.-I., 1963 [*58*].)

11. The width of a band. Another striking characteristic of the band is its thinness. A thickness of order 500 m was first suggested by VEGARD (cf. BATES 1960, p. 278 [*60*]) ELVEY 1957 [*57*], by visually comparing an arc passing through the magnetic zenith with the adjacent moon, estimated the thickness to be not more than 250 m. The first photographic determination of the thickness was made by AKASOFU (1961 [*61*]).

The best way of determining the width of a band is to take a photograph when the bottom edge of the band is crossing the magnetic zenith. In this situation, we can view the bottom edge alone, without seeing the surface. In Figs. 18a and b, the thickness at the points A and B is respectively about 340 and 140 m, assuming the height of the band to be 95 km.

We have seen in Sect. 10 that an arc can extend over 5000 km in length, but yet the thickness is of order a few hundred meters. This characteristic structure is particularly important in considering the origin of auroral particles.

12. The height and vertical extent of bands. *a) General.* Much work has been done on the determination of the height of arcs and bands. The most reliable method is by stereophotography, that is, by simultaneous photography of the same aurora from two or more points. STÖRMER was the first who conducted such an extensive study. His results and others are well summarized in his book (STÖRMER 1955 [*3*]; pp. 67—134) and by HARANG (1951 [*6*]).

A similar study was made by McEwen and Montalbetti (1958 [62]) at Churchill during the winter of 1957—1958. Their analysis showed that the mean height of weak auroras was 107.7 km and that of strong ones was 100.1 km, the mean height being 104.8 km (Fig. 19).

a

b

Fig. 18a and b. Photographs of an active corona taken at College (Alaska) at 0245 hours on 19 February 1961. They were taken with a time difference of about 5 sec. (Akasofu, S.-I., 1961 [61])

Fig. 20 shows the distribution of the heights of auroras obtained during the period 1911—1944 by Störmer (1948 [63], see also [3]); it indicates a great concentration of auroral heights around 90—110 km, but there are some heights as great as 1000 km. A similar study was conducted by Currie (1934 [64]), by stereophotography at Chesterfield, Canada (gm lat. 73.5° N), in 1932—1933. He found that although the station is located well inside the auroral cap, the height distribution is very similar to that found from northern Norway. The latitude dependence of the height of the aurora was discussed by STÖRMER (cf. HARANG 1951 [6]); there seems to be no definite variation of the height of the lower border with latitude.

As mentioned already, our discussion here is seriously limited by the fact that our eye is not sensitive to the whole spectral range of the auroral luminosity, so that the height distribution diagram will differ if the measurement is made by some instrument which is sensitive to a different spectral range from eyes or photographic plates.

b) Type A aurora (high altitude red arc). High altitude red arcs are often referred to as type A aurora (STÖRMER 1955 [3], p. 162). This is, however, a

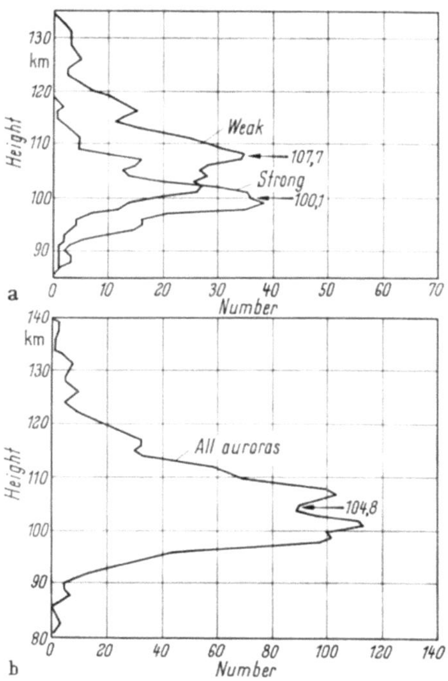

Fig. 19a and b.

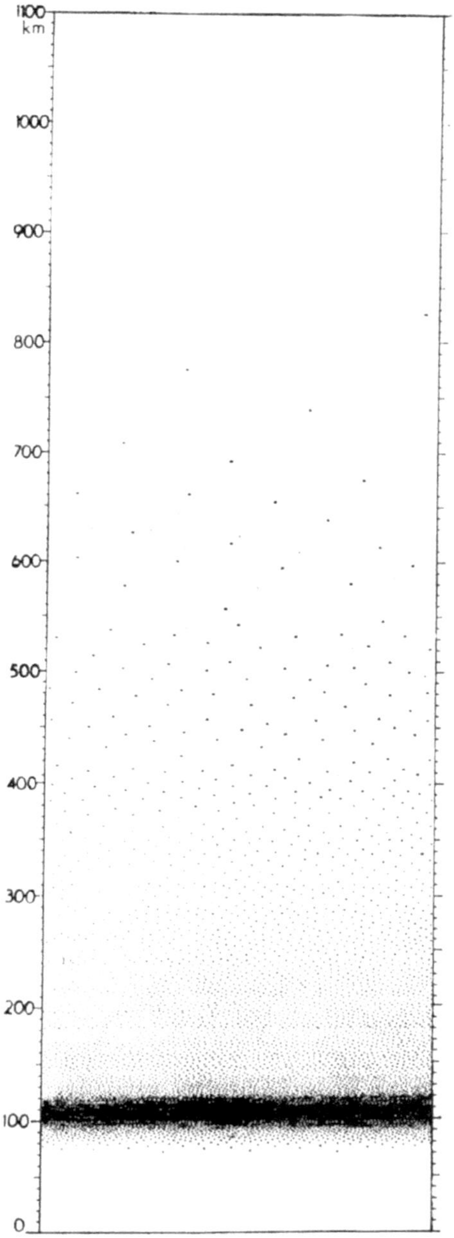

Fig. 20.

Fig. 19a and b. Height distribution of lower edges of homogeneous and rayed arcs and bands. (a) Showing weak and strong auroras only, and (b) showing all measured points. (McEWEN, D. J., and R. MONTALBETTI, 1958 [62].)

Fig. 20. Distribution of heights of 12330 auroral points measured from southern Norway 1911-44. This includes every measured point. (STÖRMER, C., 1948 [63].)

rather rare phenomenon; ELVEY (1957 [57]) reported that a red arc occurred on February 25, 1956, after the intense solar flare (accompanied by a spectacular increase in cosmic ray intensity) of February 23, 1956. It appeared to extend from the Bering Sea to the Atlantic Coast, aligned approximately along the 54° N gm latitude circle. Its height was found to be 335 km.

Another example of such a red arc was reported by REES and DEEHR (1961 [65]). It occurred over Alaska in the evening of November 27, 1959. The height was determined photographically and photometrically, and was found to be about 350 km for the region of maximum luminosity. The arc was characterized by the [OI] 6300—6364 doublet line (see also VEGARD 1938 [66]) and by the N_2^+ first

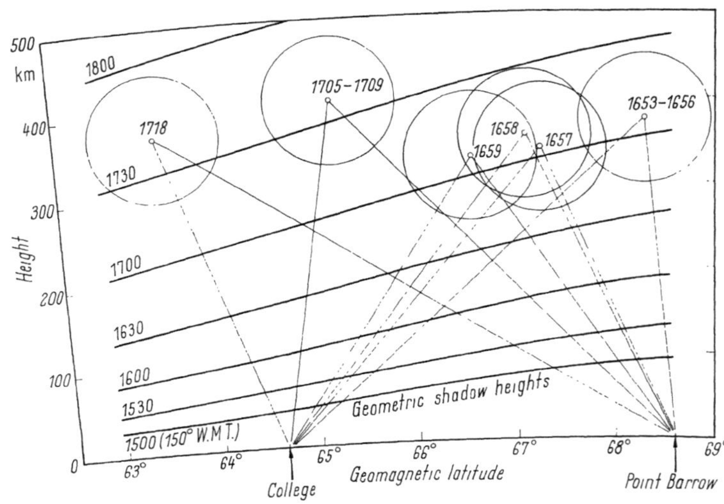

Fig. 21. The height of the red arc over Alaska, determined by the triangulation from Point Barrow and College. The height of the solid earth's shadow at various times is also shown. (REES, M. H., and C. S. DEEHR, 1961 [65].)

negative group bands, other usual auroral emission being notably absent. The maximum intensity was 66.7 kilorayleighs[1] at the zenith. The arc was approximately aligned with the geomagnetic latitude circle and moved southward. Fig. 21 shows the height and position of the red arc over Alaska.

Such an arc appears to be red all over the surface; further, both ELVEY and REES and DEEHR reported red rays superposed on the surface. One of the most spectacular examples of such a red ray was the red aurora of February 10/11, 1958. CLARK and BELON (1959 [67]) observed extremely long red rays of which the top was as high as 1100 km. The red arc is discussed further in Sect. 70.

c) Type B aurora (purplish-red lower border). When the aurora becomes active, the lower border may descend, with a particular enhancement of the first positive group of nitrogen, together with the first negative system of the ionized oxygen molecules (DAHLSTORM and HUNTEN 1951 [68]) resulting in a colorful purplish-red lower border. This phenomenon is fairly common and often appears when an arc or band breaks up.

According to HARANG and BAUER (1932 [69]), there was one occasion on which the lower border was below 70 km, almost down to 60 km. The common height,

[1] The rayleigh (R) is a photometric unit used to express the surface brightness, or intensity integrated along the line of sight; 1 R is an apparent emission rate of 10^6 photons/cm² sec. See CHAMBERLAIN ([4], pp. 569f.)

however, appears to be around 80 km. The type B aurora is discussed further in Sect. 71.

d) Sunlit aurora. STÖRMER reported also another type of aurora, *sunlit auroras* (STÖRMER 1955 [3], pp. 128—134). They may have a lower border at unusually great heights, up to 400 km; their top may extend to 800 km or even 1000 km in height. Fig. 22 shows the location of sunlit auroral rays observed on March 22/23, 1920, relative to the earth's shadow line,

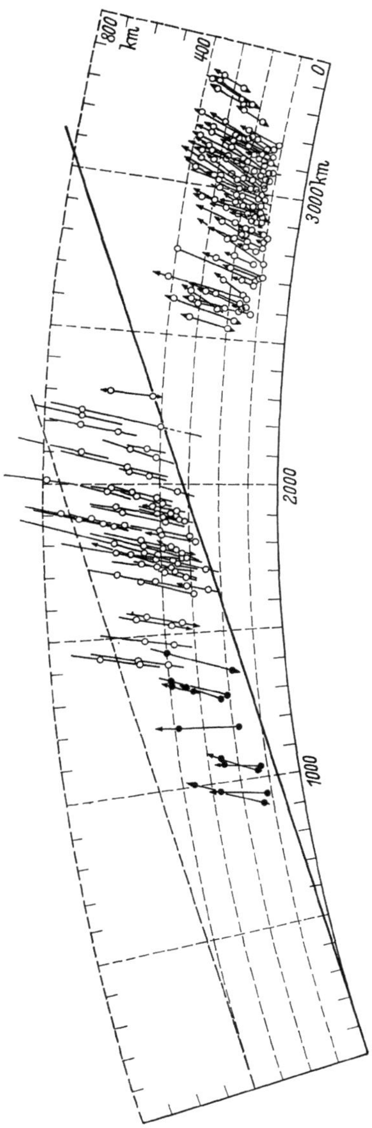

Fig. 22. Situation of the auroral rays relative to the earth's shadow line on 22-23 March 1920, and 13-14 May 1921 (to the left). (STÖRMER, C., 1955 [3].)

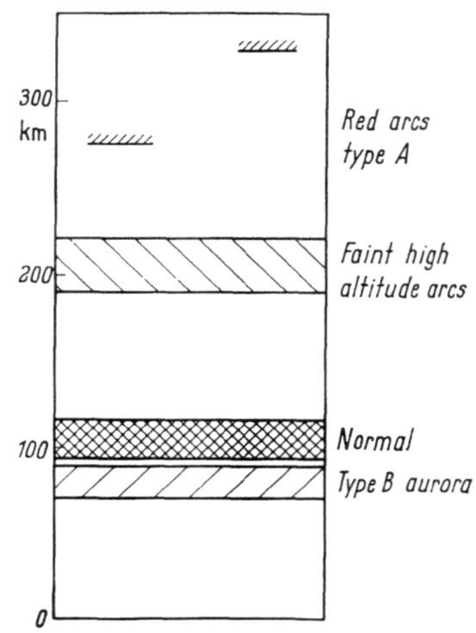

Fig. 23. Heights of various types of auroral arc. (ELVEY, C. T., 1957 [57].)

together with normal rays in the dark atmosphere. The base of the sunlit rays seems to be terminated at about the height of the shadow line. Some of them, however, penetrate it, and sometimes they are divided, one part in sunlight and the other in shadow, with a dark space between.

There is yet another type of aurora reported by STÖRMER (1955 [3], p. 96). It is a very high narrow arc of very feeble luminosity and green color. The height is estimated to be around 200 km. In Fig. 23, some of the auroras discussed here are schematically summarized (ELVEY 1957 [57]).

13. The rays. *a) Geometry.* When a homogeneous arc is striated, by "rays" of greater brightness perpendicular to its border, or nearly so, it is said to be a *rayed arc.* We can imagine straight lines which lie wholly in the surface of an arc or band and are also perpendicular to the bottom edge, which is fairly uniform in height. These lines may be regarded as the rays. One of the principal facts regarding the geometry of an arc or band is that these lines coincide very nearly with lines of force of the earth's magnetic field. Their inclination to the vertical increases with distance from the gm pole. In the auroral zone, the inclination to the vertical is of order 13°, and in the subauroral zone 16°∼27°.

The rays are nearly parallel, their apparent convergence in the corona being merely an effect of perspective. The direction of the rays can best be determined by finding their convergence point (in an ideal case) or (small) area. The measurement of the elevation and azimuth of their point of convergence shows that they lie within about 1° of the magnetic direction at the place of observation (Vegard and Krogness 1920 [70]): the point in this direction in the sky is called the *magnetic zenith*. The wandering of the converging point was discussed in the past and also recently by Abbott (1958 [71]) (Fig. 24). After analyzing magnetic records to determine the instantaneous magnetic zenith, he concluded that the position and motion of the radiant point are independent of the position and motion of the magnetic zenith when seen from the earth's surface.

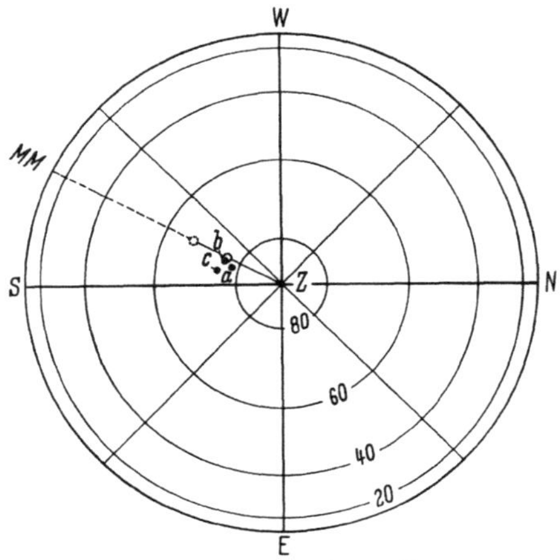

Fig. 24. Changes in the position of the radiant point during the auroral disturbance of September 22, 1957, at College, Alaska. The solid circle is the mean, "quiet", magnetic zenith. The dots a, b and c mark the position of the radiant points, and the dotted circle the magnetic zenith derived from the magnetic observation at the earth's surface. (Abbott, W. N., 1958 [71].)

Recently, Gadsden and Loughman (1960 [72]) reported some high altitude red rays not alined along the field lines.

b) Nature. The nature of the rays has not been discussed much in earlier literature, although Störmer (1955 [3], p. 326) had already speculated that the ray structure sometimes can only be apparent and may not really be a "column" of light. He suggested also that if a thin auroral curtain structure is folded into a zigzag form, the brightness at a great distance normal to the ribbon-like surface is tripled, giving rise to an apparent series of rays superposed on a weak arc or band.

One of the ways of examining this point is to observe the structure of the bottom edge of a rayed arc or band when it is located around the magnetic zenith, and if possible simultaneously, to observe the corresponding surface from different places. Fig. 25a shows a photograph of the bottom edge of a rayed band observed over Alaska on the night of March 24, 1962. When the band was homogeneous, the bottom edge was almost a straight line around the magnetic zenith. However, when the band became a little active and the ray structure appeared on other portions of the surface and quite possibly on the surface around the zenith, the bottom edge waved in various ways. Such activity was repeated many times, with intervening calm periods. During the active periods, a series of simple waves was first seen. They became quickly folded by some non-linear action upon them. Fig. 25a shows one deformed bottom edge around the zenith. It is very likely that the folded portions are seen as "rays" if the corresponding surface is seen from different places.

An analogous instability is often seen in electronic tubes, with a thin electron sheet-beam. The instability is produced by a space charge in the sheet-beam (cf. Webster 1957 [73]); see also Fig. 25b. Gould (1960 [74]) showed that the rate of growth of this particular instability is largest for "wave length" about eight times the beam thickness. For an active aurora, the thickness is of order 150~350 m (Sect. 11), so that the "diameter" of the apparent rays would be of order 1.2~2.8 km. Even for a quiet arc, it is not likely to be more than 10 km. The details are discussed in a paper by AKASOFU (1963 [75]).

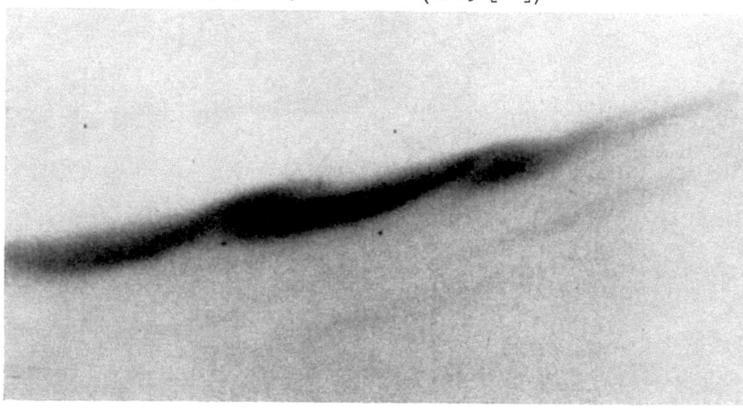

Fig. 25a. Photograph of the bottom edge of a band taken at College (Alaska) between 2300 and 2400 hours on March 24, 1961. (AKASOFU, S.-I., 1963 [75].)

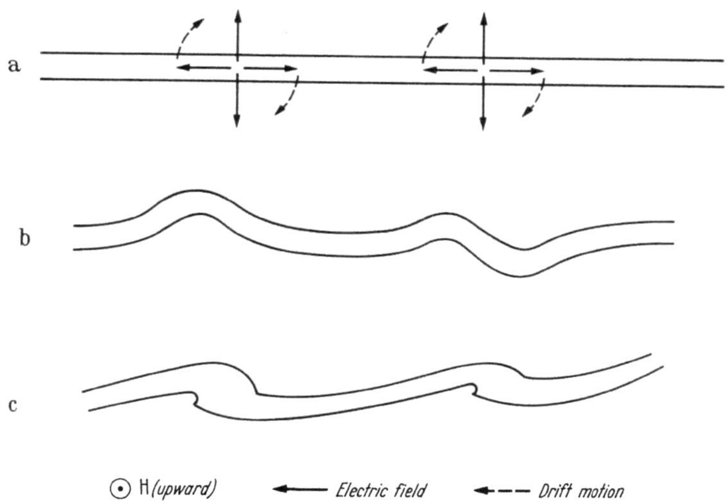

Fig. 25b. Schematic diagram to show the growth of an instability of auroral electron sheet-beam due to a positive space charge in it (looking up the bottom edge of the beam in the northern auroral zone). (AKASOFU, S.-I., 1963 [75].)

14. The orientation of bands. Another important feature of the geometry of an arc or band is its orientation. Most of the earlier studies were made within or equatorward of the auroral zone, and it has long been known that arcs lie roughly along circles of gm latitude. Its daily and seasonal variations are discussed by several authors (VEGARD and KROGNESS 1920 [70]; CURRIE and JONES 1941 [76]; JENSEN and CURRIE 1953 [77]; STÖRMER 1955 [3]). From the detailed analysis of the IGY auroral data, particularly through the works by DAVIS (1962 [36]) and DENHOLM (1961 [39]), the alinement of arcs over the entire auroral region is now available.

Using the all-sky camera records taken at ten stations in the auroral zone, DAVIS (1960 [78], 1962 [36]), obtained the local time variations in the direction of the alinement of aurora larcs and bands. His polar diagram is shown here

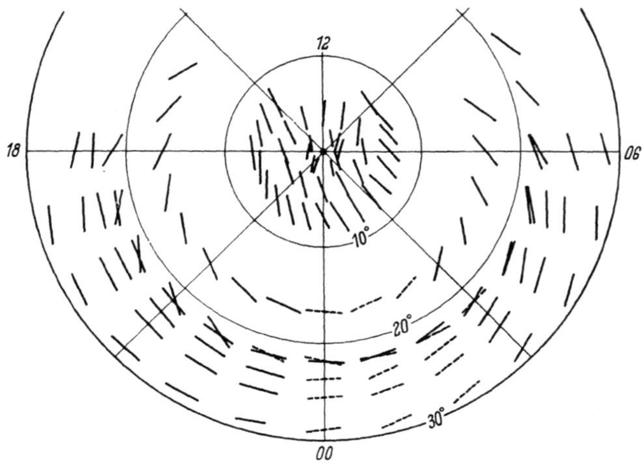

Fig. 26. Hourly median values of auroral alinements plotted on a polar coordinate system with geomagnetic colatitude and local time as polar and azimuthal angles. (DAVIS, T. N., 1962 [36].)

as Fig. 26. One of his most important findings is that arcs located well inside the auroral cap (beyond gm lat. 80°) lie approximately in the direction towards the sun.

The alinement within and to the south of the (northern) auroral zone is more or less in the east-west direction with a slight daily variation. A belt between gm lat. 68 and 80° is a transition region. A similar result was obtained by WEILL (1958 [79]) and DENHOLM (1961 [42]) in the Antarctic region (Fig. 27).

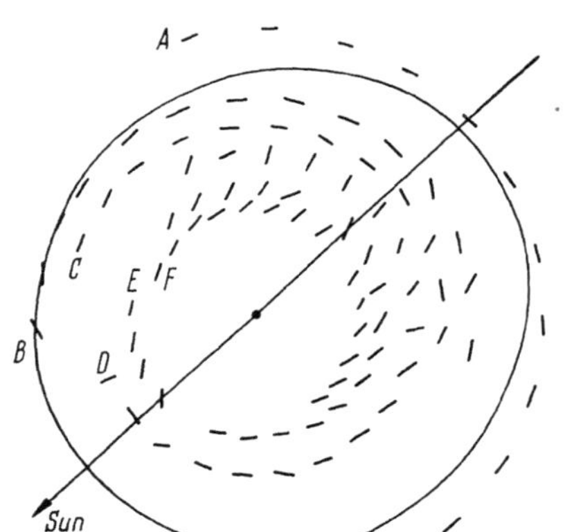

Fig. 27. Hourly mean orientations at Antarctic stations plotted in system centered on eccentric dipole axis and orientation-fixed relative to sun. Stations: A, Macquarie Island, 1957-1958; B, Mawson, 1958; C, Point 3° SE of Mawson, 1958; D, Hallett, 1957-1958; E, Scott Base, 1957-1958; F, Wilkes, 1959. The closed curve is the isoaurora of maximum frequency. (DENHOLM, J.V., 1961 [39].)

The pattern of the alinement over the entire auroral region is of great interest. It had been suggested by DAVIS (1960 [78]) and COLE (1962 [80]) that the pattern has a close resemblance to the SD current system obtained by CHAPMAN (1935 [81]). However, the relationship between the auroral zone auroras and the cap auroras is not clearly understood; they show somewhat different characteristics, as discussed in Sects. 6 and 19. Therefore, such a suggestion must be tentative until a more detailed analysis is completed; in fact, there is no case analyzed to show that individual arcs in the auroral zone are connected to those in the auroral cap.

VESTINE and SIBLEY (1959 [82]) and HULTQVIST (1962 [83]) showed that for the auroral zone auroras the average direction of the alinement of arcs is in close agreement with their theoretical curves. HULTQVIST found that there appears to be a rather discontinuous change in the direction of arcs in the morning over

Kiruna; a further extensive analysis at many other places would be necessary to confirm such a phenomenon.

15. The multiplicity. In the simplest case the display over a large area, for example over the entire Alaskan sky, may consist of only one arc ,but most displays consist of a number of arcs, lying roughly parallel to each other when they are in a quiet condition. Fig. 28 shows such an example over the Alaskan sky in the

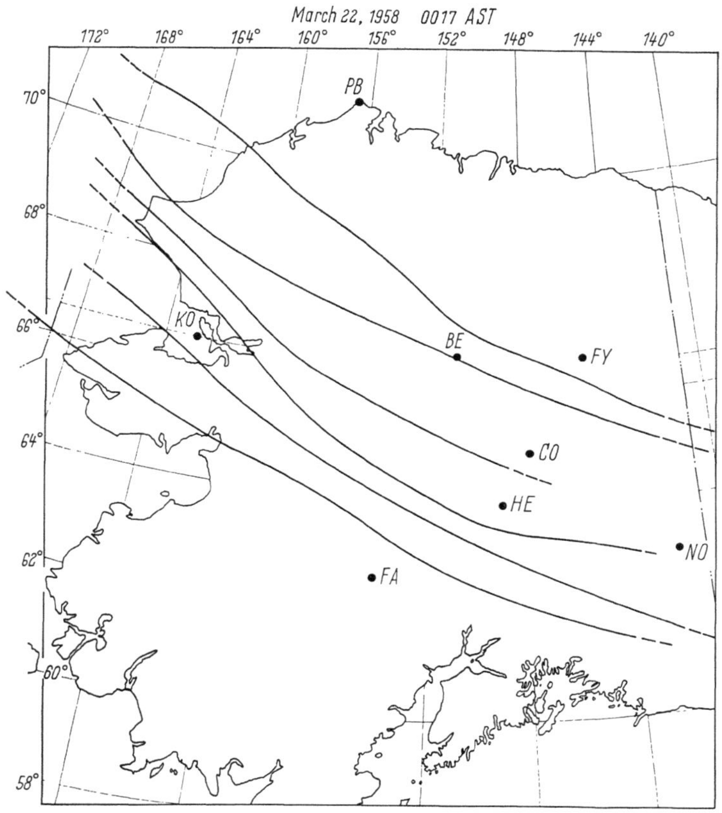

Fig. 28. The distribution of auroral arcs at 0017 AST, March 22, 1958. (AKASOFU, S.-I.)

morning of March 22, 1958. The diagram is constructed by use of data from eight Alaskan all-sky camera stations. A similar example is also seen in Fig. 15a.

ELVEY (1957 [57]) pointed out the importance of the multiple structure of an arc; there can be two to six thin arcs within a distance of 25 km. Looking up at them at the magnetic zenith, their bottom edges wave in a similar fashion.

B. The surfaces.

There are perhaps two forms of auroral surface, namely the *patches* and *veil* (JACKA and PATON 1963 [84]).

16. The patches. The patch is simply an area of luminosity with an ill-defined boundary; it may resemble an isolated cloud (Figs. 29a and b). Patches often occur after an intense break-up. Some pulsate more or less rhythmically, with a period of a few seconds, or as much as a minute. Some are rather sharply defined, particularly when they are around the magnetic zenith. Some, however, are nebulous, particularly if they are dark red patches.

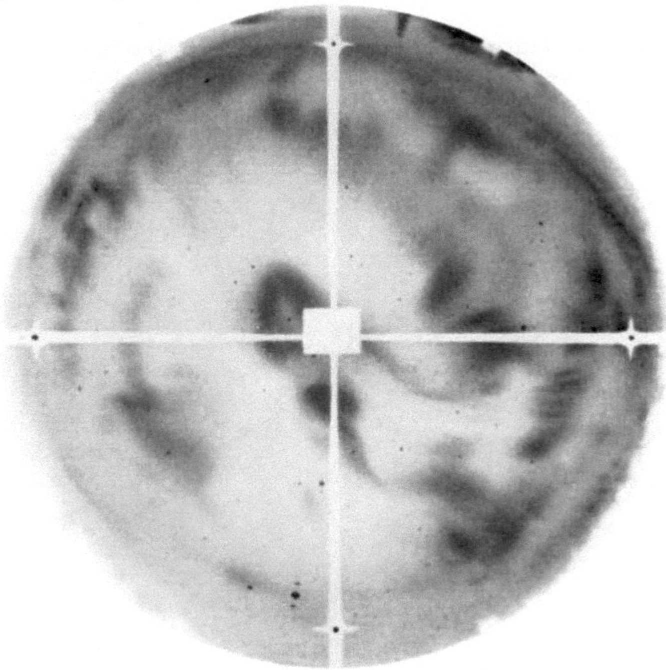

Fig. 29a. The patchy structure of a broken-up aurora observed at Meanook at 1210 GMT, February 14, 1958.

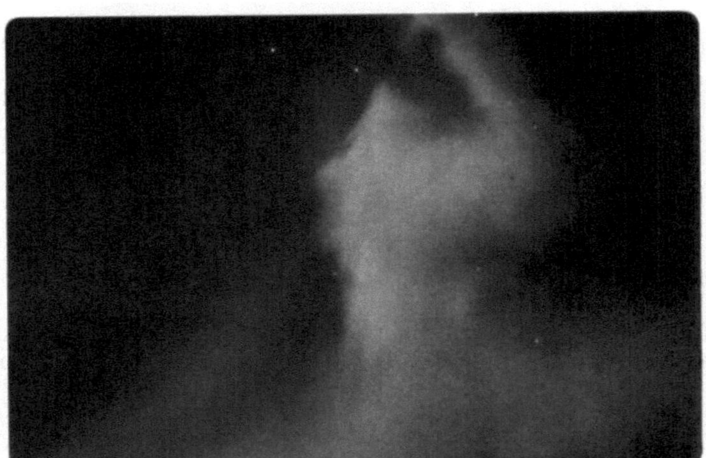

Fig. 29b. Color photograph of a patchy aurora. (AKASOFU, S.-I.)

Most of the patches are yellow-green or grey, and have a height of 100 km. STÖRMER's study (1955 [3], p. 107) suggests that some red patches have much greater heights, distributed between 150 and 800 km (Fig. 30a).

17. The veil. The veil is a more or less uniform luminosity that may cover a large part of the sky. It may appear as an abnormal enhancement of some of the airglow light over a large portion of the sky.

During great magnetic storms, some of the atomic oxygen lines are enhanced. Particularly, observing in the subauroral zone and in the minauroral belt, the λ 6300/34 [OI] doublet

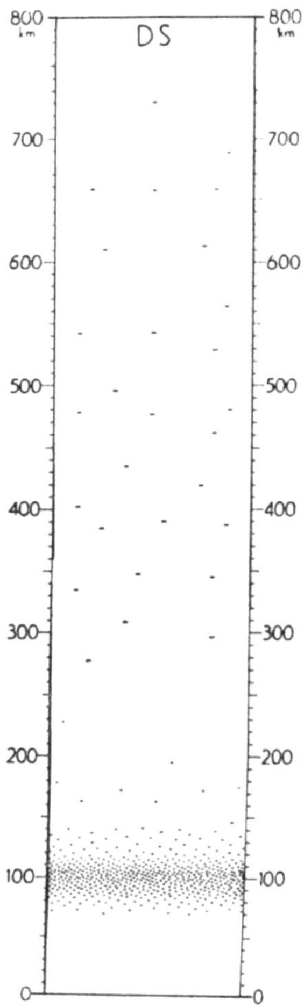

Fig. 30a. Height of ordinary cloudlike aurora and the much higher red ones. (STÖRMER, C., 1955 [3].)

Fig. 30b. The red veil. 20.18 MEZ, September 5, 1958 [BARTELS, J.]

is greatly enhanced: it is seen as a dark red luminosity covering the northern half sky (Fig. 30b).

It is not definitely known whether some of such luminosity is an upper part of bright bands located far to the north of the observing point. It has been often

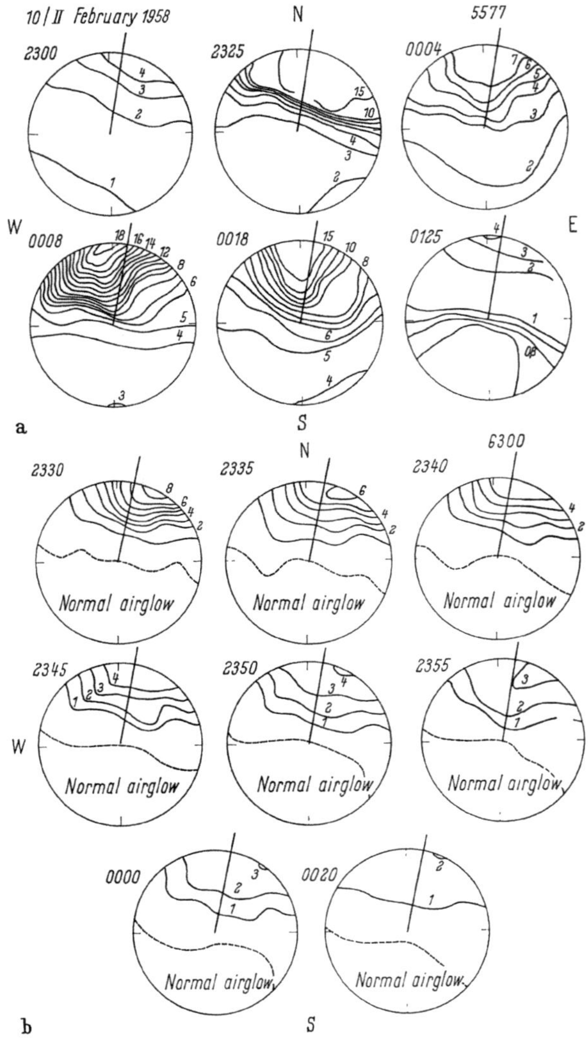

Fig. 31. (a) A detailed study of circular plots for the 5577 A emission for February 10-11, 1958. The times are in MST and indicate the start of the survey. The isophotes are in Kilorayleighs. The direction of the magnetic field is indicated. (Manring, E. R., and H. B. Pettit, 1959 [86].) (b) A detailed study of circular plots from the 6300 A emission for February 10-11, 1958. The times are in MST and indicate the start of the survey. The isophotes are in Rayleighs. (Manring, E. R., and H. B. Pettit, 1959 [86].)

reported that white-yellowish rays are seen in it, when the red luminosity is greatly enhanced. Further, such an enhancement seems to be accompanied by the break-up phase of the aurora to the north (Huruhata 1958 [85]; Akasofu 1962 [131]). In Fig. 31a and b, circular plots for the λ 5577 [OI] and λ 6300 [OI]

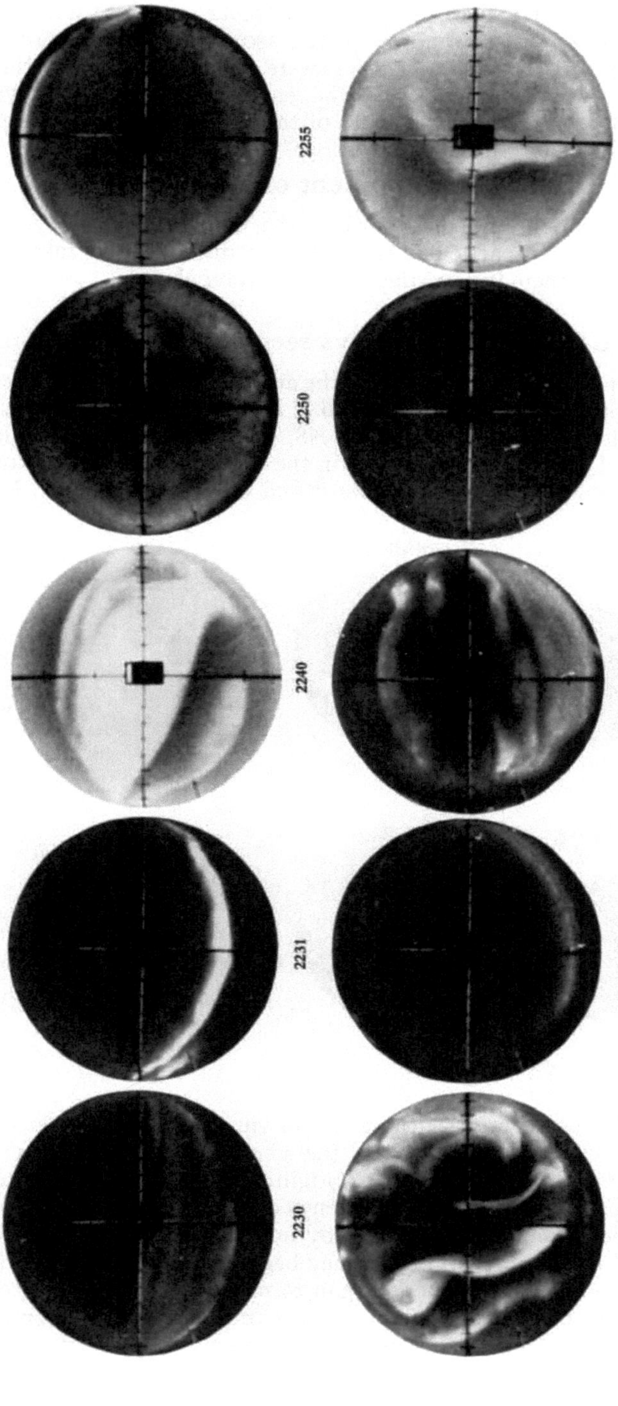

Fig. 32. The Fort Yukon all-sky camera photographs, 2230-0550 AST (= GMT-1000), February 12/13, 1958.

emissions during the great magnetic storm of February 11, 1958, are shown (Manring and Pettit 1959 [86]). Similar plots were also made for Japan by Huruhata (1958 [85]). We note also Krassovskii's report (1961 [87]) that at Zvenigorod (gm lat. 51.1° N) 35 auroras were *visually* observed during the IGY/C, but *spectrographic* records showed the enhancement of the oxygen red line λ 6300 [OI] on 157 occasions; only 14 of them were recorded *photographically*.

III. The development of auroral displays.

Many descriptions are on record of the development of typical or special auroral displays from particular stations. The displays seem infinitely various in their details, but there is nevertheless a certain regularity.

A. The typical displays seen at a single station.

18. Stations in and equatorward of the auroral zone. *a*) *The aurora of February 12/13, 1958, at Fort Yukon (gm lat. 66.7° N).* The Fort Yukon all-sky camera, during the night of February 12/13, 1958, (150° WMT=AST, Alaskan Standard Time), showed signs of an auroral arc in the northern horizon at about 2040 AST. The arc gradually advanced southward and crossed the zenith. New arcs were

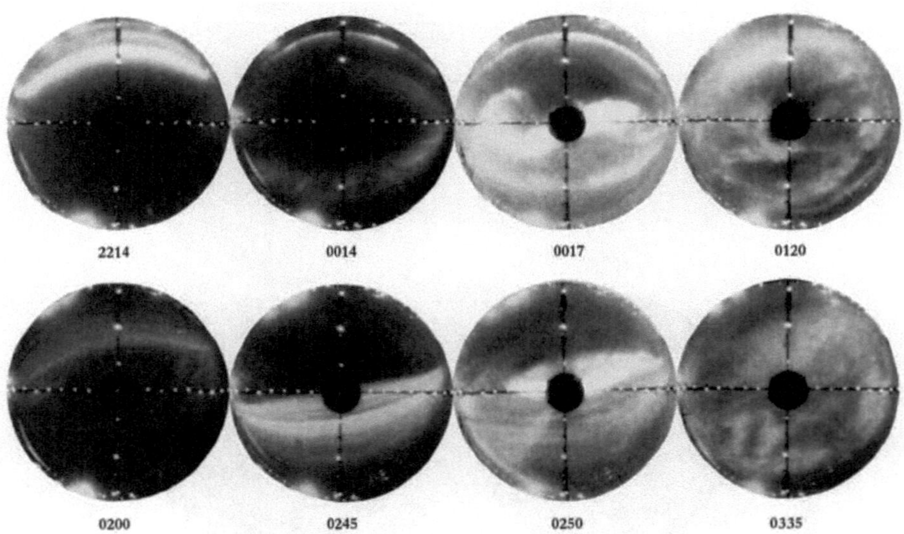

Fig. 33. The Kotzebue all-sky camera photographs, 2214-0335 AST, March 22/23, 1958.

formed to the north and south of the original arc during this period, and at 2230 AST several arcs were seen in the southern part of the sky (Fig. 32). At 2236 AST, one of the arcs began to brighten. One minute later, an arc located just north of this one suddenly brightened and moved rapidly northward.

At 2250, the whole sky was filled by luminous patches, and new arcs seemed to be formed in the northern sky. They began to brighten at about 2300; at the same time the southernmost one began to move rapidly southward. Because of violent motions of the arcs, single arc structure was lost and complicated forms were seen around 2307. These bright and irregular forms drifted westward, and then the whole sky was again covered by faint patches.

Many faint arcs were then formed in various parts of the sky, and moved southward. At 0130, bright arcs were formed near the southern horizon. They increased in brightness and began to move northward at 0140. After this particular display was over, the sky was filled by irregular luminous patches, which seemed

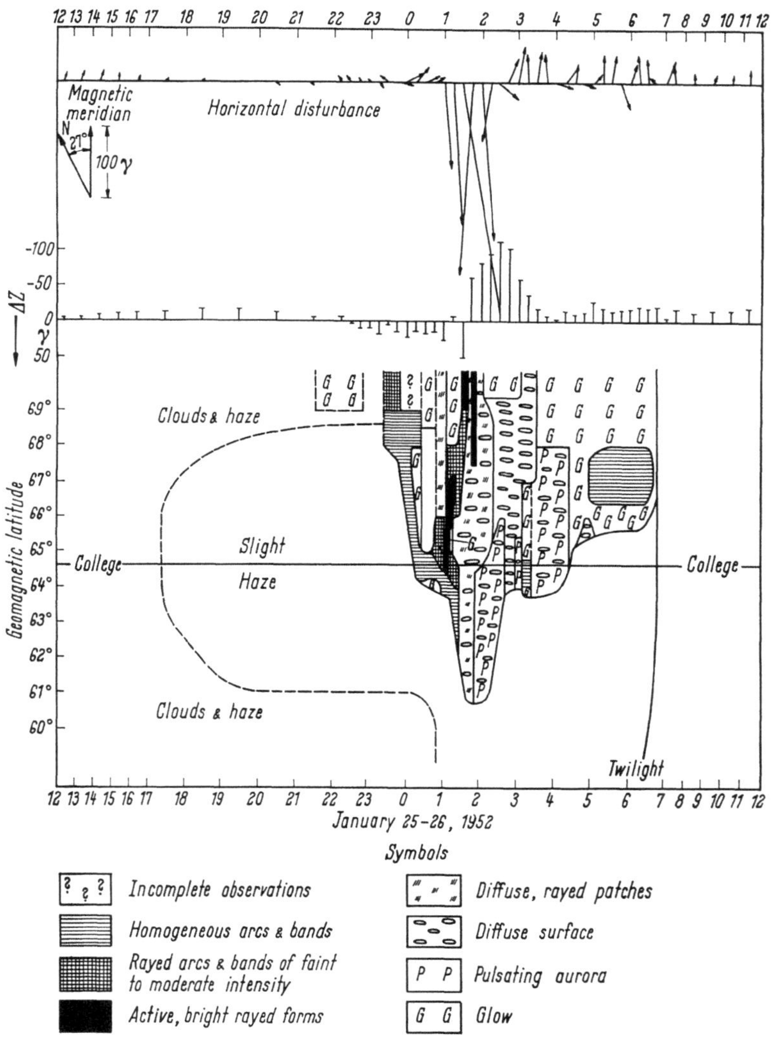

Fig. 34. Auroral display of January 25-26, 1952, observed at College, Alaska. (HEPPNER, J. P., 1954 [88].)

to drift eastward. Then a part of the sky became overcast, but a remarkable increase of brightness and of the eastward drift velocity occurred at 0420. The display lasted until the twilight obscured it.

The planetary geomagnetic indices Kp (see Sect. 33) in the 3 three-hour-intervals 1958 Febr. 13, 06 to 15 Universal Time, were 3+, 4−, and 4+, indicating rather weak geomagnetic disturbance, considered on a planetary scale.

b) The aurora of March 22/23, 1958 at Kotzebue. Fig. 33 shows a more active display of aurora during the night of March 22/23, 1958. The all-sky camera data were taken from Kotzebue, located about 200 km south of the center-line of the auroral zone. The aurora was quite active to the north from the early evening (Fig. 33, at 2234 AST). Ray structures were often seen in the arcs. The aurora over central Canada was very active during this period.

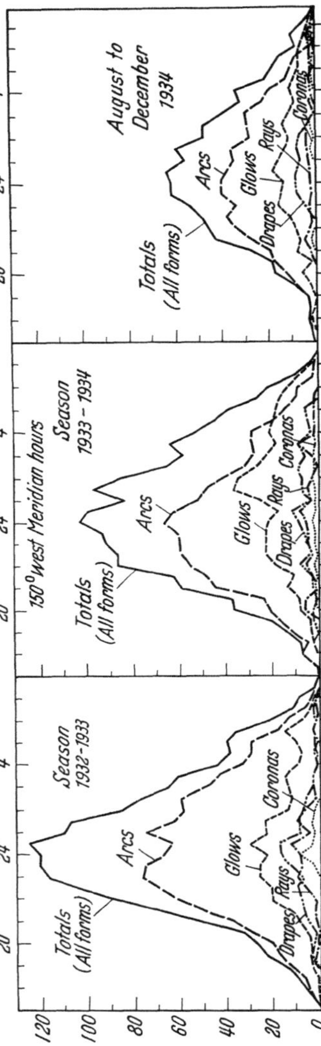

Fig. 35. Daily variation of the occurrence frequency of various types of the aurora at College, Alaska. (FULLER, V. R., 1935 [89].)

At 0014 AST, a faint arc located south of Kotzebue began to break up, and the whole sky became covered by broken auroras. Then quiet arcs began to be formed to the north (0200 AST), and slowly moved southward. At 0245, several quiet arcs were seen in the southern sky. There was no definite sign of east-west motion of the aurora at that time. At 0250, the northernmost arc began to break up, and the whole sky again became covered by luminous patches (0335 AST). This lasted until dawn.

The planetary geomagnetic indices Kp during 1958 March 23, 06 to 15 Univ. Time, were $3+$, $3+$, $3-$.

c) Summary. In the above two examples there is a sort of "cycle" of the display. Each cycle consists of three successive phases, which are characterized by the particular types of the auroras, namely typically quiet arcs, active rayed arcs, and patches, respectively.

In a simple case, we may see one such cycle during the course of a night. In such a case, observing it at a point a few degrees south of the (northern) auroral zone, a faint glow and then an arc may be seen low on the northern horizon, after darkness has fallen. The arc draws southward, rising in the sky. In most cases, the brightness is changing slowly during this phase. Around midnight, a rather sudden increase of brightness occurs, together with violent and complicated motions. The arc is sharply outlined and rayed, and may sweep as a whole to the north or south, and may form loops. The transition from the first quiet phase to this active phase (the break-up phase) may be rapid, even sudden (of order a minute or so). The transition tends to occur most often around magnetic midnight, but varies greatly from day to day, between 1000 and 0400 local time. This intense mobile phase gives place to the third phase, characterized by patches. The twilight obscures the last stage of the third phase (the *post break-up phase*). This simple case is well illustrated in one of the diagrams constructed by HEPPNER (1954 [88]), which is reproduced here as Fig. 34.

In more complicated cases there may be two or more of such cycles, perhaps together with various complexities in the details. The aurora of March 22/23, 1958, seen at Kotzebue is a typical example of two cycles. The "period" of the cycle may be of order 1/2∼2 hrs. The aurora of Feburary 12/13, 1958, seen Fig. 33, is also a typical case. Fig. 35 shows the occurrence frequency of various types of the aurora seen at College (Alaska) as a function of local time (FULLER 1935 [*89*]). We note that on the average the occurrence frequency of various types of the auroras has its maximum around midnight.

19. Stations in the auroral cap. Before and during the IGY several studies were made of the characteristics of auroras in the auroral cap. Well inside the auroral cap the occurrence frequency curve has two maxima (cf. LASSEN 1959 [*90*]; MALVILLE 1959 [*91*]). One occurs around midnight and the other in the early morning (Fig. 36). These maxima are called the "*midnight maximum*" and the "*early morning maximum*", respectively.

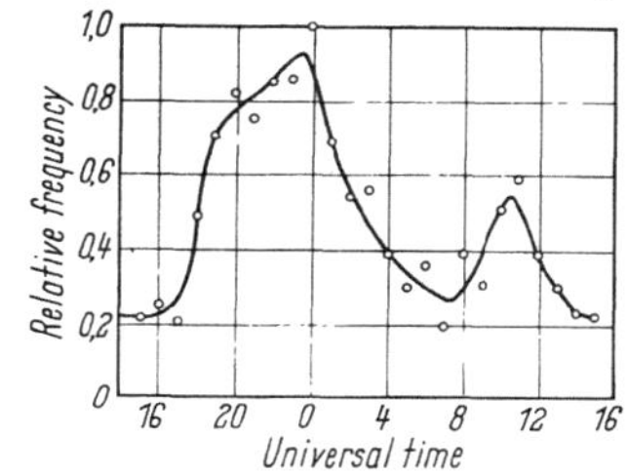

Fig. 36. Daily variation of all visible auroras at the South Pole. (MALVILLE, J. M., 1959 [*91*].)

The midnight maximum is produced by auroras which advance from the auroral zone during the break-up phase; the auroras are usually bright. Auroral displays in the early morning, unlike those around midnight, are usually very quiet, diffuse bands, sometimes with faint rays (LASSEN 1959 [*90*]). Their brightness seems to change slowly. The characteristics of the auroras observed at the South Pole are summarized by MOROZUMI (1962 [*91a*]) in Table 3.

MALVILLE (1959 [*91*]) pointed out that the hour of the midnight maximum appears to shift to earlier evening hours, as one moves polewards (Fig. 37). Further, FELDSTEIN (1960 [*28*]) showed that the hour of the morning maximum shifts to later morning hours, as one moves polewards (Fig. 38).

Table 3. *Characteristic of midnight auroras and noon auroras at the South Pole.*

	Midnight aurora	Noon aurora
Auroral brightness	Bright	Less bright
Duration of display	Long, 1∼2 hours	Sporadic
Auroral forms	Arc and bands with the break-up phase	Long rays and corona
Auroral location	Mostly northern horizon, large displays advance toward south	Mostly zenith
VLF (see Sect. 37)	Strong hiss	(Chorus)
Ionosphere	Blackout only with bright auroras	Blackout on high F minimum
Geomagnetism	Large deflection only with bright auroras	Frequent perturbations of small amplitude

Such a "spiral" feature of geophysical phenomena in the auroral cap has long been discussed in terms of STÖRMER's spiral precipitation pattern (cf. NIKOLSKI 1947 [*92*]; MEEK 1955 [*93*]; see also THOMAS and PIGGOTT 1960 [*94*]). However, it appears that such arguments are based on a mis-interpretation of STÖRMER's calculation (AGY 1957 [*95*], 1962 [*96*]).

We have already indicated some peculiarities of the polar cap auroras in Sects. 5b and 14. Further detailed analyses, based on both statistics and individual cases, are necessary to reveal the nature of the cap aurora.

Recently, DAVIS and DEWITT (1963 [97]) made an analysis of the all-sky camera data from Byrd station (gm lat. 70.6° S), Antarctica. Because of its geographic location (geogr. lat. 80.0° S), the station is in complete darkness for about two months, although it is located just inside the auroral zone. This special location makes it possible to study the auroras at the poleward edge of the auroral zone throughout the day. A typical display there is described as follows: From about 0000 UT, quiet arcs move slowly from the auroral zone to the zenith, and then displays similar to those described in Sect. 18 follow. Until 0300 UT, the predominant motion of the aurora is eastward, and then westward. In the hours between 0700 and 1400 UT, the auroral activity increases; rayed bands,

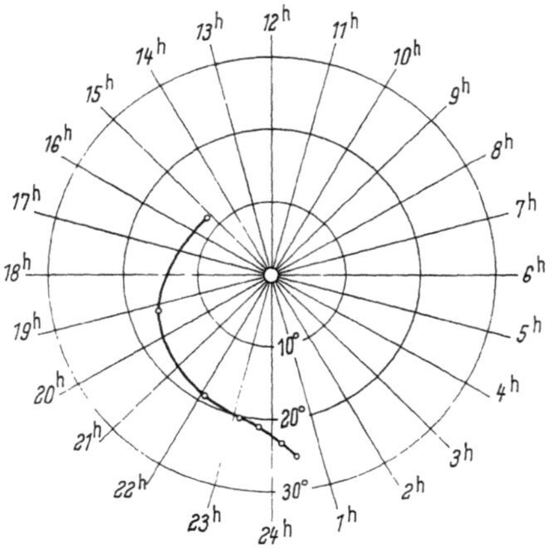

Fig. 37. Geomagnetic times of the maximum probability for the occurrence of overhead auroras in different geomagnetic co-latitudes. (MALVILLE, J. M., 1959 [91].)

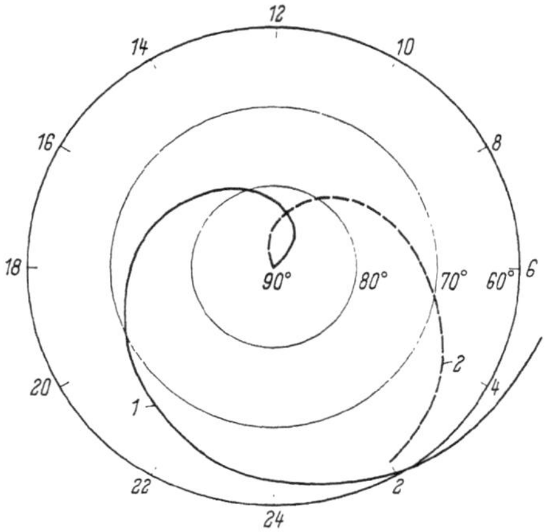

Fig. 38. Dependence of the time of morning and night extremums in the diurnal variation of aurora appearance frequency at the zenith on the corrected geomagnetic latitude. (FELDSTEIN, Y. I., 1960 [28].)

homogeneous arcs and patches are seen. At about 1400 UT, the patchy structure is overhead and drifts eastward; then it comes to a rather sudden complete stop and disappears within some tens of minutes. Between 1400 and 2200 there is but little aurora; the sky looks quite dark.

20. Stations in the tropics. The development of auroras seen in unusually low latitudes may differ greatly from that described above for stations in or near the auroral zone. There is, of course, no daily variation. It is characterized by a dark red veil or glow, occasionally permeated by white-yellowish vertical or inclined beams (rays). Its spectral characteristics are discussed in Sect. 74. Because of its uncommon nature, the description of the historic aurora of February 4/5, 1872, seen at Bombay (gm lat. 9.5° N) and elsewhere in India, is given (S. CHAPMAN 1953 [14], 1957 [98]):

Bombay (from *Times of India*, Bombay, 1872 Feb. 6, p. 2); after sunset on Sunday (Feb. 6) the aurora was slightly visible, and constantly kept changing colours, becoming deep violet when it was most intense — about three o'clock on Monday morning (Feb. 5). It was distinctly visible until sunrise on Monday ... At Aden the aurora was brilliant and extreme.

Jacobabad, Upper Sind (28° N, 68° E; 19° gm lat.); February 5; (From *Times of India*; Feb. 15, p. 2; "News letter from our own correspondent"); As I was returning home about half past 11 p.m., a sudden change from darkness to light was noticed as bright as the full moon. ... I thought it must have been a fire, no, the whole place was magically illuminated. ... I looked up towards the north; there was the phenomena of the Aurora exhibited in brilliancy. ... The light was not stationary, but seemed in motion like a flag would wave in the air; it displayed light of different hues and colours, and running undulate into one another, now disappearing and again appearing. ... I have subsequently learnt from a telegraph employee that there was a perpetual disturbance of the needle from half-past seven last night till ten this morning and that the current set up by its magnetic influence was of great intensity. They were unable to communicate with LAHORE; in fact, there was a total cessation on that line. This goes to prove that it was none other than the Aurora. Its shape was an arch, though not quite so perfect, shooting from the east horizon to the zenith and very nearly at right angles to the magnetic meridian. It seemed to have a very slow motion from E to W. The upper portion of the arch appeared like a sheet of fire of a blood red colour, and from its centre coruscations or streamers of light would issue of different hues, colours of the rainbow, and ran into one another in waves. Its grandeur was something magnificent. After an hour there appeared in the Aurora arch a little below the zenith, a bright blue light of a dome shape quite intense; after some 15 minutes it suddenly dissolved to a deep violet. The sky was clear and the phenomena was observed very distinct; it travelled towards the south and lasted till 02:15 this morning, a duration nearly 2 ($^3/_4$) hours. ... I learn that it was observed at LAHORE, MOOLTAN and SUKKUR.

Lahore is at 32° N, 74° E, 23° gm lat.; Mooltan is at 30° N, 71° E, 21° gm lat.; Sukkur is at 27.6° N, 49° E, 19° gm lat.

B. Dynamics of the auroral displays.

One most spectacular feature of auroral displays is the large-scale motion, sometimes extending over the whole sky. Though infinitely various, there are certain regularities. In this section, various motions are classified in a systematic way, and certain features are discussed.

21. The north-south motions. Auroral arcs or bands are rarely stationary; they move as a whole northward or southward, sometimes rapidly and sometimes slowly. KIM and CURRIE (1958 [*99*]) reported that in west central Canada the direction (of N-S motion) is not related to the time of night in any simple way. When observing at a station to the *south* of the northern auroral zone, the higher speeds, especially in a southward sense, occur generally at times of the break-up phase. This is opposite to what happens at a station located at the *northern* edge of the (northern) auroral zone; during the break-up phase, arcs or bands in the auroral zones move rapidly northward, then some return to the auroral zone (Sect. 30).

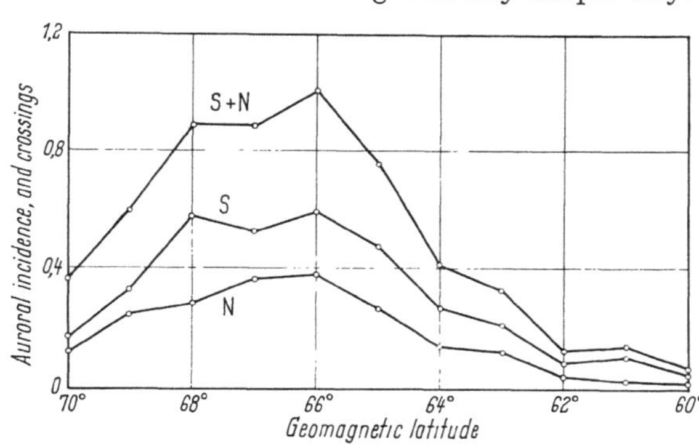

Fig. 39. Average curves of auroral incidence and north-south motions versus geomagnetic latitude. (DAVIS, T. N., 1962 [*23*].)

During a great magnetic storm, the center of the auroral activity shifts equatorward. In such special conditions, even at a station 5~15° south of the (northern) auroral zone, a rapid northward motion can be seen during the break-up phase (AKASOFU 1960 [*100*]; 1963 [*58*]; AKASOFU and CHAPMAN 1962 [*46*]). This is discussed further in Chap. IV.

KIM and CURRIE (1958 [*99*]) reported that the highest observed speed was 50 km/min or 830 m/sec. AKASOFU (1960 [*100*]) reported a similar speed for one of the remarkable northward motions which occurred during the September 23, 1957 magnetic storm.

DAVIS (1962 [*23*]) examined the number of arcs crossing to the north (N) or to the south (S) across several geomagnetic latitude circles over Alaska. He found that on the average the number of arcs crossing to the south is larger than that to the north at all times, over the entire auroral zone (Fig. 39).

22. The east-west motions; general. Since Meinel and Schulte (1953 [*101*]) found a rather systematic east-west motion of the aurora, many studies of this phenomenon have been made, particularly in connection with the SD current

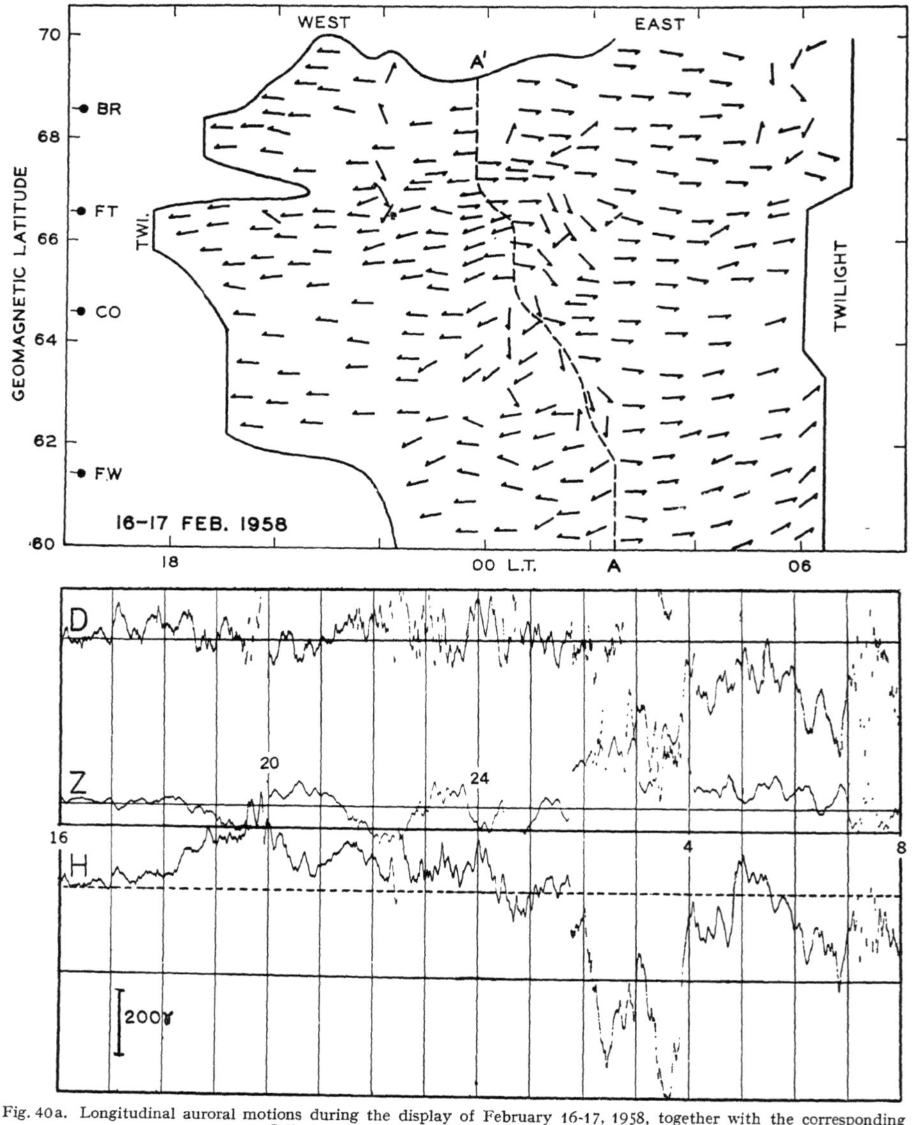

Fig. 40a. Longitudinal auroral motions during the display of February 16-17, 1958, together with the corresponding College magnetogram. (Davis, T. N., 1962 [*23*].)

system (cf. Bless, Gartlein and Kimball 1953 [*102*]; Meek 1954 [*103*]; Kim and Currie 1958 [*99*]; Malville 1959 [*91*]; Evans 1960 [*104*]; Kim and Currie 1960 [*105*]; Bhattacharyya 1960 [*106*]; Davis 1962 [*36*]).

In the auroral zone, both northern and southern, auroras move, on the average, westward in the evening and eastward in the morning (Fig. 40a). Various radio studies of the auroral motions seem to agree with this conclusion (cf. Bullough,

Davidson, Kaiser and Watkins 1957 [*107*]; Nichols 1957 [*108*]; Lyon and Kavadas 1958 [*109*]). Radio studies seem to indicate a larger average speed (of order 600 m/sec) than all-sky camera studies (of order 400 m/sec) (see Sect. 41(c)). Brown (1962 [*110*]) obtained the east-west "motions" of precipitating energetic electrons from the observation of x-ray bursts.

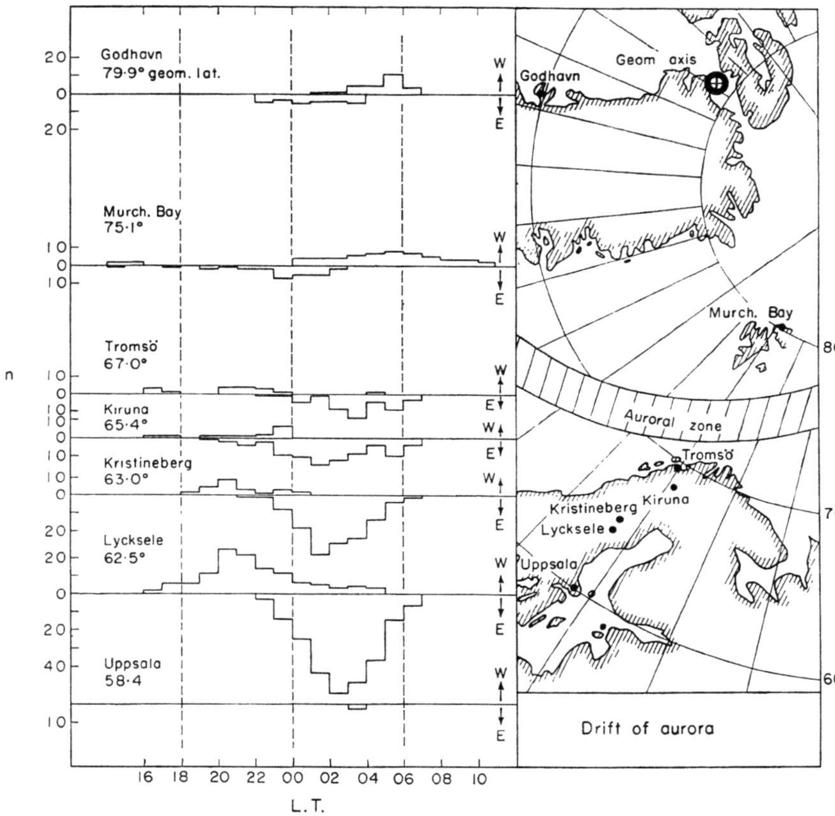

Fig. 40b. Frequency of occurrence of drift motions in E-W direction of auroral forms determined from all-sky films, versus local time (L.T.). (Stoffregen, W., 1961 [*111*].)

Recently, Stoffregen (1961 [*111*]) reported that in the auroral cap the direction of the motion is reversed, namely eastward in the evening and westward in the morning (Fig. 40b). Therefore, the whole pattern of the east-west motion in the auroral region has become clear.

In earlier literature, by "east-west motion", the authors seem to have referred to the motion of some characteristic features of the aurora, namely one or more of the following *apparent* motions of the auroras.

1. A progressive change of brightness along the arc, and the motion of rays along the arc (cf. Omholt (1962 [*112*])).
2. A westward traveling fold and the westward drift motion of loops.
3. A westward drift or progress of bands as a whole.
4. An eastward drift or progress of bands as a whole or of cloud-like patches.

It is not clear, and needs examination, whether the "east-west" motions detected by radio equipment and by photographic methods are the same. In the following, we show some of the characteristics of the above motions (cf. Akasofu 1963 [12]).

23. Westward traveling folds. In the simplest case, an arc simply waves when the fold passes along it; the arc is not seriously distorted, or displaced after the fold has passed on. An example is shown in Fig. 41 (Bettles, 0013—0029 AST,

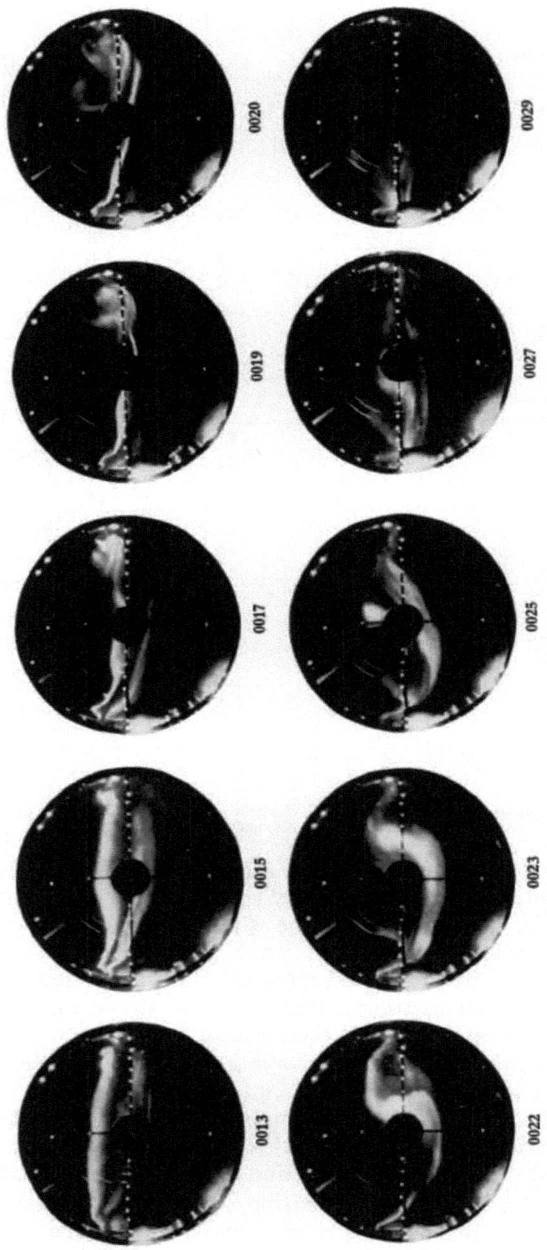

Fig. 41. An example of small westward traveling fold: Bettles, 0013-0029 AST, November 12, 1958.

Nov., 12, 1958; $Kp = 3--$). In more developed cases, the folds may greatly distort and displace the arc along which they move (Fig. 42).

Fig. 43 shows schematically the deformation of an arc due to a westward traveling fold, and also how all-sky cameras show such a fold when located as in the figure. In the northern hemisphere, the folding always develops to the north side of the arc, without exception. In the Antarctic the fold develops on

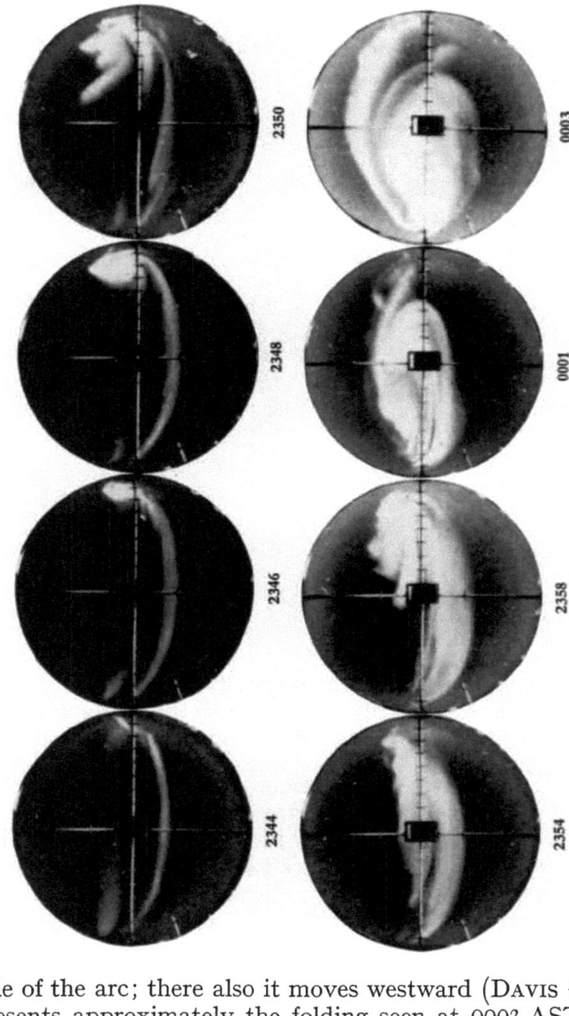

Fig. 42. The Fort Yukon all-sky camera photographs, 2344-0003 AST, February 14/15, 1958.

the south side of the arc; there also it moves westward (DAVIS 1963 [*113*]). The state C represents approximately the folding seen at 0003 AST at Fort Yukon in Fig. 43. An intense westward traveling fold may be called a surge.

24. Formation of loops. If the fold advances very rapidly the arc may become complicated, and a loop structure may be expected in the most developed stage *D* in Fig. 43. Fig. 44a shows a large loop seen at Bettles at 2211 AST, December 8, 1958; its north-south dimension exceeds 200 km. Fig. 44b shows the time sequence of its development, but (as often happens in complicated auroras) the early stages are not clearly seen. Near the western horizon, however, its form resembled stage

D in Fig. 43. The loop drifted with a speed of more than 300 m/sec. During the loop formation, we often see a sort of "bifurcation" of the band along which a fold propagates, resulting in a group of loops (Fig. 44c).

25. Pseudo break-up. Elvey (1957 [*57*]) defined the pseudo break-up as the disruption of one of the arcs near the center line of the auroral zone. The disruption is probably due to a well-developed fold traveling westward along one of the arcs from the morning side. Two examples of sudden and considerable distortion are shown in Fig. 45a and b. Although the auroras in these cases were

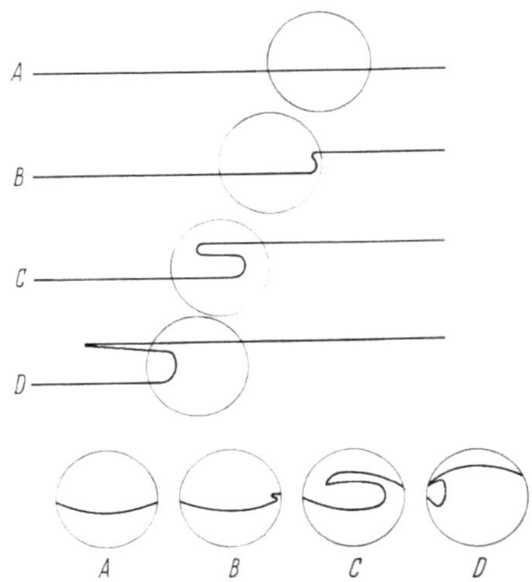

Fig. 43. The schematic representation of the deformation of an arc due to a westward traveling surge. (Akasofu, S.-I., 1963 [*12*].)

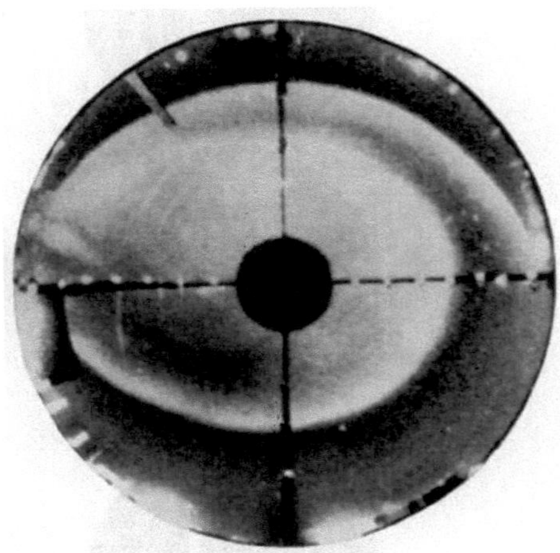

Fig. 44a. A large loop seen at Bettles at 2211 AST, December 8, 1958.

fairly bright and covered a large portion of the sky, their magnetic effects and the absorption of radio waves were small compared with those produced by a moderate break-up. Motions within the fold are usually fairly complicated.

26. Break-up. The break-up is the most violent auroral phenomenon; it is accompanied by a sudden and intense increase in brightness and motion. It often occurs within a few minutes. A complete description of the break-up over a large area is not available at present, because of the rapidity of the motions. So far as the present all-sky camera records indicate, the break-up process is characterized by a sudden increase of the brightness of all the arcs or bands over a large area on the morning side, together with north or southward motion of them as a whole, and rapid and complicated folding or wavy motions (of various characteristic lengths) of the arcs themselves. Figs. 46a and b show some examples of all-sky camera photographs during the break-up. Note the large changes of form in the all-sky camera photographs, taken every minute. It is sometimes difficult to distinguish between the pseudo break-up and the break-up from data at a *single* station (see Sects. 28—31).

If the break-up process is intense, a disruption of arcs or bands into patches occurs (the post break-up phase). The extent of the region in which the disruption occurs may correspond to the intensity of the break-up (Sects. 28—31). During the most intense break-up, the auroras on the evening side may be completely disrupted, and westward traveling folds may be seen even in the evening twilight sky.

27. Eastward drift motion. In the center and to the east of the break-up region, auroral arcs or bands move in something like a wave which progresses toward

Fig. 44b. The motion of the loop seen at Bettles, 2159-2228 AST, December 8, 1958.

the east. In most cases, their forms are so irregular that the word "wave" may not be appropriate, but in some cases their forms are simple and the characteristic length (corresponding to the wave length) is as much as a few hundred km. In general, in the auroral zone, such motion is often so rapid, also the wavy pattern changes by itself so rapidly, that it is not possible to follow the motion in any detail from all-sky camera films taken every minute. In the sub-auroral zone, however, rather simple wavy patterns are often seen (Fig. 47).

After the arc structure breaks up, cloud-like patches drift eastward. The patches can sometimes be identified in the all-sky films for at least ten minutes (Fig. 48).

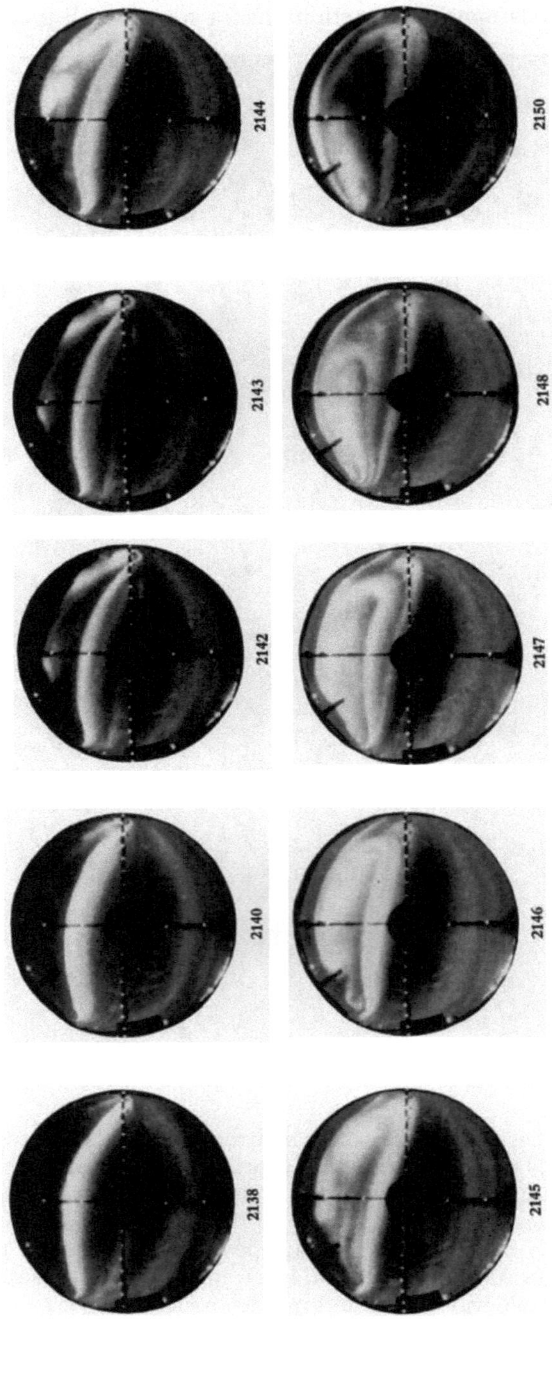

Fig. 44c. The formation of a group of loops recorded at Bettles, 2138-2150 AST, March 20, 1958.

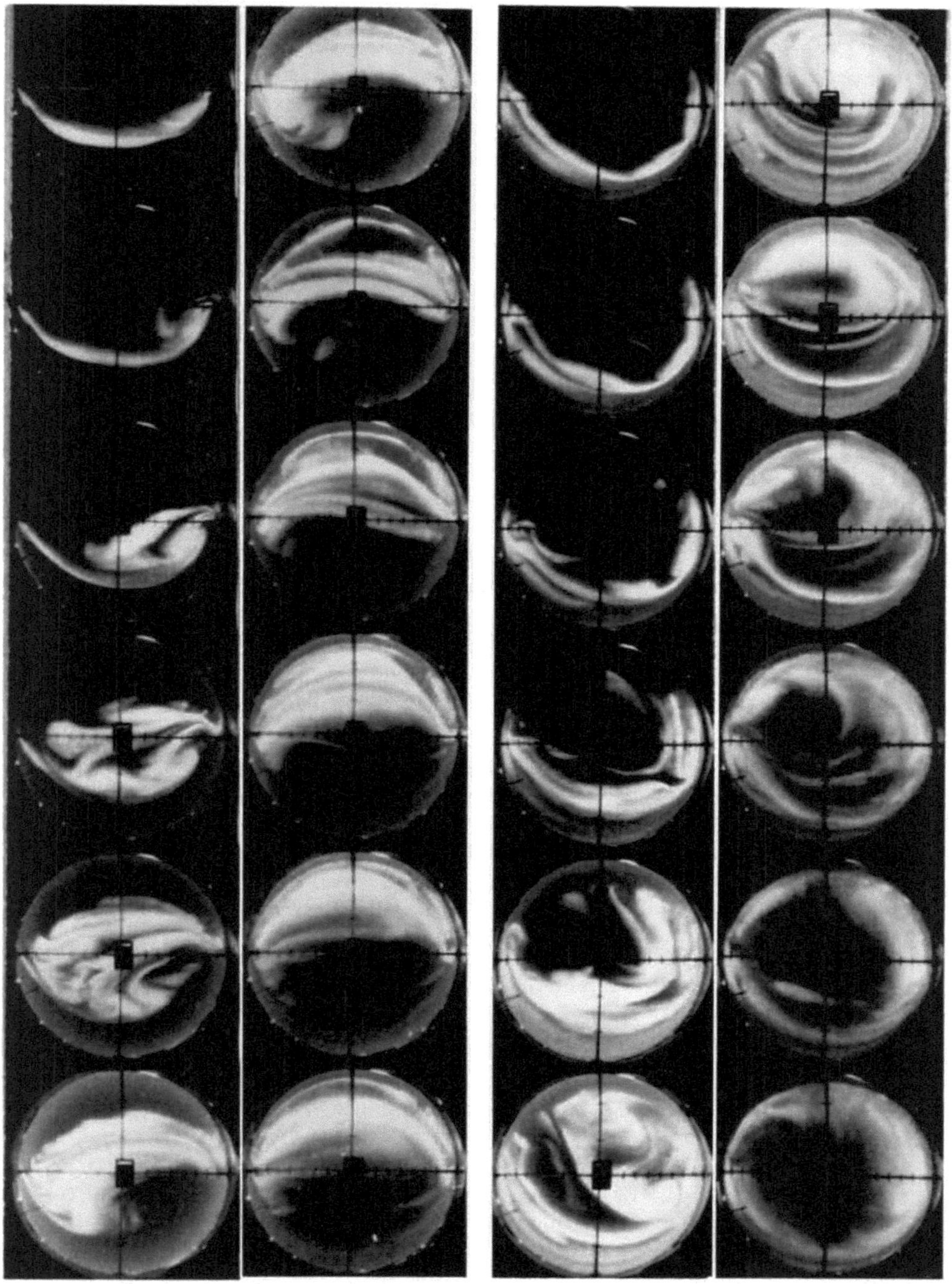

a b

Fig. 45a and b. Examples of the pseudo break-up of the auroras due to westward traveling surges.

C. Typical displays on a continental and on a planetary scale.

28. General. The aurora is a large-scale phenomenon that may extend along the entire auroral zone in the dark hemisphere. To understand the nature and origin of the aurora it is necessary to know the distribution of the arcs or bands, and their relations to other forms, over the whole polar region.

Davis (1962 [*36*]) made an extensive statistical study of the distribution and the alinement of auroral arcs over the whole polar sky. His study is based on detailed statistical studies of the daily variation of various auroral features at

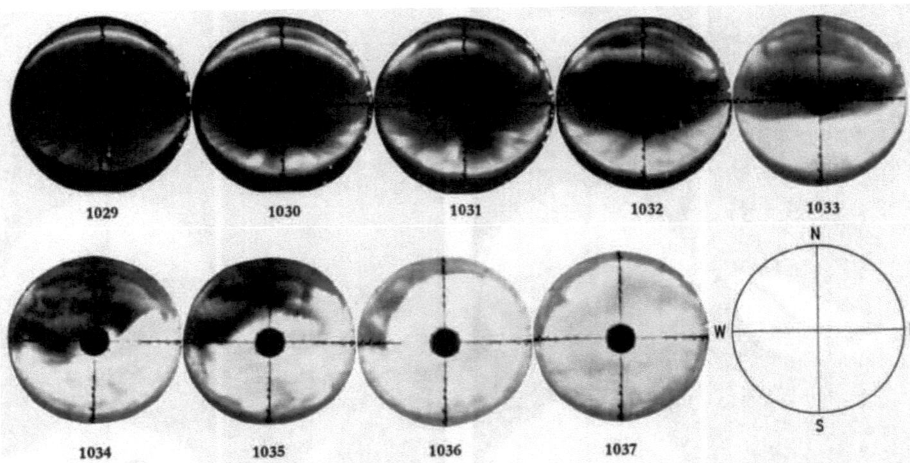

Fig. 46a. The break-up observed at Farewell, 1029-1037 GMT (0029-0037 AST), September 22, 1957.

single stations. The average characteristics thus obtained are then combined to show the morphological features of the auroral displays over the entire polar region.

Another way to approach the problem is to examine *individual* and *simultaneous* auroral displays over the whole polar sky. If this is done in a proper way, it may be possible to infer the general forms of auroral displays over such a large area. Further, various dynamical features of the aurora, such as the break-up, traveling folds, and loop formation, occur on a scale far beyond the view of a single station. This section presents a brief summary of an analysis made in this way (Akasofu 1963 [*12*] and [*58*]). (See also Khorosheva (1961 [*58a*])).

29. The aurora of February 13, 1958. The distribution of auroral forms on February 13, 1958 is shown by IGY all-sky camera data taken simultaneously over a large area, including eastern Siberia, Alaska, and western and central Canada (see also Sect. 18a).

At 0830 GMT on February 13, 1958 (Fig. 49a), at least three arcs extended across Alaska and a part of Canada, but only one was seen to the north of east Siberia. At 0842 GMT (Fig. 49b), about five minutes after the break-up, the auroras in the auroral zone in Alaska were completely disrupted. A few arcs rapidly moved northward there, and one of them went to the north of Point Barrow. In the south of the auroral zone, patchy structures are seen.

To the west of this region there was a remarkable loop formation (Sect. 24), which seemed to start at the time of the break-up of the auroras over the Bering Strait.

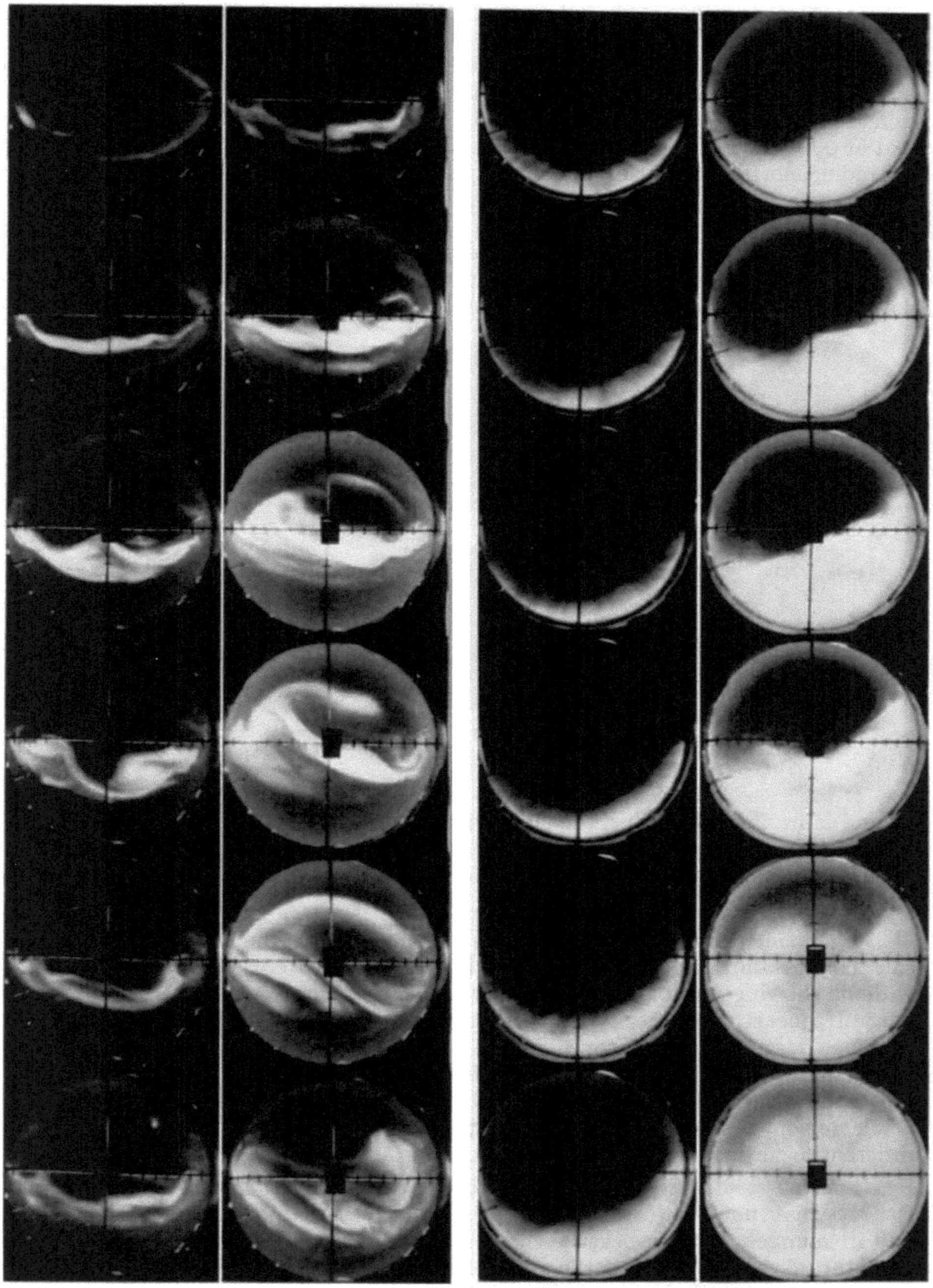

0004-0015 0030-0041

Fig. 46b. Examples of the break-up phase observed at Fort Yukon. (2236-2247 AST, February 12, 1958) (0055-0106 AST, February 15, 1958). Examples of the break-up phase observed at Fort Yukon. (0004-0015 AST, February 16, 1958) (0030-0041 AST, February 11, 1958).

Fig. 49b shows a great change at 0842 GMT in the latitude range of the aurora between Alaska and Canada. The strip in which overhead auroras are seen did not expand in latitude simultaneously all along the zone; over Alaska the expansion occurred earlier than over central Canada, although Canadian arcs increased in brightness when the break-up started in Alaska. Fig. 49c indicates that at 0900 GMT the expansion had spread to central Canada; the strip was still expanding northward when the auroras in Alaska became less active, and new regular arcs had been formed. By 0900 GMT the loops in the western sky of Alaska drifted away westward.

At 0907 GMT (Fig. 49d), a new break-up started in Alaska, and the broken bands became much brighter also in central Canada. At 1030 GMT (Fig. 49e), the auroras in Alaska and the Siberian Arctic Sea became quiet, and there were

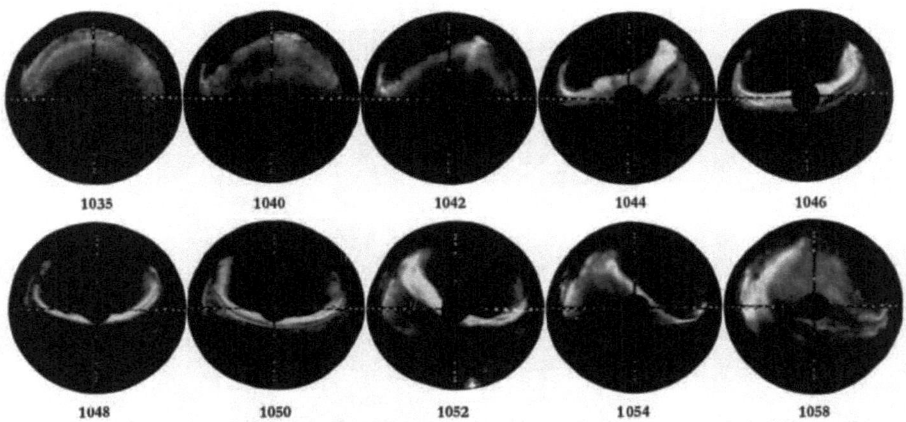

Fig. 47. The eastward progress of a wavy pattern of the auroral band observed at Choteau, 1035-1058 GMT, September 23, 1957.

several faint arcs. However, the aurora in Canada was still irregular and did not recover to this quiet phase, although it became fainter. The strip was still expanding to the north, and one of the arcs crossed the zenith of the Baker Lake station (gm. lat. 75° N).

30. A model case. The arcs or bands distributed over the whole auroral zone can remain faint and diffuse, keeping their (approximately) east-west direction without showing any particular activity. This situation occurs when the whole auroral zone is magnetically very quiet for several days, and sometimes even during a large magnetic storm, in the intervals between two DP sub-storms (Sect. 34). Such a quiet condition is disrupted by a sudden brightening of the arcs, mostly located in those on the early morning side of the auroral zone. The whole region of the auroral zone then becomes active, but in different ways at different local (magnetic) times (see Fig. 50).

a) Early evening side. With weak break-up around the midnight meridian the arcs may not be affected at all, or may increase only slightly in brightness and in number. This is because most of the disturbances associated with the break-up degenerate during their travel towards the early evening side. With moderate break-up, the arcs increase both in brightness and number. Their location seems to be confined to the vicinity of the auroral zone or slightly north of it. The H component of the magnetic field is slightly increased, or is not affected appreciably.

However, during a violent break-up, the arcs show ray structure or westward traveling folds. In such active cases, the arcs spread all over the auroral zone.

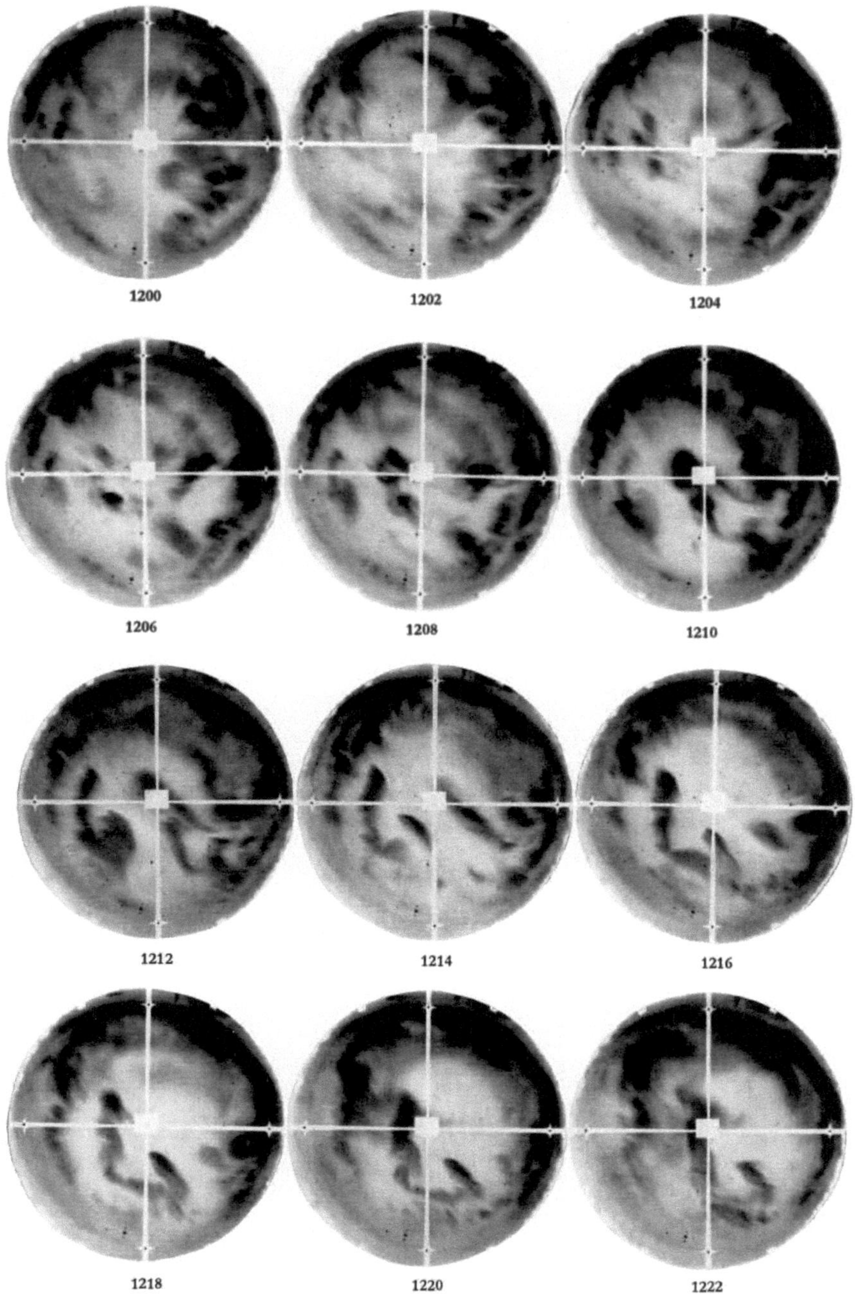

Fig. 48. The eastward drift of the broken-up patches observed at Meanook at 1210 GMT, February 14, 1958.

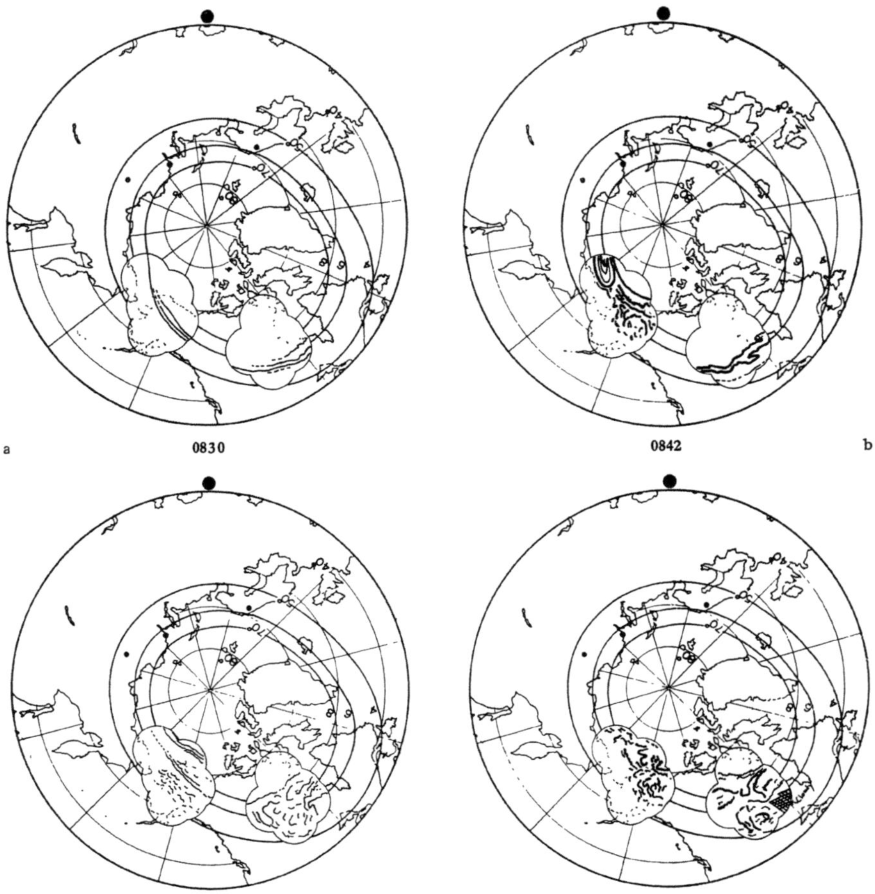

a 0830 0842 b

c 0900 0907 d

e 1030

Fig. 49a–e. The distribution of the auroras on February 13, 1958. The field of view of each all-sky camera station is indicated by a circle of radius 500 km. The dark dot on the top of each map shows the direction of the sun. McILWAIN's L curves for L 4, 6, and 8 are also shown. (AKASOFU, S.-I., 1963 [58].)

During the largest break-ups, even corona-type activity can be seen to the south of the auroral zone, e.g., College (gm lat. 64.7° N), in the evening twilight sky.

b) Evening side. With moderate break-up around the midnight meridian, the arcs increase not only in brightness and number, but also in motion. Westward

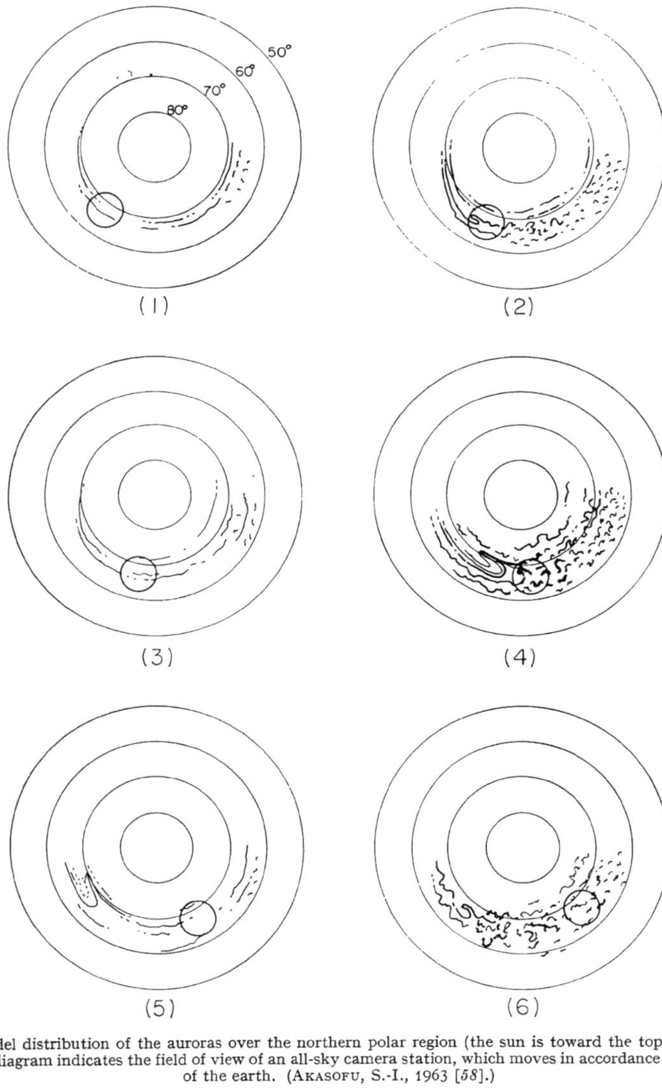

Fig. 50. A model distribution of the auroras over the northern polar region (the sun is toward the top of the diagram). A circle in the diagram indicates the field of view of an all-sky camera station, which moves in accordance with the rotation of the earth. (AKASOFU, S.-I., 1963 [58].)

traveling folds are often seen. If the intensity of the folds is weak, the arcs wave. When more intense, the arc is seriously folded. If the folding action is powerful, a group of loops may be formed.

After a loop is formed, it drifts westward at various velocities. If it covers an appreciable portion of the sky, the phenomenon is called the pseudo break-up. However, it does not produce a large change of the earth's field. With an intense break-up, the arcs are disrupted even on the evening side, and broken-up arcs drift rapidly westward. To the north of the normal auroral zone, auroral displays are much less intense than within the auroral zone except during intense break-up.

c) Morning side. With moderate break-up, most of the arcs in the auroral zone break up. Various structures thus formed progress rapidly eastward. In the northern part of the active region, new arcs are formed and rapidly advance poleward. Some of them go well inside the polar cap.

d) Late morning side. In a weak break-up, the arcs simply increase in brightness. In moderate and large activations, most of the arcs break up in the auroral zone and on its equatorial side. The broken-up patches drift eastward. In the northern part of the strip, however, the arcs tend to keep their ribbon-like structure.

The lifetime of the break-up phase is of order one or two hours. During a moderately active period, there is a succession of break-ups, every 2~3 hours around the midnight meridian. In the evening sector, the effect of the break-up is seen only as an increase in the brightness of arcs there. Thus the break-up is seen only when an observer is close to the midnight meridian, even though break-ups are occurring successively around the midnight meridian. Between two successive break-ups, the aurora becomes quiet all along the auroral zone.

31. The auroral display on a planetary scale. The aurora is produced by energetic particles entering the auroral ionosphere along the earth's magnetic field lines. The close agreement between the calculated auroral zone and the observed maximum isoaurore strongly supports this view (Sect. 4). Hence it is to be expected that the auroras seen at conjugate points (the points, one in each hemisphere, linked by a field line) will have a certain similarity. (Note that if the heights of the northern and southern mirror points differed significantly, for example, if they were at 300 km in the southern hemisphere, instead of at 100 km as in the northern hemisphere, the auroras at the conjugate points might not look the same.)

DE WITT (1962 [*114*]) has shown that the auroras seen from Farewell (gm lat. 61.4° N) and from Campbell Island (gm lat. 57.3° S) have similar forms and motions. Further, the conjugate auroras undergo similar variations in brightness, and break up simultaneously. According to the calculation by VESTINE and SIBLEY (1960 [*32*]), Campbell Island and Farewell are not exactly conjugate, Farewell being about 90 km north of Campbell Island's conjugate point. This displacement was seen in the aspect of the observed auroras. It may be noted that ANDERSON, ANGER, BROWN and EVANS (1962 [*115*]) observed simultaneous electron precipitations for the College-Macquarie Island conjugate pair (Sect. 51 c).

Other geophysical phenomena at conjugate points have recently been extensively studied (cf. NAGATA and KOKUBUN 1960 [*116*]; WESCOTT 1962 [*117*]; HOOK 1961 [*118*]; LEINBACH and BASLER 1963 [*119*]). From their studies, it seems certain that there are conjugate areas at which at least some of the geophysical phenomena are closely though not exactly correlated (see also WILSON (1963 [*120*])).

IV. The auroras and magnetic storms.

32. Solar relationships with the auroras and magnetic storms. *a) Three types of solar plasma flow.* DE MAIRAN (1754 [*1*], p. 264) discussed the possible association between auroras and sunspots, indicating series of years when both were abundant, and other series of years when both were few; but he noted that there were years when the correspondence was not good. In 1851 SCHWABE discovered the sunspot cycle of about 11 years. In 1852 SABINE (see [*4*], § 26.15) found parallel variations in disturbances of the magnetic declination at Toronto. But the auroral record was much less systematic than the magnetic record. As late as 1870, LOVERING (1870 [121]) doubted any close correlation between the sunspot number and auroral frequency.

MEINEL, NEGAARD and CHAMBERLAIN (1954 [50]) have shown, using a rather homogeneous series of auroral observations at Yerkes Observatory (gm lat. 53° N) for 55 years, 1897—1941, that the frequency of auroras visible at this subauroral latitude clearly shows an 11-year cycle, and that its minima coincide with the minima of the sunspot number; but the auroral maxima are delayed by about two years after the sunspot maxima (Fig. 51).

There is, however, considerable independence in the variations of the annual mean sunspot number and the annual auroral frequency. At Yerkes, their correlation coefficient is not more than 0.28 around the sunspot minima; for the years of sunspot maxima, it is only 0.05[1].

NEWTON and MILSON (1954 [122]) have shown that the average annual number of (a) giant sunspots, (b) intense solar flares, (c) great magnetic storms, and (d) magnetic storms with sudden commencements (ssc), all closely follow the sunspot cycle as defined by the annual mean sunspot number. The magnetic storms without sudden commencements, which on the whole are less intense than ssc storms, show a 2-years lag.

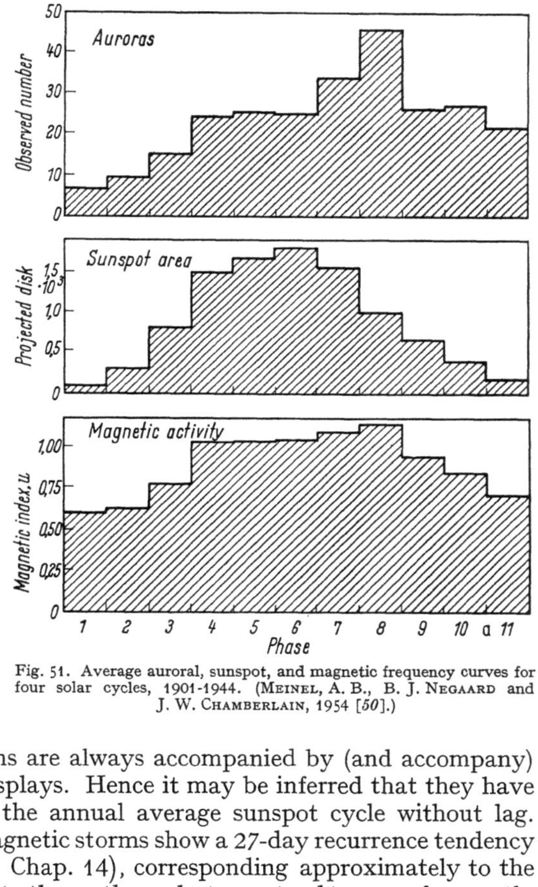

Fig. 51. Average auroral, sunspot, and magnetic frequency curves for four solar cycles, 1901-1944. (MEINEL, A. B., B. J. NEGAARD and J. W. CHAMBERLAIN, 1954 [50].)

Great and ssc magnetic storms are always accompanied by (and accompany) great or above normal auroral displays. Hence it may be inferred that they have an average frequency following the annual average sunspot cycle without lag. The moderate to weak non-ssc magnetic storms show a 27-day recurrence tendency (cf. CHAPMAN and BARTELS [5], Chap. 14), corresponding approximately to the rotation period of the sun relative to the earth; such storms tend to occur frequently during the declining phase of the sunspot cycle.

Therefore, it seems that there are two types of storm-time solar plasma flow; one originates from the region of a large solar flare (or a bi-polar magnetic region) and the other from an M region (or a uni-polar magnetic region). The former flow may occur sporadically with a solar flare and may be intense, but occupies a rather limited area at a particular instant. The flow may well be regarded as an advancing "shell"; on the other hand, the latter may be a rather continuous and wide-spread flow, a "stream", with less intensity than the former.

[1] Similarly poor relations between annual or monthly averages for sunspots and magnetic activity in certain selected groups of years have been indicated before (see [4], p. 374).

These two types of flow may be superposed on weaker but continuous flow from all over the sun; this is called the *solar wind*. PARKER (1958 [*123*]) originally suggested such a flow based on his argument that there could be no static situation in the interplanetary gas. Mariner II, one of the Venus probes, observed almost continuous plasma streaming from the sun (NEUGEBAUER and SNYDER 1962 [*124*]). The solar plasma flow was also studied by BONETTI, BRIDGE, LAZARUS, ROSSI and SCHERB (1963 [*124a*]). The relations between the solar plasma flows and the auroras are discussed in the following sections.

b) Possible geometrical relations. In the auroral regions, the long days and light nights in summer cancel or reduce the possibility of seeing the aurora; but the auroral statistics for such latitudes show an increase of auroral frequency at the equinoxes as compared with winter. In lower latitudes (e.g., for Yerkes) the auroral statistics extend throughout the year, and can be corrected for the varying duration of nightly darkness throughout the year. Like the statistics of magnetic activity, they then show two statistical minima and two equinoctial maxima of auroral frequency (MEINEL, NEGAARD and CHAMBERLAIN (1954 [*50*])).

It is not yet known with any certainty whether this is due to the changing orientation of the earth relative to the sun, or to the changing orientation of the sun, with its sunspot zones or M regions, relative to the earth. It is unlikely, however, that the two equinoctial maxima are purely due to some terrestrial effect.

33. Magnetic storm fields. The greatest and most widespread auroral displays always coincide with great magnetic storms. In order to examine the relationships between the aurora and magnetic storms, we analyze the magnetic storm field, denoted by D, into several components differently caused. Each component field is then correlated with detailed features of the auroral displays.

At magnetically quiet times the magnetic state of the earth and its near surroundings seems to be as follows. There is the main field M, proceeding from within. In the ionosphere, at a height of order 100 km, there are electric current systems (Sq and L) whose magnetic field at the earth's surface produces the solar daily and lunar daily magnetic changes. Further, there is the field R of the quiet-time ring current (AKASOFU, CAIN and CHAPMAN 1962 [*125*]) and the field CF of the solar wind.

When an advancing shell from a solar flare region or a continuous flow from an M region comes in contact with the earth's magnetic field, the field CF is enhanced. This enhanced component of CF may be denoted by DCF. The letter D signifies magnetic disturbance. When this enhancement occurs suddenly, it is noted as a *storm sudden commencement* (ssc). Then, in a typical case, the storm ring current begins to grow in the magnetosphere, resulting in the growth of the storm time R, denoted by DR (the field of the storm-time ring current). As we see in Sect. 35, there is a simultaneous growth of the field DP of *polar magnetic substorms* with the DR field. The letter P signifies *polar*. The symbol DP is used because this field and the currents that produce it are strongest in the polar regions, especially near the auroral zone. The currents are driven by electromotive forces rapidly set up there. Thence, the currents spread all over the earth, so that the DP field is world wide.

There may be a fourth addition to the pre-existing field during the storm. The solar plasma may carry with it a magnetic field transported away from the sun. This field may be denoted by DSM (disturbing solar magnetism). As yet, little is known about such fields, except that the space probe Pioneer V appeared to detect such a field in the interplanetary space (COLEMAN, SONETT and DAVIS 1961 [*125*]).

Thus, in all, the disturbing field D during a magnetic storm may be divided into the following four main parts.

$$D = DCF + DR + DP + DSM.$$

The first three components, DCF, DR and DP, are most clearly seen in the horizontal component magnetic records. Therefore, hereafter, by DCF, DR and DP, we refer to the horizontal component of DCF, DR and DP, respectively.

The field DCF increases the field at the earth's surface. CHAPMAN and FERRARO (1931 [*127*]) showed that assuming an advance of solar plasma, with a plane frontal surface, towards the earth, the field DCF can be approximated by the field of an image dipole located at a point at geocentric distance $2r$ along the earth-sun line, r being the geocentric distance of the frontal surface.

The DR field is negative and reduces the earth's surface field in low latitudes. It is likely to be the same, or nearly so, all round the earth. This field grows and then decays relatively gradually compared with the DP field.

Expressing D as the combination of familiar terms, namely an axially symmetric part Dst and the remainder DS, we may write approximately

$$Dst \sim DCF + DR \quad \text{in low latitudes}$$

$$DS \sim DP \quad \text{in the auroral zone}.$$

Typically, magnetic storms consist of three phases. The storm sudden commencement is followed by a rather steady positive phase. This phase is called the *initial phase*. The major component of D during this phase is the DCF field. The initial phase is terminated by the growth of the ring current field DR. Because DR is negative at the earth's surface, the Dst values become negative, if DR becomes larger than DCF. The second phase, the *main phase*, starts when the DR field overtakes the DCF field. After the Dst value attains its largest negative value, the DR field begins to decay. This phase may be called the recovery phase. In Sect. 34, we examine the relationships between the aurora and the DP fields, and in Sects. 35 and 36 those between the aurora and the Dst field.

The growth and decay of the combined field ($DCF + DR$) can be seen approximately from the Dst *curve*, although there is no way to separate DCF and DR from the ground magnetic observations. The activity of polar magnetic substorms DP for the whole planet earth, indicating some characteristics in the stream of solar corpuscles, is expressed approximately by the Kp index for intervals of three hours. It is a *semi-logarithmic* index, and is expressed in a scale of 28 steps, 0_0, 0+, 1−, 1_0, 1+, 2−, — to 8+, 9−, 9_0. In order to obtain a linear measure, the Kp index can be converted into the ap index according to the following table.

$Kp =$	0_0	0+	1−	1_0	1+	2−	2_0	2+	3−	3_0	3+	4−	4_0	4+
$ap =$	0	2	3	4	5	6	7	9	12	15	18	22	27	32
$Kp =$	5−	5_0	5+	6−	6_0	6+	7−	7_0	7+	8−	8_0	8+	9−	9_0
$ap =$	39	48	56	57	80	94	111	132	154	179	207	236	300	400

At a standard station at about 40° distance from the geomagnetic pole, ap may be regarded as the range of the most disturbed of the three field components expressed in the unit of 2 gammas (cf. BARTELS 1962 [*128*]). The series 1932—1961 has been discussed by BARTELS [*129*]; a shorter series has been previously interpreted with respect to the possible occurrence of aurora in lower latitudes (BARTELS and CHAPMAN 1958 [*130*]).

34. The aurora and polar magnetic substorms. *a) The break-up phase and auroral electrojets.* The break-up phase of the aurora is in general accompanied by an intense and concentrated electric current, an auroral electrojet. The

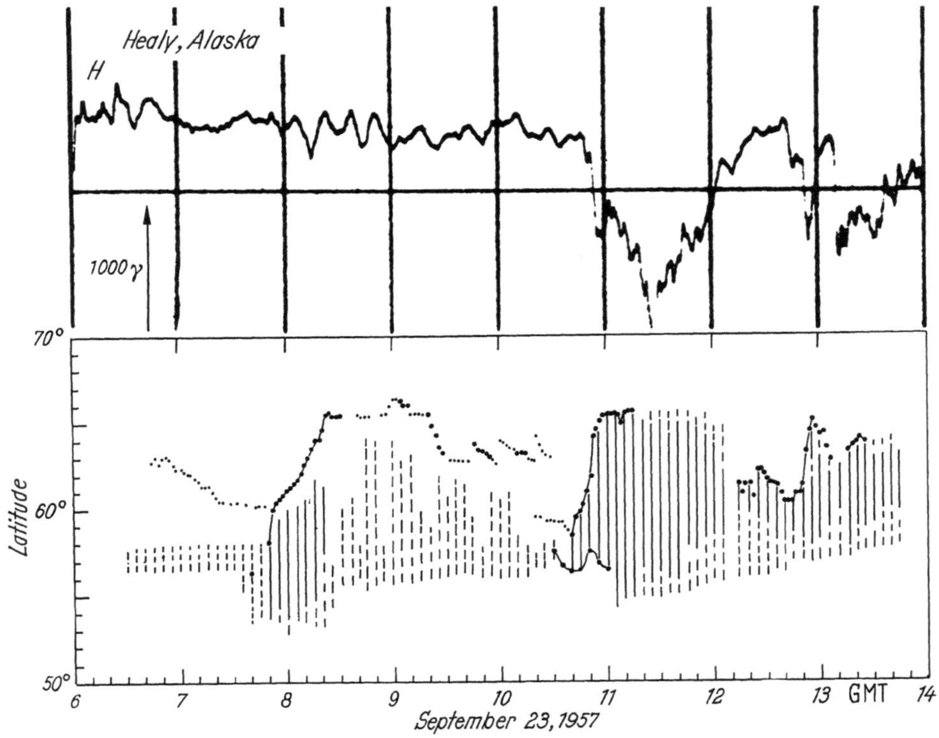

Fig. 52. The H component magnetogram of Healy (Alaska) station, 0600-1400 hours GMT, 23 September 1957. The successive positions of the auroras in the Alaskan sky (approximately in gm. longitude 250°) between 0600 and 1400 hours GMT September 1957. The positions of bright arcs are indicated by large dots, faint arcs by small dots. Solid lines indicate the latitude range in which bright, broken-up auroras were observed. (AKASOFU, S.-I., 1962 [*131*].)

magnetic field DP produced by such a current system is largest under breaking-up auroras, and can be as large as $2000\,\gamma$ or more, far greater than the fields DCF ($<200\,\gamma$) and DR ($<600\,\gamma$).

Fig. 52 shows the range of geomagnetic latitude in which overhead auroras were seen at different times in the Alaskan sky (approximately in gm long. 250°) on September 23, 1957. Large dots indicate the positions of bright arcs, and small dots faint ones. Solid lines indicate the latitude range in which bright broken-up auroras were observed. Broken lines indicate the latitude range in which broken-up auroras were overhead, or the approximate position of a faint arc or band.

The period (0600—1400 GMT, September 23, 1957, when Kp was 8— in all three intervals) witnessed three successive intense magnetic sub-storms (AKASOFU 1962 [*131*]). As mentioned already in Sect. 21, the center of auroral activity shifts equatorwards during the main phase of large magnetic storms. In Fig. 52, both the northern and southern borders of the aurora shifted equatorward, and the usual region of most frequent aurora (between gm lat. 66 and 67° N) was often completely deserted.

In addition to the equatorward shift, there were also large and rapid changes of the range of latitude in which auroras appeared overhead. Rapid northward

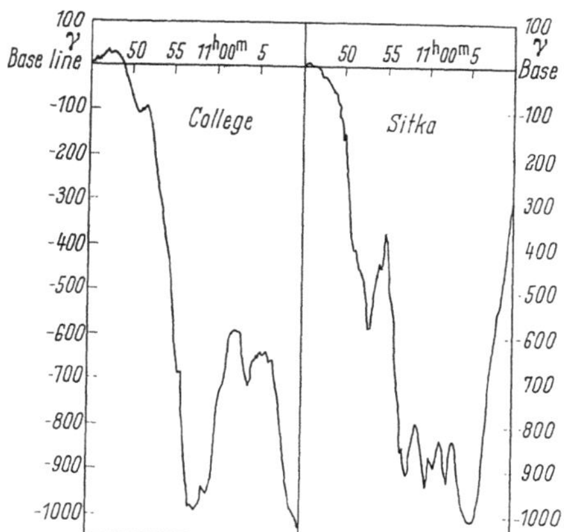

a H-Component $10^h45^m-11^h10^m$ GMT Sept. 23, 1957

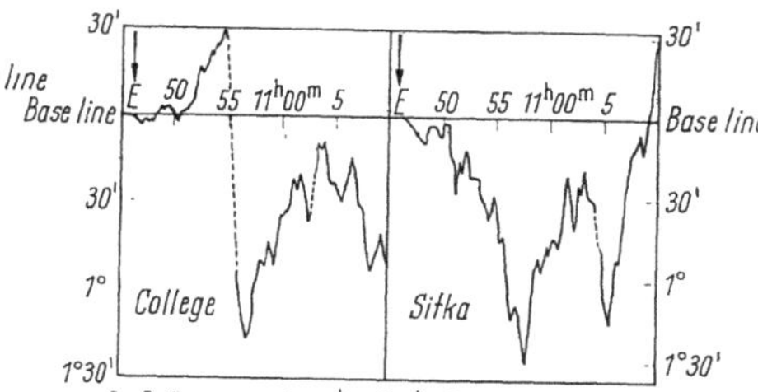

b D-Component $10^h45^m-11^h10^m$ GMT Sept. 23, 1957

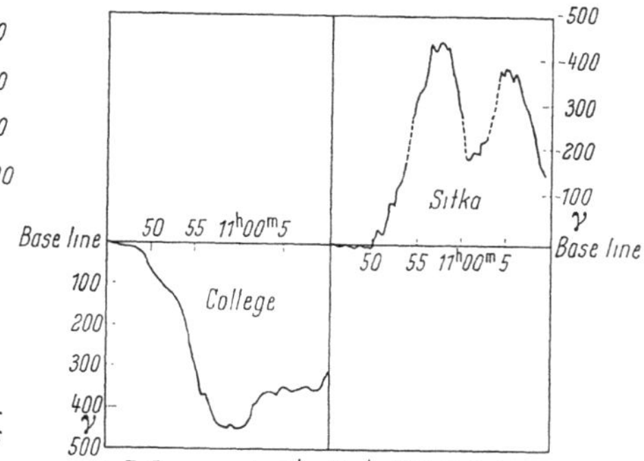

c Z-Component $10^h45^m-11^h10^m$ GMT Sept 23, 1957

Fig. 53a–c. Magnetic records constructed from the rapid-run and the low-sensitive magnetograms for the polar magnetic storm of 1050 GMT, 23 September 1957. (AKASOFU, S.-I., 1960 [*100*].)

expansions led by northward motions of bright auroras were followed by subsequent coverage of broken-up auroras in the area swept by the bright auroras.

The first expansion, starting at about 0750 GMT, was accompanied by the break-up of the aurora in central Canada. Because the Alaska area was on the evening side (1000 GMT = 0000 AST), there was only a little indication of the development of the DP field there. The second expansion, starting at about 1040 GMT, was accompanied by a large development of the DP field, exceeding 1000 γ. Fig. 53a–c shows the magnetic records from College (gm lat. 64.7° N) and Sitka (gm lat. 60.0° N). Detailed analyses show that an electrojet was flowing approximately along the circles of gm latitude. In such a rather simple case, it may be possible to find the location of the jet by triangulation, using records like those of Fig. 53a–c, assuming that the electrojet was essentially a line current. The geometry of the triangulation is given in Fig. 54. This method was first used by BIRKELAND (1913 [*132*]), but his estimated heights are too great, as CHAPMAN (1935 [*81*]) pointed out, because no account was taken of the induced currents within the earth (for details, see AKASOFU 1960 [*100*]).

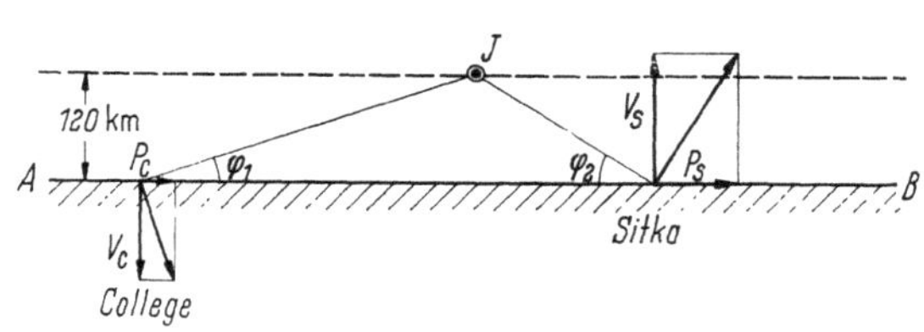

Fig. 54. Geometry showing parameters determining a position of auroral electrojet. (AKASOFU, S.-I., 1960 [*100*].)

Fig. 55 shows the position of the auroral jet at each minute for three successive intervals during the second expansion. Clearly the jet moved rapidly northward or southward. The speed of the first northward motion, which was accompanied by the rapid northward motion of a bright aurora (Fig. 52), was 1 km/sec. During this period the jet grew rapidly at about 1054 GMT; the total current intensity was 2.5×10^6 amp.

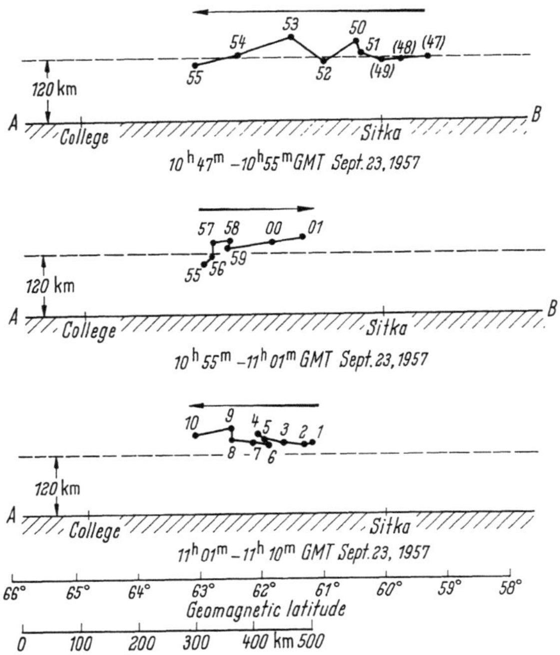

Fig. 55. Positions of the auroral electrojet between College and Sitka at each minute for three successive stages (1047-1055, 1055-1101, 1101-1110 GMT, 23 September 1957.) (Akasofu, S.-I., 1960 [100].)

As mentioned in Sect. 18, there is a normal sequence of three phases in an auroral display. The second active phase, the break-up phase, is accompanied by the growth of auroral electrojets. In a simple case, there may be only one such sequence during the course of a night, so that the jet grows once: Heppner's diagram (Fig. 34) shows such an example. Fig. 56 shows the H component magnetogram from College, about 200 km south of Fort Yukon, on February 12/13, 1958, corresponding to the auroral display at Fort Yukon (Sect. 18a) on that night (Fig. 32). (See also Fig. 63 in Sect. 35.) Polar magnetic substorms are shaded for emphasis. There were three major DP's, and corresponding break-ups of the auroras as seen from Fort Yukon (Sect. 18a), illustrating the intermittent and sporadic nature of polar magnetic substorms.

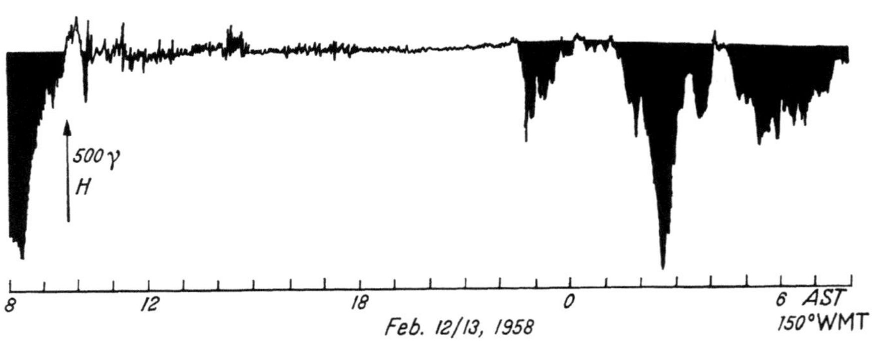

Fig. 56. The H component College magnetogram, February 12/13, 1958. Negative DP's are shaded for emphasis. (Akasofu, S.-I., 1963 [12].)

The aurora is not necessarily accompanied by a polar magnetic storm. It is the second (break-up) phase that is associated with DP's. During the third phase also the *DP*'s can be intense: this phase is usually accompanied by rapid fluctuations of the magnetic field (Sect. 37).

b) The electromotive force. The detailed distribution of electric current in the auroral region cannot as yet be deduced from the available magnetograms because of its complexity. But a reasonable illustrative model can be given, indicating some probable features of the disturbance. To obtain the current density we treat the auroral electrojet as a uniform sheet with current density J and width $2w$. The analysis in Sect. 34a suggests that w is less than about 500 km, the distance between College and Sitka (along the geomagnetic meridian line). Corresponding to the conditions at 1056 hours, we may take h, the height of the layer, to be 120 km, $w=120$ km, and l the horizontal distance from the center of the layer 240 km. These give a horizontal intensity of the magnetic field P due to this current sheet (CHAPMAN 1951 [*133*]):

$$P = 2J \tan^{-1} \frac{2w/h}{1 + (l/h)^2 - (w/h)^2}.$$

Thus, for the disturbance $P = 515\,\gamma$, two-thirds of the actual value of $770\,\gamma$ at 1056 GMT, J would be approximately 5.6×10^{-3} e.m.u./cm $= 5.6\times 10^{-2}$ A/cm. This might be compared with the equatorial noon current density of the solar daily variation S_q of about 1.9×10^{-4} A/cm (CHAPMAN and BARTELS 1940 [*5*], p. 233) and the equatorial electrojet of 7.86×10^{-4} A/cm ($=8650$ A/°lat., ONWUMECHILLI [*134*] 1959, his Table 3). The total current is 5.6×10^{-2} A/cm $\times$ 2.4×10^{7} cm $= 1.3\times 10^{6}$ A.

The electric current equations in the ionosphere (BAKER and MARTYN 1953 [*135*], p. 285) may be written:

$$J_x = K_{xx} E_x + K_{xy} E_y,$$
$$J_y = -K_{xy} E_x + K_{yy} E_y;$$

K_{ii} and K_{ij} respectively denote the height integrals of the direct and transverse conductivities: the x and y axes coincide with geomagnetic south and east. The conductivities in the ionosphere are not properly known, but it has been concluded (MAEDA 1953 [*136*]), that in the polar region K_{xy} exceeds K_{xx} and K_{yy}. As a basis for later numerical discussion, FEJER's values (1953 [*137*]) for the geomagnetic latitude 60° are adopted.

$$K_{xx} = K_{yy} = 4\times 10^{-9} \text{ e.m.u.}$$
$$K_{xy} = 1.4\times 10^{-8} \text{ e.m.u.}$$

In the disturbances here considered, the currents were westward ($-y$ direction), so that $|J_y| \gg |J_x|$.

Thus the westward electric current might be caused by a southward ($+x$ direction) electric field. If so, the current equations become

$$J_x \simeq K_{xx} E_x$$
$$J_y \simeq -K_{xy} E_x.$$

It is not known how much the electron density is increased in the ionosphere at the time of an auroral display. Assuming an average twenty-fold increase ($K_{xy} = 2.8\times 10^{-7}$ e.m.u.), the last equation gives

$$E_x = \frac{J_y}{K_{xy}} = \frac{5.6\times 10^{-3}\ (\text{e.m.u.})}{2.8\times 10^{-7}\ (\text{e.m.u.})} = 2\times 10^{4}\ (\text{e.m.u.})$$
$$(= 2\times 10^{-4}\ \text{V/cm} = 20\ \text{V/km}).$$

This might be compared with the electric field responsible for the S_q variation, produced by a dynamo e.m.f. of about 1.7×10^{-3} e.m.u. ($=1.7\times10^{-5}$ V/cm), induced by a wind velocity of order 30 m/sec (= 108 km/hr).

One of the outstanding features of the display of 23 September 1957 was the sudden brightening and simultaneous appearance of rapid eastward motion of the auroras. A sudden considerable negative deviation of the H-component from its normal value (which in most polar magnetic disturbances is the most prominent feature of the magnetograms) is generally accompanied by mainly eastward auroral motion. This has been studied by Kim and Currie (1958 [*99*]), using all-sky camera data, and by Kaiser (1955 [*138*]), Bullough and Kaiser (1954, 1955 [*139*]), Bullough, Davidson, Kaiser and Watkins (1957 [*107*]), Dagg (1957 [*140*]) and Nichols (1959 [*141*]), using radio data.

Table 4. *Typical values of ν_i, ν_e, α_e, and α_i at heights 110, 130 and 200 km.*

Height	110 km	130 km	200 km
ν_e (/sec)	3×10^4	5×10^3	10^3
ν_i (/sec)	2×10^3	3×10^2	10
α_e	89° 50′	89° 58′	90°
α_i	9°	48°	88

$\omega_e=9.6\times10^2$/sec, $\omega_i=3.3\times10^6$/sec.

Another quite independent estimate of the electric field can be made from the velocity of auroral drift motions (Akasofu 1960 [*100*]; Chamberlain, Kern and Vestine 1960 [*142*]). Assuming that the magnetic field is vertical (which in auroral latitudes is nearly true, though strictly true only at the poles), Martyn (1953 [*143*]) gave the following horizontal drift velocities of charged particles (v_e for electrons, v_i for ions) due to an applied electric field:

$$|v_e|=(E/F)\sin\alpha_e,$$

$$|v_i|=(E/F)\sin\alpha_i.$$

Here

$$\tan\alpha_{e,i}=\omega_{e,i}/\nu_{e,i},$$

$$\omega_{e,i}=eF/m_{e,i}$$

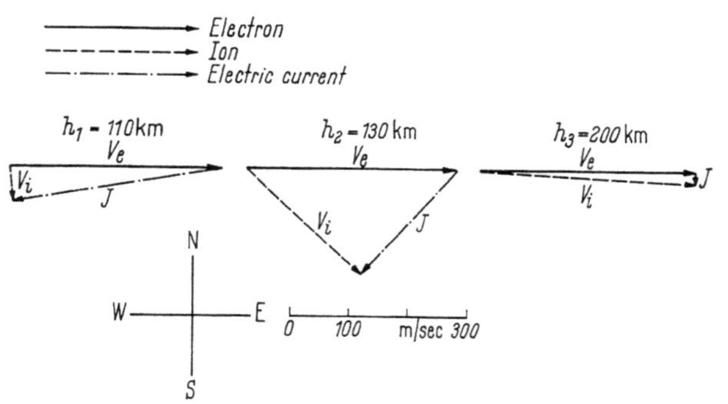

Fig. 57. Electrons and ions drift velocity vectors at 110, 130 and 220 km in height due to the southward electric field of 2×10^4 e.m.u. (= 20 V/km) in the polar region. (Akasofu, S.-I., 1960 [*100*].)

where e and $m_{e,i}$ are the charges and the masses, and ν_e and ν_i denote the collision frequencies of electrons and ions with neutral atoms, respectively; F denotes the total intensity of the magnetic field. Table 4 gives typical values of ν_e, ν_i, and α_e, α_i. Fig. 57 gives corresponding values of v_e and v_i due to the southward electric field of 2×10^4 e.m.u. (= 20 V/km). It is rather surprising that the eastward drift velocity of electrons is almost constant over the height range.

Fig. 57 also shows the electric current vectors. Clearly the eastward auroral electrojet discussed in the above section is mainly due to the eastward motion of electrons in the lower part of the ionosphere with speeds of order $E/F=360$ m/sec. This drift of electrons is almost insensitive to the physical state in the ionosphere region, because of the high gyrofrequency ω_e, compared with the collision frequency ν_e. Therefore this estimation has a great advantage compared with the discussion in the above section, which strongly depends on the conductivity of the ionosphere and so on the electron density.

c) The current system of polar magnetic substorms. The auroral jet currents are limited to certain longitude sectors of the auroral zone. They set up a polarization field at their eastern and western ends. This causes the electric current to complete its circuit over the polar cap, and also in lower latitudes. Thus, changes of electromagnetic condition in the auroral zone are communicated even to the equatorial zone, without delay (within the time accuracy of the observation).

The Second International Polar Year

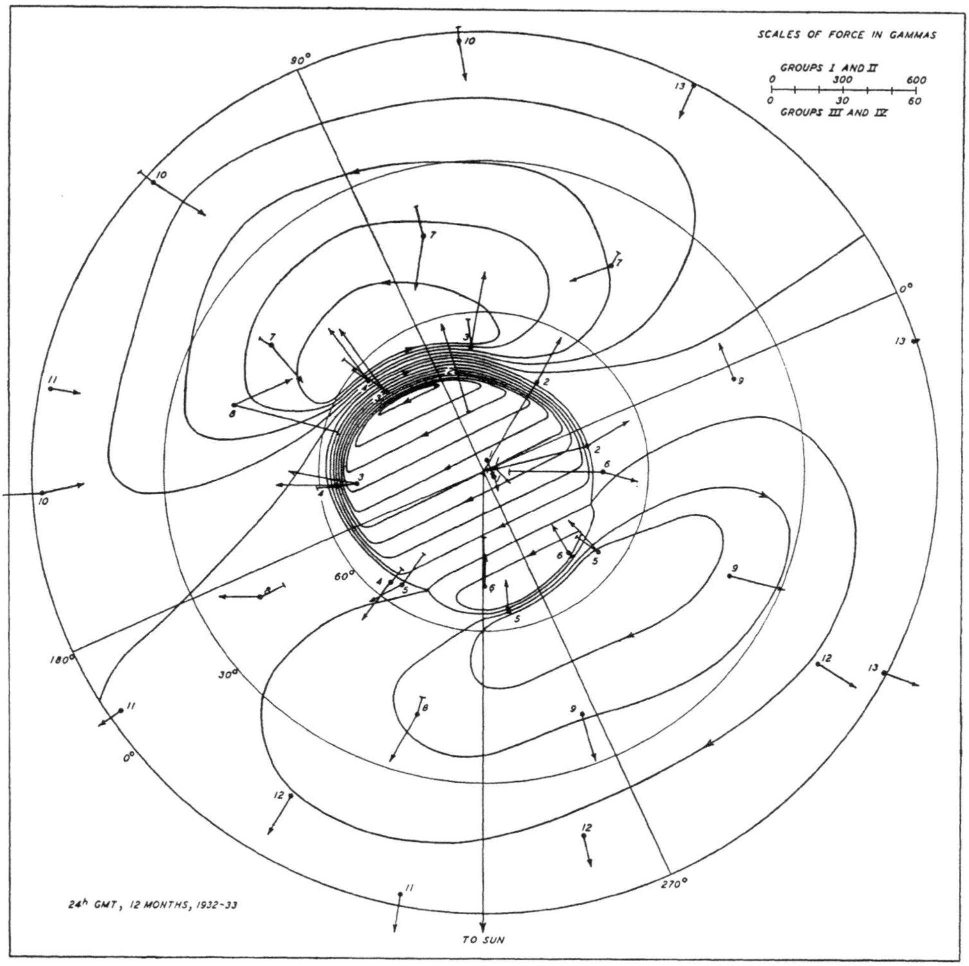

Fig. 58. Mean 3-hour disturbance vectors of magnetic bays centered at 21 hr, 24 hr and 3 hr GMT (tentatively corrected for induced currents) and corresponding average electric-current system of height 150 km. View from above Geomagnetic North Pole. (50000 A flow between successive current lines). (SILSBEE, H. C., and E. H. VESTINE, 1942 [*145*].)

The *DP* disturbance, originating in the polar regions, becomes a worldwide disturbance. Current diagrams for this kind of disturbance have been drawn by CHAPMAN (1935 [*81*]), VESTINE (1940 [*144*]), SILSBEE and VESTINE (1942 [*145*]) and FUKUSHIMA (1953 [*146*]). (See also WHITHAM, LOOMER and NIBLETT 1960 [*147*]; BHATTACHARYYA 1961 [*148*]). The current diagram given by SILSBEE and VESTINE is reproduced here as Fig. 58. The current is greatly concentrated on the early morning side of the auroral zone. It is the area where the most intense break-up occurs (Sect. 30). AKASOFU and CHAPMAN (1963 [*149*]) have shown also that the equatorial electrojet may be greatly enhanced when auroral electrojets appear in the auroral zone. Like active auroras, *DP* disturbances are intermittent and sporadic, but their scale is more extensive, often worldwide. They are impulsive, and their life time is a few hours at most. Between the disturbances the magnetic condition is rather quiet.

d) The latitudinal variation of the aurora with the Kp-index. The variation of the distribution of overhead auroras with magnetic activity has long been discussed. It appeared that at a station in the subauroral zone the occurrence of overhead auroras increases with the increase of magnetic activity. GARTLEIN (C. W. and H. E.) and SPRAGUE (1960 [*150*]), using IGY visual observations, compared the latitudinal distribution of overhead auroras with the planetary 3-hourly index *Kp*.

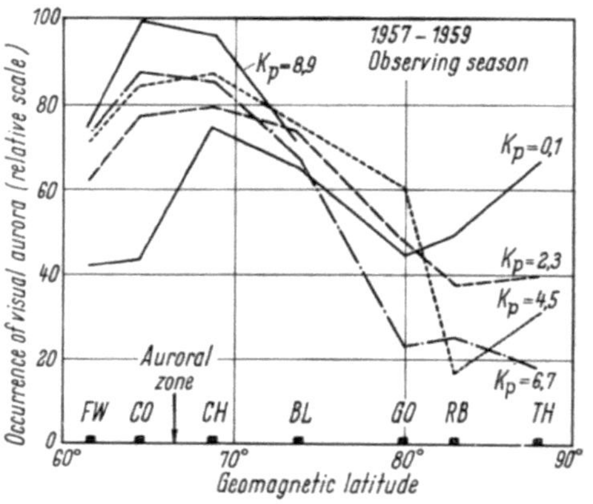

Fig. 59. The occurrence of visual aurora plotted as a function of geomagnetic latitude for various levels of planetary magnetic activity. Observations come from Farewell, Alaska (FW), College, Alaska (CO), Churchill, Canada (CH), Baker Lake, Canada (BL), Godhavn, Greenland (GO), Resolute Bay, Canada (RB), and Thule, Greenland (TH). (DAVIS, T. N., 1963 [*151*].)

DAVIS (1963 [*151*]) compared *Kp* with the occurrence of auroras between gm lat. 60° and 90°, using all-sky camera data from seven Canadian stations and Alaskan stations (Fig. 59). Between gm lat. 60° and 72°, the auroral occurrence frequency increases in general with increase of the *Kp* index. Further, the latitude of maximum frequency shifts equatorwards as *Kp* increases. On the other hand, above gm lat. 80°, the occurrence frequency tends to decrease as *Kp* increases. The zone between 72° and 80° is a transition region between the two. This negative correlation is discussed in Sects. 5b and 19 (see also AKASOFU and KIMBALL (1964 [*152*])).

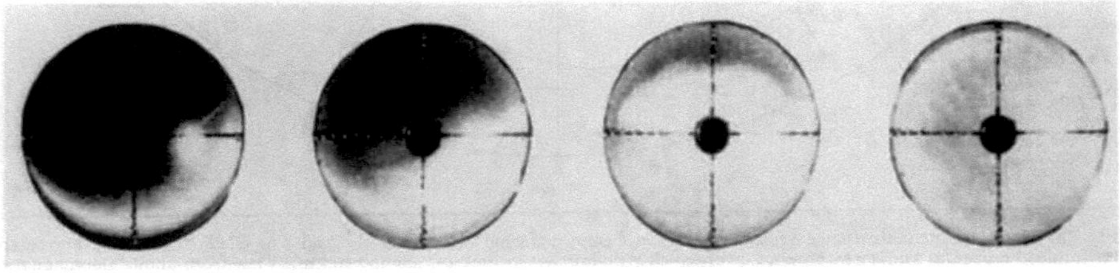

Fig. 60. Pullman (gm lat. 53.5° N) all-sky camera photographs, 0813-0820 GMT, 13 September 1957. (AKASOFU, S.-I., and S. CHAPMAN, 1963 [*152*].)

35. The aurora and the main phase. *a) The lower limit of latitude (U.S. sector) of northern quiet arcs, and its relation to Dst(H).* A striking feature of magnetic storms with an appreciable main phase is that auroras then become visible in middle or low latitudes. One major cause of this tendency is closely related to the main phase decrease. This can easily be examined, because even during an exceptionally large magnetic storm there may be relatively calm intervals, free from intense polar magnetic substorms, but yet low latitude auroras are seen during such intervals.

A typical example occurred during one of the largest magnetic storms in the IGY period. Around 0810 GMT, September 13, 1957, almost at the maximum epoch of the main phase (of great magnitude, $-390\,\gamma$), there was a fairly quiet interval free from polar magnetic storms. At that time a quiet arc was overhead in gm lat. as low as 49.5° N. Fig. 60 shows the all-sky camera photograph taken between 0813—0820 GMT, September 13, 1957, at Pullman (gm lat. 53.5°). At 0814 GMT, a new polar magnetic substorm started, and the arc became quite active.

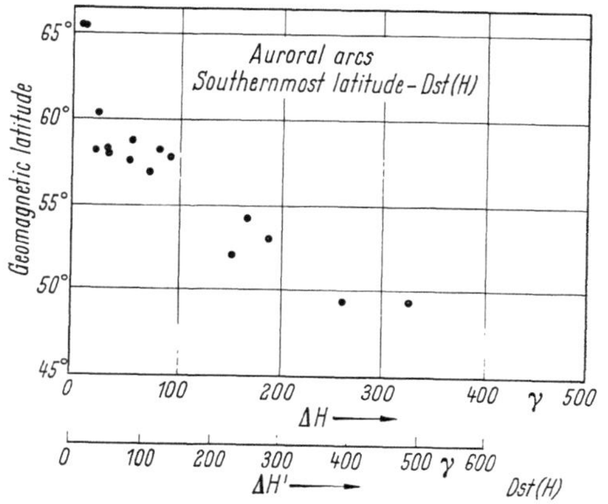

Fig. 61. Relation between the lower limit of latitude attained by quiet arcs in the US sector during 16 magnetic storms during the IGY, and the intensity of the magnetic field produced by the ring current, denoted by $\Delta H'$ and also ΔH, where $\Delta H = 2\Delta H'/3$. (AKASOFU, S.-I., and S. CHAPMAN, 1963 [*48*].)

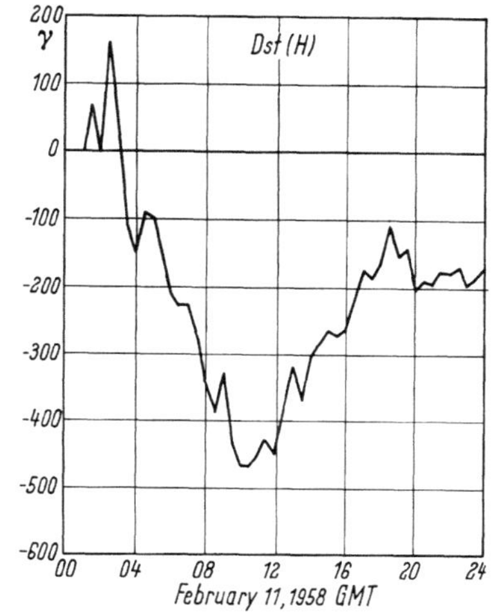

Fig. 62. The Dst (H) curve for the storm of 11 February 1958. It is based on the records of eight low latitude magnetic observatories. (AKASOFU, S.-I., and S. CHAPMAN, 1962 [*46*].)

AKASOFU and CHAPMAN (1963 [*48*]) showed that the lower limit of latitude (U.S. sector) of northern quiet arcs is rather simply related to the magnitude of the main phase decrease. In Fig. 61, the southernmost latitudes are plotted against the *Dst* (H) decrease, denoted by $\Delta H'$; a scale is given also for ΔH, which is $^2/_3$ of $\Delta H'$. We assume that $^1/_3$ of $\Delta H'$ is due to induced earth current, so that ΔH represents approximately the external magnetic field responsible for the main phase. The analysis was based on data from the all-sky camera stations Choteau (gm lat. 55.5° N), Pullman (gm lat. 53.5° N), Rapid City (gm lat. 53.1° N) and Redmond (gm lat. 50.5° N), during 16 IGY magnetic storms. We show later (see Sects. 51a and 55) that during the main phase of a large magnetic storm, the zone of strong x-ray bursts shifts equatorwards; similarly the outer boundary of the outer radiation belt, beyond which there are no appreciable trapped electrons, contracts and shifts equatorwards.

The main phase decrease is ascribed to the growth of the ring current encircling the earth in the radiation belt region (cf. AKASOFU and CHAPMAN 1961 [*153*]; AKASOFU, CAIN and CHAPMAN 1961 [*154*]; AKASOFU 1963 [*155*]). It may be inferred that the formation of quiet arcs is closely related to the ring current.

b) The aurora of February 11, 1958. The magnetic storm of February 11, 1958 had a large sudden commencement (ssc) at 0125 GMT; a second ssc occurred shortly afterwards, at 0159 GMT. A large Dst developed, with maximum epoch at 1000 GMT; the peak decrease of the Dst was about 480 γ (Fig. 62). From 00 to 12 GMT, the four intervals gave $Kp=9_0$, 8+, 9−, 8+.

As the ring current grew and the Dst decrease developed, *both the northern and southern borders* of the zone in which overhead auroras were seen moved equatorwards. A remarkable feature was the absence of overhead aurora over a large region on the poleward side of the zone. At times the aurora was completely absent from the region of the auroral zone, where normally it is most frequent and intense.

However, during several intense polar magnetic substorms that occurred in the course of the growth and decline of the Dst decrease, the zone expanded

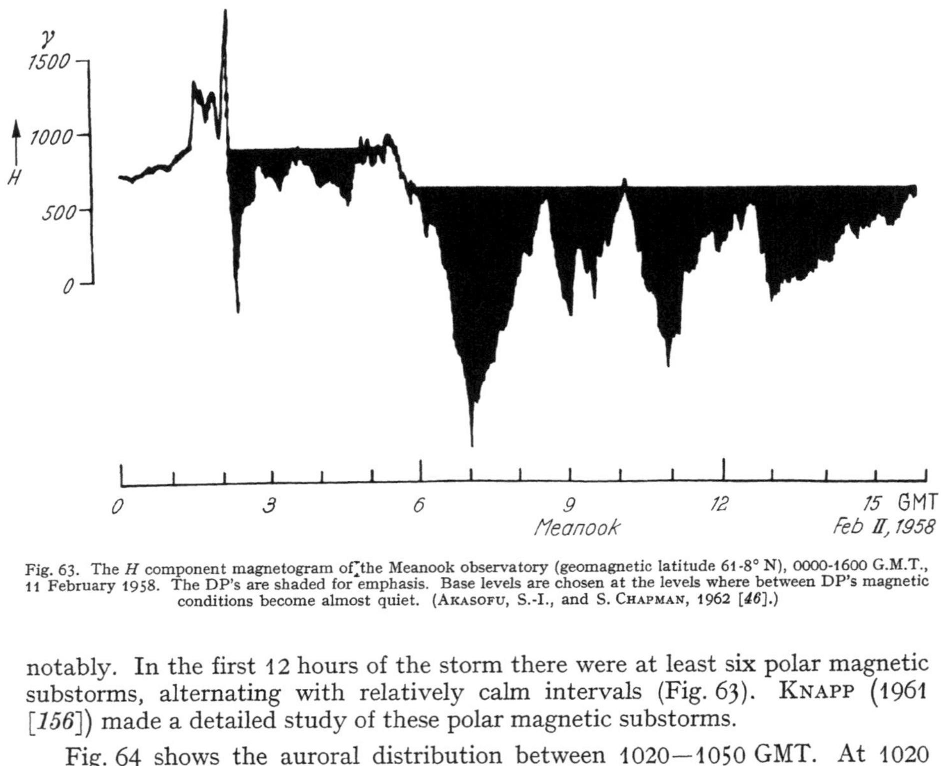

Fig. 63. The H component magnetogram of the Meanook observatory (geomagnetic latitude 61·8° N), 0000-1600 G.M.T., 11 February 1958. The DP's are shaded for emphasis. Base levels are chosen at the levels where between DP's magnetic conditions become almost quiet. (AKASOFU, S.-I., and S. CHAPMAN, 1962 [*46*].)

notably. In the first 12 hours of the storm there were at least six polar magnetic substorms, alternating with relatively calm intervals (Fig. 63). KNAPP (1961 [*156*]) made a detailed study of these polar magnetic substorms.

Fig. 64 shows the auroral distribution between 1020—1050 GMT. At 1020 GMT, just preceding a new polar substorm, the zone was located between the curves $L=2$ and $L=4$ (for the L curves, see Sect. 4). Except for faint arcs seen from Tixie and Saskatoon, there was no indication of auroras to the north of the zone. Between 1020 and 1030 GMT the zone began to expand rapidly northwards, and at 1030 GMT the northern border crossed the curve $L=6$. At 1050 GMT, it reached just south of the curve $L=10$. The region swept during the northward expansion was covered by bright luminosity. During this interval the auroras showed violent eastward motion throughout the expanding strip.

This example shows the combined effects of polar magnetic substorms (Sect. 34a) and of the main phase decrease (Sect. 35). The storm of September 13, 1957 showed similar features. These two storms had an extraordinary growth of the main phase (AKASOFU, CHAPMAN and VENKATESAN 1963 [*157*]).

36. The simultaneous development of the main phase (DR) and of polar magnetic substorms (DP). Akasofu and Chapman (1963 [*158*]) pointed out that the main phase may develop in considerably different ways in different individual magnetic storms, although the average features are rather simple. Some storms show a remarkable sudden commencement (ssc) and a long initial phase (3∼10 hours),

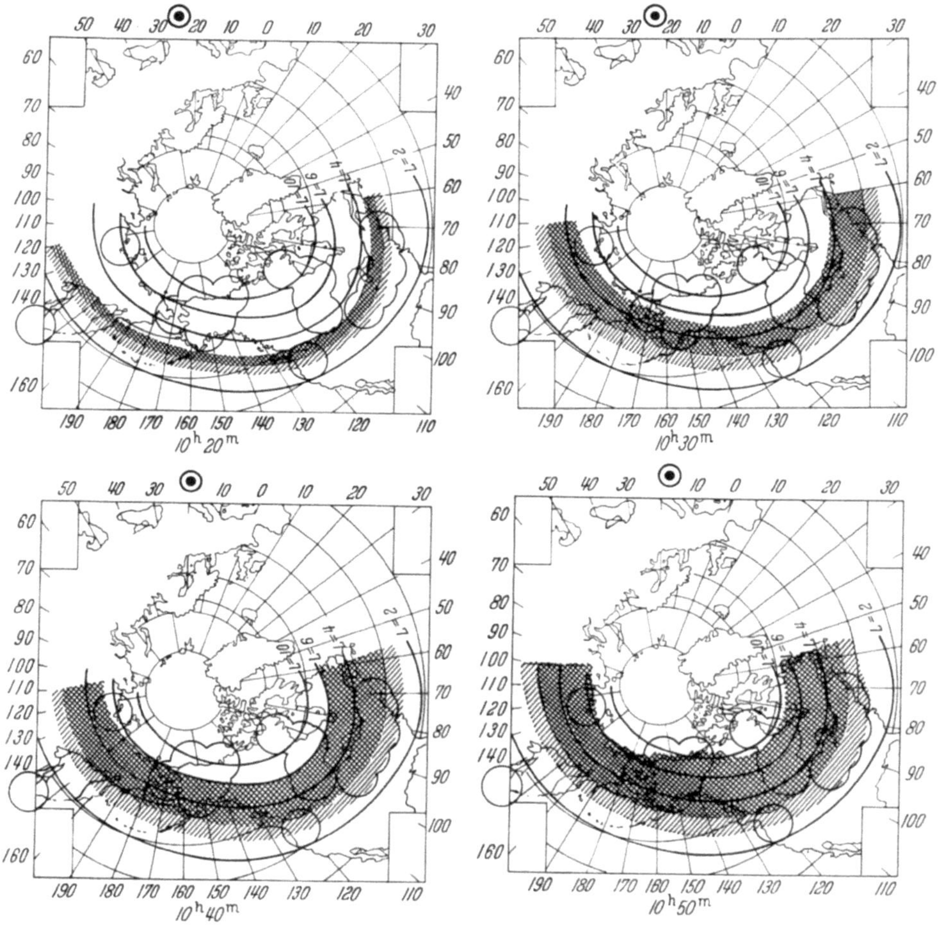

Fig. 64. The distribution of the auroras, 1020-1050 G.M.T., 11 February 1958 from twenty-three all-sky camera stations. (Akasofu, S.-I., and S. Chapman, 1962 [*46*].)

yet show no significant main phase, whereas other storms show a very small ssc followed soon after (20∼30 minutes) by a fairly large main phase. They showed also that magnetic storms with a large main phase (DR) in general exhibit intense activity of polar magnetic substorms (DP), whereas those whose main phase is small or inappreciable have but little DP activity.

They pointed out further that the existence of storms with a large and long initial phase, but with no appreciable main phase, may imply that the simple solar plasma flow hitherto considered in theoretical studies may not be capable of generating the DR and DP fields.

Fig. 65 (AKASOFU and CHAPMAN 1963 [*159*]) gives the H component records for a highly contrasted pair of storms, of October 17/18, 1948 and March 29/30, 1950, from Honolulu, a typical low latitude station (gm lat. 21° N). The 1948 storm showed a ssc of order 25 γ; the initial phase lasted for only 30 minutes, and was soon followed by a large main phase decrease, of about $-150\,\gamma$. The 1950 storm showed a comparable ssc; the initial phase lasted for at least a few hours, but there was no appreciable main phase.

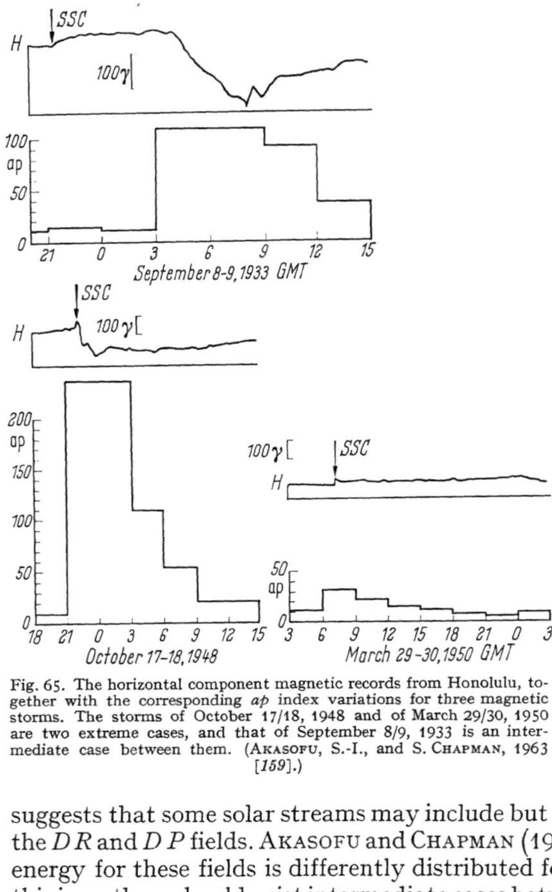

Fig. 65. The horizontal component magnetic records from Honolulu, together with the corresponding *ap* index variations for three magnetic storms. The storms of October 17/18, 1948 and of March 29/30, 1950 are two extreme cases, and that of September 8/9, 1933 is an intermediate case between them. (AKASOFU, S.-I., and S. CHAPMAN, 1963 [*159*].)

To indicate the activity of polar magnetic substorms, they showed also the *ap* indices during the above two storms. In general, polar storms are localized, on the early morning side of the auroral zone; thus it is more appropriate to examine *planetary* indices, rather than their activity at a single station (see also KREMSER, footnote at the end of Sect. 48).

Fig. 65 shows that the DP activity, manifested by *ap* indices, is drastically different for the above two storms, though the details of the DP development are not seen, because the *ap* index is a 3-hourly one[1]. The amplitude in γ (twice *ap*) was at least 7 times larger for the 1948 than for the 1950 storm. The total energy deposited in the auroral zone could be at least 10 times greater for the former.

The occurrence of storms like that of March 29/30, 1950 suggests that some solar streams may include but little energy available to generate the DR and DP fields. AKASOFU and CHAPMAN (1963 [*159*]), have suggested that the energy for these fields is differently distributed for different solar plasma flows. If this is so, there should exist intermediate cases between the above two extreme cases.

The September 8/9, 1933 storm in Fig. 65 is a typical intermediate case. The Honolulu record clearly shows the ssc (of order 25 γ), indicating a sudden increase of the solar plasma pressure. The initial phase then lasted for at least six hours, and was terminated by the growth of the main phase at about 4 GMT on September 9. There was little sign of sudden changes in the solar plasma pressure. The auroral zone was fairly quiet during this long initial phase. A sudden burst of DP activity seemed to start between 3 and 6 GMT, September 9, apparently coinciding with the growth of the main phase.

---

[1] The Q-index, for quarter-hourly intervals, is derived by a few polar stations like Kiruna; it may provide the desired sub-division of the 3-hour interval (BARTELS and FUKUSHIMA 1956 [*160*]; BARTELS 1957 [*161*]). However, nothing can replace the information provided by an inspection of simultaneous magnetograms from many observatories (Ed.).

Fig. 66 shows this and two other such examples. In both the H records and the ap diagrams the initial phase is hatched for emphasis. In all three cases the first large increase in the ap index coincided with the rapid growth of the main phase. The August 7, 1950 storm showed a remarkable ssc and a large initial phase, which lasted more than 10 hours; this suggests an intense flow of the solar plasma, yet throughout this time the DP activity was insignificant. The initial phase ended when a large DP developed and the main phase rapidly grew.

The August 12, 1937 storm showed a double ssc; the first ssc was followed by a long initial phase (about 7 hours). Quickly after the second ssc there was rapid

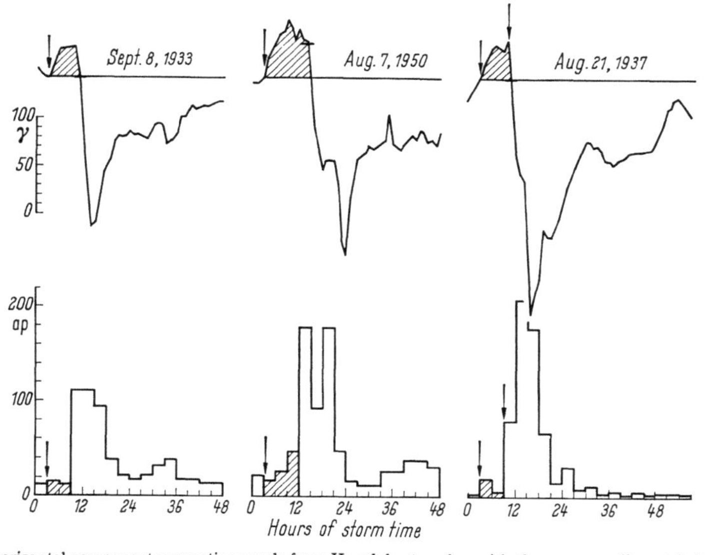

Fig. 66. The horizontal component magnetic records from Honolulu, together with the corresponding ap index variations for three magnetic storms. Notice that a large and long initial phase (hatched) is not accompanied by any significant DP activity. (AKASOFU, S.-I., and S. CHAPMAN, 1963 [*159*].)

growth of the main phase, and, as the ap index showed, great development of DP. This resembles a combination of the 1950 and 1948 magnetic storms of Fig. 65f

These are but a few examples among many. The variety of development o. magnetic storms is considerable; it seems to show no definite relation between the magnitude of the ssc, the duration of the initial phase, or the DR, DP activities.

All these examples demonstrate that the enhancement of the solar plasma pressure, indicated by the ssc and the DCF field of the initial storm phase, does not necessarily imply the growth of the DR and DP fields. The DR field is associated with an increase in the particle kinetic energy density in the ring current belt, and the DP substorms are associated with enhanced injection of energetic particles from the magnetosphere into the auroral ionosphere . It seems that in order to initiate these DR and DP development there must be some additional property associated with the solar plasma flow.

37. The aurora and geomagnetic micropulsations. CAMPBELL (1960 [*162*]) and CAMPBELL and REES (1961 [*163*]) have reported that rapid fluctuations of the intensity of the λ 3914 N_2^+ First Negative bands (see Sect. 63 d) in the auroral spectrum are accompanied by an intense activity of geomagnetic micropulsations. The rapid fluctuations of the former tend to occur in the last phase of the cycle (Sect. 18 c), namely when the aurora becomes patchy — the post break-up phase. CAMPBELL and LEINBACH (1961 [*164*]) showed that such rapid fluctuations of the λ 3914 emission coincide with rather heavy absorption of radio waves that pass through the auroral ionosphere. Fig. 67 shows an example of well correlated changes in the geomagnetic field, the intensity of the λ 3914 bands and the cosmic radio wave absorption recorded by a riometer (see Sect. 40b).

TROITSKAYA (1961 [*165*]) reported that a particular type of geomagnetic or earth-current micropulsation accompanies the break-up phase and post break-up

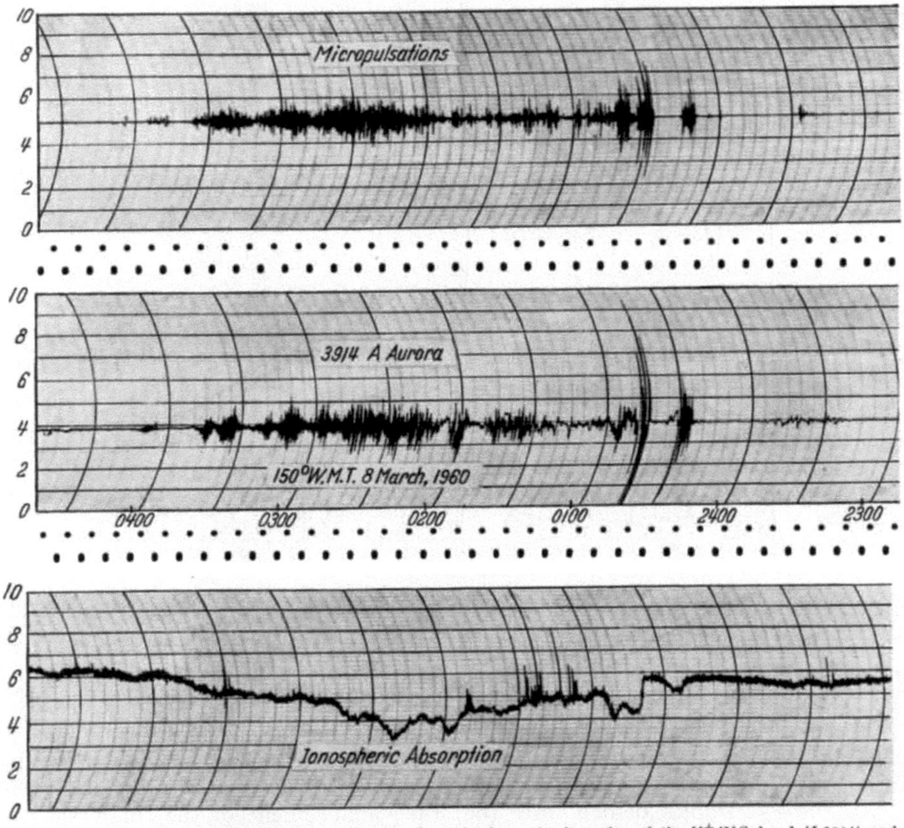

Fig. 67. Simultaneous observations of geomagnetic micropulsations, the intensity of the N_2^+ 1NG band (λ 3914) and cosmic noise absorption. (CAMPBELL, W. H., and H. LEINBACH, 1961 [*164*].)

phase of the display. Such micropulsations are irregular, with periods of 1 to 15 seconds at the beginning, followed by pearl series of diminishing period. FLOCK, BELON and HEACOCK (1962 [*166a*]) reported also that a sudden increase in auroral luminosity is accompanied by the onset of earth-current micropulsations. Because of heavy absorption of hydromagnetic waves in the ionosphere, however, the correlation between a certain type of micropulsations and the auroral luminosity is likely to be reduced (HEACOCK 1962 [*166b*]).

Some VLF (very low frequency) emissions also accompany a strong auroral display. ELLIS (1959 [*167*]) reported such an association in the 2—40 kc/s range, and that the increased emission was accompanied by simultaneous enhancement of the red oxygen lines. MOROZUMI (1962 [*91a*]) reported that at the South Pole strong VLF hiss is accompanied by arcs and bands that appear around midnight.

V. Radio studies of the aurora.

38. Introduction. In high latitudes ionospheric changes are closely associated with auroras. These were first studied during the Polar Year 1932/33 by APPLETON NAISMITH and INGRAM (1937 [*168*]), using the vertical incidence pulse method of ionospheric exploration at Tromsö (gm lat. 67°) in the auroral zone. Similar

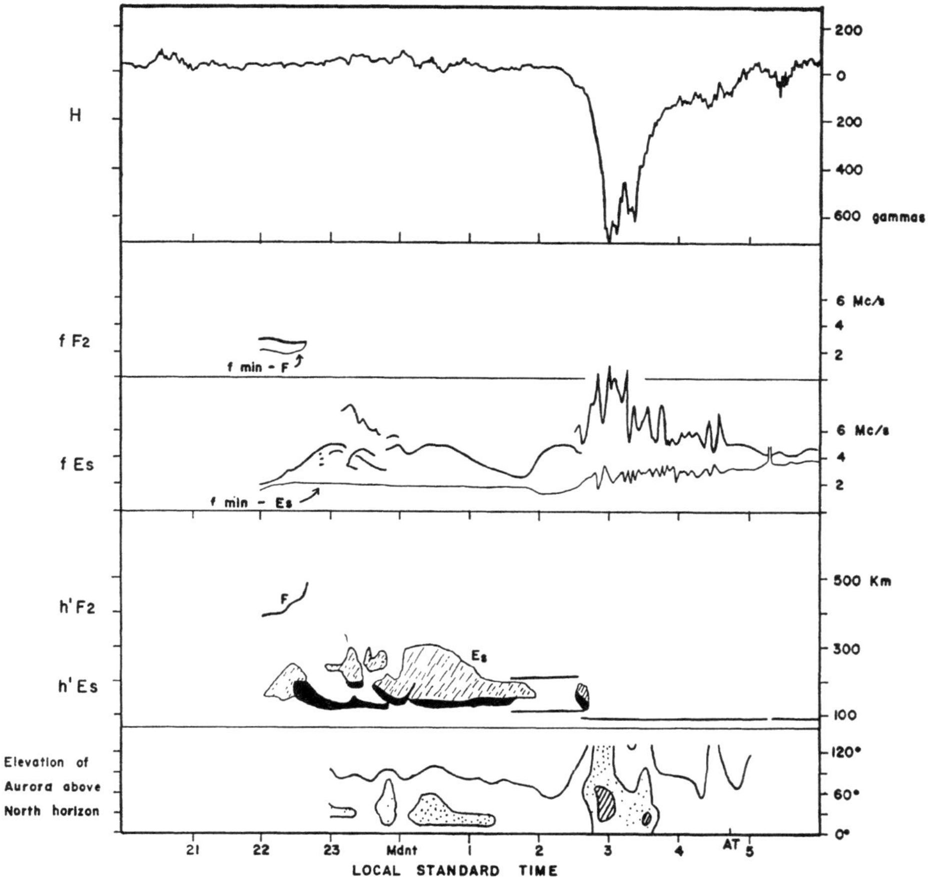

Fig. 68. The simultaneous magnetic, ionospheric and auroral records on March 6/7, 1952. (MEEK, J. H., 1954 [*170*]).

studies were also made by HARANG (1936, 1938 [*169*]) and others. They are well summarized by HARANG (1951 [*6*], Chap. 8) and MEEK (1953, 1954 [*170*]). MEEK'S diagram is reproduced here as Fig. 68.

At a station several degrees south of the (northern) auroral zone, a faint arc may appear after darkness has fallen. When such an arc (sometimes multiple) ascends towards the zenith, an ionospheric sounder detects an abnormal ionization around 100 km or above. As noted in Sects. 18 and 34a, this phase of the aurora may not be accompanied by any particular magnetic activity, except that the increase of brightness may sometimes be associated with a small positive increase in the H component. The onset of the break-up phase, however, is accompanied by a polar magnetic substorm (Fig. 68; see also Sect. 34); the aurora spreads over a large portion of the sky, and the maximum frequency fEs of the sporadic E trace increases. On occasions when the auroral and magnetic activity become strong, their effects soon cease to be observable in the ionogram because a *polar black-out* develops. This is ascribed to the absorption of radio waves in a layer, not normally appreciably ionized, below the usual ionospheric levels; in this lower layer the electron collision frequency is high and the electrons quickly lose the energy they absorb from the radio waves. Heppner, Byrne and Belon (1952 [*171*]) showed that the absorption is largest for the broken-up auroras (patches).

Non-vertical beams of radio waves may also be reflected, scattered or absorbed by ionization associated with auroras. Harang and Stoffregen (1940 [*172*]) investigated 40 Mc/s echoes, and observed discrete fluctuating echoes of small amplitude coincident with the appearance of an auroral display. Since then this radio method, the *auroral radar* technique, has been extensively used. The results are summarized by Booker (1960 [*173*]). However, we are still far from the answer to the question, "What is the radar aurora?"; the mechanism of auroral echoes needs to be examined further both experimentally and theoretically.

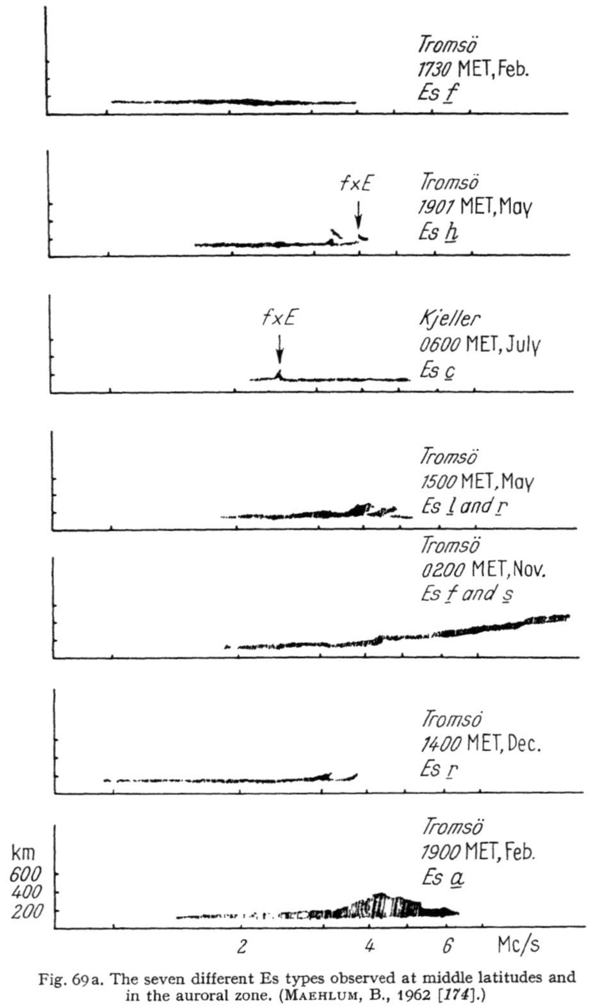

Fig. 69a. The seven different Es types observed at middle latitudes and in the auroral zone. (Maehlum, B., 1962 [*174*].)

39. The sporadic *E*-layer associated with the aurora. There are several types (Fig. 69a) of high latitude sporadic E layers, all denoted by Es: they are represented by the following lower-case letters (IGY Annals, Vol. 3, pp. 94—95, 1957):

r (retardation): Non-blanketing Es with group retardation at the high-frequency end.

l (low): Es below the regular E layer.

c (cusp): Es between the minimum virtual height of the regular E layer and the height of maximum ion density. This appears as an Es trace that is usually continuous with the regular E reflection except for a rather symmetrical cusp around foE.

h (high): *Es* above the height of maximum ion density of the *E* layer.
f (flat): *Es* that shows no increase in height with frequency.
s (slant): A steady rising diffuse trace in the ionogram.
a (auroral): The trace has a rather well-defined flat or gradually rising lower edge, with stratification and diffuse (spread) echo presented above it. This sometimes extends several hundred kilometers in virtual height.

MAEHLUM (1962 [*174*]) made an extensive study of arctic *Es* and its relation to auroral and magnetic phenomena (Fig. 69a). MONTALBETTI and MCEWEN (1962 [*175*]) showed that the daily curve for *Es* type *r* observed at Churchill (gm lat. 68.7° N) resembles that of the hydrogen emissions, whereas type *f Es* has a maximum around midnight, like the auroral intensity. This is also the case for type *a* (Fig. 69b). At College (gm lat. 64.7° N), however, both $H\alpha$ emission and *Es* type *r* have a single maximum around midnight. Thus although their daily appearance may differ from place to place, depending on the latitude, the close relation between them has been well established.

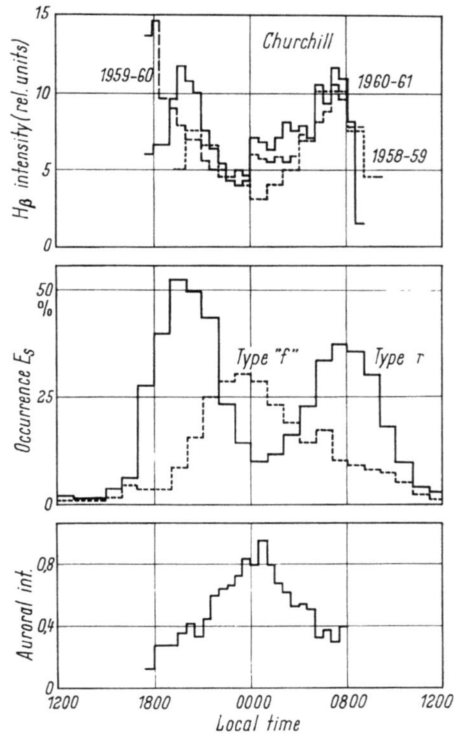

Fig. 69b. Daily variation of H_β, and aurora measured at Churchill, Manitoba. (MONTALBETTI, R., and D. J. MCEWEN, 1962 [*175*].)

OMHOLT, STOFFREGEN and DERBLOM (1962 [*176*]) showed that the evening aurora consists of a weak but persistent glow showing an unusually strong $H\alpha$ line, and of distinct, brighter forms showing no trace of *H*; between the brighter forms and the glow there is a dark area. They suggested that this "hydrogen glow" is excited purely by proton impact, and that the retarded type sporadic *E* is closely associated with the glow: agreeing with earlier work by MEEK and MCNAMARA (1954 [*177*]), who concluded that a "fringe" of sporadic *E* ionization moves southward in advance of the auroral light (see also POTAPOV, RAPPOPORT and BARSUK 1960 [*178*]).

The relationships between the slant *Es* and the aurora were studied by BATES (1961 [*179*]). He showed that whenever the slant *Es* echo was strong on the College oblique-incidence, sweep-frequency sounder (1—25 Mc/s), the College 41 Mc/s auroral radar recorded an echo on the sweep-frequency record at the same range as that at 25 Mc/s. He suggested that the high latitude slant *Es* echo is an aspect-sensitive auroral echo (see also HUNSUCKER and OWREN (1962 [*260*])).

40. The absorption of radio waves. *a) The polar blackout.* Absorption of radio waves at vertical or oblique incidence in high latitudes is often associated with the occurrence of auroras. It can interrupt long-distance short-wave radio communication in such regions. During severe absorption the wave energy sent out by an ionospheric sounder is almost completely absorbed in the ionospheric region, so that there is no trace of echo in the ionogram: such total absorption is called a *polar blackout*.

PIGGOTT (1953 [*180*]), AGY (1954 [*181*], [*182*]) and MEEK (1953 [*170*]) are among those who have studied the polar blackouts. AGY found that these blackouts occur only when there is magnetic disturbance, but the relation between the polar blackouts and the aurora is not simple.

Some of the apparent complexities, however, were removed by PIGGOTT and THOMAS (1960 [*183*]). They examined the relation between the Kp index and the percentage of the total number of hours of blackout, at three ionospheric stations, Ottawa (gm lat. 56.9° N), Winnipeg (gm lat. 59.6° N) and Resolute Bay (gm lat. 82.9° N). They found that as Kp increases, at Ottawa the blackout percentage increases, whereas at Resolute Bay and Winnipeg the percentage increases with Kp up to $Kp=3\sim4$ and then decreases. This seems to imply that the zone of polar blackout shifts equatorward when Kp becomes more than 4. Such an equatorward shift may be related to that of the center of auroral activity during large magnetic storms (Sect. 35a).

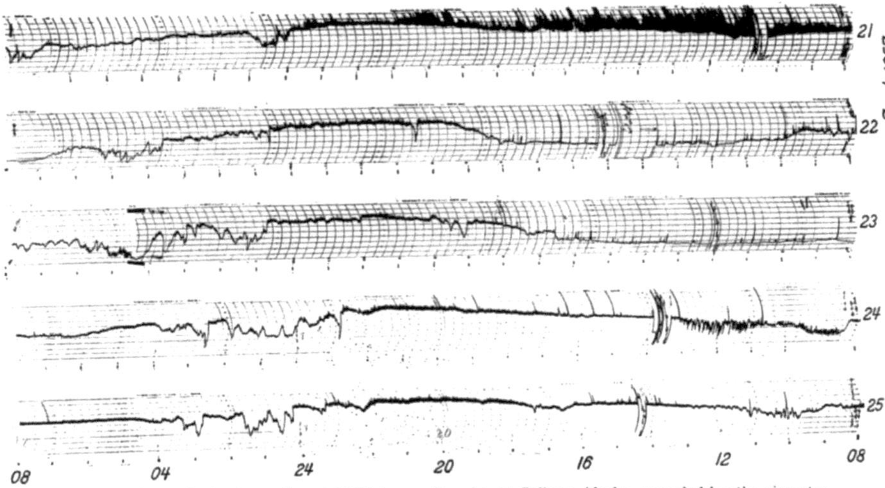

Fig. 70. Ionospheric absorption of 30 Mc/s cosmic noise at College, Alaska, recorded by the riometer. (LEINBACH, private communication, 1959.)

One peculiarity of the polar blackout is that it appears to occur along a particular curve for a particular Universal Time. The similarity between such a curve with STÖRMER's spiral has long been discussed (AGY 1957 [*184*]; THOMAS and PIGGOTT 1960 [*94*]; PIGGOTT and THOMAS 1960 [183]). This may be related to the problem discussed in Sect. 19, but at present there seems to be no obvious solution.

Intense ionization in the D-region was studied in detail by STOFFREGEN, DERBLOM and OMHOLT (1960 [*185*]). They showed that a sudden increase of ionization in the D-region is associated with sudden increases of auroral activity. The lowest measured height of this D layer was about 65 km. They inferred that both primary electrons and secondary x-rays contribute to the formation of the auroral D ionization down to 65 km.

b) Riometer studies of radio absorption. α) Auroral type absorption. Studies of the polar blackout have given much important information on abnormal ionization in the auroral ionosphere. However, during a polar blackout the radio wave absorption is so large as to defeat the common type of ionospheric sounder. LITTLE (1957 [*186*]) and LITTLE and LEINBACH (1958 [*187*]), by recording cosmic noise at relatively high frequencies, demonstrated that it is thus possible to measure the amount of absorption during the polar blackout; such a device is called the *riometer*. An example of a riometer record is shown in Fig. 70. (Note the close similarity between the riometer record on September 23 and the magnetic record in Fig. 52).

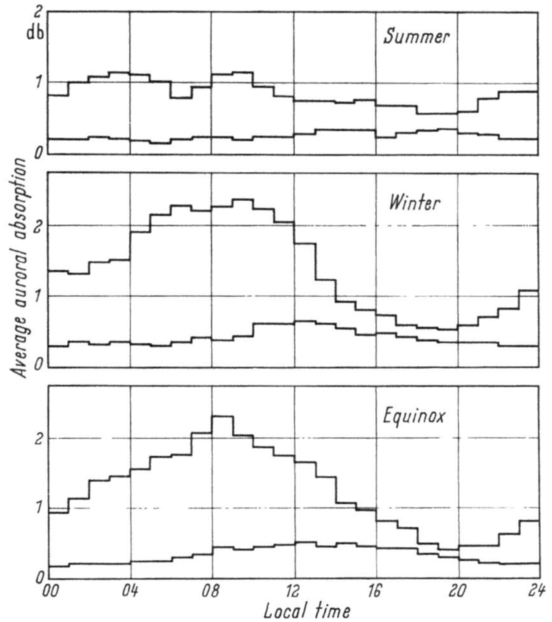

Fig. 71. The daily variation of auroral absorption at College observed during the five-year period from September 1957 through August 1962. (BASLER, R. P., 1963 [*188*].)

An extensive study of radio wave absorption measured by the riometer was made by BASLER (1963 [*188*]), using data from six riometer stations distributed over Alaska. Fig. 71 shows the daily variation of auroral absorption at College, for the five magnetically disturbed days (above) and five quiet days (below) (September 1957—August 1962). It shows that during the winter and the equinoxes the absorption has a pronounced daily variation, with a rather broad maximum between 0600 and 1300, and a minimum between 1900 and 2200 local time. During the summer months, however, there seems to be no such marked daily variation; further, on the average the absorption is less during the summer than in winter.

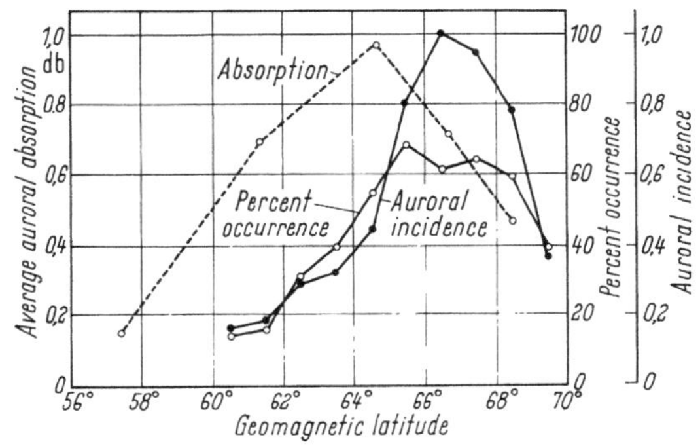

Fig. 72. A comparison of the auroral absorption zone with the visual auroral zones determined from Alaskan all-sky camera data by DAVIS [*23*]. (BASLER, R. P., 1963 [*188*].)

Fig. 72 shows the latitude dependence of the absorption over Alaska, together with DAVIS' auroral incidence (Sect. 3). The absorption has a well-defined peak at College, a few degrees south of the center of the visual auroral zone. (See also AGY 1954 [*182*]; MONTBRIAND, HARTZ and VOGAN 1962 [*189*]; HOLT, LANDMARK and LIED 1961 [*190*]). As HEPPNER, BYRNE and BELON (1952 [*171*]) pointed out, the greatest percentage of blackout time observed at College corresponded with periods of pulsating patchy auroras. HEPPNER (1954 [*88*]) indicated that the maximum frequency of occurrence of pulsating aurora was at about 64°, a few degrees south of the center of the auroral zone (see also Sects. 28 to 30). These two effects seem to be closely related to the fact that the peak absorption caused by the aurora occurs a few degrees south of the visual auroral zone. BASLER [*188*] pointed out also that at Thule (gm lat. 88° N) there has been no detectable absorption since 1958.

Recently multi-frequency riometers have been used to determine the height at which the absorption takes place (ZIAUDDIN and FORSYTH 1961 [*191*]; PARTHASARATHY, LERFALD and LITTLE 1963 [*192*]). However, there seems to have been not much progress in the theoretical studies of the absorption, since CHAPMAN and LITTLE (1957 [*193*]) proposed their theory of non-deviative absorption (see also ANSARI (1963 [*194*]) and JELLY, MATHEWS and COLLINS (1961 [*195*])).

β) Polar cap absorption. The evidence for the entry of non-relativistic solar protons over the entire polar cap was first set out in 1959 by LEINBACH and REID [*196*]. Their riometer records, obtained simultaneously at Thule (gm lat. 88° N), Barrow (gm lat. 70° N), and College (gm lat. 64.7° N), showed that after a large solar flare the ionosphere is heavily bombarded by particles guided by the earth's magnetic field toward the polar regions. Subsequently, from ionosonde data, HAKURA and GOH (1959 [*197*]) showed also that the entire polar ionosphere is blacked out after large solar flares. Preceding these studies, BAILEY (1959 [*198*]) made an extensive theoretical study of the ionization of the polar upper atmosphere by solar protons during the February 23, 1956 event.

The earliest direct observations of non-relativistic solar protons were made by ANDERSON (1958 [*199*]), by use of balloons, and by ROTHWELL and MCILWAIN (1959 [*200*]) and VAN ALLEN and LIN (1960 [*201*]), by the detectors carried by Explorers IV and VII, respectively. These studies confirmed the earlier conjecture that many non-relativistic protons are produced during intense solar flares. Further, the Explorer XII satellite observations and others (BRYANT, CLINE, DESAI, and MCDONALD 1962 [*202*]), make it likely that a considerable part of the inner interplanetary space is filled by such energetic solar proton "gas" for many hours after large solar flares, and that the earth and its magnetosphere are enclosed in it.

During the years 1958—1962, many direct observations of such solar protons have been made at relatively low altitudes (30—1500 km) by detectors carried by balloons, rockets and satellites (cf. BROWN and D'ARCY 1959 [*203*]; WINCKLER and BHAVSAR 1960 [*204*]; WINCKLER, BHAVSAR, and PETERSON 1961 [*205*]; ANDERSON, ARNOLDY, HOFFMAN, PETERSON, and WINCKLER 1959 [*206*]; PFOTZER, EHMERT, and KEPPLER 1962 [*207*]; OGILVIE, BRYANT, and DAVIS 1962 [*208*]; DAVIS and OGILVIE 1962 [*209*]; LIN and VAN ALLEN 1963 [*210*]; MAEHLUM and O'BRIEN 1962 [*211*]; PIEPER, ZMUDA, BOSTROM, and O'BRIEN 1962 [*212*]). A close comparison of the satellite data and the riometer records has been made by VAN ALLEN (1962 [*213*]).

When non-relativistic solar protons were discovered, it was thought that they followed Störmer trajectories (Sect. 82) from the sun or from its vicinity, as relativistic solar protons do approximately. However, it has become apparent that many solar protons impinge upon regions where, according to STÖRMER's theory (for the earth's dipole field), their entry is forbidden because their energy is too low. This abnormal entry of non-relativistic solar protons was first indicated by FREIER, NEY and WINCKLER (1959 [*214*]) and NEY, WINCKLER and FREIER (1959 [*215*]).

The deviation of the trajectories of galactic cosmic rays from the STÖRMER paths was studied well before the abnormal entry of non-relativistic solar cosmic rays was discovered. TREIMAN (1953 [*216*]) and RAY (1956 [*217*]) made a detailed study of the effect of the ring current: see also STÖRMER (1955 [*3*], p. 338).

Similarly, assuming various types of distortion of the earth's magnetic field, several attempts have been made to investigate this anomalous reduction of the cut-off energy: cf. ROTHWELL (1959 [*218*]) and OBAYASHI and HAKURA (1960 [*219*]). However, their models are either unrealistic or not sufficiently accurate for a quantitative study of this problem, in the light of present theoretical and observational knowledge.

KELLOGG and WINCKLER (1961 [*220*]) have shown that the observed anomaly can be explained if the magnetic moment M_R of the ring current exceeds M_E, that of the earth. Using more realistic ring current models, AKASOFU and LIN (1963 [*221*]) have shown that the ratio M_R/M_E cannot exceed unity.

AKASOFU, VAN ALLEN and LIN (1963 [*222*]) and AKASOFU (1963 [*155*]) have shown that the combined effect of the limitation of the earth's magnetic field (the Chapman-Ferraro effect) and of the ring current can reasonably explain the anomalous entry of such non-relativistic solar protons. The excitation of auroral type glow over the polar cap during polar cap absorption events is discussed in Sect. 76 (see also WEBBER (1963 [*223*])).

41. Auroral radar echoes. *a) Aspect sensitivity.* Following radar studies of the aurora by HARANG and STOFFREGEN [*172*], much effort has been devoted to this problem. However, serious controversy has developed as to the interpretation of auroral echoes. BULLOUGH and KAISER (1954 [*224*]) and KAISER (1957 [*225*]) concluded that auroral echoes come from the surface of an arc lying along a parallel of geomagnetic latitude. UNWIN (1958 [*226*]), SEED (1958 [*227*]), POGORELOV (1960 [*228*]), and BARKER, SUTCLIFFE and WATKINS (1962 [*233*]) demonstrated that the echoes are reflections from aspect-sensitive "columns" lying along the geomagnetic field lines; they showed further that KAISER's results can be consistently interpreted as due to reflection from aspect-sensitive columns.

UNWIN's argument was based on calculations made by CHAPMAN (1952 [*229*]), who outlined the geometry of aspect-sensitive reflection from auroral ionization alined along geomagnetic field lines, for various radar locations. For this particular mechanism the echoes are strongest when the ionized "column" of ionization, the ray structure, is seen perpendicularly by radar radio beams. The theory of this reflection was developed by MOORE (1951 [*230*]), BOOKER, GARTLEIN and NICHOLS (1955 [*231*]), and BOOKER (1956 [*232*]). The diagram by BARKER, SUTCLIFFE and WATKINS (1962 [*233*]), here reproduced as Fig. 73, suggests that the zone where echoes are most probable closely follows the curve of aspect-sensitive reflection.

However, this required perpendicularity is not attained at high latitude regions above gm lat. 62°, if the ionization is located at or above an altitude of 100 km (cf. BOWLES 1954 [*234*]; DYCE 1955 [*235*]; BOOKER 1960 [*173*], pp. 366—368; OWREN 1961 [*236*]). Nevertheless, there seems to be tacit agreement among the workers that this aspect-sensitive reflection plays the major role in auroral radar echoes. The echoes do not come merely from the foot P of the perpendicular from the radio station to the ionized column; a column above P, but extending nearly down to it, can still give a good echo; the Fresnel zones discussed in meteor-trail reflections are important here also (cf. HERLOFSON 1947 [*237*]). FRICKER, INGALLS, STONE and WAUG (1957 [*238*]) showed that echoes may return from regions as much as 20 to 30 degrees off perpendicular.

Further, DYCE's study (1955 [*235*]) at Point Barrow (gm lat. 68.5° N) with a 51.9 Mc/s radar suggested that the rather poor correlation between overhead visual auroras and auroral radar echoes mainly depends on the (near) perpendicularity, and not on the absorption (CURRIE, FORSYTH and VAWTER 1953 [*239*]) or on insufficient ionization to refract the waves (HARANG and LANDMARK 1954 [*240*]). However, although aspect-sensitive reflection is an important mechanism, there seems to be no definite one-to-one correlation between visual auroras and radar echoes. GADSDEN (1959 [*241*]) showed that visual auroral rays in the right

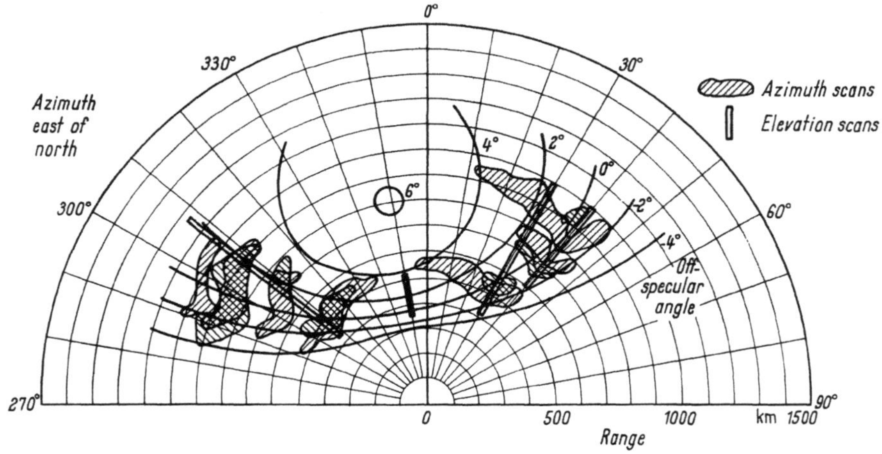

Fig. 73. The locations of auroral echo regions detected at 500 Mc/s between 1600 and 1855 U.T. on 10 April 1959, together with off-specular angles for a height of 110 km. (BARKER, D., H. K. SUTCLIFFE and C. D. WATKINS, 1962 [*233*].)

position are not necessarily "seen" by auroral radar. His conclusion agrees with some earlier critical studies (cf. KAISER and BULLOUGH 1955 [*242*]; HELLGREN and MEOS 1952 [*243*]; BULLOUGH 1962 [*244*]). The mechanism of auroral echoes needs to be re-examined, both theoretically and experimentally. Unfortunately because

Table 5. *Characteristics of VHF auroral echoes.*
(F = 55 Mc/s; Prf = 50 cps; Pulse width = 18 μsec; Peak power output 45 kw) (gm lat. 51.5° S.)

	Diffuse	Diffuse with structure	Short discrete	Long discrete
Extension				
In Range	30∼1000 km	30∼1000 km	<40 km	∼20 km
In Time	Minutes to hours	Minutes to hours	<1 min	∼1 min
Strength	Generally weak	Weak to strong	Generally strong	Weak
Time of occurrence	Afternoon and early evening	From late afternoon until dawn	Night time	Chiefly from midnight until dawn
Antenna lobe structure	Always clearly defined	Partially or wholly suppressed	Almost always suppressed	Suppressed
Remarks			Frequently clustered in both range and time	Usually grouped over several hundred km

of uncertainty as to the mechanism of auroral echoes, auroral radar cannot be regarded as a simple quantitative tool for the exploration of the aurora. Some other mechanisms of auroral echoes have been proposed by FORSYTH (1960 [*245*]), LYON (1960 [*246*]), MOORCRAFT (1961 [*247*]), KELLY, HANSEN and FORSYTH (1961 [*248*]), LYON and FORSYTH (1962 [*249*]), and VERSHININ (1962 [*251*])

b) Types of auroral echoes. KAISER (1955 [*250*]) showed that there are two basic types of auroral echoes, "diffuse" and "discrete" (see also CURRIE, FORSYTH and VAWTER 1953 [*239*]; ASPINALL and HAWKINS 1950 [*252*]; and HELLGREN and MEOS 1952 [*243*]). UNWIN (1959 [*253*]) showed that each of the basic types may further be subdivided into two subtypes. They are "diffuse", "diffuse with structure", "short discrete" and "long discrete" types. Their characteristics, as seen on film running at 12 cm/hr, are summarized in Table 5 (UNWIN 1959 [*253*]).

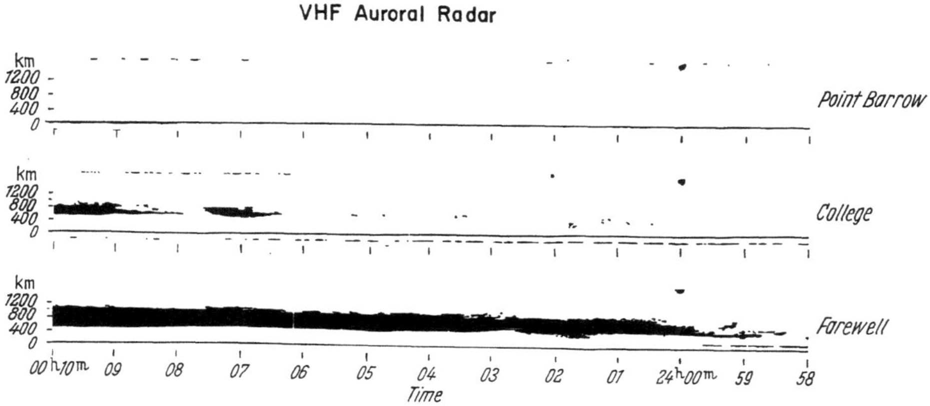

Fig. 74. VHF auroral radar data taken at Point Barrow (Alaska) (gm lat. 68.6° N), College (Alaska) (gm lat. 64.7° N) and Farewell (Alaska) (gm lat. 61.4° N) on the night of 5 to 6 December 1958. (LEONARD, private communication, 1960.)

42. The characteristics of auroral radar echoes. *a) Radar auroral zone.* LEONARD (1962 [*254*]) made an extensive study of the echoes by use of four stations alined approximately along gm long. 250°, namely Point Barrow (gm lat. 68.5° N), College (gm lat. 64.7° N), Farewell (gm lat. 61.4° N) and King Salmon (gm lat. 57.5° N). This array of stations covered the gm latitude range of approximately 58∼68°. Fig. 74 shows a typical example of simultaneous records from Point Barrow, College and Farewell. As expected for aspect-sensitive reflection (slightly modified), the Farewell station observed mostly the echoes in the range of 600∼700 km north of the station. This is the zone of maximum occurrence of visual aurora.

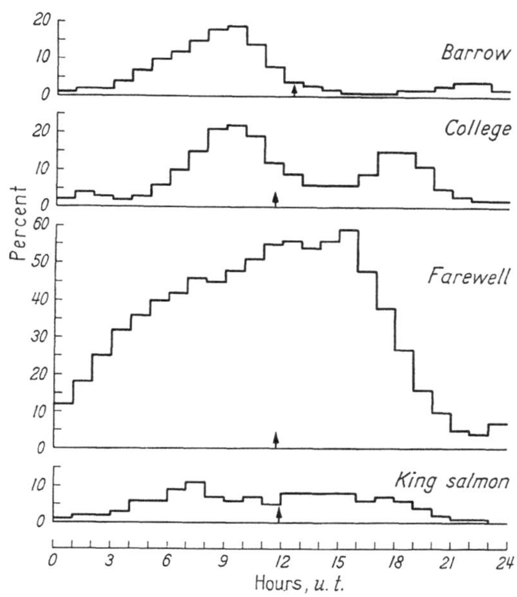

Fig. 75. Daily variation of auroral echoes at Point Barrow (gm lat. 68.6° N), College (gm lat. 64.7° N), Farewell (gm lat. 61.4° N) and King Salmon (gm lat. 57° N). (LEONARD, R. S., 1962 [*260*].)

b) Daily characteristics. The daily characteristics of radar echoes have been studied by CURRIE, FORSYTH, VAWTER (1953 [*239*]), BULLOUGH and KAISER (1955 [*255*]), UNWIN (1959 [*253*]), LYON (1960 [*246*]), BYRFELD (1960 [*256*]),

GRATCHEV (1960 [*257*]), POGORELOV (1960 [*228*]), YARIN (1960 [*258*]), YARIN (1961 [*259*]), LEONARD (1962 [*254*]), and LANGE-HESSE [*261*—*263*]. LEONARD'S diagram for the above four stations is reproduced here as Fig. 75; magnetic midnight at each station is indicated by an arrow (note also that UT is 10 hours ahead of Alaska standard time). FAREWELL shows a large single maximum a few hours after magnetic midnight, whereas both Point Barrow and College show a double peak, one before magnetic midnight and one well after midnight. LEONARD plotted the daily peaks for different geomagnetic latitudes, together with other results (Fig. 76). Clearly the second peak tends to shift from the midnight maximum as one moves from the auroral zone towards the gm pole. A similar tendency of visual auroras is discussed in Sect. 19.

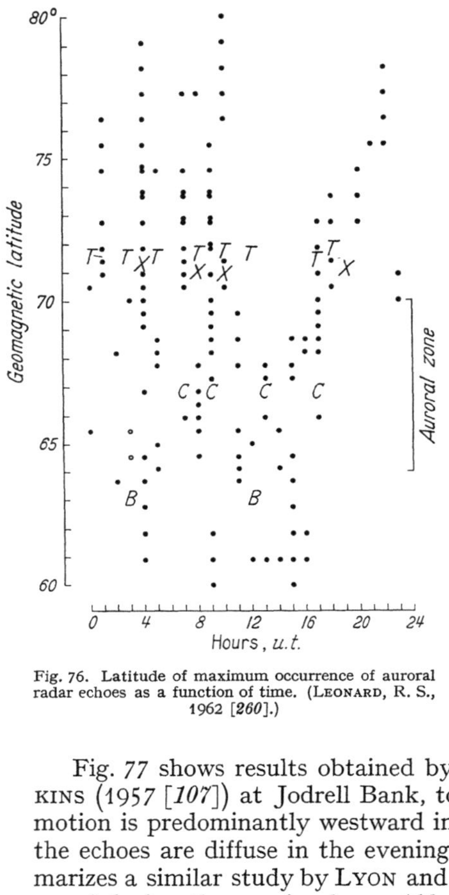

Fig. 76. Latitude of maximum occurrence of auroral radar echoes as a function of time. (LEONARD, R. S., 1962 [*260*].)

LANGE-HESSE (1963 [*261*]; 1962 [*262*]; 1963 [*263*] and [*263a*]) discussed a large material of amateur radio-contacts on 145 Mc/sec in Middle-Europe via aurora. An interesting feature is a pronounced daily variation, with a strong minimum around 21 h local time, at about the time of the change of the ionospheric DS-current system (Sect. 33) from the direction towards east to the direction towards west.

c) Motion. Apparent "motions" of auroral radar echoes have been extensively studied by BULLOUGH and KAISER (1954 [*224*]), NICHOLS (1957 [*108*]), BULLOUGH, DAVIDSON, KAISER and WATKINS (1957 [*107*]), LYON and KAVADAS (1958 [*109*]), UNWIN (1959 [*253*]) and DOVGER (1961 [*267*]). There are several ways of detecting apparent motions, including Doppler shifts of the echoes, and the shift in the range of radar echoes.

Fig. 77 shows results obtained by BULLOUGH, DAVIDSON, KAISER and WATKINS (1957 [*107*]) at Jodrell Bank, together with the H component record. The motion is predominantly westward in the evening and eastward in the morning; the echoes are diffuse in the evening and discrete in the morning. Fig. 78 summarizes a similar study by LYON and KAVADAS (1958 [*109*]). The mean westward speed during the evening hours (6 km/min or 100 m/sec) is less than during the morning hours (15 km/min or 250 m/sec).

When treated *statistically*, the speed variation during the night gives a rather smooth curve showing a gradual decrease prior to the change to eastward motion. Individual cases often do *not* show such a tendency (Fig. 77); the motions are intermittent. KIM and CURRIE (1958 [*105*]) pointed out that the statistical study is rather misleading, because one tends to imagine a simple fixed pattern of motion under which the earth rotates.

Fig. 79 shows the relation between the velocity of the auroral echoes and the horizontal magnetic field changes. The westward moving echoes are associated with negative magnetic disturbances. However, there seems to be no simple relationship between the speed shown by individual echoes and the magnitude of the magnetic disturbance.

PRESNELL, LEADABRAND, BERG and DYCE (1959 [*268*], [*269*]) made a study of the auroral motion at College, Alaska, using UHF radar at 400 Mc/s. Their Doppler data showed predominantly east-west motion, with speeds of order 500 m/sec, and direction independent of time of day. The reason why they did not find the systematic reversal of motion found by other workers

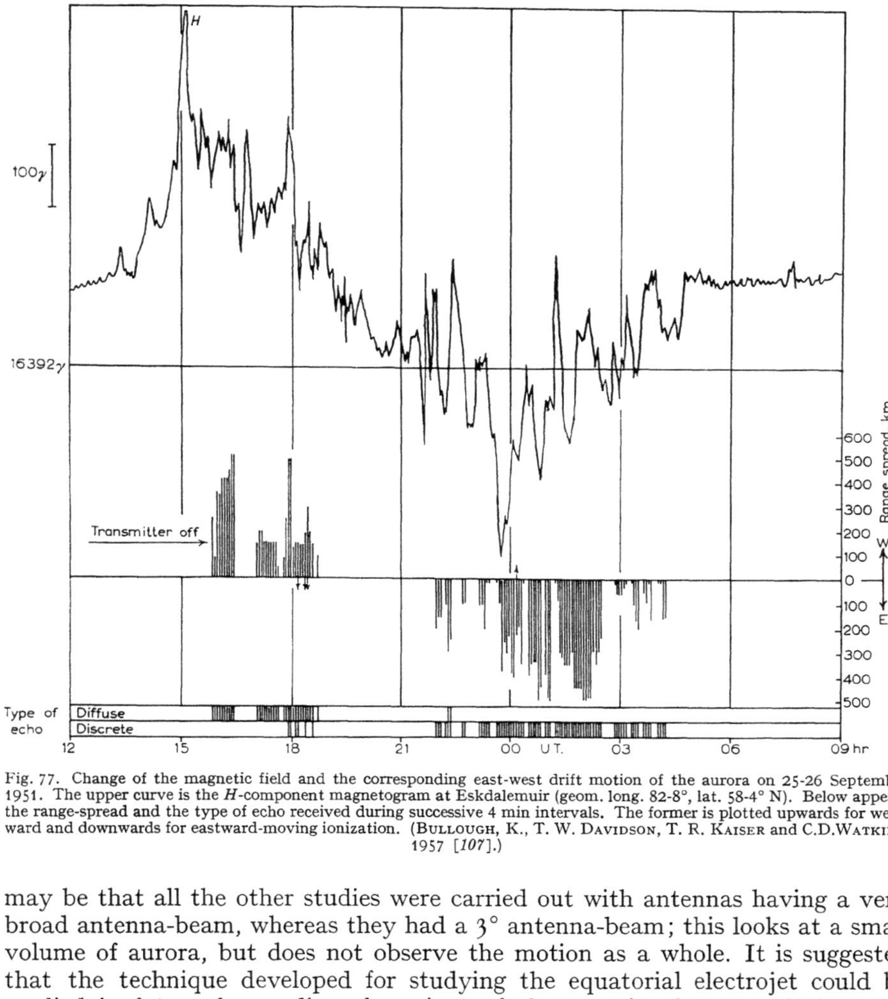

Fig. 77. Change of the magnetic field and the corresponding east-west drift motion of the aurora on 25-26 September 1951. The upper curve is the H-component magnetogram at Eskdalemuir (geom. long. 82·8°, lat. 58·4° N). Below appears the range-spread and the type of echo received during successive 4 min intervals. The former is plotted upwards for westward and downwards for eastward-moving ionization. (BULLOUGH, K., T. W. DAVIDSON, T. R. KAISER and C.D.WATKINS, 1957 [*107*].)

may be that all the other studies were carried out with antennas having a very broad antenna-beam, whereas they had a 3° antenna-beam; this looks at a small volume of aurora, but does not observe the motion as a whole. It is suggested that the technique developed for studying the equatorial electrojet could be applied in future for studies of motions of electrons in the auroral electrojet (cf. COHEN and BOWLES 1963 [*264*]; BOWLES, BALSLEY and COHEN 1963 [*265*]; COHEN and BOWLES 1962 [266]).

d) Daytime echoes. An advantage of auroral radar, provided that the mechanism of auroral echoes is known, is that it can detect aurora in the sunlit hours. PRESNELL, LEADABRAND, PETERSON, DYCE, SCHLOBOHM and BERG (1959 [*268*]) studied the aurora at College, Alaska, using VHF and UHF radar.

Their UHF radar detected echoes in the early morning and in the late afternoon. They showed that daytime aurora is distributed over a larger region of space than the more commonly observed night-time aurora. The daytime echoes are diffuse and last longer than the night-time echoes.

LEADABRAND (1961 [270]) further analysed the daytime echoes, and confirmed that they occur at times of high magnetic activity, and are field-alined.

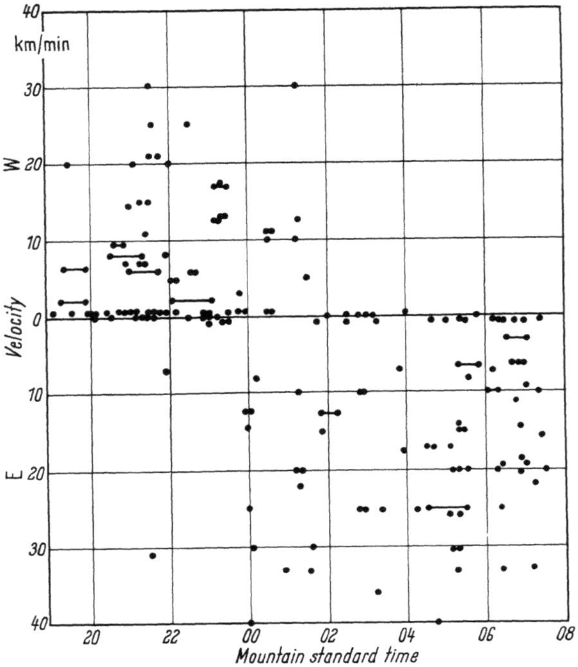

Fig. 78. Scatter diagram; individual echo-velocity versus time. (LYON, G. F., and A. KAVADAS, 1958 [109].)

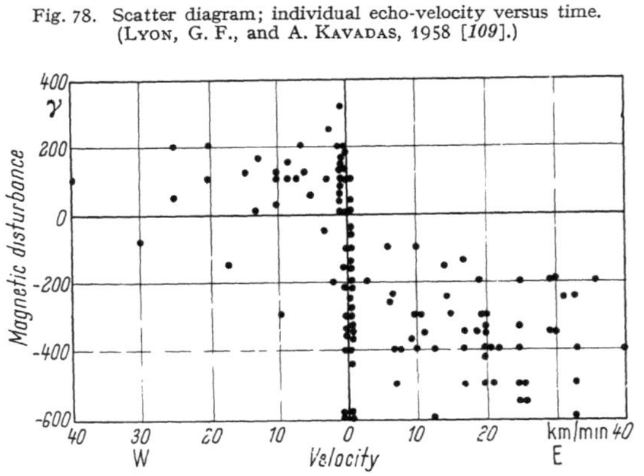

Fig. 79. Scatter diagram; individual east antenna echo velocity versus magnetic disturbance. (LYON, G. F., and A. KAVADAS, 1958 [109].)

e) Auroral type echoes in the minauroral zone. DYCE, DOLPHIN, LEADABRAND and LONG (1959 [271]) regularly observed anomalous echoes by a shipborne radar located at Antigua, British West Indies (gm lat. 30°). These echoes were observed at ranges from about 200 to 600 km. They have many of the characteristics of echoes from auroras observed in the arctic regions. They resemble E-region echoes from magnetic field-alined ionization observed at Stanford (PETERSON, VILLARD, LEADABRAND and GALLAGHER 1954 [272]). The echoes occurred every night, and the occurrence frequency was more than 10% of the time. However, visible auroras were not observed even when the night-time radar echoes were unusually strong.

NAKATA (1958 [273]) observed slant-range auroral echoes by an ordinary ionospheric sounder located in Tokyo (gm lat. 24.4° N). The echoes coincided with the visual auroras of September 13, 1957, September 21, 1957, and February 11, 1958. The echoes appeared first at frequency ranges of 3—10 Mc/s at a range of about 700 km; they moved with speed of order 25 m/s, and disappeared at 400 km. The appearance of such an echo was repeated.

On February 11, 1958. SCHLOBOHM, LEADABRAND, DYCE, DOLPHIN and BERG (1959 [276]) observed radar echoes at 106.1 Mc/s at an altitude of 300 km from about 900 km north of Stanford (gm lat. 43° N).

43. Radio star scintillations and the aurora. When cosmic radio waves penetrate the auroral ionosphere, their propagation is considerably disturbed. At ground level it is seen as a "scintillation" of the radio waves; see HEWISH (1952 [*273a*]), LITTLE and MAXWELL (1952 [*274*]), COSTAIN (1955 [*275*]), HARTZ (1959 [*277*]), BRIGGS (1959 [*278*]), LITTLE, REID, STILTNER and MERRITT (1962 [*279*]), BENSON (1960 [*280*]), FORSYTH and PAULSON (1961 [*281*]), MOORCRAFT and FORSYTH (1963 [*282*]).

The correlation between radio star scintillation and auroral activity is not necessarily positive (BRIGGS 1961 [*283*]). BENSON (1960 [*280*]) by comparing the radio star scintillation data with simultaneous all-sky camera films, showed that the presence of aurora in the line of sight to the radio star can directly affect the scintillations of the radio signal. MOORCRAFT and FORSYTH (1963 [*282*]) concluded that the scintillations are characterized more by fluctuations in the phase than in the amplitudes of the waves from the radio stars.

VI. Auroral particles and the radiation belts.

44. Introduction. The history of the observational discovery of energetic charged particles trapped in the geomagnetic field goes back to 1952 and 1953, when VAN ALLEN and his associates launched *rockoons* (small rockets carried aloft by balloons and started in the stratosphere) over a latitude range from Baffin Bay, near the geomagnetic North Pole, to the North Atlantic Sea. They found intense and continuous radiation over the latitude range in which visible auroras are most often seen (Fig. 80). The radiation was later identified as bremsstrahlung from electrons with energy $\sim$100 kev; the flux was of order 10^6 particles/cm^2 sec (MEREDITH, GOTTLIEB and VAN ALLEN 1955 [*284*]; VAN ALLEN 1957 [*285*]).

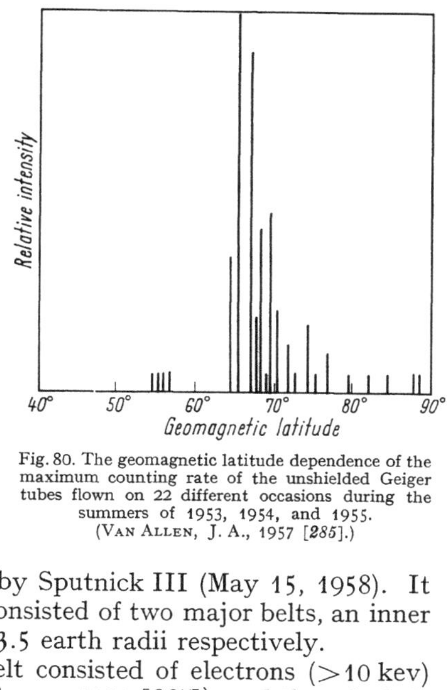

Fig. 80. The geomagnetic latitude dependence of the maximum counting rate of the unshielded Geiger tubes flown on 22 different occasions during the summers of 1953, 1954, and 1955. (VAN ALLEN, J. A., 1957 [*285*].)

The first sketch of the radiation belts was based on the *satellite* data obtained by Geiger counters carried by Explorer I (launched on January 31, 1958), Explorer III (March 26, 1958), Pioneer I (October 11, 1958) and Pioneer III (December 6, 1958). VERNOV and GORCHAKOV, LOGACHEV and CHUDAKOV (1959 [*286*]) confirmed the existence of the belts by apparatus carried by Sputnick III (May 15, 1958). It was then believed that the radiation zone consisted of two major belts, an inner and an outer belt, located at about 1.5 and 3.5 earth radii respectively.

It was first concluded that the outer belt consisted of electrons ($>$10 kev) with a flux of order $10^{10} \sim 10^{11}$/cm^2 sec (VAN ALLEN 1959 [*287*]), and that the belt had ample energy to produce the aurora according to the "leaky bucket" model. This conclusion was based on the assumption that the satellite Geiger counters responded to bremsstrahlung in their shielding, produced by electrons with energy of order 20 kev, rather than to direct penetration by higher energy electrons ($>$1.5 mev).

Various detectors carried by Injun I (June 29, 1961) and Explorer XII (August 16, 1961) changed the earlier concept of the belts and led to a better one, not yet complete (O'BRIEN, LAUGHLIN, VAN ALLEN and FRANK 1962 [288]; O'BRIEN 1963 [289]). Fig. 81 and the table below summarize the present view of the major features of the belts of geomagnetically trapped particles (VAN ALLEN 1962 [290]).

Hereafter, by the outer radiation belt, we refer to the belt consisting of protons and electrons which extends from 2 to about 10 earth radii; (see Figs. 81 b and d). The outer boundary of the outer belt in the equatorial plane is in general sharply defined, and coincides with the termination of the geomagnetic field, namely the outer boundary of the magnetosphere.

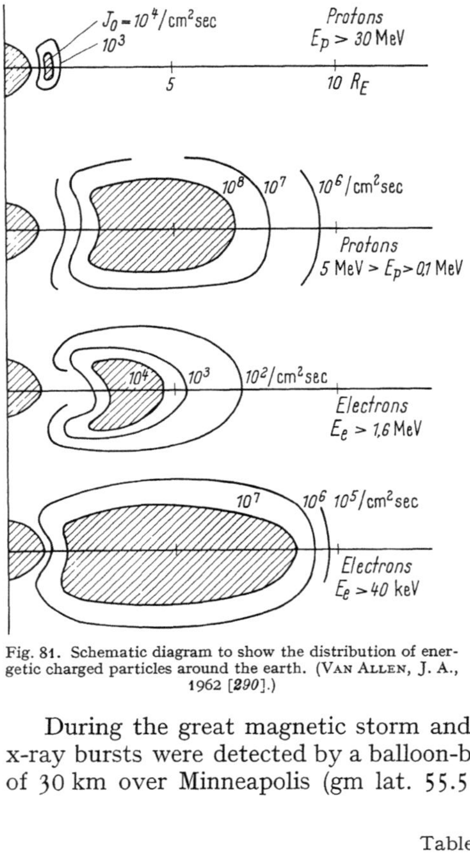

Fig. 81. Schematic diagram to show the distribution of energetic charged particles around the earth. (VAN ALLEN, J. A., 1962 [290].)

At present it is impossible for any one detector to "see" the whole belt system. Each detector has its own characteristic response for different energies and for different particles, and thus for different components of the belts. Our present knowledge of particles of lower energy than those in the table, namely protons with energy less than 100 kev and electrons less than 40 kev, is incomplete and fragmentary, although for auroral studies such information is urgently needed.

ANDERSON (1959 [291]) at Churchill (gm lat. 68.7° N) observed soft radiation by balloon-borne apparatus during the magnetic storm of August 29/30, 1957. He concluded that soft radiation is produced by electron bremsstrahlung in the atmosphere.

During the great magnetic storm and aurora of September 23, 1957, intense x-ray bursts were detected by a balloon-borne ionization chamber at an altitude of 30 km over Minneapolis (gm lat. 55.5° N) (WINCKLER, PETERSON, ARNOLDY and HOFFMAN 1958 [292]). The bursts occurred when a bright rayed band passed overhead (Fig. 82). The large burst that started at about 1050 GMT, September 23, 1957, was associated with the break-up of the aurora and with the polar magnetic storm observed in Alaska (Sect. 34a).

Table 6.

	Energy	Location*	Flux
Protons: inner	$E > 40$ Mev	$L = 1.4 \sim 1.5$	2×10^4/cm² sec
Protons: outer	100 Kev $< E <$ 4.5 Mev	$L > 2$	6×10^7/cm² sec at $L = 3.5$
Electrons: inner	$E > 1.5$ Mev	$L \sim 3.5$	$10^3 \sim 10^6$/cm² sec
Electrons: outer	$E > 40$ Kev	$L > 2$	10^7/cm² sec

* L = MCILWAIN's L parameter, see Sect. 4.

The bremsstrahlung is produced when primary electrons enter the ionosphere from above. They come to rest at a height of 80~100 km. The photons travel down to a height of 30~40 km, and are detected by various balloon-borne

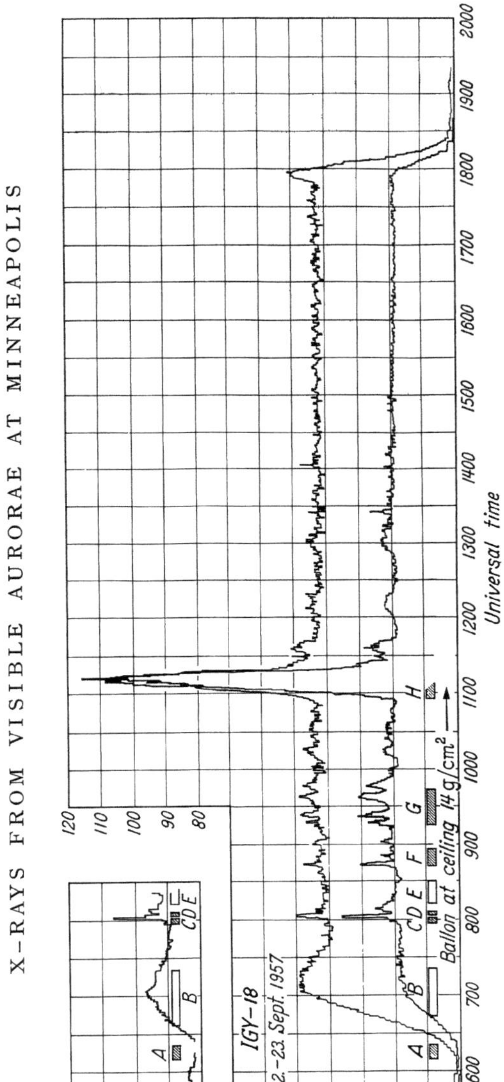

Fig. 82. Ion chamber (bottom), single counter (top), and photon counter (upper left, separate curve) response during the intense auroral storm of September 22-23, 1957. The balloon was launched about midnight CST during a lull in auroral activity, which had been very high in the earlier evening. The letters refer to observed auroral phenomena as follows: A, flaming auroral rays reaching zenith from all points of compass; red color. B, 1 curtains in north but activity generally decreasing overhead. B-C (0735 UT), arcs in narrow bands extending across zenith and almost to horizon in east and west directions; north dark; no rays. C, extremely strong ray buildup from all directions; intense red patches at 30° elevation in west. D, red patches intense again after decrease. E, flaming rays; no red; activity decreased. F, strong rays in northwest; red color; strong flaming in west; general activity increasing. G, major ray buildup in west, north, and east, with corona; red color reappeared at 0945 UT; following G, general intensity decreased with flaming rays. H, very strong ray structure at 30° elevation in east with intense red color; visible against predawn sky light. (Winckler, J. R., L. Peterson, R. Arnoldy and R. Hoffman, 1958 [292].)

devices. They are absorbed during their passage, mainly by the photoelectric process (Winckler 1962 [293], see also Sects. 50 and 51).

Rockets have been used to explore the particles that produce visible auroras, and have provided some of the most direct information as to the nature of these particles.

A. Motions of charged particles in a dipole field.

45. General. The first extensive study of the trajectories of a charged particle in a dipole field was made by STÖRMER, whose results are summarized in his book "*The Polar Aurora*" (1955 [*3*]). If STÖRMER's integration constant γ exceeds unity, there are two separate regions in which the particle can move. One

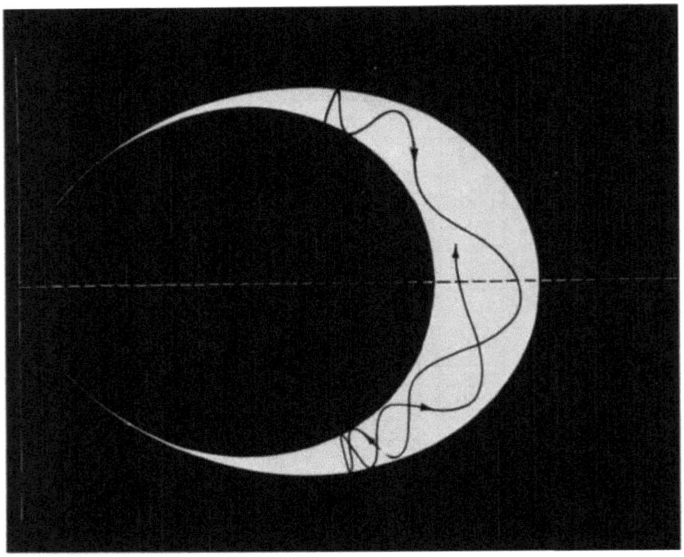

Fig. 83. A diagram after STÖRMER [*3*], illustrating the meridian projection of the spatial trajectory of an electrically charged particle in the field of a magnetic dipole and the boundaries of the rigorous trapping region (VAN ALLEN, J. A.).

region extends to infinity and is called the outer allowed region. The other is connected to the origin and is called the inner allowed region (STÖRMER 1955 [*3*], pp. 229—239).

Our concern here is with the particles confined in the inner allowed region (Fig. 83). They are now called "*trapped particles*", because in a steady state their trajectories do not extend to infinity; they remain indefinitely in the inner allowed region.

In 1940 ALFVÉN (1940 [*294*], 1950 [*13*]) devised the "guiding center" approximation, which greatly simplifies the description of the motions of charged particles in a sufficiently strong magnetic field. By this approximation the motion of a charged particle in the inner allowed region is analyzed into three parts (Table 7), namely a circular motion (the *gyration*), in a plane perpendicular to the magnetic field vector $\boldsymbol{B}$, a motion along $\boldsymbol{B}$ (the *oscillation*), and a *drift motion* along the direction of $\pm\boldsymbol{B}\times\nabla^2 B$ and $\pm\boldsymbol{B}\times[(\boldsymbol{B}\cdot\nabla)\boldsymbol{B}]$ (the plus sign refers to a positive charge and the minus sign to a negative charge; see Eq. (47.1) and (47.2). Notation:

a = the earth's radius;
r = the radial distance from the earth's center O to a point P;
φ = the latitude;
λ = the longitude;
c = velocity of light;
m = the mass of a trapped particle;
e = its charge, in esu;
w = its velocity;

E = its energy, in kev, namely $\frac{1}{2}mw^2/1.602\times 10^{-9}$;
n = the number density of the particles;
$\boldsymbol{B}$ = the magnetic vector, of magnitude B;
ϑ = the angle, called the *pitch angle*, between w and $\boldsymbol{B}$;
w_s = the component of w along $\boldsymbol{B}$, so that $w_s = w\cos\vartheta$;
w_n = the component of w normal to $\boldsymbol{B}$, so that $w_n = w\sin\vartheta$.

46. Gyration and oscillation between the two mirror points M, M'. The radius R and the period T of the gyration of the particle are given by

$$R = \frac{m w_n}{eB}, \qquad (46.1)$$

$$T = \frac{2\pi m}{eB}. \qquad (46.2)$$

In the rapid oscillation of a trapped particle between the mirror points M and M', the magnetic moment is invariant; it is given by

$$\mu = \frac{\frac{1}{2}m w_n^2}{B} = \frac{\frac{1}{2}m w^2}{B_m}. \qquad (46.3)$$

Here B_m denotes the value of B at the mirror point, where $w_s = 0$, $w_n = w$. Thus the particle is reflected at the point where

$$B_m = \frac{w^2}{w_{ne}^2} B_e \qquad (46.4)$$

where the suffix e refers to values at the point P_e where the field line crosses the equatorial plane. Because of the relation $w_{ne} = w\sin\vartheta_e$, the Eq. (4) may be rewitten as

$$B_m = \frac{B_e}{\sin^2\vartheta_e}. \qquad (46.5)$$

The time T_0 required for one complete oscillation between the two mirror points (from P_e to M, thence through P_e to M' and back to P_e) is given by

$$T_0 = 4l/w. \qquad (46.6)$$

Here l denotes the arc length of the *spiral* path of the particle from P_e to M or M'. Its value has been calculated by WENTWORTH, MACDONALD and SINGER (1959 [*295*]; their Fig. 2). For the case $w_{ne}/w = \sin\vartheta_e = 0.1$, l is approximately given by

$$l = 1.25\, r_e. \qquad (46.7)$$

In calculating the illustrative data of Table 7, the distance r_e is taken to be $6a$. The field intensity H_e at $6a$ is about $148\,\gamma$. Thus Eq. (5) shows that a particle for which $\sin\vartheta_e$ is 0.1 will be reflected at a point M where the field intensity on the line of force crossing the equatorial plane at 6 earth radii is $14800\,\gamma$. In this case M is 3190 km above the ground. The corresponding mirror latitude φ_0 is approximately 60°. For such a particle, according to (7),

$$l = 7.50\, a \qquad (46.8)$$

and

$$T_0 = 30\, a/w = 1.91\times 10^{10}/w \qquad (46.9)$$

if w is expressed in cm/sec and T_0 in sec.

For a particle that crosses the equatorial plane at a given distance r_e the latitude φ_0 and height of its mirror points depend only on its pitch angle ϑ_e, not on its speed w.

47. The slow east or west drift motion. During the oscillations between the mirror points, the protons drift westward and the electrons eastward. The guiding centers move on a surface of revolution containing the lines of force r_e.

The drift motion results from two causes, (a) the inhomogeneity of the earth's field and (b) the centrifugal force associated with the motion along the curved lines of force. The corresponding parts $\boldsymbol{u}_1$ and $\boldsymbol{u}_2$ of the drift velocities are respectively given (cf. PARKER 1957 [*296*]) by the general formulae

$$\boldsymbol{u}_I = (\tfrac{1}{2} m w_n^2 / e B^4)\, \boldsymbol{B} \times \nabla B^2/2\,, \tag{47.1}$$

$$\boldsymbol{u}_2 = (m w_s^2 / e B^4)\, \boldsymbol{B} \times (\boldsymbol{B} \cdot \nabla)\, \boldsymbol{B}\,. \tag{47.2}$$

During each complete oscillation between the two mirror points the drift produces a longitudinal displacement $\delta\lambda$ given (ALFVÉN 1950 [*13*], p. 28) by

$$\delta\lambda = 4 \left(\frac{r_e}{C_{st}}\right)^2 I_1(\varphi_0)\,. \tag{47.3}$$

Here C_{st} denotes STÖRMER'S unit length (STÖRMER 1955 [*3*], p. 217) given by

$$C_{st} = \left(\frac{M e}{m w}\right)^{\frac{1}{2}} = \left(\frac{M}{R_M}\right)^{\frac{1}{2}}. \tag{47.4}$$

Here M denotes the dipole moment of the earth (8.1×10^{25} gauss cm^3), and

$$R_M = m w / e\,. \tag{47.5}$$

R_M is called the *magnetic rigidity* of the charged particle. This length C_{st} was introduced by STÖRMER in his calculations of the paths of *single* particles entering the earth's magnetic field from outside, or trapped in the field. Also I_1 is the integral (see also HAMLIN, KARPLUS, VIK and WATSON (1961 [*314*]))

$$I_1(\varphi_0) = \int_0^{\varphi_0} \left(\frac{r_e \cos^4\varphi}{R_c}\right) \frac{1 - \frac{1}{2} B/B_m}{(1 - B/B_m)^{\frac{1}{2}}}\, d\varphi\,, \tag{47.6}$$

where R_c denotes the length of the radius of curvature of the line of force. For a dipole field line, R_c at a point is given by

$$\frac{1}{R_c} = \frac{3(1 + \sin^2\varphi)}{r_e \cos\varphi\, (1 + 3\sin^2\varphi)^{\frac{3}{2}}}\,. \tag{47.7}$$

Here r_e denotes, as above, the radial distance of the point P_e where the line of force through P crosses the equatorial plane. The integral $I_1(\varphi_0)$ has been evaluated by ALFVÉN (1950 [*13*], p. 29) for a series of values of the mirror latitudes φ_0. When φ_0 exceeds 45°, $I_1(\varphi_0)$ is nearly independent of φ_0, with the approximate value 76°. Thus, the corresponding longitudinal displacement per complete oscillation is given by

$$\delta\lambda = 4 \times 76° \left(\frac{r_e}{C_{st}}\right)^2. \tag{47.8}$$

The number of oscillations made by the particle per revolution is $(360°/\delta\lambda)$. The time T_R required for one revolution is given by

$$T_R = (360°/\delta\lambda)\, T_0\,. \tag{47.9}$$

Table 7 illustrates these formulae numerically, for protons and electrons that cross the equatorial plane at $r_e = 6a$ with a pitch angle $\vartheta_e = \sin^{-1} 0.1$; the latitude of the mirror points for these particles is approximately 60°. Various energies are considered, corresponding to a series of values of the magnetic rigidity R_M that are included in a Table given by STÖRMER ([*3*], p. 294, Table 1).

Table 7. *Numerical particulars relating to the gyratory, oscillatory and drift motions of electrons and protons of various energies, that cross the equatorial plane with pitch angle* $\vartheta_e = \sin^{-1} 0.1$, *and are associated with the lines of force* $r_e = 6a$ *of the earth's dipole field. In this Table a number expressed as* X^y *signifies* $X \times 10^y$.

	R_M	w (cm/sec)	E (Kev)	C_{st} (km)	R (km)	T (sec)	$\delta\lambda$ (°)	$360°/\delta\lambda$	T_0 (sec)	T_R
electrons	1^2	1.76^9	0.879	9.00^6	6.76^{-2}	2.42^{-4}	0.0055	65455	10.95	8.4 days
	3^2	5.28^9	7.85	5.20^6	2.03^{-1}	2.46^{-4}	0.0185	19459	3.67	19.5 hours
	6^2	9.95^9	30.7	3.67^6	4.06^{-1}	2.57^{-4}	0.0356	10121	1.92	5.4 hours
	1^3	1.52^{10}	81.4	2.85^6	6.76^{-1}	2.81^{-4}	0.0578	6233	1.27	2.2 hours
	5^3	2.84^{10}	1.07^3	1.27^6	3.38	7.48^{-4}	0.2845	1265	0.673	14.2 minutes
	1^4	2.96^{10}	2.53^3	9.00^5	6.76	1.44^{-3}	0.5673	635	0.645	6.8 minutes
protons	5^3	4.79^7	1.2	1.27^6	3.38	0.433	0.2845	1265	3.99^2	6.6 days
	1^4	9.58^7	4.79	9.00^5	6.76	0.433	0.5673	635	1.99^2	35 hours
	3^4	2.87^8	43.1	5.20^5	2.03^1	0.433	1.712	210	66.6	3.9 hours
	6^4	5.75^8	112	3.67^5	4.06^1	0.433	3.429	105	33.2	58 minutes
	1^5	9.57^8	479	2.85^5	6.76^1	0.433	5.688	63.3	20.0	21 minutes
	1^6	9.13^9	4.79^4	9.00^4	6.44^2	0.433	57.06	6.3	2.09	16 seconds

In calculating R and T, the relativistic correction is taken into account for high energy electrons.

R_M = the magnetic rigidity (47.5);
w = the speed;
E = the energy;
C_{st} = Störmer length unit (47.4);
R = the radius of gyration at $r_e = 6a$ (46.1);
T = the period of gyration ar $r_e = 6a$ (46.2);
$\delta\lambda$ = the longitudinal displacement per oscillation (47.8);
$360°/\delta\lambda$ = the number of oscillations per revolution;
T_0 = the time of oscillation (46.9);
T_R = the time of revolution (47.9).

Three "abiabatic" invariants (Northrop and Teller 1960 [*297*]) are associated with the above three motions, the gyration, oscillation and drift motion. They are

(a) the magnetic moment μ of the gyratory motion, given by Eq. (46.3);
(b) the longitudinal invariant I of the oscillatory motion defined in Eq. (4.1);
(c) the third invariant Φ = the magnetic flux through a surface bounded by a magnetic surface (defined below).

"Adiabatic" here means that the above three quantities μ, I and Φ change so slowly in a steady state that they may be taken as constants of the motion to a high degree of accuracy (Kulsrud 1957 [*298*]) (see also Northrop 1963 [*299*]).

The invariance of μ has been an important guide to the characteristics of particle motions in a dipole field. The Eq. (46.3) is simply a direct consequence of this. The longitudinal invariant I makes it possible to define a surface or shell on which the guiding center oscillates and drifts. It is the locus of the lines of force which have the same value of I, and is called a *magnetic surface*. In a dipole field, because of its axial symmetry, the shell can be generated by rotating the lines of force around the dipole. If the first two invariants are modified in such a way that μ is decreased and I is increased, the altitude of the mirror point is reduced; if the reduction is significant, it increases the rate of loss of the particle by collisions in the denser atmosphere. Change in the third invariant signifies the migration of a particle from one magnetic surface to another.

There are two functions that describe the group characteristics of a set of trapped particles, namely

$n(r, \varphi, \lambda; w)\,dw =$ the number density per cm³ with velocities in the range dw at w at the point (r, φ, λ)

$F(r, \varphi, \lambda; \vartheta; w)\,d\vartheta =$ the fraction of the particles of velocity w with pitch-angle in the range $d\vartheta$ at ϑ at the point (r, φ, λ).

The latter is called the *pitch-angle distribution function*. In the simplest case it is a function of ϑ only, and

$$\int_0^\pi F(\vartheta)\, d\vartheta = 1 .$$

B. Balloon observations of auroral particles.

48. X-ray bursts, auroras, auroral electrojets, cosmic radio-noise absorption. X-ray bursts in the auroral zone have been much studied (ANDERSON 1958 [*291*], 1960 [*300*]; ANDERSON and ENEMARK 1960 [*301*]; BROWN 1961 [*302*], 1962 [*303*]; BARCUS and BROWN 1962 [*304*]; BROWN and CAMPBELL 1962 [*305*]; EVANS 1963 [*306*]; PFOTZER et al. 1962 [*307*]; KEPPLER et al. 1962 [*308*]; KREMSER 1963 [*309*]), also in the subauroral zone (WINCKLER, PETERSON, ARNOLDY and

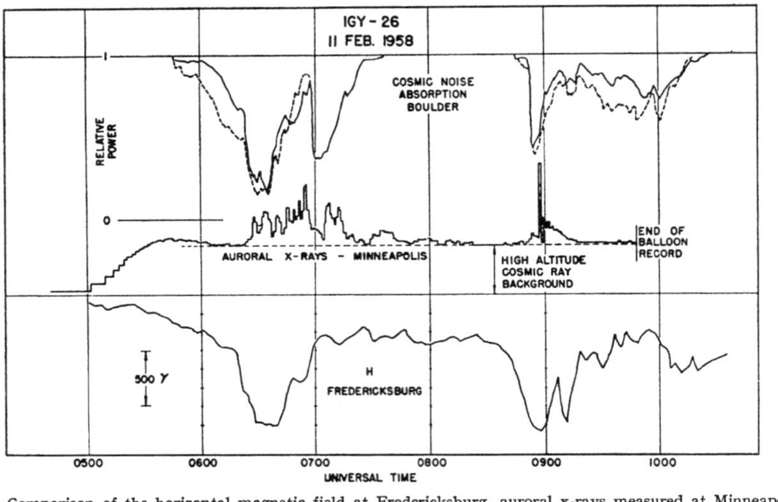

Fig. 84. Comparison of the horizontal magnetic field at Fredericksburg, auroral x-rays measured at Minneapolis, and cosmic noise absorption measured at Boulder. Two cases of intense ionospheric ionization apparently occurred, reaching maxima at approximately 0630 and 0900 UT. The dotted Boulder curve refers to the total interferometer power and the solid curve to the vertical beam power. The balloon curve shows the x-rays compared with the normal high-altitude cosmic rays background. (WINCKLER, J. R., L. PETERSON, R. HOFFMAN and R. ARNOLDY, 1959 [*310*].) See also Fig. 63.

HOFFMAN 1958 [*299*]; WINCKLER, PETERSON, HOFFMAN and ARNOLDY 1959 [*310*]; WINCKLER 1960 [*311*]; MAY 1960 [*312*]; BHAVSAR 1961 [*313*]); WINCKLER 1962 [*293*]) summarized some of the earlier results.

The auroras over the subauroral zone are usually associated with medium or great magnetic storms, and are fairly active (except during such periods as are discussed in Sect. 35a). This simplifies correlation studies between the x-ray bursts and the aurora or other geomagnetic phenomena. The x-ray bursts are mostly accompanied by active auroras with rays, and by polar magnetic substorms (bay disturbances); they occur simultaneously during the night hours (WINCKLER 1960 [*311*], 1962 [*314*]; MAY 1961 [*312*]). Fig. 84 shows a typical example of closely related x-ray bursts and polar magnetic substorms on February 11, 1958 (WINCKLER, PETERSON, HOFFMAN and ARNOLDY 1959 [*310*]).

In the auroral zone, however, the relation between auroral phenomena and x-ray bursts is more complicated. Auroral displays and polar magnetic substorms

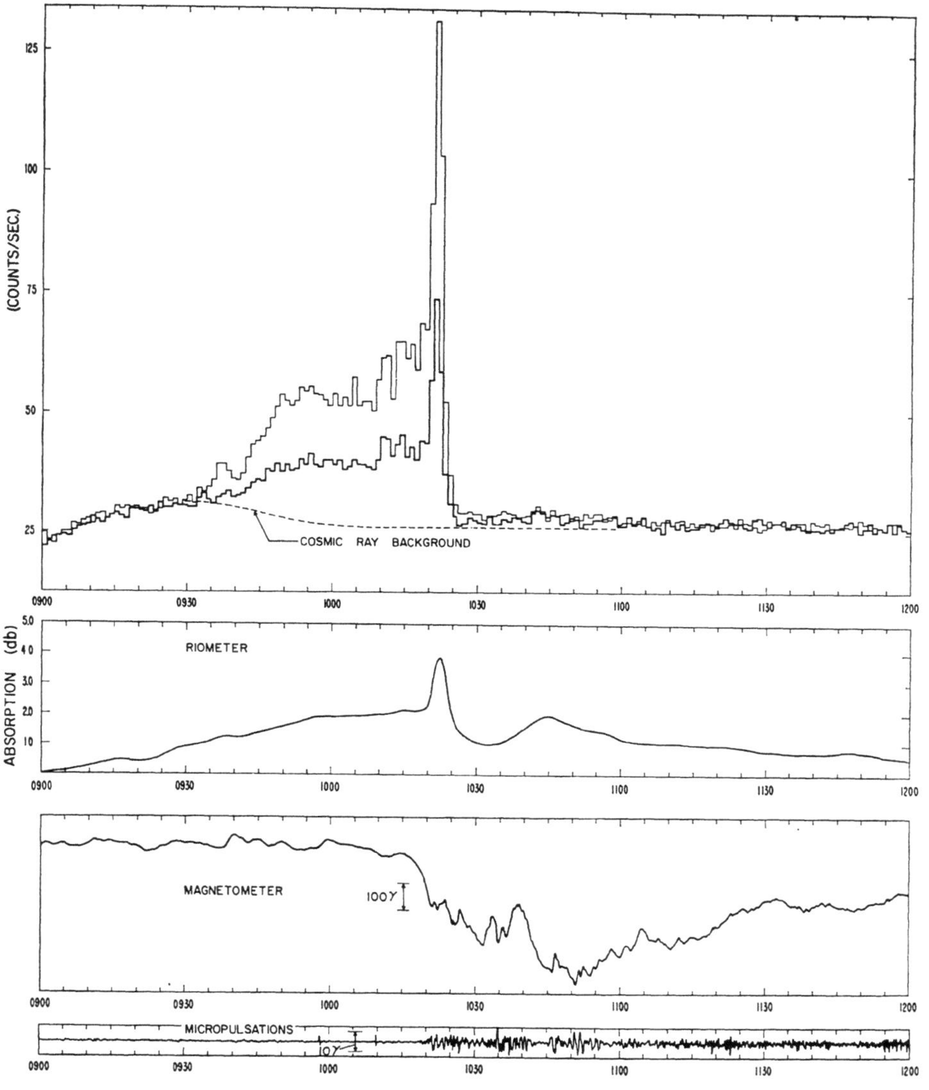

Fig. 85. The magnetic bay of June 25, 1961. Top: counting rates of balloon-borne counters (B1 counter, light line; A1 counter, dark line). Second from top: ionospheric absorption as recorded by a 27.6-Mc/s vertical riometer. Third from top: rapid-run magnetogram trace of the horizontal field component. Bottom: geomagnetic micropulsations of horizontal field. (BROWN, R. R., and W. H. CAMPBELL, 1962 [*305*].)

are not always closely correlated; as discussed in Sect. 25, the pseudo break-up of the aurora, however bright it may be, produces less magnetic effect and less cosmic noise absorption than a medium break-up. Fig. 52 shows a typical example; the expansions of the zone in which overhead auroras were seen occurred at 0750 and 1045 GMT, September 23, 1957; they were similar in their rapidity and in the scale of the motions. The former was accompanied by a marked polar magnetic substorm over the border between Canada and the USA, but not in Alaska. During intense magnetic storms the latitude of maximum auroral activity shifts towards the equator (Sect. 35); the equatorward shift of the zone of x-ray bursts is discussed in Sect. 51. In correlation studies with the polar aurora, such complexities must be kept in mind.

ANDERSON (1960 [*300*]), ANDERSON and ENEMARK (1960 [*301*]) and BROWN (1961 [*302*]) reported that x-ray bursts are poorly correlated with magnetic and auroral activity. ANDERSON and ENEMARK (1960 [*301*]) showed that x-ray bursts did not coincide with the time of maximum visible auroral activity, but instead precede it by about half an hour. However, further detailed study is necessary before forming definite conclusions on their correlation, because of the above-mentioned complexities of the polar aurora (see also Sect. 53).

CAMPBELL and MATSUSHITA (1962 [*315*]), BROWN and CAMPBELL (1962 [*305*]), BARCUS and BROWN (1962 [*304*]) found a rather close relation between x-ray bursts, cosmic noise absorption and the auroral electrojets. Fig. 85 shows a typical example of simultaneous observations of an intense x-ray burst, cosmic noise absorption and the polar magnetic substorm of June 25, 1961. As discussed in Sect. 18, polar magnetic storms are undoubtedly accompanied by the break-up phase of the aurora.

Auroral electrojets appear in a localized narrow zone where a strong electron precipitation and thus an intense x-ray burst occurs. X-ray bursts can, however, be detected only in a rather limited area close under such a narrow zone, whereas the magnetic effect of the jet can be recorded at a distance of a few hundred km away from the zone, without much reduction of magnitude. WINCKLER (1962 [*314*]) gave two fine examples in which x-ray bursts and cosmic noise absorption are well correlated.

Another characteristic difference of x-ray bursts in the auroral zone from that in the subauroral zone is that their appearance is not limited to the nighttime. ANDERSON (1960), ANDERSON and ENEMARK (1960), BROWN (1961), PFOTZER (1962 [*307*]) and KEPPLER (1962 [*308*]) reported that x-ray bursts occur in the daytime and that their intensity can be greater than that in the nighttime[1].

49. Electron energy spectrum and flux. By balloon-borne NaI scintillation spectrometer, BHAVSAR (1961 [*313*]) obtained the electron energy spectrum in the energy range 22 to 263 kev for the intense x-ray bursts of May 13, 1959, over Minneapolis (gm lat. 55.5° N). Assuming that x-rays are the bremsstrahlung of electrons striking the top of the atmosphere, and that the energy spectrum of these electrons follows an integral number power law, BHAVSAR obtained the

[1] KREMSER (1963 [*309*]) discussed x-ray bursts observed over Kiruna (Lapland, gm lat. 65.3°) by PFOTZER [*307*] and KEPPLER [*308*] and compared them with geomagnetic records from many stations. He found that in some cases x-ray bursts over Kiruna were accompanied by bay-disturbances (here called polar sub-storms) in the Kiruna magnetic records, while in other cases the x-ray burst had no counterpart in the geomagnetic variations. However, while such simultaneous geomagnetic variations were absent at Kiruna, they were quite distinct in the magnetic records at observatories in other longitudes, where they could be recognized even at stations near the equator. Such cases were found at the two local daytimes when the horizontal field variations of the polar sub-storms (what was formerly called *SD* now *DS*, the solar diurnal variation on disturbed days) change sign. At such times, an x-ray burst over Kiruna may not be accompanied by a magnetic bay at Kiruna; but the more or less world-wide geomagnetic effects associated with the burst show clearly at other stations where the local time is favorable for the production of a bay, positive or negative, in the horizontal component. (Ed.)

spectrum $N(>E) = 0.66 \times 10^{15} E^{-5}$ (22 kev $< E <$ 263 kev)

with E in kev. He indicated, however, that for energies less than 40 kev, N is proportional to E^{-7}; thus the spectrum becomes steeper for lower energies.

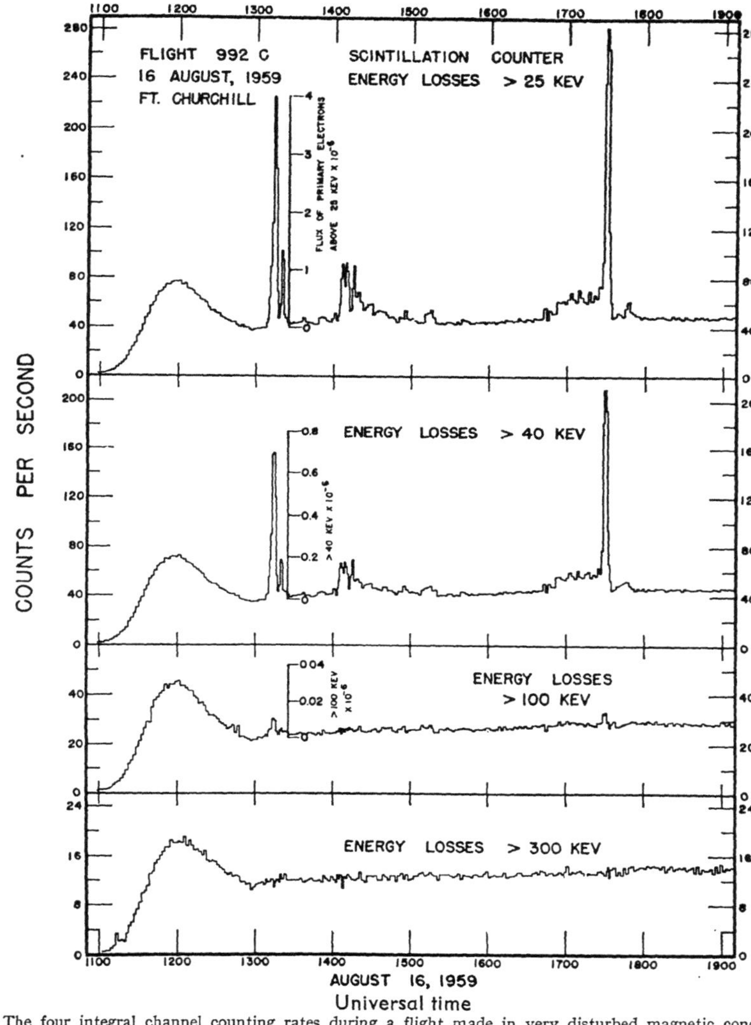

Fig. 86. The four integral channel counting rates during a flight made in very disturbed magnetic conditions. (ANDERSON, K. A., and D. C. ENEMARK, 1960 [301].)

The flux of electrons of energies above 40 kev varies between 10^9 and 10^{12} electrons/cm^2 sec; the typical burst time is approximately one hour. The peak flux in Fig. 82 was 8×10^9/cm^2 sec (WINCKLER 1960 [311]). BHAVSAR (1961 [313]) reported that the peak flux of electrons of energies above 22 kev was 10^9/cm^2 sec during the event of May 14, 1959.

Using a four channel scintillation counter over Churchill (gm lat. 68.7° N) ANDERSON and ENEMARK (1960 [301]) obtained the spectrum given by

$$N(>E) = 1.1 \times 10^6 \, e^{-0.04E} \quad (25 \text{ kev} < E < 100 \text{ kev}).$$

Fig. 86, an example of their observations, shows a rapid decrease of the intensity for higher energy electrons.

They indicated also that the auroral zone x-ray bursts are present about 40% of the time, and that the daily integrated flux of electrons above 25 kev into the auroral ionosphere is of order $10^{10}/cm^2$ (a daily average flux $10^5/cm^2$ sec). In Sects. 54—56 these numbers are compared with those obtained by satellites.

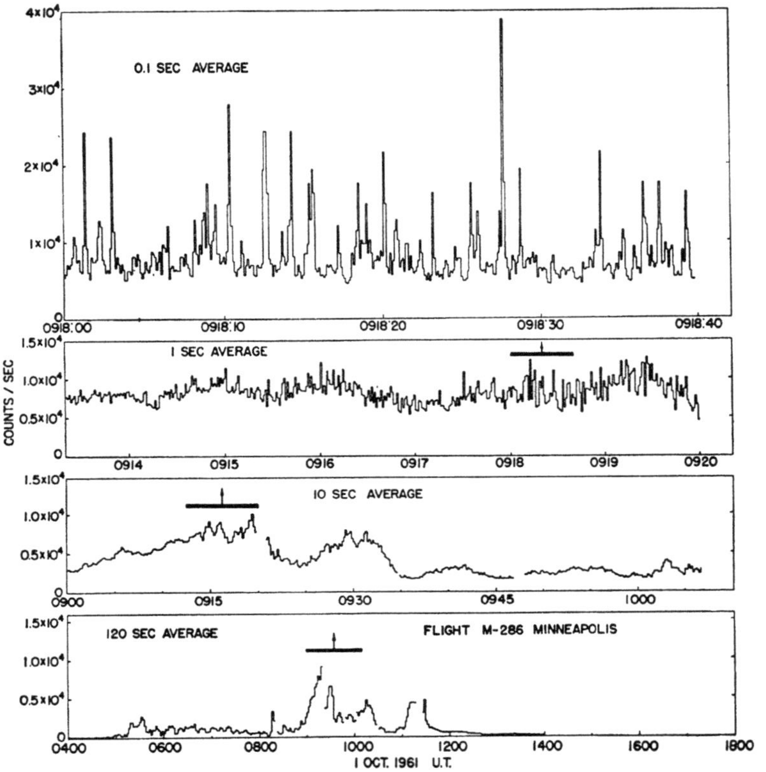

Fig. 87a. Samples of the time structure of the electron precipitation measured at Minneapolis on October 1, 1961, for the 20 keV X-ray channel. The intensity scale is linear, and beginning with the bottom section the region covered by the black bar is expanded into the section just above. Note that the maximum variance seems to occur in the very short intervals where extreme intensity variations occur in 0.1 sec. (WINCKLER, J. R., P. D. BHAVSAR and K. A. ANDERSON, 1962 [*316*].)

50. Time structure of the precipitation. During the magnetic storm of September 29/30, 1961 (Fig. 87a) WINCKLER, BHAVSAR and ANDERSON (1962 [*316*]) studied the structure of x-ray bursts (time scale 0.1 sec).

The bottom part of Fig. 87a is the standard plot of a 2-minute average, like that in Figs. 82 and 84. The portion under the black line is expanded to show about a 10-second average plot. Repeating twice more such an expansion of the time scale, the top part shows the plots of 0.1-second average during a short interval of the black line in the bottom part. Clearly the rather gradual increase of intensity in the bottom part is due to a succession of short bursts of life time of order 0.1 sec.

Their power spectrum analysis showed precipitation with predominant periods of 0.8, 1.6 and 3.2 seconds and further higher multiples. They suggested that this constitutes direct evidence that mono-energetic electrons of order 60 kev oscillate between conjugate points. They inferred also that such periodic bursts account for pulsating patches that often appear during the post break-up phase.

Evans (1963) reported fairly periodic x-ray bursts of period 100 seconds; the observation was made around 0200 AST on June 17, 1961, at College, Alaska (gm lat. 64.7° N). The electron energy spectrum had the form $E^{-4.3}$, with the

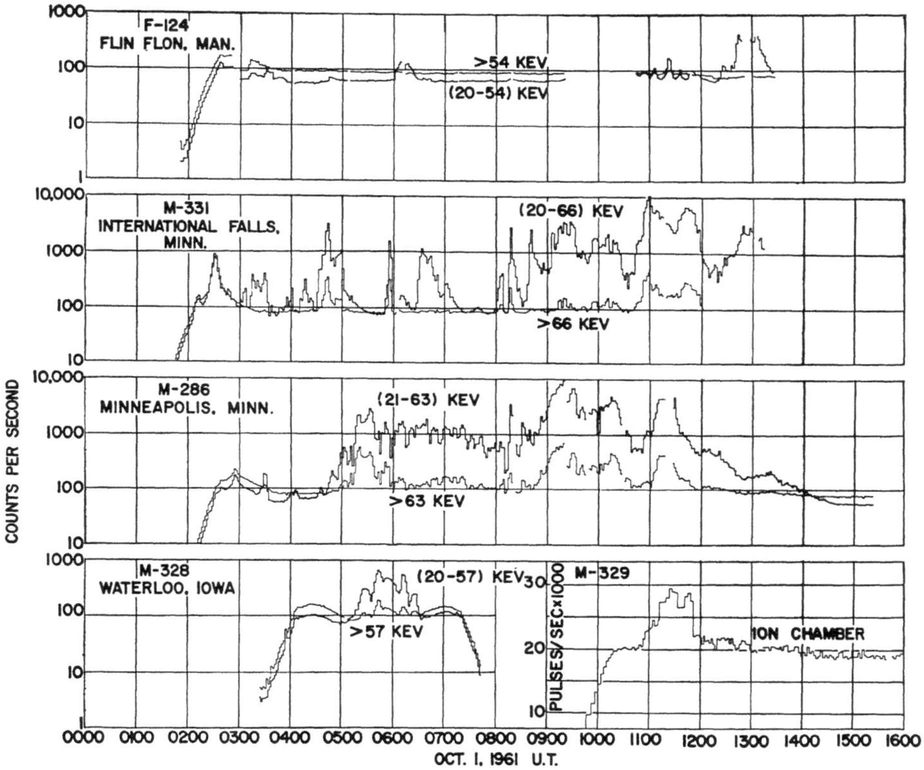

Fig. 87b. Simultaneous balloon flights at the four locations on the occasion of the large flare-induced storm on October 1, 1961. Note that the maximum precipitation occurs at intermediate latitudes and is sometimes simultaneous over a wide range of latitudes and sometimes unrelated. (Winckler, J. R., P. D. Bhavsar and K. A. Anderson, 1962 [*3 16*].)

most intense flux $9\times10^6/\mathrm{cm}^2$ sec for electrons with energies above 40 kev. Because of arctic daylight the pulsations could not be compared with auroral features, but he suggested that they might be associated with pulsating patches.

51. Large-scale precipitation of electrons. *a) Equatorward shift of the electron precipitation zone during a large magnetic storm.* Winckler, Bhavsar and Anderson (1962 [*316*]) studied the large-scale movements of the electron precipitation zone, by simultaneous balloon observations at four places, Flin Flon (gm lat. 64.5° N), International Falls (gm lat. 58.7°), Minneapolis (gm lat. 55.4° N), and Watheroo (gm lat. 52.8° N) (Fig. 87b). The network spans a range of L from 2.4 to 6.

During weak magnetic activity there was a general increase of x-ray intensity with latitude, the maximum intensity being at Flin Flon. During the large magnetic storm of September 29/30, 1961 (with maximal $Kp=9-$), however,

Table 8.

Geomagnetic Latitude L	52.8° 2.4	55.4° 3	58.7° 4	64.5° 6
Satellite Total energy/cm² (in kev) > 60 kev	7×10^{8}	2.6×10^{9}	7.50×10^{10}	2×10^{11}
Balloons (Oct. 1, 1961) Total energy/cm² (in kev) > 60 kev	1.5×10^{11}	5×10^{12}	4×10^{12}	1×10^{11}

the zone of maximum intensity of x-ray bursts shifted equatorward; at Flin Flon the x-ray bursts were few, but they were intense at International Falls. During the maximum epoch of the main phase of the magnetic storm of February 11, 1958 (attaining $Kp=9_0$), the precipitation was certainly at times (0730—0830, 0940—1020 GMT) south of Minneapolis, though the precipitation zone expanded and contracted rapidly (Sect. 35a and AKASOFU and CHAPMAN 1963 [*152*]).

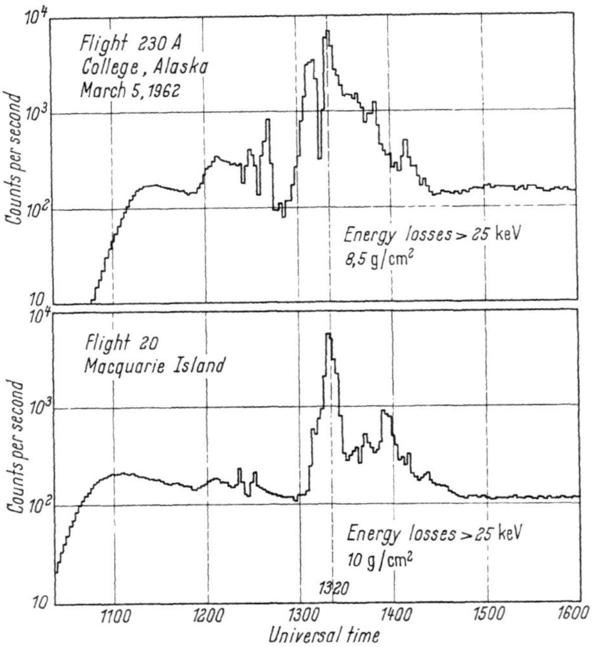

Fig. 87c. Counting rates of scintillation detectors launched from College and Macquarie Island. The two stations are approximately magnetically conjugate. (ANDERSON, K. A., C. D. ANGER, R. R. BROWN and D. S. EVANS, 1962 [*320*].)

Over a 10-hour period during the storm of October 1, 1961 the total energy precipitation of 1 cm² at height 100 km is compared in Table 8 with the total energy in a magnetic tube of force of 1 cm² cross-section at 100 km in a static outer radiation belt as inferred from the Explorer XII satellite observations.

Thus the total energy of trapped electrons, namely the stored energy in the belt, was at least ten times less than that precipitated during a 10-hour period. The static outer belt seemed to have insufficient energy to produce the aurora. O'BRIEN (1962 [*317*]) proposed to replace the leaky bucket model of the outer belt by a "*splash catcher*" model. The implication of this is discussed further in Sect. 55 and Chapter VIII.

b) Large-scale precipitation of electrons at the time of sudden commencement of magnetic storms. BROWN, HARTZ, LANDMARK, LEINBACH and ORTNER (1961 [*318*]) reported that the geomagnetic sudden commencement at 0146 GMT, June 26, 1960, was accompanied by a brief x-ray burst at Happy (near College), Alaska, and by cosmic noise absorption at College (gm lat. 64.7° N, gm long. 256.5°) and Kiruna (gm lat. 65.1° N, gm long. 115.9°). They suggested that an intense electron bombardment occurred in a zone at least 100 km wide, all around the polar cap, near the latitude of the auroral zone. The estimated electron flux was of order $2.6 \times 10^7/\text{cm}^2$ sec for energies greater than 50 kev. They inferred also that the rapid geomagnetic field variation at the time of the sudden commencement induced a "dumping" of electrons from the trapping region.

EVANS (1963 [*307*]) describes a 100-second periodicity in x-ray intensity at balloon altitudes observed over College (Alaska), 1961 June 17, 1148 to 1210 UT, in the course of a pronounced x-ray event (see also ANGER, BARCUS, BROWN and EVANS (1963 [*219*])).

c) Simultaneous precipitation at conjugate points. ANDERSON, ANGER, BROWN and EVANS (1962 [*320*]) simultaneously observed x-ray bursts at College and Macquarie Island during February and March 1962, one, at 1320 GMT, March 5, while Kp rose to 5+, was intense (Fig. 87c). They suggested that the zone of precipitation was at least 2° wide in latitude and 14° in longitude.

The large-scale precipitation of electrons with nearly equal intensity at the ends of the same bundle of field lines suggests that the electron precipitation has its cause in a region far out along the lines of force near the equatorial plane. The precipitation in the northern hemisphere cannot be a consequence of back scattered electrons from the southern hemisphere.

C. Rocket observations of auroral particles.

52. Auroral protons. The first direct observation of the nature of auroral particles was made by rockets by MEREDITH, DAVIS, HEPPNER and BERG (1958 [*321*]) at Churchill (gm lat. 68.7° N). MCILWAIN (1960 [*322*]) and MCDIARMID, ROSE and BUDZINSKI (1961 [*323*]) made further detailed similar studies.

The entry of protons does not seem to be related to visible auroras. MEREDITH et al., and MCILWAIN, observed no protons in one of their flights, and when they were detected, they entered both within and outside visible auroras; they were distributed fairly uniformly above 140 km in height.

The integral-number energy spectrum of protons incident upon the atmosphere is approximately expressed by

$$N(>E) = 2.5 \times 10^6 \exp(-E/30) \text{ protons/cm}^2 \text{ sec ster.} \quad (80 \text{ kev} < E < 250 \text{ kev})$$

where E denotes the proton energy in kev. The total flux calculated from the above equation is of order 1.6×10^7 protons/cm² sec, assuming isotropic incidence of protons upon the atmosphere. The simultaneous ground measurement of the $H\beta$ intensity indicated about 6×10^7 quanta/cm² sec, suggesting that each proton produces $4 H\beta$ quanta (MCILWAIN 1960 [*322*]; CHAMBERLAIN 1954 [*324*]).

53. Auroral electrons. MEREDITH et al. [*321*], showed that both auroral rays and glows are primarily due to electrons. Unlike the proton flux, the electron flux does not remain constant during the rocket flight. (Note, however, that it is not possible to separate spacial variations and time variations in rocket or satellite studies). MCILWAIN [*322*] indicated that in one of the flights, the electron flux exhibited very rapid and irregular fluctuations, and the light intensity measured by the rocket-borne photometer varied similarly.

The integral-number energy spectrum obtained by MCILWAIN [*322*] is expressed (Fig. 88) by

$$N(>E) = 2.5 \times 10^9 \exp(-E/5) \text{ electrons/cm}^2 \text{ sec} \quad (3 \text{ kev} < E < 30 \text{ kev}).$$

The total flux was approximately 2×10^{11} electrons/cm² sec, assuming the average electron energy of electrons to be 6 kev. This flux corresponds to a total energy of about 200 ergs/cm² sec. He showed that at least 75% of the light in this case was produced by such electrons. The total energy of ionization deposited in a bright aurora (1 BCIII) is about 400 ergs/cm² sec (CHAMBERLAIN 1961 [4], p. 283; see also Sect. 68).

Both balloon and rocket observations suggest a rather steeply graded energy spectrum for auroral electrons, indicating far more of low energy (a few kev) than of higher energy (>100 kev). It has been suggested that the polar aurora is mainly produced by electrons of energy less than 20 kev. But type *B* aurora (Sects. 12a and 71) is often seen during the break-up phase, at 80 km height. If it is produced by electrons, their estimated energy may be of order 200 kev or more (MALVILLE 1959 [325]). One obvious question here is whether a really active aurora in the auroral zone has yet been observed by balloons or rockets. A close inspection of the aurora penetrated by MCILWAIN's rocket (his Fig. 17) suggests that it showed the pseudo break-up, or a westward traveling fold (Sects. 23 and 25), rather than the break-up. Further extensive studies are needed before reaching a definite conclusion on the nature of auroral primary electrons.

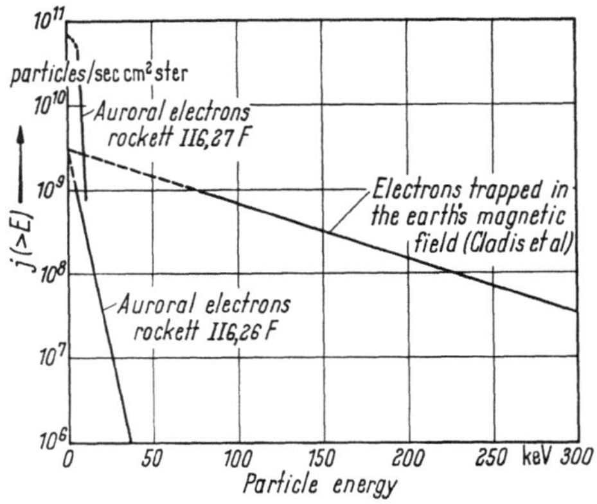

Fig. 88. The integral electron energy spectra obtained by two rocket flights. (MCILWAIN, C., 1960 [322].)

D. Satellite observations of auroral particles.

54. Auroras and energetic electrons (E > 1.5 mev) in the outer belt. A large decrease of the counting rate of energetic electrons in the outer belt (E>1.5 mev; see Fig. 81c) during a magnetic storm was first shown by ARNOLDY, HOFFMAN and WINCKLER (1960 [326]). During the main phase of the August 16, 1959 magnetic storm (attaining $Kp = 8+$), counters carried by the Explorer VI satellite indicated a remarkable reduction of the counting rate over a large portion of the energetic electron belt, particularly in the outer part. Then, during the recovery phase of the storm, the flux started to increase rapidly, and eventually far exceeded the pre-storm level.

The reduction of the counting rate during the main phase was interpreted as a "dumping" of energetic electrons into the zone where active auroras were seen (gm lat. 55∼60° N). Later other possibilities, such as a softening of the energy spectrum (a deceleration), have been discussed.

ARNOLDY, HOFFMAN, WINCKLER and AKASOFU (1962 [327]) showed that a rapid decrease in the counting rate coincided with a strong break-up of the aurora seen at Fargo (gm lat. 56.7° N). The satellite was about 4.6 earth radii away from the earth's center in the evening sector.

FARLEY and ROSEN (1960 [328]) and FAN, MEYER and SIMPSON (1961 [329]) studied the variations of the belt during the storm. The latter showed that the initial reduction and the later build-up cannot be simply due to the betatron effect. FORBUSH, VENKATESAN and MCILWAIN (1961 [330]) further extensively studied the belt during a number of other storms, and confirmed the above general behavior of the belt during storms. They found also a close relation between the magnitude of the main phase decrease and the reduction of the counting rate.

It does not seem likely that such high energy electrons ($E > 1.5$ Mev) contribute much to the auroral luminosity. However, the close relation between the decrease of the counting rate and the auroral break-up suggests that the decrease is a part of the process that increases the injection of auroral electrons (perhaps of $E < 200$ kev) into the auroral ionosphere. These electrons may be a high energy tail of the auroral spectrum. The process that produces an intense flux of auroral electrons seems to affect such a high energy tail. At present both the initial reduction of the counting rate and the later build-up are not understood.

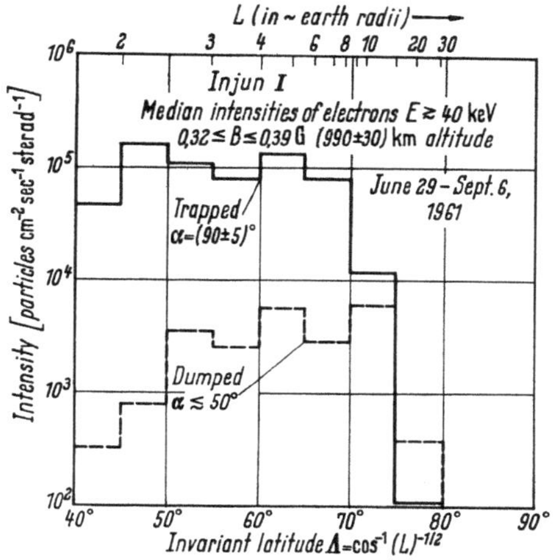

Fig. 89. The median intensities of trapped and dumped electrons as a function of the invariant latitude. (O'BRIEN, B. J., 1962 [334].)

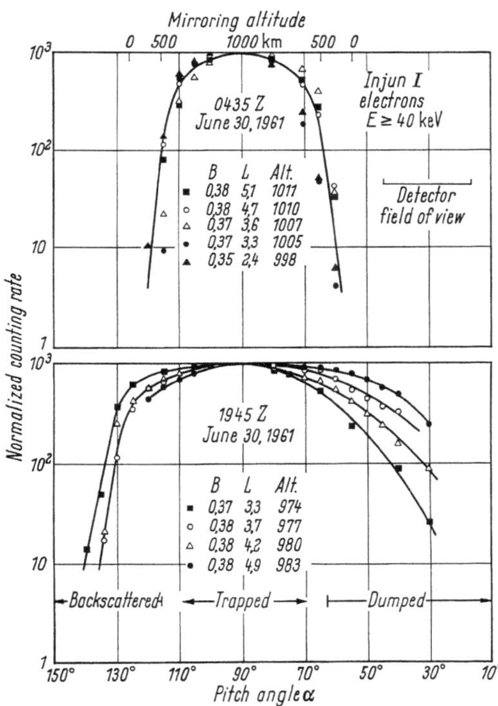

Fig. 90. Contrast of the two typical types of angular distributions, showing that dumping and backscattering occurred in the northbound pass at 1945 UT but not in the southbound at 0435 UT. Both passes were over the same region of longitude from 250° E to 270° E. Units of B are gauss, and of altitude are kilometers. (O'BRIEN, B. J., 1962 [334].)

55. Auroras and lower energy electrons (1.5 mev > E > 40 kev). The direct relationships between visual auroras and energetic electrons were first studied by O'BRIEN, VAN ALLEN, ROACH and GARTLEIN (1960 [331]). During a large magnetic storm on November 28, 1959 (attaining $Kp = 8_0$), two Geiger counters carried by the Explorer VII satellite found narrow zones of radiation over a visible aurora at an altitude of about 1000 km; the zone of radiation was traversed by the satellite in only 3 seconds. They suggested that the peak flux of electrons ($E > 40$ kev) was of order $5 \times 10^{10} \sim 5 \times 10^{11}/\text{cm}^2$ sec, but O'BRIEN (1963 [332]) later revised the estimate (see Sect. 44) to about $5 \times 10^7/\text{cm}^2$ sec. The Explorer VII satellite also passed over a subvisual red arc (Sect. 74).

ARNOLDY, HOFFMAN, WINCKLER and AKASOFU (1962 [327]) reported that an increase of soft electrons at a geocentric distance of 42000 km during the August 16, 1959 storm coincided with the break-up of the aurora over Churchill (gm lat. 68.7° N) (see also O'BRIEN (1963 [333]) and O'BRIEN and TAYLOR (1963 [334])).

New sets of counters carried by the Injun I and Explorer XII have greatly changed our view of the structure of the outer radiation belt (Sect. 44; see Fig. 81 d). O'BRIEN (1962 [332]) successfully separated the spacial distribution of *trapped* electrons and *dumped* electrons, by counters carried by the Injun I

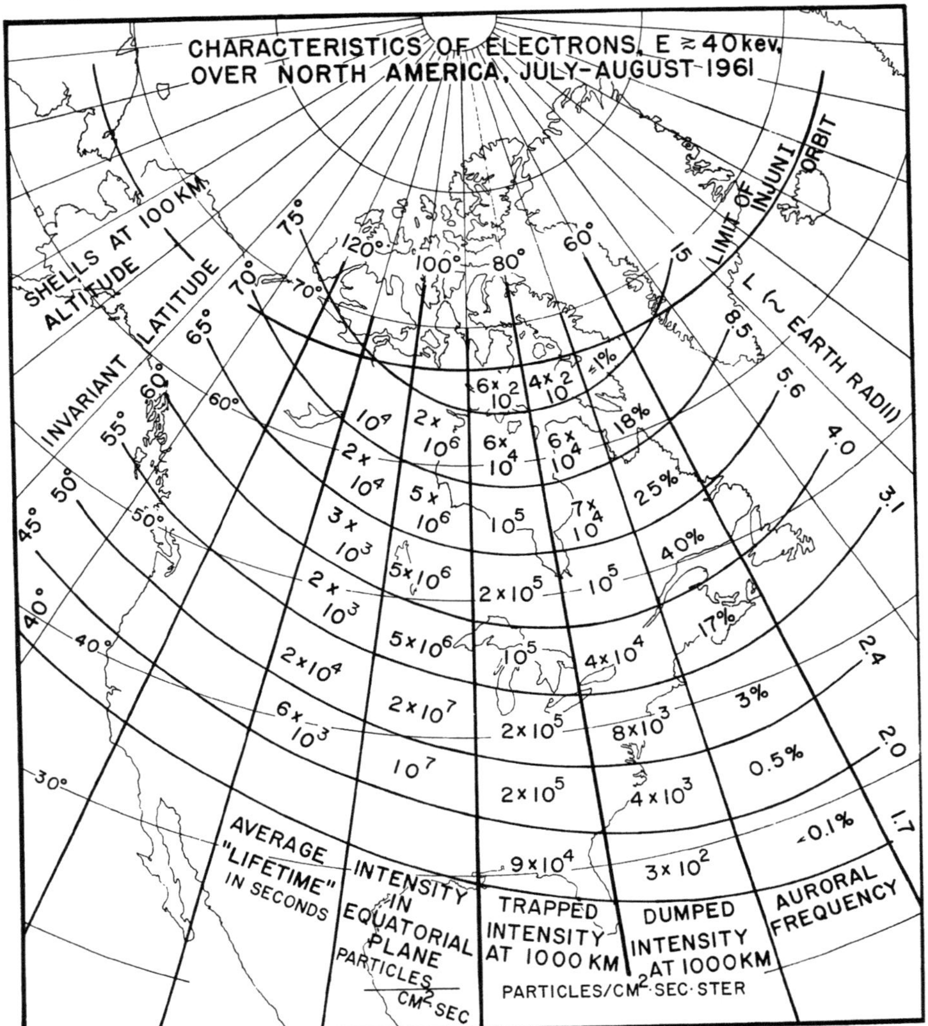

Fig. 91 a. Summary of the data of O'BRIEN's paper projected on a map of North America for comparison with balloon data. Auroral data by courtesy of W. C. GARTLEIN. (O'BRIEN, B. J., 1962 [*334*].)

satellite (Fig. 89). If the pitch-angle distribution (Sect. 46) of the electrons is known at a certain point in space at a certain instant, it is possible to classify electrons into three groups, namely dumping, trapped and backscattering electrons. Fig. 90 shows an example of the pitch-angle distribution, together with the three classifications. Electrons whose pitch-angle is less than 60° at the point of observation (at about 1000 km height, over the North American continent) have mirror heights located in the ionosphere, so that most of them are expected to be lost in the dense atmosphere. Electrons with pitch-angle close to 90° are mirroring there and would not be lost to the belt unless their pitch-angle is seriously changed by some mechanism and the mirror height is reduced. Some of the electrons which have mirror points between 1000 km and the ionosphere move upward after they are reflected at the mirror point or scattered back in the ionosphere. The trapped electrons were found to have a rather uniform distribution over a large range of latitude (gm lat. 40~70°), whereas the maximum flux of dumped electrons is near the boundary, beyond which the flux of both trapped and dumped electrons decreases sharply. Their fluxes near the boundary are of similar magnitude. The implication of this is important. The dumped electrons there cannot be a small portion of the trapped ones; they must be produced continuously. This agrees with the observation by WINCKLER, BHAVSAR and ANDERSON (1962 [*316*]), that the total energy of the trapped electrons is not enough to produce the observed x-rays (Sect. 51). It can rather be said that trapped electrons are a small portion of dumped electrons; the geomagnetic field acts as a "splash catcher" of electrons continuously produced there by some mechanism. O'BRIEN (1962 [*334*]), using the Explorer XII and Injun I data obtained the life time of low energy electrons. It is shown in Fig. 91 a, together with some other related quantities.

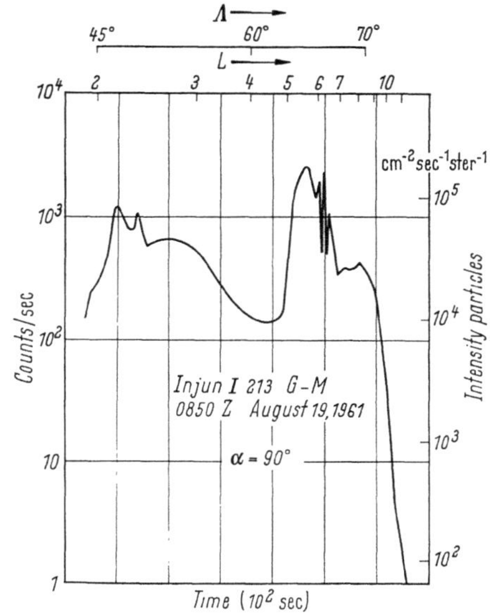

Fig. 91 b. Counting rate of the 213 GM during a northbound pass when the detector was always pointing normal to the magnetic field. The electrons were therefore mirroring at altitudes between about 800 and 1000 km. (O'BRIEN, B. J., C. P. LAUGHLIN, J. A. VAN ALLEN and L. A. FRANK, 1962 [*335*].)

The analysis of the Injun I data for individual passes shows irregular structures at times beyond gm lat. 60° (Fig. 91 b). O'BRIEN, LAUGHLIN, VAN ALLEN and FRANK (1962 [*335*]) suggested that some of such anomalous fluxes may be related to the radiation detected by MEREDITH, GOTTLIEB and VAN ALLEN (1955 [*284*]).

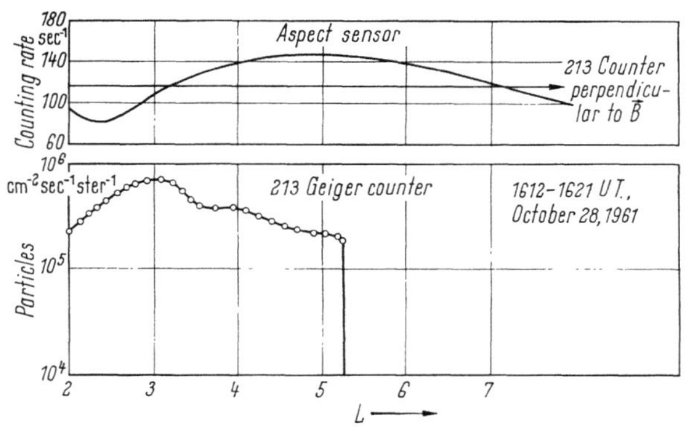

Fig. 92. Typical latitudinal variation of the directional flux of electrons of greater than 40 keV energy during geomagnetically disturbed conditions in October 1961. The aspect sensor indicated the orientation of the 213 G-M counter relative to the local magnetic field line. High counting rates of the aspect sensor refer to periods when the 213 G-M counter is looking up the geomagnetic field lines in the northern hemisphere. (MAEHLUM, B., and B. J. O'BRIEN, 1962 [*336*].)

No direct relationship between the aurora and the electrons detected by the Injun I satellite has yet been found, so it is not possible to infer what role the electrons of energy 40 kev in the outer belt play in the production of the auroral luminosity. However, it has become clear that the auroral electrons are most likely to be produced near the outer boundary of the outer radiation belt.

During the main phase of magnetic storms, the center of auroral activity shifts equatorward (Sect. 35), and so does the precipitation zone of electrons ($E > 60$ kev) (Sect. 51 a). MAEHLUM and O'BRIEN (1963 [*336*]) found that the outer boundary of the outer belt shifted drastically towards the equator; the observations were made by the Injun I satellite at an altitude of about 1000 km (Fig. 92). The equatorward shift of the outer boundary of the outer belt near the earth's surface is ascribed by AKASOFU (1963 [*337*]) to the growth of the ring current which is imbedded in the outer belt.

The outer belt shown in Figs. 81 c and 89 is an average feature; the belt undergoes considerable changes, even within a few hours. This variability of the belt bears on the origin of the electrons in the outer belt.

KRASOVSKII, SHKLOVSKI, GAL'PERIN, SVETLITSKII, KUSHNIR and BORDOVSKII (1962 [*338*]) also found an intense flux of electrons of energy above 10 kev over the auroral zone. Their suggested flux was at times 400 ergs/cm sec ster; however O'BRIEN (1963 [*339*], p. 123) criticized their interpretation.

56. Auroras and electrons of energy less than 10 kev. The most detailed study of the relation between visual auroras and auroral primary electrons was made by a joint experiment between the Geophysical Institute, University of Alaska, and the Lockheed Aircraft Corporation (EVANS and BELON 1963 [*340*]). Various counters able to detect electrons of 2 kev were carried by one of the polar orbiting satellites. When the satellite passed over Alaska, approximately in geographic longitude 156°, a stable homogeneous arc was overhead near Fort Yukon. Two photometers located at College and Fort Yukon scanned the sky and "triangulated" the aurora. As the satellite moved from south to north, incident particles were detected within the auroral form. They indicated an integral-number energy spectrum, $N(>E)$ proportional to $E^{-3.6}$, E the electron energy being in kev. The luminosity profile of the N_2^+ First Negative group bands at λ 3914 (see Sect. 68) obtained by the two scanning photometers seems to be reasonably reproduced by the observed energy spectrum, except in the upper part of the observed arc. They suggested that the upper part of the arc is produced by electrons of energy less than 2 kev.

It now seems that the main portion of the auroral luminosity is produced by electrons which have a wide range of energy, perhaps extending from 100 ev to 200 kev. The process that produces auroral primary electrons seems to affect the entire range of the electron spectrum, namely from 100 ev$\sim$2 mev; it seems to operate usually near the outer boundary of the outer radiation belt.

VII. The auroral spectrum.

57. Introduction. The auroral spectrum has been extensively studied by a number of workers from the middle of the last century (*cf.* CHAMBERLAIN and MEINEL [*341*]). Because the auroral luminosity is faint, variable and complex, and parts of the spectrum are difficult to produce in the laboratory, the identification of the intermingled atomic and molecular features is difficult. Key lines are often obscured by strong molecular bands. As the wave-length tables do not make this readily apparent, Fig. 93 a—j reproduce microphotometer tracings of the auroral spectrum of an ordinary aurora (PETRIE and SMALL 1952 [*346*]).

The early history of the identification of the lines and bands is a fascinating part of this subject. After ÅNGSTRÖM, VEGARD was the pioneer, and made many important contributions to it. Our present knowledge of the auroral spectrum owes much also to MCLENNAN, KVIFTE, NICOLET, BATES, MASSEY, RENSE,

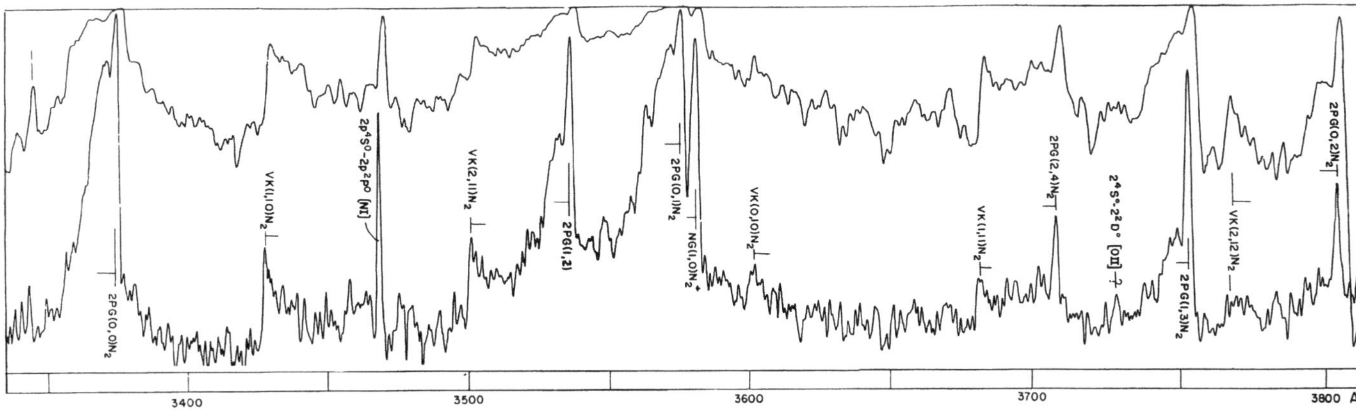

Fig. 93. Microphotometer tracings of the spectrum of the aurora by PETRIE and SMALL, by permission of the University of Chicago Press. The lower tracing is of a moderately exposed strip of the step-slit spectrum while the upper is of a heavily exposed strip. The upper edge of the graph represents dark while the lower is approximately a clear or unexposed reading. Identifications have been marked by CHAMBERLAIN and MEINEL (1953). Additional identifications are found in CHAMBERLAIN and OLIVER (1953). Half-tone reproductions of these spectra are found in PETRIE and SMALL (1952) at 43 A/mm dispersion.

Fig. 93a. Region 3340-3800 A.

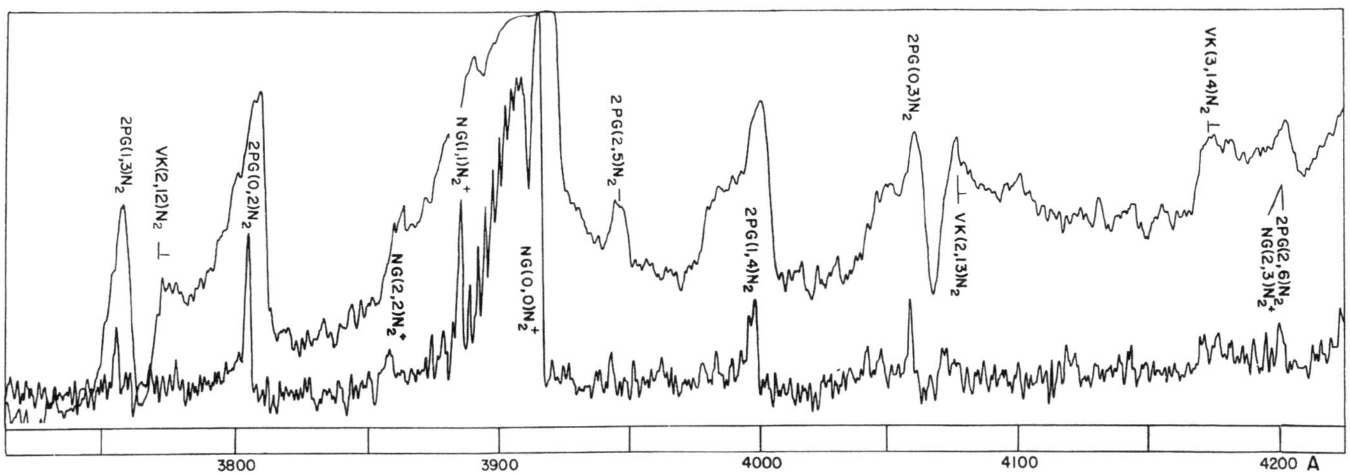

Fig. 93b. Auroral spectrum 3750-4225 A.

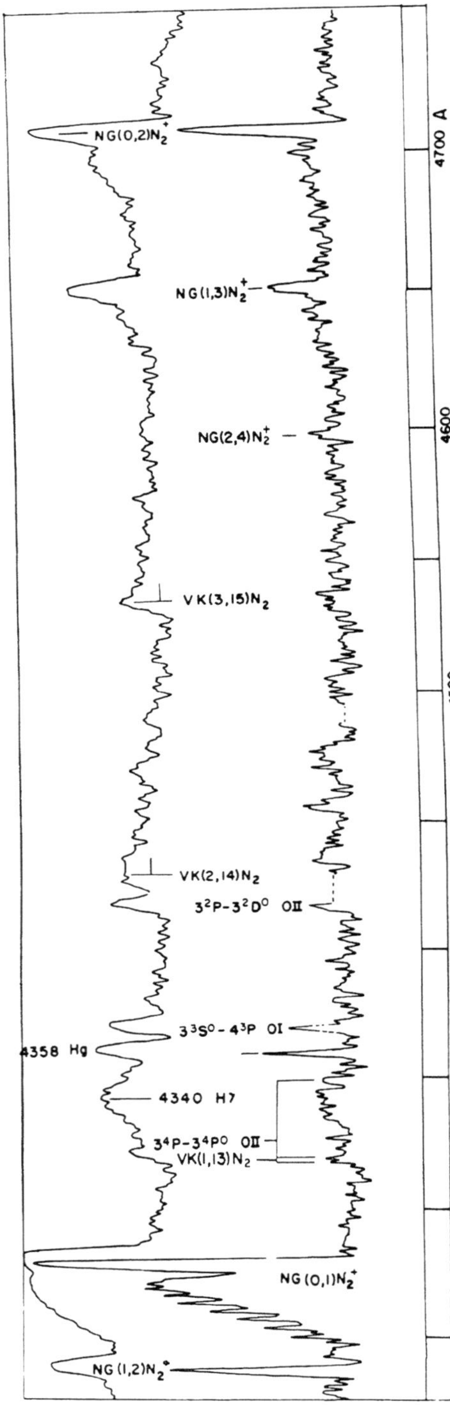

Fig. 93c. Auroral spectrum 4225-4750.

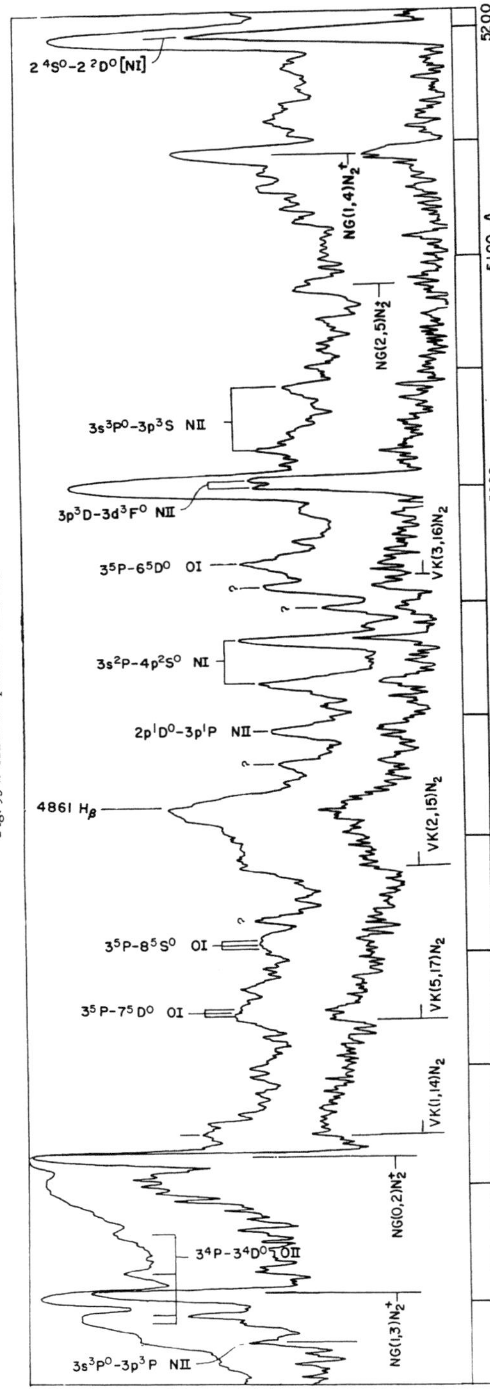

Fig. 93d. Auroral spectrum 4620-5210.

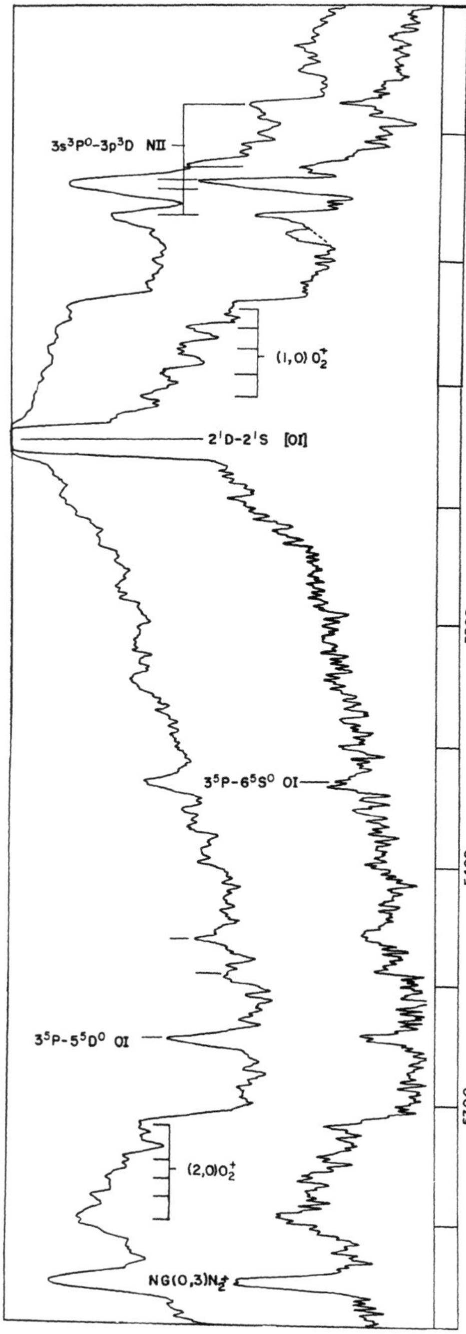

Fig. 93e. Auroral spectrum 5210-5750.

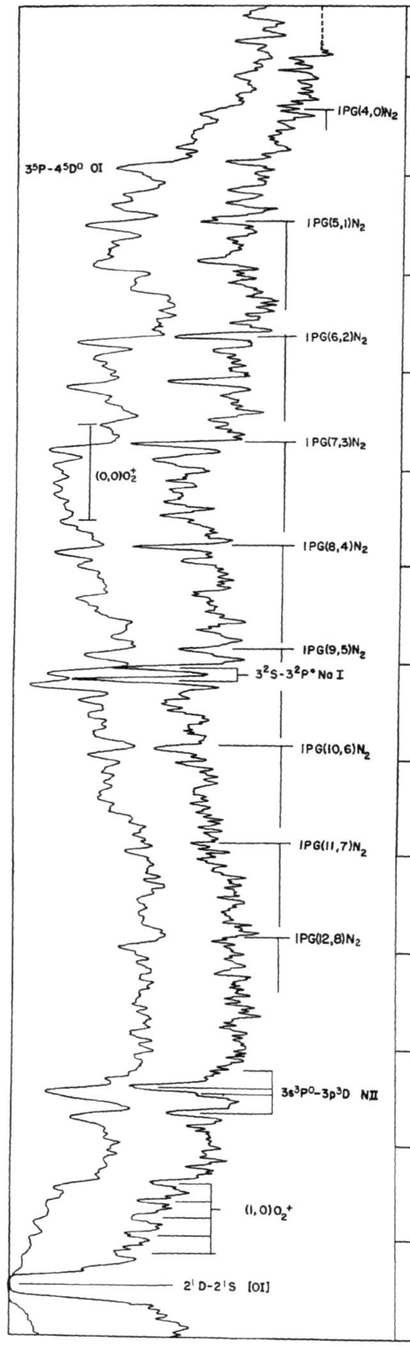

Fig. 93f. Auroral spectrum 5550-6240.

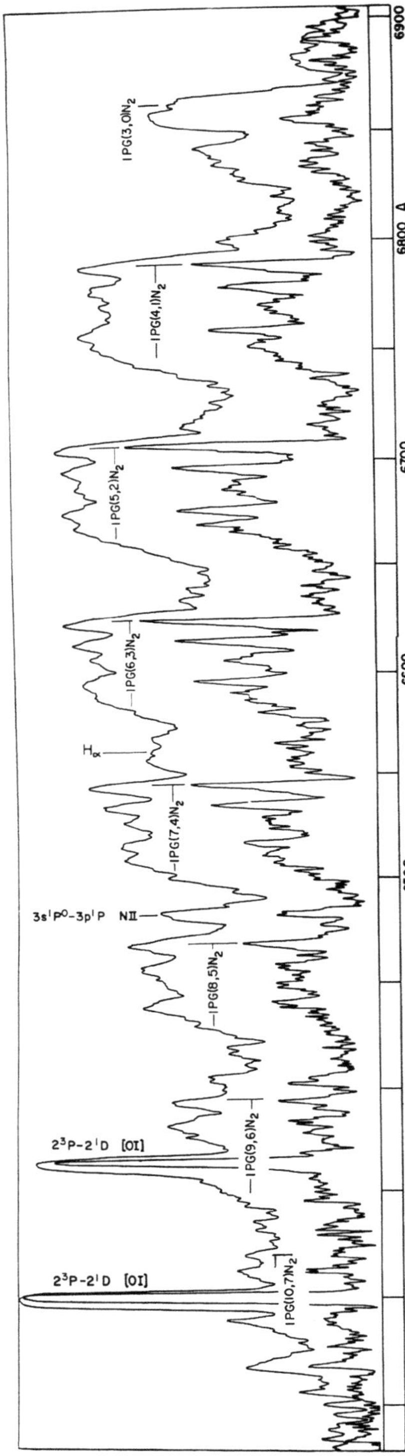

Fig. 93g. Auroral spectrum 6240-6900.

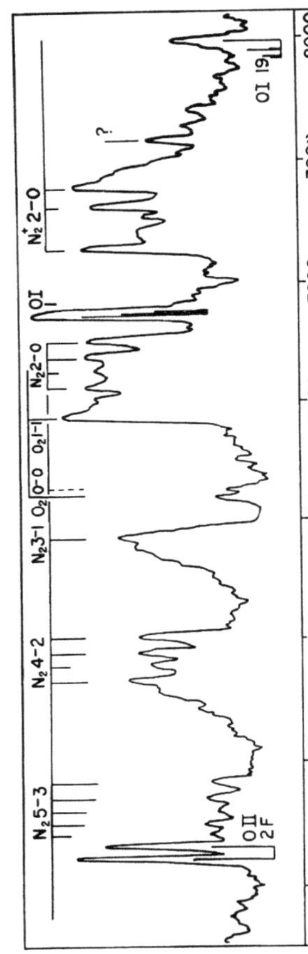

Fig. 93h. Microphotometer tracings of the spectrum of an auroral by A. Vallance Jones, 7250-8200 A, courtesy of Academic Press.

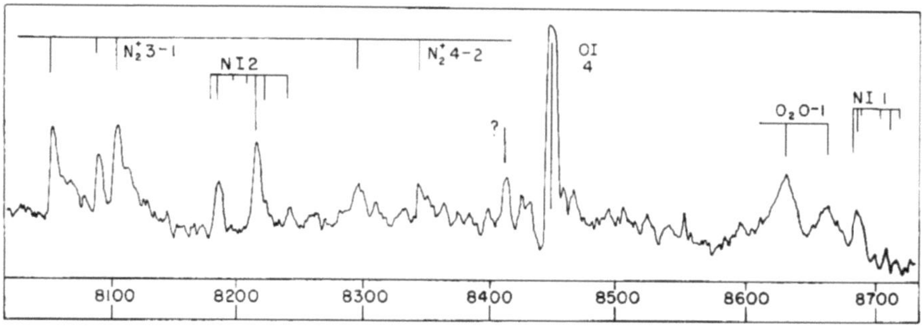

Fig. 93i. Auroral spectrum, 8000-8730 A.

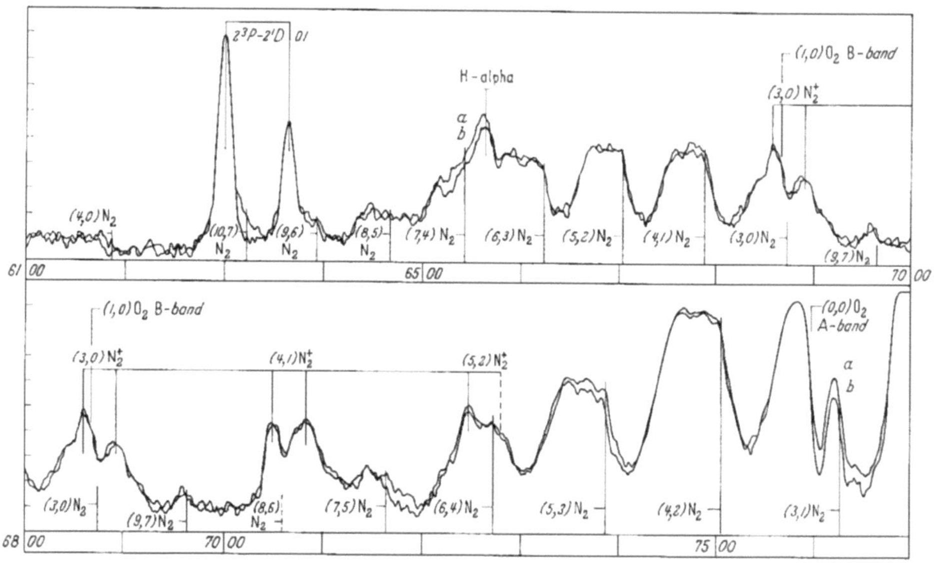

Fig. 93j.

Fig. 93 j. Microphotometer tracings of the spectrum of an auroral from two regions of the sky 30° apart from a plate by MEINEL (1951) at low dispersion (250 A/mm). Spectrum a is from the portion nearest to the zenith, showing the Doppler, shifted violet wing on H_α 6563 A, courtesy of the Astrophysical Journal

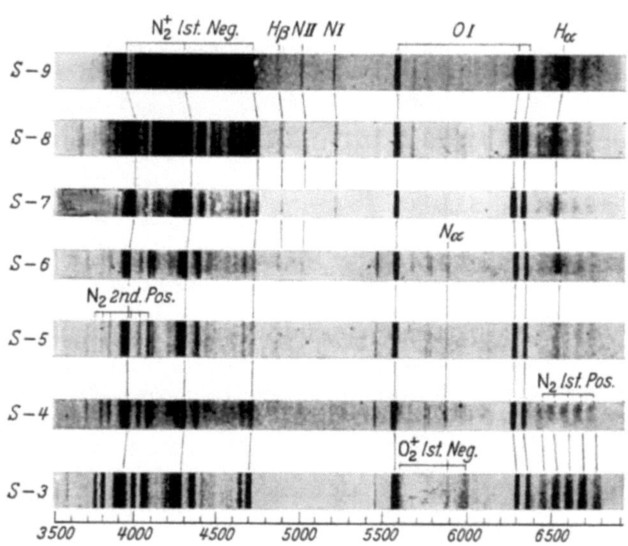

Fig. 94. Various types of the auroral spectra. (Geophysical Institute, University of Alaska.) C.T. ELVEY, private communication.

Swings, Barbier, Williams, Meinel, Petrie, Small, Chamberlain, Oliver, Vallance Jones, Hunten, Harrison, Wallace, Krasovskii, Shefov, and others.

Another important field is the theoretical and experimental study of the mechanism for the production of the auroral spectrum. It seems now that most of the visible features of the polar aurora are produced by energetic electrons injected into the auroral ionosphere from outside. There are some special types of aurora, however, whose excitation mechanism cannot simply be attributed to such electrons.

Still another aspect is the determination of the temperature of the atoms or molecules emitting the auroral light.

Fig. 94 shows several types of auroral spectrum. The first three (*S—9, S—8, S—7*) are for high altitude red auroras (Type *A*); the next two are for the usual yellow-green type; the last two (*S—4, S—3*) are for Type *B* auroras. The great variety of auroral spectral types indicates the complexity of the aurora.

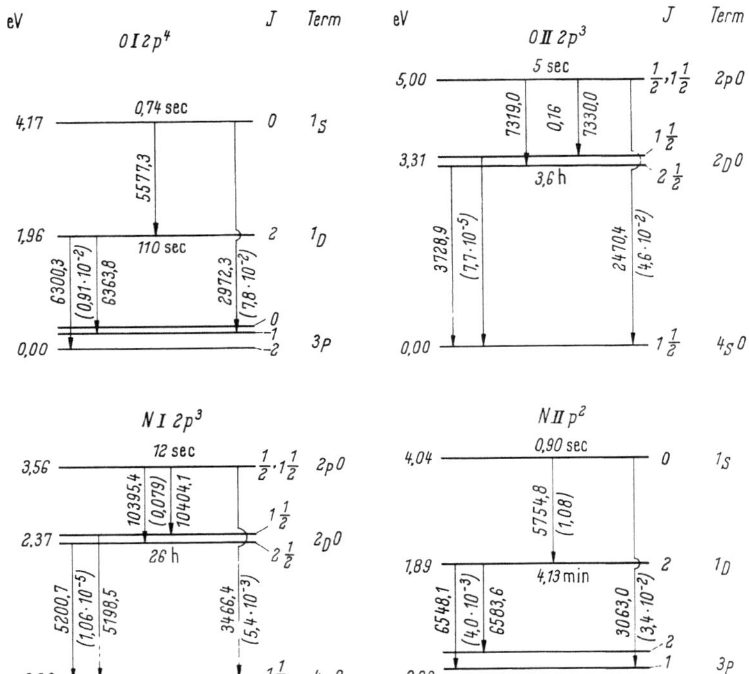

Fig. 95. Level structure and multiplets of the ground configurations of O I, O II, N I, and N II. The transition probabilities and the term lifetimes are given in parantheses. (Chamberlain, J. W., and A. B. Meinel, The Earth as a Planet, Chap. II, 1954.)

A. Description and analysis of the auroral spectrum.

58. Oxygen: Forbidden atomic lines. *a) Neutral atom [OI].* The famous green line of the aurora, while measured by means of a direct vision spectroscope by Ångström in 1860, remained a puzzle for many years. Only after the production of this line in the laboratory by McLennan and Shrum in 1925 [*342*] was it identified as a forbidden line of the oxygen atom. The production of forbidden lines in the upper atmosphere is favored by the low density. Some atoms can remain in a metastable excited state for seconds or even hours, rather than radiating in 10^{-6} to 10^{-8} seconds. In laboratory experiments, an atom in a metastable state is usually de-excited by collisions with particles or with the wall of the vessel. At the low density prevailing above 100 km, however, the atom has a larger probability of remaining undisturbed long enough to radiate a forbidden quantum.

The neutral oxygen atom has two forbidden lines in the visible region of the spectrum; the green line λ 5577 [OI], $^1S \rightarrow {}^1D$, is generally the strongest, although in some low latitude auroras the red doublet λ 6300—6364 [OI], $^1D \rightarrow {}^3P$, may be stronger (Fig. 95). The third forbidden line arising from the upper state of the λ 5577 [OI] transition is at λ 2972, $^1S \rightarrow {}^3P$. Although it is approximately 20 times weaker than λ 5577, it would be a strong feature if observed above the ozone layer, which cuts off all wave lengths less than λ 3000.

The height distribution of the [OI] lines has been discussed by a number of workers. Fig. 96 shows one of HARANG's diagrams (1957 [*343*]); the intensity of

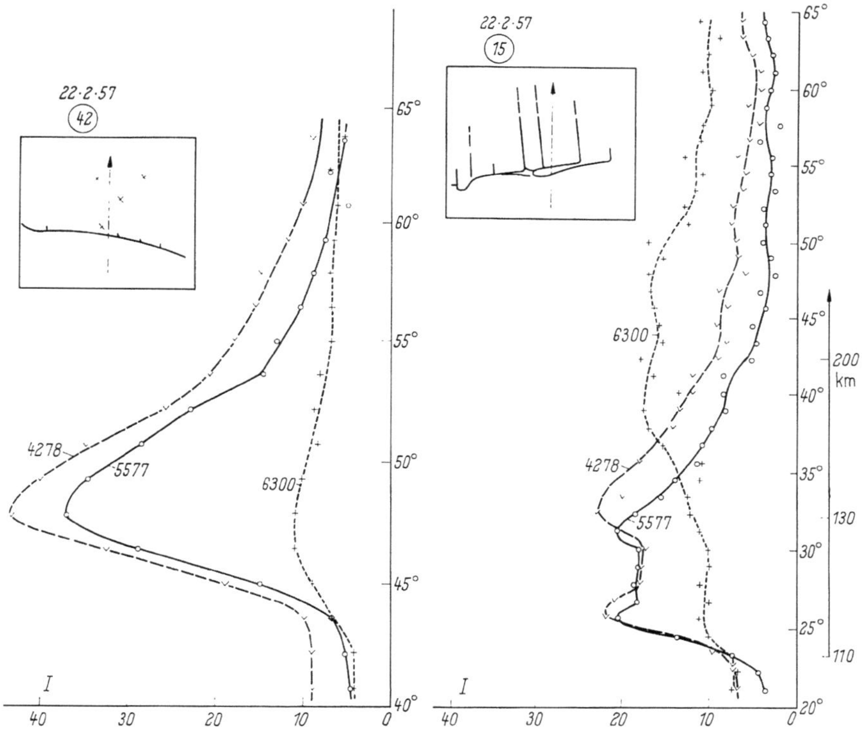

Fig. 96. Luminosity curves of an arc and a drapery. (HARANG, L., 1957 [*343*].)

λ 6300 [OI] shows a considerably slower decrease with height than λ 5577 [OI] line and λ 4278 of N_2^+ (0,1) bands. Because of the long life time of the ^1D state (110 seconds), a strong de-excitation takes place around 90—100 km height, diminishing greatly the intensity of the λ 6300 [OI] line there. JORGIO (1960 [*344*]) examined the intensity ratio $I(6300)/I(5577)$ as a function of $I(5577)$ for different types of aurora, and found a significant growth of de-activation with increase of auroral brightness, especially in the ray forms of the aurora.

b) Ionized atom [*OII*]. The presence of the principal forbidden line from ionized atomic oxygen, λ 3727 [OII], $^2D - {}^4S$, was considered doubtful; however higher dispersion spectra show it as a sharp feature between two strong N_2 bands (BARBIER and WILLIAMS 1950 [*345*]; PETRIE and SMALL 1952 [*346*]; OLIVER, WOLNIK, SCANLON and CHAMBERLAIN 1953 [*347*]; WALLACE 1959 [*348*]; BELON and CLARK 1959 [*349*]).

The other forbidden lines of [O II] at λ 7319—7330, $^2P \rightarrow {}^2D$, were identified by OMHOLT (1957 [*350*]) and M. DUFAY (1957 [*351*], 1959 [*352*]). They come from the transitions from the next higher level (Fig. 95). There is no evidence for the [O III] lines $\lambda\lambda$ 5007 and 4959, $^1D - {}^3P$, and λ 4363, $^1S - {}^1D$, which are so prominent in astronomical emission regions.

59. Oxygen: Permitted atomic lines. As higher dispersion studies have progressed, many permitted atomic lines have been found. Many of these appeared on earlier lists by VEGARD and KVIFTE (1945 [*353*]), but their acceptance was not general until confirmed by later observations, particularly by MEINEL (1948 [*354*]) and PETRIE (1950 [*355*]). MEINEL found O I multiplets at λ 8446 and λ 7774 in the infra-red region, and the prominence of their strong lines led to the first general acceptance of the presence of the permitted atomic lines in the aurora.

The intensity of these infra-red transitions is high because they arise from the first observable levels lying above the three metastable ground configuration levels. The atomic lines in other regions arise from higher levels and are correspondingly weaker.

While many atomic lines are now identified, examination of the electron transitions from which these lines result discloses that most arise from $(np \rightarrow 3s)$ transitions, where $n = 3$, 4, and 5, with 3 much the strongest. Some examples are:

$(3p \rightarrow 3s)$	7774.7, 8446.4
$(4p \rightarrow 3s)$	3947.5
$(5p \rightarrow 3s)$	3692.4

A few weak lines are perhaps due to the $(ns \rightarrow 3p)$ and $(nd \rightarrow 3p)$ transitions; among these are:

$(5s \rightarrow 3p)$	6454.7
$(4d \rightarrow 3p)$	6157.0

HARANG (1962 [*356*]) showed that λ 8446 O I has a height distribution like that of λ 5577 [O I]. SHEFOV (1962 [*357*]) reported however that the intensity of λ 8446 O I seems to have no definite relation with characteristics of auroras.

60. Oxygen: Molecular bands. *a) O_2 bands* $(b\,{}^1\Sigma_g^+ \rightarrow x\,{}^3\Sigma_g^-)$. The only confirmed bands of this system in the aurora are the Kaplan-Meinel band (0,1) at λ 8645 and the (1,1) band at λ 7688 (MEINEL 1950 [*358*]; CHAMBERLAIN, FAN and MEINEL 1954 [*359*]). The bands show two heads, one shading to the red and the other to the blue. The Kaplan-Meinel band is present in the airglow, but much stronger in the aurora. On one occasion, however, MEINEL (1950 [*358*]) observed this band to be greatly enhanced, although other auroral emissions were absent.

b) O_2^+ First Negative bands $(B\,{}^4\Sigma_g^- \rightarrow A\,{}^4\Pi_u)$. Second multiple-headed bands at λ 5603 and λ 5234 were identified by VEGARD (1950 [*360*]) and GARTLEIN and SHERMAN (1952 [*361*]) as due to this system. DAHLSTROM and HUNTEN (1951 [*68*]) found the O_2^+ bands to be greatly enhanced in type *B* auroras. This is easily explained by the fact that type *B* auroras appear below the normal auroras, where O_2 is not dissociated. However, this band is not responsible for the red border, since the visual stimulus for the O_2^+ was observed by FAN to be yellow.

61. Nitrogen: Forbidden atomic lines. *a) Neutral atom [N I].* The presence of forbidden atomic nitrogen lines in the aurora remained doubtful for many years. λ 3466 [N I], ${}^2P \rightarrow {}^4S$, is strong, but in a region that is difficult to study because of the numerous intense N_2 bands (Fig. 95). It was first identified by BERNARD (1939 [*363*]) and confirmed by BARBIER and WILLIAMS 1950 ([*345*]). λ 5200 [N I], ${}^2D \rightarrow {}^4S$, identified by DUFAY and TCHENG (1946 [*365*]), is of particular interest because the lifetime of the parent term is approximately 26 hours. Because of collisional de-excitation this line must originate in the extreme upper atmosphere. GÖTZ (1947 [*366*]) observed that the λ 5200 [N I] emission was still quite strong one hour after the end of a low latitude aurora, while the green line had resumed its auroral airglow intensity. MIRONOV (1960 [*367*]) showed that the intensity of the λ 5200 [NI] line increases with the Kp index.

The third [N I] line at λ 10404, ${}^2P \rightarrow {}^2D$, and also the line λ 10395 [N I], were confirmed by BAGARIATSKII and FEDOROVA (1955 [*368*], 1956 [*369*]) and HARRISON and VALLANCE JONES (1957 [*370*]).

b) Ionized atom [N II]. The N [II] line at λ 5755, ${}^1S \rightarrow {}^1D$, was reported by PETRIE (1952 [*371*]). However, because it falls in a region filled with N_2 First Positive bands, its identification is not easy. Later BELON and CLARK (1959 [*349*]) and MIRONOV, PROKUDINA and SHEFOV (1961 [*373*]) found the [NII] lines at λ 5755 and λ 6548 (${}^1D \rightarrow {}^3P$) during the red aurora of February 11, 1958.

62. Nitrogen: Permitted atomic lines. The most prominent permitted atomic nitrogen line is from the ($3p \rightarrow 3s$) transition at λ 8712 N I. The intensity is high because it arises from the first observational level. BAGARIATSKII (1957 [*374*]) reported the ($3d \rightarrow 3p$) transition at λ 9060 N I. Two of the strongest permitted atomic multiplets are at λ 5003 N II, $3p \rightarrow 3s$, and at λ 5686 N II, $3p \rightarrow 3s$. These multiplets are present in all well-exposed spectra. λ 5003 N II was once ascribed to λ 5007 [O III]; however, the absence of an equally strong line at λ 4950 showed this identification to be incorrect. The high dispersion spectra by PETRIE and SMALL (1952) enabled each emission to be identified as due to N II. The levels from which these lines arise are about 23 ev above the ground level of N II. This result seemed to constitute a puzzle; however, "laboratory auroras" produced by FAN (1956) show these same lines to be strong. WALLACE (1959 [*348*]) found the ($4f \rightarrow 3d$) transition of λ 4026 N II (see also YEVLASHIN (1962 [*372*])).

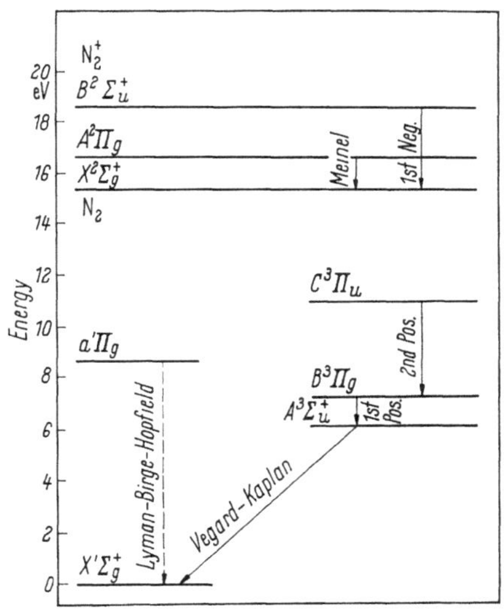

Fig. 97. Electronic states and band systems of N_2 and N_2^+. The dashed transition lies in the far ultraviolet and has not been detected in aurora. (CHAMBERLAIN, J. W., 1961 [*4*], Physics of the Aurora.)

63. Nitrogen: Molecular bands. By far the greatest amount of radiative energy released in the aurora is from molecular species. The relative intensity of all bands to λ 5577 [OI] is perhaps of order 200 ~ 500 between 1 μ and the atmospheric cut-off at λ 3000. The most prominent band systems are due to neutral and ionized nitrogen molecules (Fig. 97).

a) N_2 First Positive bands ($B^3\Pi_g \rightarrow A^3\Sigma_u^+$). This band system extends from beyond the (0,0) band at λ 10440 to the yellow region of (10,6) at λ 5854, and accounts for some fifty observed bands. The infra-red bands drop off rapidly in intensity with increasing v', while bands in the visible drop off slowly, in accordance with the Franck-Condon principle. Some infra-red bands, once ascribed to large v' values, were re-identified by MEINEL (1951 [*378*]) as due to a new infra-red N_2 band system. The First Positive bands are distinguished by their complex structure; each band, under low to moderate dispersion, shows two or three maxima about 30 A apart. The bands are composed of nine strong branches and 18 satellite branches. In Fig. 94 some of these are clearly seen on the infra-red side of λ 6300 [O I].

b) N_2 Second Positive bands ($C\,^3\Pi_u \rightarrow B\,^3\Pi_g$). These bands fall in the ultra-violet, and extend from the atmospheric cut-off at λ 3000 to approximately λ 4000, namely (0,0) at λ 3371, (0,1) at λ 3577, (0,2) at λ 3805, (0,3) at λ 4059. The appearance is much simpler in structure, consisting of a strong head, and a weak shading towards the blue. This system blends almost imperceptibly with the ultraviolet N_2^+ system near λ 4000, and high dispersion is required to separate them.

c) N_2 Vegard-Kaplan bands ($A\,^3\Sigma_u^+ \rightarrow X\,^1\Sigma_g^+$). Unlike the other systems in the blue and ultra-violet, the Vegard-Kaplan bands shade from violet toward the red, with a gradual diminution in intensity, quite unlike the sharp gradients of the Second Positive N_2 bands. The bands are considerably weaker than the other ultra-violet systems of N_2 and N_2^+; consequently the Vegard-Kaplan bands can only be identified in regions free from those bands. The presence of progressions for $v'=1$, 2, and 3 is well established, namely (1, 10) at λ 3425, (2, 11) at λ 5202 and (3, 14) at λ 4171.

d) N_2^+ First Negative bands ($B\,^2\Sigma_u^+ \rightarrow X\,^2\Sigma_g^+$). Although the (0,0) band of N_2^+ at λ 3914 is the strongest feature in the blue and ultra-violet, the intensities in this system drop off rapidly with increasing v'. The bands all show a strong P — branch head and a weak R wing extending to the violet. The presence of progressions for $v'=0$, 1, 2, and 3 is confirmed (see also WEISSBERG (1962 [*364*])).

e) N_2^+ Meinel bands ($A\,^2\Pi_g \rightarrow X\,^2\Sigma_g^+$). The bands of this system fall in the infra-red and are characterized by double maxima; they degrade to the red (MEINEL 1951). The infra-red region of the Meinel bands has been explored by BAGARIATSKII and FEDOROVA (1956 [*379*]) and by HARRISON and VALLANCE JONES (1959 [*380*]). Recently NOXON and VALLANCE JONES (1960 [*381*]) reported an enhancement of the auroral emission at 2.1 μ during an intense auroral display, and identified it as the (0,2) Meinel band (see also FEDOROVA (1961 [*384*])).

64. Hydrogen: Review. The hydrogen Balmer lines have been an elusive feature of the auroral spectrum; they have been listed by VEGARD since 1939, but the identifications were considered doubtful by most investigators, since the lines were clearly absent from many spectra. A study of the relative intensities of the Balmer lines from spectrum to spectrum by GARTLEIN (1950 [*382*]) led to the general acceptance of the identification, namely $H\alpha$ line at λ 6562.8, $H\beta$ at λ 4861.3 and $H\gamma$ at λ 4340.5. In particular, GARTLEIN found the hydrogen lines to be strong early in an auroral display. MEINEL (1951 [*383*]), and DAHLSTROM and HUNTEN (1951 [*68*]), found that strong hydrogen lines are observed during the quiet homogeneous arc phase, but are absent when the arc acquires rayed structure.

The discovery of the Doppler broadening of the $H\alpha$ line by GARTLEIN (1950 [*382*]) and of its Doppler shift by MEINEL (1951 [*383*]), marked an important epoch in the development of auroral physics. The $H\alpha$ profile obtained from an aurora in the magnetic zenith indicated a Doppler shift towards the violet side, showing that the emitting hydrogen atoms are descending along the geomagnetic field lines.

Later FAN and SCHULTE (1954 [*385*]), ROMICK and ELVEY (1958 [*386*]), FAN (1958 [*387*]), GALPERIN (1959 [*388*]) and others confirmed the earlier observations that the hydrogen emissions are seen in the early phase of the aurora (quiet arc), but not in the active rayed forms. They thought that the protons were responsible for the production of quiet arcs, but not for active forms.

Many theoretical studies were then devoted to the relation between auroras and incoming protons (see CHAMBERLAIN 1961 [*4*]; Chapt. *7*). The main effort was to calculate theoretically the luminosity profile for the $H\alpha$ intensity, and the intensity ratio $I(3914)/I(H\beta)$ for monoenergetic protons, or for protons with a certain energy spectrum. It was found, however, that the predicted intensity ratio $I(3914)/I(H\beta)$ is much less than is observed. CHAMBERLAIN (1961 [*4*], p. 256) concluded that protons are not responsible for auroral arcs. Although the theories of proton bombardment do not seem to explain complicated features of the aurora, the basic studies in this approach must not be forgotten.

65. Hydrogen glow. Later observations indicated that the hydrogen emissions appear in a large part of the sky, and that the "zone" of the emissions moves equatorward in the early evening; there is no conclusive evidence of the concentration of hydrogen emissions in any distinct auroral forms (REES, BELON and ROMICK 1961 [*389*]; OMHOLT, STOFFREGEN and DERBLOM 1962 [*390*]; STOFFREGEN and DERBLOM 1962 [*391*]; MONTBRIAND and VALLANCE JONES 1962 [*392*]; GALPERIN 1963 [*393*]). The diagram given by OMHOLT, STOFFREGEN and DERBLOM [*390*] is reproduced here as Fig. 98; it shows a dark space between the hydrogen zone and the visible auroras.

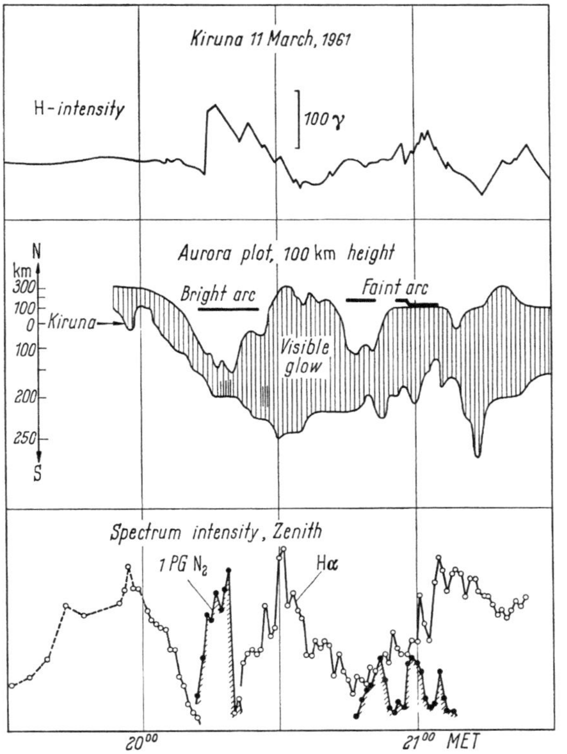

Fig. 98. Geomagnetic data and latitudinal occurrence of the aurora with time at Kiruna, 11 March 1961. (OMHOLT, A., W. STOFFREGEN and H. DERBLOM, 1962 [*390*].)

In summary, the important features of the hydrogen zone (REES, BELON and ROMICK 1961 [*389*]) are:

(1) The hydrogen emissions are associated with a broad diffuse arc.

(2) The arc is barely visible to the eye on the all-sky auroral photographs.

(3) The $H\alpha$ emission is definitely not associated with *all* auroral arcs; the emission is not associated solely with quiet arcs; after an auroral display has reached full development the $H\alpha$ emission may occur in any type of auroral forms, or in essentially formless aurora. This agrees with rocket studies of auroral protons (Sect. 52).

(4) The appearance of the hydrogen zone and its motion at least to the zenith (at College, gm lat. 64.7 N; a few degrees south of the auroral zone) always

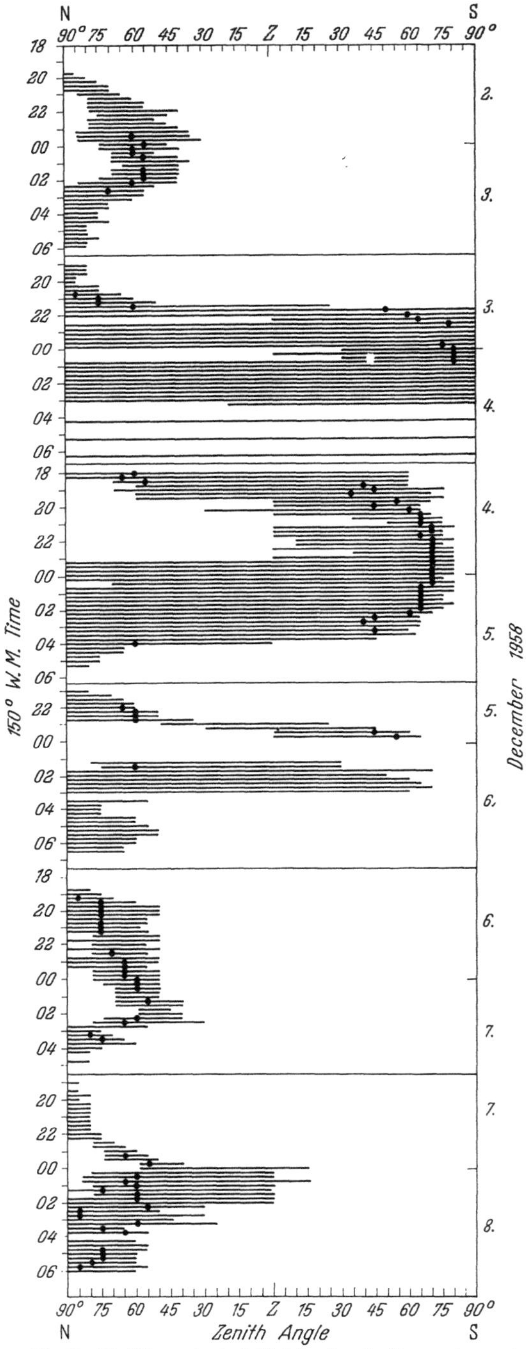

Fig. 99. Meridian extent of H-alpha line in the spectrum, at 15 minute intervals throughout 6 consecutive nights. The circles indicate the position of the H-alpha intensity peak when this was clearly evident. (Rees, M. H., A. E. Belon and G. J. Romick, 1961 [389].)

precedes the bright aurora or break-up phase of an auroral display; in the majority of cases, the $H\alpha$ emission peak moves well south of the magnetic zenith before break-up (Fig. 99).

(5) Recession of the hydrogen zone may occur at any time after it has reached its southern extent.

(6) At the time of onset of polar magnetic storms, the zone seems to shift rapidly equatorward. The southern extent during the disturbed period is not certain.

(7) Before midnight the hydrogen zone lies to the south of the distinct auroral forms, but after midnight to the north of them (see also Montalbetti and Vallance Jones 1957 [394]; and Stoffregen and Derblom 1962 [391]).

(8) During magnetically quiet periods, the southward advance and northward recession seem to be a regular feature. It seems therefore urgently necessary to extend the network of observations for the $H\alpha$ emission so as to cover a larger latitude extent than now. It is not known where the hydrogen zone first appears, nor how far south it shifts during polar magnetic storms.

Malville (1959 [395]) and Ingram (1962 [396]) found a narrow $H\alpha$ emission in the aurora and airglow. Ingham concluded that it is of geocentric origin. Galperin (1963 [393]) indicated that in most cases the $H\alpha$ profiles are practically identical within observational errors (Fig. 100).

66. Helium. Shefov (1961 [397], 1962 [398], 1963 [399]) reported an emission at λ 10830 in the sunlit atmosphere both during auroral displays and in ordinary twilight with absence of auroras. The intensity of the emission was as large as 10 kR. He suggested that the emission is produced by a fluorescent scattering of solar radiation by He I atoms in the ^3S metastable state. The metastable atoms are likely to be produced by electron impacts. MALVILLE (1961 [*400*]) showed that under reasonable assumptions solar particles would produce only 300 R of the λ 10830 He I line.

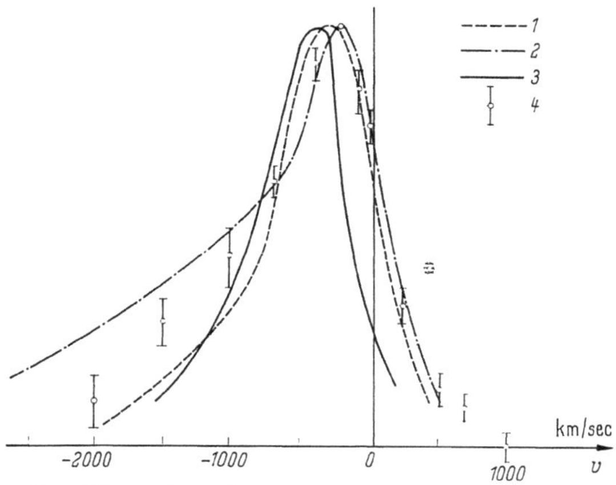

Fig. 100. Typical hydrogen H_γ and H_β velocity profiles without instrumental broadening. 1. Mean H_β profile from spectro-electrophotometer. 2. H_γ profile. 3. Mean H_γ and H_β profile. 4. Mean H_γ profile after reductions for 1 *PG* H_α blending. (GALPERIN, Y. I., 1963 [*393*].)

B. The production of the auroral spectrum.

67. Introduction. Rocket and satellite observations of the aurora have shown that there exists an intense flux of energetic electrons *within* visible auroras (Chap. VI). Further, it has been shown that the energy carried by such electrons is of order a few hundred ergs/cm² sec; this is sufficient to produce various auroral spectra. Therefore there is no doubt that the electrons carry a large portion of the energy that produces various auroral lines and bands. Electrons that penetrate into the ionosphere from outside may be called the *primary electrons*.

Most of the primary electrons have ample energy to ionize the constituents in the auroral ionosphere. The excitation by such energetic particles occurs after they are greatly slowed down by repeated ionization processes. An appreciable fraction of the secondary electrons are also energetic enough to ionize the constituents once; their contribution to the excitation is considerable. Here the ionization and the excitation are discussed separately.

68. Ionization. *a)* N_2^+ *First negative bands.* One most important process that takes place in the aurora is the simultaneous ionization and excitation of N_2 molecules by energetic electrons, both primary and secondary. Fig. 101 shows four spectrograms taken at College (gm lat. 64.7° N) on the night of 5/6 December, 1958. Two, (a) and (c), were just before break-ups; the others, (b) and (d), were taken just after break-ups. A strong enhancement of the N_2^+ First Positive group bands obscured the existence of the $H\alpha$ line, although there seemed to be no particular enhancement of it, corresponding to other lines and bands. The process is thus likely to be

$$N_2(X^1\Sigma_g^+) + e \rightarrow N_2^+(B^2\Sigma_u^+) + e + e.$$

STEWART (1956 [*401*]) has determined the cross-sections of this process for the (0,0), (0,1), and (0,2) bands. The cross-section for the (0,0) band increases rapidly for electrons of energy E increasing beyond 20 ev, and reaches the maximum value 6×10^{-18} cm² at $E = 80$ ev; it decreases slowly with further increase of E (Fig. 102). OMHOLT (1959 [*402*]) has shown theoretically that each primary electron loses about 35 ev kinetic energy for each ionization. Since it is very unlikely that ionospheric electrons attain such a high energy by any plausible acceleration mechanisms, the λ 3914 emission has been regarded as an indication

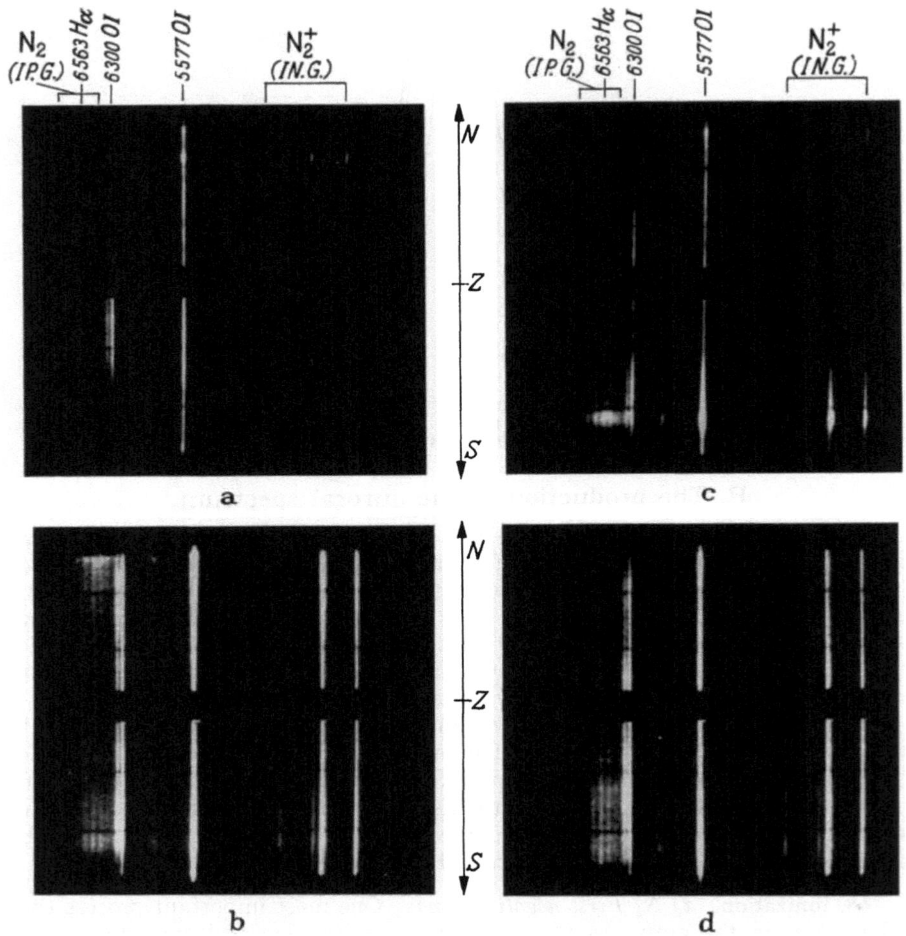

Fig. 101 a—d. Four spectrograms taken at College (Alaska) on the night of 5 to 6 December 1958. Two of them, (a) and (c), were taken just before the auroral break-ups shown in Fig. 5.2. The others, (b) and (d), were taken just after the break-upso (REES, private communication, 1960.)

of the direct bombardment of the auroral ionosphere by primary particles, although the solar ultra-violet radiation will also ionize some N_2 molecules (cf. CHAMBERLAIN and SAGAN 1960 [*403*]; HUNTEN 1963 [*404*]).

It is commonly assumed that in electron bombardment about one ionization in 50 leads to the emission of a λ 3914 photon. If in an aurora the intensity of λ 3914 emission is 1000 kR (= an apparent emission of 10^{12} photons/cm² sec), the ion production rate must be of order $10^6 \times 10^6$ (photons/cm² sec)/0.02 $= 5 \times 10^{13}$ (ions/cm² sec). The total energy involved is then 5×10^{13} (ions/cm² sec) $\times 35$ (ev) $=$ 1.75×10^{15} ev/cm² sec. If this amount of energy is carried by 5 kev electrons, the flux must be of order 1.75×10^{15} (ev/cm² sec)/5×10^3 (ev) $= 3.5 \times 10^{11}$ electrons/cm² sec.

The N_2^+ ions are most likely to be lost by dissociative recombination with ionospheric electrons,

$$N_2^+ + e \rightarrow N^* + N^* .$$

The rate of this dissociative recombination is of order 10^{-6} cm³/sec (cf. DALGARNO 1961 [*405*]). LYTLE and HUNTEN (1962 [*406*]) showed that N_2^+ ions are lost much more rapidly below 150 km in height.

b) O_2^+ bands. The ionization process is expressed by

$$O_2(X^3\Sigma_g^-) + e \rightarrow O_2^+(b^4\Sigma_g^-) + e + e .$$

The cross-section, measured by TATE and SMITH (1932 [*407*]) and FITE and BRACKMANN (1959 [*408*]), is of order 3×10^{-16} cm².

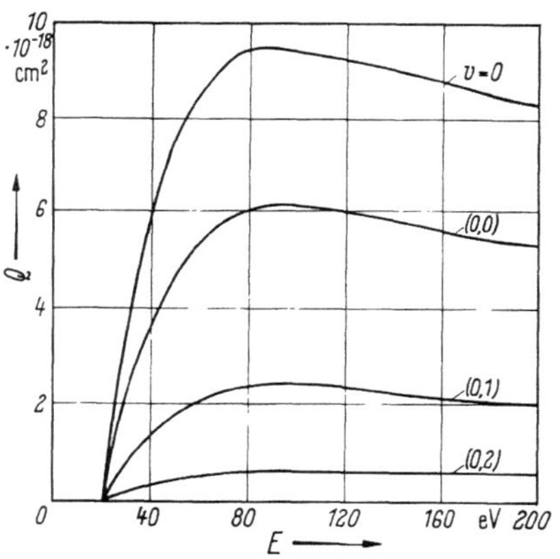

Fig. 102. Cross sections for the excitation of the first negative systems of nitrogen by electron impact. (STEWART, D. T., 1956 [*401*].)

c) [OII] and [NII] lines

$$O(^3P) + e \rightarrow O^+(^2P) + e + e$$

$$N_2(X^1\Sigma_g^+) + e \rightarrow N^+(^1S) + N + e + e$$

69. Excitation. *a) [OI] forbidden lines.* The most familiar lines λ 5577 [OI] and

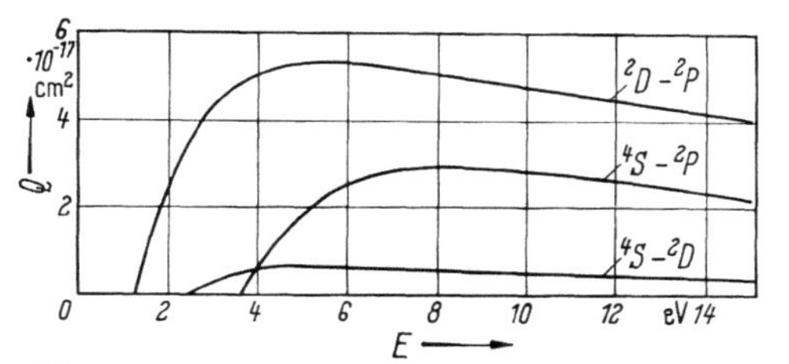

Fig. 103. Cross sections for the excitation of metastable levels of atomic nitrogen by electron impact. (SEATON, M. J., 1956 in [*409*].)

λ 6300 [OI] in the auroral spectrum are produced by secondary electrons;

$$O(^3P) + e \rightarrow O(^1S) + e$$

$$O(^3P) + e \rightarrow O(^1D) + e .$$

SEATON (1956 [*409*]) computed the cross-sections for the above two processes; for the first one, the maximum cross-section (0.4×10^{-17} cm²) is attained by electrons of $E \sim 8.5$ ev; for the second one, the maximum cross-section (3.3×10^{-17} cm²) is attained by electrons of $E \sim 4.2$ ev, and it decreases rapidly with increase of E. The lifetime of undisturbed $O(^1S)$ is of order 0.71 sec, and that of $O(^1D)$ is of order 110 sec. Fig. 103 shows the cross section for the atomic nitrogen.

Many studies have been devoted to find the relationships between the emission due to the direct ionization and the above excitations. By simultaneous observation of I(3914) and I(5577), ASHBURN (1955 [*410*]) showed that major pulsations of the N_2^+ First Negative group band were almost synchronous with pulsations of λ 5577 [OI], with a time of about 0.5 sec. MURCRAY (1959 [*411*]) also reported a close relation between I(5577) and I(3914). JORGIO (1962 [*412*]) shows the rapid and simultaneous fluctuations of the intensities of several lines and bands (Fig. 104).

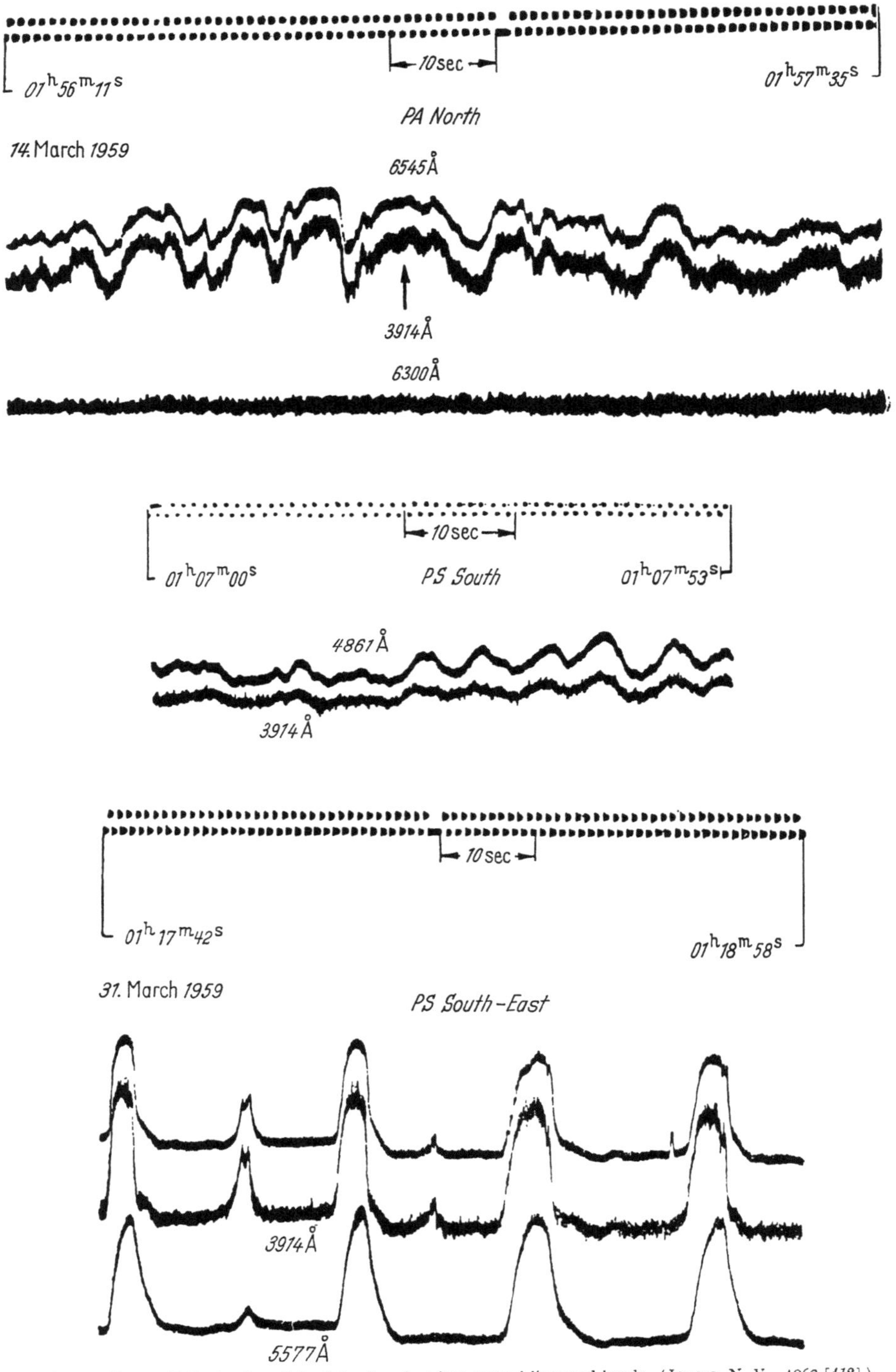

Fig. 104 The rapid fluctuation of the intensity of various auroral lines and bands. (JORGIO, N. V., 1962 [412].)

b) [NI] forbidden lines:

$$N(^4S)+e\to N(^2D)+e$$
$$N(^4S)+e\to N(^2P)+e$$
$$N(^2D)+e\to N(^2P)+e$$

c) OI permitted lines:

$$O(^3P)+e\to O(^3p\,^3P)$$
$$O(^3P)+e\to O(^3p\,^5P)$$

d) N_2 second positive group bands:

$$N_2(X^1\Sigma_g^+)+e\to N_2(C^3\Pi_u)+e$$

DALGARNO (1951 [*405*]) has summarized the cross-sections for (b), (c), and (d).

C. Special types of aurora.

70. Type A aurora (high altitude red arc). The spectrum of a high altitude red arc, obtained by REES and DEEHR (1961 [*65*]) on the night of November 27, 1959, was characterized by λ 6300 [OI] and the N_2 First Negative group bands, but not by the other usual auroral emissions. The intensity of λ 6300 [OI] was 66.7 *kR*, and the intensity ratio $I(6300)/I(5577)$ was 25 (see also YEVLASHIN (1961 [*413*])).

The arc was located in the sunlit atmosphere at an average height of 350 km. After a detailed examination of several plausible processes, REES (1961 [*414*]) concluded that protons of energy of order a few kev with a flux of order 10^{10}/cm²sec were involved. He proposed a succession of charge-exchange and ionization processes;

$$H^+ + O\to H + O^+$$
$$H + O\to H + O^+ + e$$

and

$$H + O(^3P)\to H + O(^1D).$$

71. Type B aurora (purplish-red border). An active aurora is often characterized by the so-called purplish-red lower border. The purplish-red emission is mainly due to the enhancement of the N_2 First Positive group bands. MALVILLE (1959 [*325*]) proposed the following processes:

$$N_2 + e\to N_2^+ + e + e$$
$$N_2^+ + O^-\to N_2(B^3\Pi_g) + O$$
$$N_2^+ + O_2^-\to N_2(B^2\Pi_g) + O_2.$$

To ionize N_2 molecules around 80 km, the energy of the primary electrons must exceed 200 kev. Rocket or satellite studies of such active auroras are needed (see also VEGARD and TÖNSBERG (1937 [*415*]); CHAMBERLAIN [*4*] p. 314).

72. Sunlit aurora. Most auroras in the *dark* atmosphere do not extend above 500 km. STÖRMER (1942 [*416*]) found some faint arcs and rays extending up to 1000 km, in the *sunlit* atmosphere.

BATES (1949 [*417*]) ascribed the sunlit aurora to the strong enhancement of the N_2^+ First Negative group bands;

$$N_2^+(X^2\Sigma_g^+) + h\nu\to N_2^+(B^2\Sigma_u^+)$$
$$N_2^+(B^2\Sigma_u^+)\to N_2^+(X^2\Sigma_g^+) + h\nu.$$

To produce a detectable intensity of the N_2^+ First Negative group bands the number density of N_2^+ molecules at 1000 km height must be at least $10^2/cm^3$. It is not known whether N_2^+ molecules at this high altitude are a permanent feature of the auroral ionosphere. Some heating mechanisms have been proposed, but none is convincing. However, KRASOVSKII (1961 [*87*]) reported an upward movement of fluorescent N_2^+ molecules, first in separate narrow filaments, and then in larger forms. As such observations are valuable for studies of the very high atmosphere and for the heating mechanism, further effort in this direction is desirable (see also SHISKAYA (1961 [*418*])).

73. Red veil in the minauroral zone. Major spectral features of auroras seen from the minauroral zone are the great enhancement of the N_2^+ First Negative group bands and of the 6300 [OI] line, together with some enhancement of the λ 5577 [OI] line. SEATON (1956 [*419*]) made a detailed theoretical study of the spectrum of the red veil, and attributed the enhancement of the N_2^+ First Negative group bands to resonance scattering in the sunlit atmosphere (Sect. 72). Such enhancement occurs at a great height (> 200 km).

To explain the enhancement of λ 6300 [OI] line and the large intensity ratio $I(6300)/I(5577) \sim 13$, SEATON suggested excitation of O atoms by electrons associated with electric current systems in the high atmosphere. During polar magnetic substorms, the auroral electrojet produces a large-scale electric polarization field throughout the ionosphere. AKASOFU and CHAPMAN (1963 [*149*]) showed that even the equatorial electrojet is greatly enhanced by the polarization field. If this is the case, the intensity $I(6300)$ should change in accordance with the activity of polar magnetic substorms around the auroral zone. AKASOFU (1962 [*131*]) found that on February 11, 1958 the red line was greatly enhanced at Memanbetsu, gm lat. 34.1° N) during a large polar magnetic storm to the north. Another possibility is that such a red veil is an upper part of arcs. CLARK and BELON (1959 [*67*]) reported that in February 11, 1958 the top of the red arc over Alaska was at times 1000 km in height; an upper part of such a high red arc would be visible from Memanbetsu.

HIKOSAKA and YANO (1961 [*421*]) proposed that the above spectral features of the minauroral zone are produced by incident protons of energy of order 3 kev. Such protons can produce strong emission of the N_+^2 First Negative group bands; and the secondary electrons produced can have enough energy to populate the ^1D state of the O atoms (but not the ^1S state), resulting in the high intensity ratio $I(6300)/I(5577) \sim 20$ (see also HIKOSAKA (1958 [*420*])).

74. Subvisual red arcs (λ 6300 [OI]). BARBIER (1958 [*422*]) reported a rather stable and persistent subvisual red arc over Haute Provence (gm lat. 40° N) during large magnetic storms. A similar study by ROACH and MAROVICH (1959 [*423*]) suggested that the red arc observed by them in the northern USA on the night of September 29, 1957 was an extension of the arc then observed by BARBIER at Haute Provence.

One most characteristic feature of the subvisual red arc is the large intensity ratio $I(6300)/I(5577)$. BARBIER (1960 [*424*]) reported that the ratio was at times $125\ kR/1.5\ kR = 80$. ROACH and MAROVICH (1959 [*423*]) reported also that when the λ 6300 [OI] was enhanced to about 50 times the normal airglow intensity there was no detectable enhancement of λ 5577 [OI] (Fig. 105). ROACH (F. E. and T. R.) (1963 [*425*]) reported no definite sign of the enhancement of the N_2^+ First Negative group bands.

Another characteristic feature of the red arc is its alinement along MCILWAIN'S L lines. Many of the arcs appear along L between 2.0 and 3.0, and seem to be on a global scale; there is an indication of a corresponding conjugate arc in the southern hemisphere (ROACH, BARBIER and DUNCAN 1962 [*426*]). Hence the arcs are likely to be related to trapped charged particles.

ROACH and MAROVICH (1959 [*423*]) reported that the arc is not affected by the break-up of the λ 5577 [OI] aurora to the north, so that polar magnetic substorms do not seem to be related to it. After reaching its maximum intensity, the red arc decays slowly, without much irregular change of intensity. On the other hand, REES and AKASOFU (1962 [*427*]) reported a close relation between the intensity $I(6300)$ and the magnitude of the main phase decrease, $Dst(H)$.

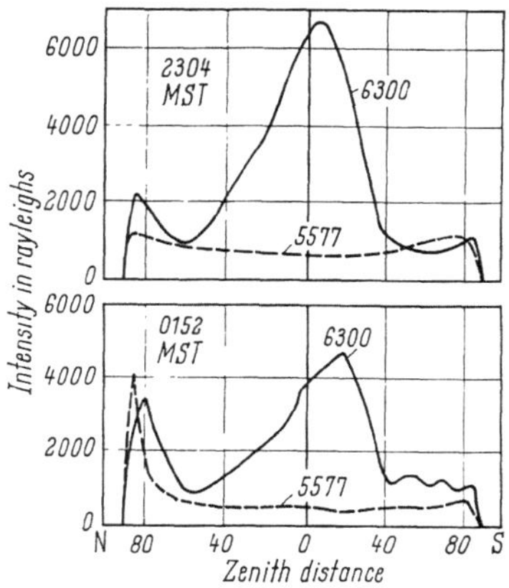

Fig. 105.

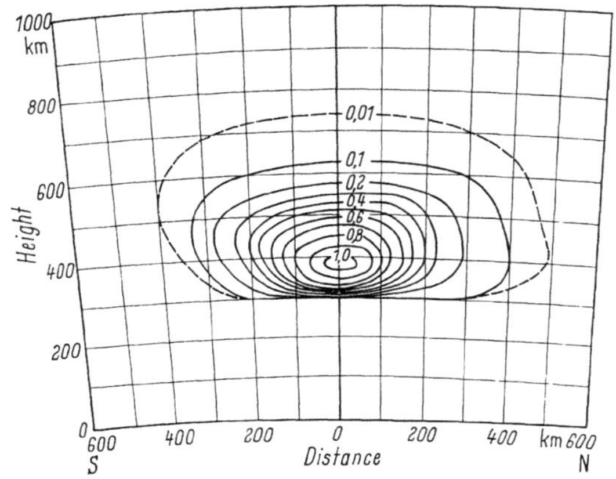

Fig. 106.

Fig. 105. Sample horizon to horizon traces made at Fritz Peak on September 29/30, 1957, at times of a strong red arc near the zenith. The 5577 A traces are nearly normal except for an enhancement in the extreme north at 0152 MST. The 6300 A traces show enhancements of several thousand rayleighs as the photometer traverses the sky near the zenith. (ROACH, F. E., and E. J. MAROVICH, 1959 [*423*].)

Fig. 106. Cross section isophotal representation of a red arc. (TOHMATSU, T., and F. E. ROACH, 1962 [*428*].)

Other characteristic features are the height and thickness. Fig. 106 shows the meridian cross section of a red arc obtained by TOHMATSU and ROACH (1962 [*428*]). The height of the brightest portion is of order 380 km; it extends from 300 km to 700 km in height, and also several hundred km in latitude north and south of the center. Both BARBIER (1958 [*422*]), and ROACH and MAROVICH (1959 [*423*]) reported southward motion of the arc, with speed of order 5~50 m/sec.

Because of the large ratio $I(6300)/I(5577)$, the excitation of the red arc must operate preferentially to populate $O(^1D)$ but not $O(^1S)$. The absence of any definite sign of the enhancement of the N_2^+ First Negative group bands suggests that the arc is not due to a direct bombardment of the upper atmosphere by energetic particles.

KING and ROACH (1961 [*429*]) suggested the following processes:

$$N_2 + O^+ \rightarrow NO^+ + N$$

$$NO^+ + e \rightarrow N + O(^1D).$$

They noted however that these require an abnormal increase of N_2 molecules above 300 km.

Assuming an electrostatic field from outside the ionosphere, MEGILL, REES and DROPPELMAN (1963 [*430*]) suggested a heating of ionospheric electrons; they estimated the energy distribution of the electrons by solving the Boltzmann equation, including all the inelastic collisions of importance in the upper atmosphere and then obtained the intensity — height profile and the intensity of the λ 6300 [OI] emission. These agree fairly well with the observations. However, the required intensity of the electric field is rather large, of order 200 volts/km, because their proposed field is perpendicular to the magnetic field; the efficiency of the field is extremely small in such a gyro-free region for electrons.

75. Tropical subvisual red arcs (λ 6300 [OI]). During the intense magnetic storm of November 12/13 and 14/15, 1960, BARBIER, WEILL and FAFIOTTE (1961 [*431*]) and BARBIER (1961 [*432*]) observed tropical red arcs, of about 500 km half width, over Tamanrasset (gm lat. 25.4° N). BARBIER (1961 [*432*]) found a semi-empirical formula for the relation between the zenith intensity Q of the λ 6300 [OI] line in rayleighs, f_0 the critical frequency of the F layer in kc, and h' its minimum virtual height in km:

$$Q = 2.35\, f_0^2 \exp\left(\frac{-h' - 200}{41.2}\right) + 87.$$

BARBIER (1963 [*433*]) suggested the following processes:

$$O^+ + O_2 \rightarrow O + O_2^+$$
$$O_2^+ + e \rightarrow O(^1D) + O(^1S).$$

There is no enhancement of the N_2^+ First Negative bands in the arc.

76. Polar glow aurora in polar cap absorption (PCA) events. During polar cap absorption events SANDFORD (1962 [*434*], 1963 [*435*]) observed an extensive auroral type glow, the polar glow aurora. The glow shows no discrete structure. The geographical extent and time variation vary in parallel with the polar cap absorption; their minimum gm. latitude is approximately 60°. The glow is primarily excited at heights below 100 km (down to 60 km) by protons in the energy range 0.5~100 mev.

The most prominent feature of the glow spectrum is the enhancement of the N_2^+ First Negative group bands (of order 10 kR), namely,

$$N_2 + p \rightarrow N_2^+(B^2\Sigma_u^+) + p + e$$
$$N_2 + p \rightarrow N_2^+(B^2\Sigma_u^+) + H^*$$
$$N_2 + e \rightarrow N_2^+(B^2\Sigma_u^+) + e + e$$

where p denotes solar protons and e secondary electrons.

The $H\beta$ emission observed by MONTALBETTI and MCEWEN (1961 [*436*]) was of order 22R. SANDFORD (1963 [*435*]) indicated also some enhancement of the λ 5577 [OI] line, produced by secondary electrons. He showed also that the metastable Vegard-Kaplan bands of N_2 and the emission from atomic species [OI], [OII], [NI] and [NII] are all relatively weaker than they are in common auroral forms.

77. Artificial auroras. Recently "auroras" produced by high altitude nuclear explosions have been studied in detail. The explosion "Teak" over Johnston Island in the Pacific on August 1, 1958 produced an arc from the exploding mushroom cloud (Fig. 107). STEIGER and MATSUSHITA (1960 [*437*]) indicated that it was produced by β decay electrons, ejected from the nuclear device, directed along the geomagnetic field lines. Bright auroral rays were seen also at Apia (CULLINGTON 1958 [*438*]). The rays lasted approximately six minutes.

At first they were violet covered with red, and gradually changed to green. The rays were followed by a red glow which continued for another eight minutes. In addition, a crimson arc was observed north of the conjugate point.

MALVILLE (1959 [*439*]) suggested that the N_2^+ First Negative group bands gave the observed violet color over Apia, and that the [OI] lines are probably due to (a) direct collisional excitation by electrons at 100 km or above and (b) to the following processes at heights below 100 km:

$$O^+ + N_2 \rightarrow NO^+$$
$$NO^+ + e \rightarrow O + N$$
$$O_2^+ + e \rightarrow O(^1D) + O(^1D).$$

The red color covering the violet rays may result from

$$N_2^+ + O^- \rightarrow N_2(B\Pi_g)$$
$$N_2^+ + O_2^- \rightarrow N_2(B^3\Pi_g) + O_2.$$

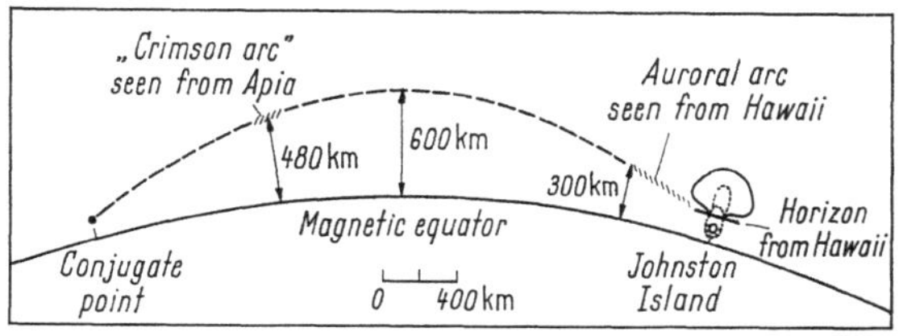

Fig. 107. Diagram showing the location of two artificial auroral displays seen at the time of the 1 August 1958 nuclear bomb blast at Johnston Island. A mushroom cloud was visible from Hawaii as was the auroral arc shown at the right. Electrons created by the blast traveled along lines of magnetic force of the earth's field and created the crimson auroral arc seen at Apia Island south of the earth's magnetic equator. (STEIGER, W. R., and S. MATSUSHITA, 1960 [*437*].)

During the high altitude nuclear detonations Argus I, II, and III in August to September, 1958, artificial auroras were seen at both the launching and the conjugate areas (NEWMAN 1959 [*440*]). Around the launching area, one of the explosions produced a complicated pattern of luminosity, a white patchy glow surrounded by red, lemon, and dull red, together with green-yellow and electric-blue streams. The photometric record indicated an intensity ratio $I(5577)/I(3914)$ of order 10. The aurora over the conjugate area was colored blue-green and red.

A sudden rise of the λ 3914 emission from the N_2^+ First Negative group bands and a subsequent slow increase of the emission λ 6300 [OI] were observed at China Lake, California by ODENCRANTZ, SAINT-AMAND and MOORE (1962 [*441*]) soon after the nuclear explosion over Johnston Island. They suggested that x-rays or a beam of neutrons from the explosion area was responsible for the enhancement. The same explosion produced a brilliant artificial aurora over Christchurch, New Zealand (NEFF 1963 [*442*]). The peak intensities of λ 6300 [OI], λ 5577 [OI], and λ 4278 emission from the N_2^+ First Negative group bands were of order 200, 20, and 1 kR, respectively. NEFF suggested that λ 5577 [OI] and λ 4278 were excited by the same mechanism, namely electron bombardment; however the excitation of λ 6300 [OI] is still little understood. The decay rate of the intensity λ 6300 [OI] was 0.0087/sec (see also CARMAN *et al.* (1963 [*442a*])), close to the theoretical value 0.0092/sec.

D. Upper atmospheric temperatures inferred from auroral spectra.

The determination of upper atmospheric temperatures from auroral spectra is difficult for numerous reasons. Any assumption of thermodynamic equilibrium is doubtful, particularly when the populations of atomic or molecular *vibrational* levels are considered. Thermodynamic equilibrium is attained only within the *rotational* levels of a molecule. The energy changes required to re-establish equilibrium after excitation are small, and the large inertia of the molecules resists abnormal excitation in collisions with an electron gas that is not in equilibrium at the kinetic temperature of the molecules. Unfortunately a clear distinction between valid and non-valid temperature criteria has not always been made in the literature.

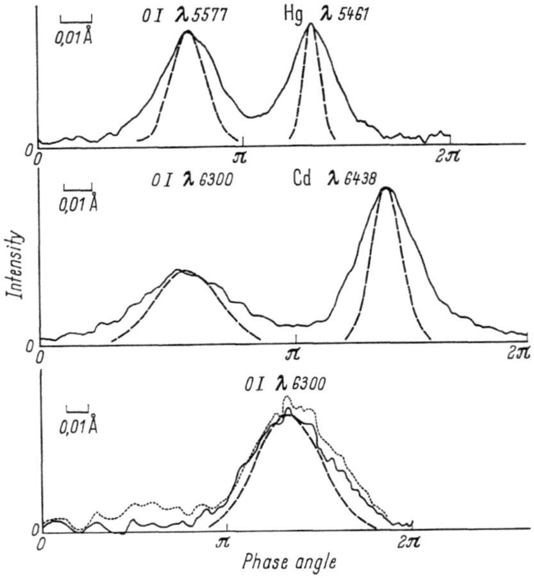

Fig. 108

Fig. 108. Representative airglow line contours. Solid curves are observed profiles of interference fringes, and dashed lines are intrinsic profiles; the dotted line is the observed profile before subtraction of neon fringes. Top: night sky λ 5577; center: twilight λ 6300; bottom: night-sky λ 6300. Dispersions are indicated. (Wark, D. Q., 1960 [443].)

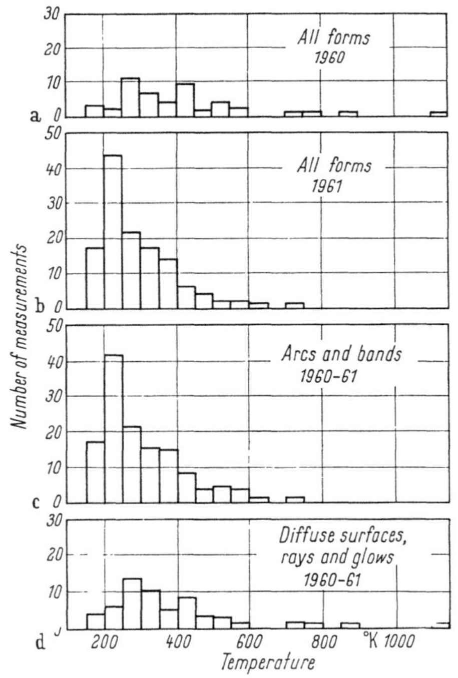

Fig. 109 a–d

Fig. 109 a–d. Histograms showing the number of temperature measurements from the OI 5577 A line falling in 50° intervals. (a) All forms observed in the fall of 1960; (b) all forms observed in the spring and summer of 1961; (c) arcs and bands observed during 1960 and 1961; (d) rays, coronas, diffuse surfaces, and glows during 1960 and 1961. (Turgeon, E. C., and G. C. Shepard, 1962 [445].)

78. Doppler temperature. The least ambiguous temperature determination is obtained by measurement of the thermal Doppler broadening of atomic lines. The very small natural width of the forbidden lines makes the strong λ 5577 [OI] line ideal. Since these measurements require great precision, a Fabry-Perot interferometer has been used. Wark (1960 [443]) showed that the lines λ 5577 [OI] in the night sky and λ 6300 [OI] in the twilight, the aurora, and the night sky yielded temperature of 184°, 710°, and 980° K, respectively (Fig. 108).

Further extensive studies of emitting [OI] atoms have been made by Nilson and Shepherd (1961 [444]), Turgeon and Shepherd (1962 [445]) and Mulyarchik and Scheglov (1963 [446]). Fig. 109a–d gives histograms showing the number of temperature measurements from the λ 5577 [OI] emission (Turgeon and Shepherd 1962 [445]). They differ greatly from one type of aurora to another, and also from year to year.

Turgeon and Shepherd (1962 [445]) estimated the temperature variation with height; T is about 300° K at 100 km and increases to 600° K at about 160 km. They obtained T also from the λ 6300 [OI] emission. It is quite high compared with that obtained from λ 5577 [OI], and ranges from 1000 to 1900° K. It is likely that the emission was from a rather high atmospheric level (above 175 km). Mulyarchik and Shcheglov (1963 [446]) indicated that the temperature obtained from λ 6300 [OI] can vary between 1200° K, on a night when there is no aurora, to 3500° K in bright red aurora.

79. Vibrational temperature. Some early attempts to measure temperatures from the observed band intensities within a given system, but originating from different vibrational levels, as by ROSSELAND and STEENSHOLD (1933 [*447*]), immediately showed that such determinations are difficult. To be a valid temperature, the excitation must be by thermal collisions. Any other mechanism will only give a fictitious temperature.

HUNTEN and SHEPHERD (1955 [*448*]) obtained the intensity ratios (0,2) : (1,3): (2,4) for the N_2 Second Positive group bands, and from them they estimated the temperature. In most cases the temperature was below 1000° K.

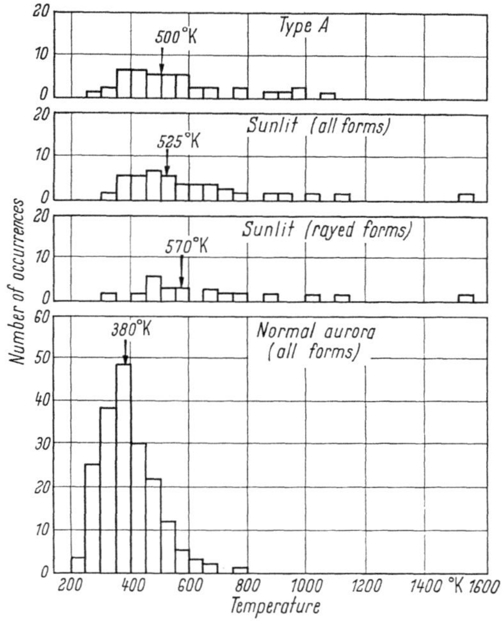

Fig. 110a. Distribution of measured temperatures obtained with narrow-field lens for different types of the aurora. (JOHANSON, A. E., and A. VALLANCE JONES, 1962 [*449*].)

80. Excitation temperature. In the case of atomic lines, we again have a case where a study of the populations of different electronic levels will yield a valid temperature only when the excitation is by thermal processes. If the mechanism of excitation could be carefully defined, a temperature could be deduced. For instance, if the excitation were by electrons, then an effective electron temperature could be derived. There would, however, be no *a priori* reason for this to represent the kinetic temperature of the air molecules. The point is clear when one considers the physics of interstellar gases or planetary nebulae. The kinetic temperature of the atoms and molecules is low, perhaps a few tens of degrees, while the electron temperature may measure tens of thousands of degrees. Clearly where thermodynamic equilibrium is not established between all species of particles, various "effective" temperatures will result.

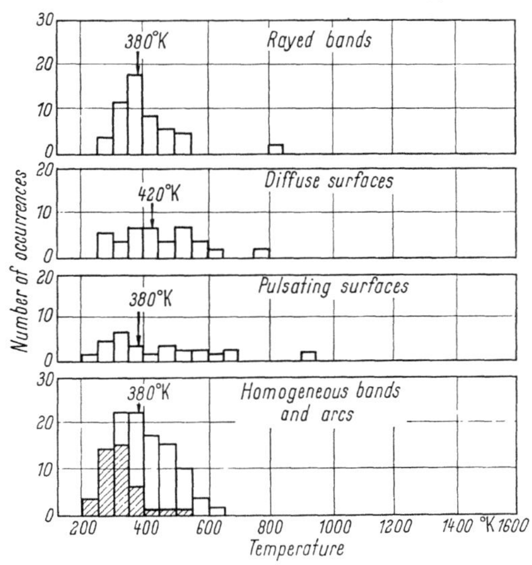

Fig. 110b. Distributions of measured temperatures obtained with narrow-field lens for different auroral forms. In the bottom histogram for homogeneous arcs and bands, the distribution of temperatures from the regions of maximum brightness found from the long slit spectra are shown by the cross-hatched areas. (JOHNSON, A. E., and A. VALLANCE JONES, 1962 [*449*].)

81. Rotational temperature. In rotational temperatures we find the nearest approach to valid kinetic temperatures. A molecule has a large moment of inertia, so that an electron gas at a very high temperature has little effect: the momentum of an electron is small compared with that of the molecule. Collisions with other heavy particles are alone effective in producing an equilibrium distribution. As a consequence, if the molecule existed as a molecule *before* excitation, a valid temperature can be obtained from the rotational line intensities. (If the molecule results as a part of the excitation process, as in chemical reactions, very anomalous temperatures can result.) Therefore in the aurora the rotational temperature can be determined with reasonable accuracy; it should be noted however that for a more rigorous treatment the lifetime of the excited state must be taken into account.

Rotational temperatures have been determined with great accuracy by Canadian and USSR workers. JOHANSON and VALLANCE JONES (1962 [*449*]) determined the N_2^+ rotational temperature and its gradient, in various forms of the aurora. Their results are shown in Fig. 110a and b. The temperature for type *A* red aurora is distinctly higher than that of normal auroras; this is consistent with the view that type *A* red aurora is caused by a mechanism providing efficient excitation of the upper atmosphere at heights up to 200 km or more. Diffuse and pulsating patches also show a higher temperature than do bands and arcs. There is a definite indication of an upward increase of temperature; the gradient is of order 6° K/km. Rotational temperatures have also been obtained by SHUISKAYA (1961 [*450*]) and YARIN (1961 [*451*]); see also MULYARCHIK and SHCHEGLOV (1963 [*451*]). They indicated that the N_2^+ rotational temperature in type *A* and sunlit auroras is much higher than in others.

VIII. Theories of the aurora.

A. Historical reviews.

82. Early speculations; Mairan to Lemström. The beauty and mystery of the aurora have long stimulated speculation about its causes. As early as 1733 de MAIRAN [*1*] presented to the French Academy an able treatise (republished with additions in 1754) on the history and physics of the aurora. He gave reasons for rejecting two hypotheses then current, (a) that the aurora is caused by reflexion of sunlight from polar ice and snow, (b) that is due to terrestrial exhalations; why, he asked, would these be specially luminous in high latitudes?

His own view was that the cause of the aurora is extra-terrestrial, due to gas belonging to the solar atmosphere, impinging by free fall (under the earth's gravitational pull) upon the upper atmosphere. He believed that the solar corona, seen during eclipses, extends at times up to and beyond the earth's orbit, and that its illumination by sunlight is seen as the zodiacal light. He felt confirmed in this view by his estimates that the height of the aurora is several hundred km, not merely a few miles, as then often supposed.

He ascribed the variable frequency of auroral occurrence to variability in the extension of the solar atmosphere (he wrote more than a century before SCHWABE discovered the sunspot cycle, in 1851, see [*5*], § 15, 26). His explanation of the polar situation of the aurora, which he ascribed to greater density of the atmosphere in high latitudes, was vague and inadequate.

LOOMIS [*15*], who discovered the existence of the auroral zone, rejected the extra-terrestrial theory because of its weakness in this respect. Like the chemists DALTON and SCHÖNBEIN before him, he regarded the aurora as an electrical phenomenon. He ascribed it to positive electricity carried to high levels with vapors arising from the oceans, and thence transported by upper air currents towards the polar regions. But he recognized the variation of auroral frequency with the sunspot cycle, and ascribed it to the influence of the sun upon the electrical state of the earth (and other planets). LOVERING, however, in 1870 [*121*], still did not regard the connection of auroral frequency variations with the sunspot cycle as proved. LEMSTRÖM supposed that the sun affects the auroral frequency because its vaporizing influence changes from time to time.

83. Single particles. In the latter part of the nineteenth century came the discovery of the electron and of free gaseous ions; it was found that the motion of such charged particles is affected by the presence of a magnetic field. BIRKELAND (1908, 1913 [*132*]) saw that this might explain the polar situation of the aurora, because particles coming from outside, if electrically charged, could be deflected towards the polar regions; he suggested that the particles are electrons (cathode rays) coming from the sun. He became confirmed in these ideas by experiments he made, in which cathode rays were projected in a partly evacuated chamber towards a magnetized sphere, a "*terrella*". This was coated with fluorescent paint so as to reveal where the electrons impinged on the sphere; he found that they did so along two zones (one round each magnetic pole), which he likened to the auroral zones; the zones extended round the rear as well as on the near side of the sphere. He also studied the geomagnetic disturbances in the polar regions and elsewhere, and inferred that they are associated with strong electric currents flowing high in the atmosphere, and especially above the auroral zone; he regarded these currents as streams of free electrons coming from the sun and descending to the auroral zone (and sometimes passing outwards again). His conclusions, though imperfect, marked an important step towards a true theory of the aurora; his hypothesis supplied a natural link between the changing frequency of auroras and sunspots.

Stimulated by BIRKELAND'S ideas, STÖRMER undertook the study of the motion of electric charges projected from a distance, into a magnetic dipole field such as that of the earth. His work, continued with great ability and energy for several decades, dealt with the motion of a *single* charged particle, subject only to the influence of the magnetic field; as the force is perpendicular to the direction of motion, the speed remains constant. His work is summarized in his book "*The Polar Aurora*" (1955 [*3*]).

STÖRMER'S work led to results of much mathematical interest and physical importance. He showed that there is an immense variety of possible paths for such charged particles moving in a magnetic dipole field. Many of these paths have been beautifully demonstrated experimentally by BRÜCHE (1931 [*453*]), and later by BENNETT (1958 [*454*]), using narrow beams of electrons in a high vacuum. Some of the paths from the sun towards the earth were found by STÖRMER to have a perigee point considerably distant from the earth, whence they bent back into space; particles moving in such paths would not enter the atmosphere and contribute to the aurora. Hence in many of his computations, STÖRMER calculated the orbits back from the earth towards the sun (see also BLOCK 1955 [*452*])).

From his mathematical equations he derived some important and elegant conclusions regarding the regions accessible to isolated charged particles near the earth. One was that there is a natural unit of length C_{st} in his problem, defined by

$$C_{st} = \sqrt{M e / m v}$$

where e is the charge, m the mass and v the speed of the particle, and M is the earth's magnetic dipole moment. He showed that particles coming from infinity cannot enter certain "forbidden" regions near the earth. Their boundary meets the earth in latitude circles at angular distance ϑ from each magnetic pole, where ϑ is given by

$$\sin^2\vartheta = \frac{2a}{C_{st}} + \frac{a^2}{2C_{st}^2} + \frac{a^4}{4C_{st}^4}.$$

Here a denotes the radial distance from the earth's centre to the upper atmosphere. When a/C_{st} is small, approximately

$$\sin^2\vartheta = 2a/C_{st}.$$

STÖRMER interpreted this as giving the outer angular radius of the observed auroral zones, measured from the geomagnetic poles. We now know that the polar aurora is mainly initiated by primary electrons of energy less than 30 kev. According to the above equation, however, such electrons can only enter within a circle $\vartheta = 2.9°$ from the magnetic poles. Even for electrons of energy as high as 1 mev, the value of ϑ is only 5.8°. In order that electrons may reach the auroral zone, their energy must be as high as 300 mev (STÖRMER 1955 [*3*], p. 296).

STÖRMER proposed an auxiliary hypothesis to remove the discrepancy. Some of the orbits he calculated in the equatorial plane bend round the night side of the earth; some of them encircle the earth. A succession of such charges moving along such orbits would constitute an electric current, flowing westwards partly or wholly round the earth. STÖRMER located this "*ring current*" at a great distance from the earth, beyond the moon. He identified its magnetic field with that observed on the earth during a magnetic storm, and inferred the corresponding strength of the ring current. He showed that other charged particles moving in the combined field of the earth and of the ring current (so located and estimated) could reach the earth in the observed auroral zone, and also (as observed) at lower latitudes, when the magnetic disturbing field is as strong as in great magnetic storms.

This explanation, however, meets the difficulty that near the ring current so located, its own field much exceeds that of the earth, so that the calculation of its location is not valid. Its magnetic moment would exceed that of the earth.

STÖRMER also showed that particles projected within a narrow cone from the sun could impinge on the earth at points spread along a narrow band of latitude. This he interpreted as explaining the formation of auroral arcs and draperies; the particles would enter the atmosphere nearly along the magnetic lines of force (possibly spiralling around them), thus agreeing with the observed direction of auroral rays. As shown in Sect. 45, motions of auroral particles lie in STÖRMER's inner allowed region, rather than in the outer allowed region. Therefore except for energetic solar protons, auroral primary particles can not directly come from the sun to the auroral zone in the way discussed by STÖRMER.

Despite the promising analogy of some of STÖRMER's results with important auroral features, it seems certain that BIRKELAND's hypothesis and STÖRMER's calculations make only a first distant approach to a true auroral theory. SCHUSTER (1911 [*455*]) showed that magnetic storms cannot reasonably be explained by streams of solar particles of one sign, and therefore carrying a current; their mutual electrostatic repulsion would disperse them to negligible density before they could reach the earth.

LINDEMANN (1919 [*456*]) similarly criticized a one-sign stream theory of magnetic storms proposed by CHAPMAN (1918 [*377*]). LINDEMANN added a constructive suggestion, namely that magnetic storms and auroras are due to the impact upon the earth of a cloud or stream of solar gas that is neutral but ionized—consisting mainly of electrons and ionized hydrogen atoms (that is, protons).

84. Neutral ionized solar plasma. Based on LINDEMANN'S suggestion, CHAPMAN and FERRARO (1931 [*457*], 1940 [*458*]) initiated a new approach towards the problem of the interaction between the solar gas and the geomagnetic field. Their solar gas consisted of a neutral ionized stream in which the number of positive and negative particles is nearly equal. They showed that such a solar gas can be treated as a good conductor. The problem is therefore analogous, in some sense, to a solid conductor approaching a magnet. The electric currents are induced first near the surface of the conductor, and shield the interior of the conductor from the field. At the same time, the Lorentz force tends to retard the advance of the conductor.

They estimated the addition to the geomagnetic field by the induced currents in the solar gas by treating these currents as flowing in an infinite plane perfectly conducting sheet; in this particular case, the additional magnetic field due to the current is that of an image dipole located at a geocentric distance of $2r_0$, r_0 being the distance of the sheet. This is called the *image dipole approximation*. The space on the rear side of the sheet is completely shielded from the geomagnetic field. This phenomenon is now commonly expressed by the word "compression" of the geomagnetic field by the solar gas.

Further, by use of a "cylindrical sheet" model, CHAPMAN and FERRARO (1940 [*458*]) attempted to infer by mathematical deduction the detailed motions of the particles of the neutral ionized gas advancing towards the axis of the cylinder; the field lines were taken to be parallel to the axis, with the field intensity increasing towards the axis. In particular cases positive particles in the cylindrical sheet may advance ahead of the electrons. The electrons are dragged on by the protons and come much closer to the axis than they could do individually; the protons are held back by the electrons and travel less far towards the axis than they would do alone. Kinetic energy is transformed into magnetic energy in the space within the sheet.

They inferred that this could be the case when the solar gas is approaching towards the earth. They showed that STÖRMER'S treatment of solitary particles can only be applied for sufficiently energetic and rarified ionized gas, like cosmic rays. For the solar neutral ionized gas, the electrostatic interaction between the charged particles is enormous, so that each particle cannot move independently of the others.

When this phenomenon is examined microscopically, each charged particle is reflected back or scattered sideways from the front, but the deflection of the protons is more eastward and of the electrons more westward. Consequently, a net eastward current is produced at the front, increasing the magnetic field within the cylinder (cf. CHAPMAN 1960 [*459*]). The sheet problem was further elucidated by CHAPMAN and KENDALL (1961 [*460*]).

At about the time when CHAPMAN and FERRARO were analyzing the motion of a neutral ionized medium, LANGMUIR (1929 [*461*]) was independently studying neutral ionized gas, to which he gave the name *plasma*, now commonly used.

In their most developed model, the advancing solar plasma was replaced by successive sheets of ionized gas, and the geomagnetic field by a unidirectional field (FERRARO 1952 [*462*]). In this case the front sheet is first most retarded, and thus overtaken by the sheet to the rear, previously only partly retarded because of the shielding effect of the frontal sheet.

After being overtaken the front sheet moves backwards. This process is successively repeated. This model enabled FERRARO to infer several important results on this problem, such as the minimum distance attained by the sheets, the rise time of the disturbance field and the shielding distance. For the number density $n=1/\text{cm}^3$ and velocity 10^8 cm/sec, the plasma is stopped at a distance of $8.86\,a$ (a = the earth's radius), and produces a magnetic field of order $21.5\,\gamma$ within about 2 minutes. This problem has recently been extended to a two or three dimensional dipole field by many workers (Sects. 85—87). CHAPMAN and FERRARO (1940 [*458*]) suggested also the growth of a toroidal ring current, electrically polarized, within the cavity carved in the solar plasma. Its polarization field expels some of the charged particles along the geomagnetic field lines towards the ionosphere. MARTYN (1951 [*463*]) suggested that this is the cause of the aurora, and that the expelled charged particles convey the polarization field into the ionosphere to drive the electric current there. However, present knowledge on the radiation belts does not confirm the existence of such a toroidal ring in the magnetosphere.

85. Trapped particles. Although STÖRMER found a group of trajectories in the inner allowed region (Sect. 45), he paid little attention to them. This was because in a pure dipole field, particles from infinity cannot enter the inner allowed region.

When ALFVÉN (1940 [*294*]) proposed the guiding center approximation for motions of charged particles in a strong magnetic field, he showed the usefulness of his simple "smoothed out" version of the path of such a gyrating particle by comparing it with STÖRMER's complicated trajectories in the inner allowed region (see ALFVÉN 1950 [*13*], p. 30). The significance of such orbits for the aurora was first recognized by SINGER (1957 [*464*]). He suggested that the ring current is produced by such trapped particles.

The discovery of the radiation belts confirmed the existence of such trapped particles (Chap. VI), although the problem of their origin is still little understood, except perhaps for some of those produced by neutron decay.

Part of the motion of auroral primary electrons can be described with sufficient accuracy by the guiding center approximation. The agreement between the observed and theoretical auroral zones supports this view (HULTQUIST 1959 [*31*]; VESTINE and SIBLEY 1960 [*33*]).

Obviously, however, in a region where such electrons are produced or introduced, the three invariances (Sect. 47) are likely to be violated, whatever the mechanism may be.

Further, the studies mentioned in Sect. 83, together with recent satellite observations (particularly Explorer X, XII, and XIV), have enabled us to recognize that the geomagnetic field is confined in the Chapman-Ferraro cavity. This region is now commonly called the *magnetosphere* (GOLD 1959 [*465*]).

It is a region carved in the solar wind during quiet periods, and in an intense plasma flow during storm times. The size of the magnetosphere continuously changes with the intensity of the solar plasma flow. Trapped particles can exist only in this region. It is the region where auroral primary particles circulate, and where the ring current grows.

The extensive IGY network of all-sky camera stations has confirmed that the aurora is a more or less permanent feature of the planet earth, at least during sunspot maximum. The aurora was almost always seen somewhere in the auroral region, regardless of whether or not magnetic storms were in progress. Thus even during quiet periods the supply of auroral energy from the sun seems to continue, though at times it is weak.

However, the fact that strong auroral displays are always accompanied by magnetic storms indicates that the supply is larger during magnetic storms than in quiet periods. Thus it is likely that two types of solar plasma flow discussed in Sect. 32a, namely the shell and the stream, contain *in general* a considerable amount of energy, enough to cause the aurora. But this is *not always* the case; as seen in Sect. 36, the enhancement of the solar plasma pressure, indicated by the storm sudden commencement and *DCF* field of the initial storm phase, does not necessarily ensure the enhancement of the supply of the auroral energy. It can only be said *statistically* that intense solar plasma flows tend to contain substantially more energy than weaker ones.

Thus, as correctly recognized by BIRKELAND, the origin of the aurora is closely related to magnetic storms. Without understanding the mechanism involved in magnetic storms, it is not possible to solve the auroral problem, and vice-versa. In this sense there is no definite separation between the theory of the aurora and of magnetic storms.

B. The geomagnetic field bounded by the solar plasma.

86. The shape of the boundary surface. In the Chapman-Ferraro mathematical models (see Sect. 84), the front surface of the solar plasma was treated as plane or cylindrical. However, they recognized that the cavity carved in the solar plasma flow by the geomagnetic field has a complicated three-dimensional boundary surface.

The exact boundary surface for a two-dimensional dipole has been obtained by ZHIGULEV and ROMISHEVSKII (1959 [*466*]), DUNGEY (1961 [*467*]), HURLEY (1961 [*468*]), SPREITER and BRIGGS (1962 [*469*]), SPREITER and HYETT (1963 [*470*] MIDGLEY and DAVIS (1963 [*407a*])). In their treatments, the current was assumed to flow in a thin layer close to the boundary surface, so that in the cavity

$$\operatorname{div} \boldsymbol{B} = 0,$$

$$\nabla \times \boldsymbol{B} = 0.$$

At the boundary surface,

$$B_n \text{ (the normal component of the magnetic field)} = 0,$$

$$B^2/8\pi = 2\, n m v^2 \cos^2\chi,$$

where n is the number density in the plasma, m the ion mass, v the velocity of the plasma at infinity, and χ the angle between the normal to the boundary and the direction of incidence. The magnetic field can then be expressed by a scalar function Φ. The problem thus becomes one of finding a function Φ satisfying the conditions that its gradient expresses the intensity of the two-dimensional dipole field near the origin, and that it satisfies the above boundary condition. The problem is transformed to find Φ in a complex w plane in which the surface is expressed by a circle. Then, using a conformal transformation, the circle is distorted until the pressure balance condition is also satisfied.

In a three-dimensional case, the pressure-balance equation is given by

$$4 f^2 H^2/8\pi = 2 m n v^2 \cos\chi,$$

where f is taken to be a constant (see FERRARO 1960 [*471*]). MEAD (1962 [*472*]) has shown that the three-dimensional boundary surface and the magnetic field within the cavity can be obtained with reasonable accuracy by an iteration method.

Fig. 111a shows the relation between the equatorial distance r_0 of the boundary along the sun-earth line and the velocity v, for several values of the number density n of protons in the incident plasma and for two different values of the parameter f. It shows also the northern half of the boundary surface. Fig. 111b shows an isometric drawing of the boundary of the equatorial plane and the meridian plane for various values of the ratio of the static pressure (p_0) to the dynamic pressure (p_d). AXFORD (1962 [*473*]) and KELLOGG (1962 [*474*]) proposed that there is a standing shock wave outside the cavity (see also FRANK, VAN ALLEN and MACAGO (1963 [*474a*]), HONE 1963 [*470b*], CAHILL and AMAZEEN (1963 [*474b*]), FREEMAN, VAN ALLEN and CAHILL (1963 [*474c*]))

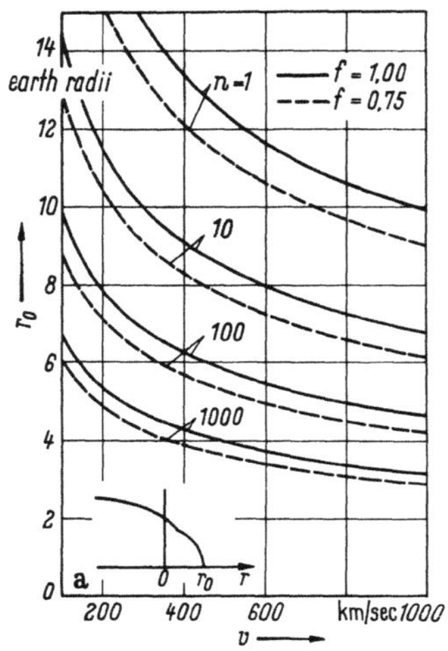

Fig. 111a. Values for r_0 for various velocities and number densities of the corpuscular stream. (SPREITER, J. R., and B. R. BRIGGS, 1962 [*469*].)

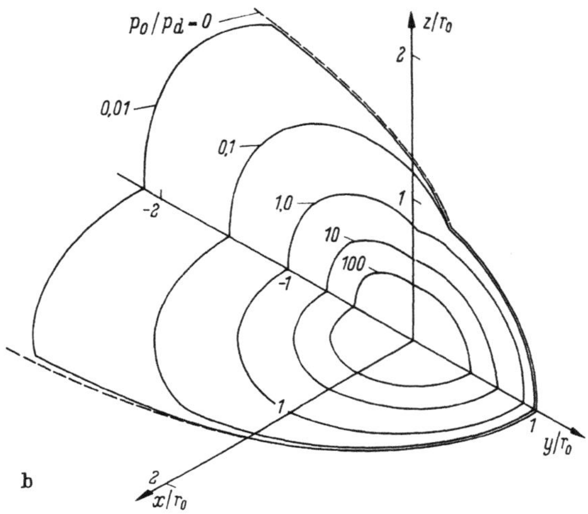

Fig. 111b. Isometric drawing of the boundary of the geomagnetic field in the equatorial plane and the meridian plane containing the sun-earth line, for various p_0/p_d. (SPREITER, J. R., and B. J. HYETT, 1963 [*470*].)

87. Transmission of magnetic changes from the solar plasma front to the earth's surface. In the studies by CHAPMAN and FERRARO and others (cf. Sect. 86) the space between the earth and the frontal surface was treated as emply. In that case the magnetic field produced by the surface current is propagated towards the earth with the speed of light *in vacuo*.

STOREY's theory of whistlers (1953 [*475*]) gave the first strong evidence of the presence of ionized gas out to distances of a few earth radii. His study was later refined by ALLCOCK (1959 [*476*]), POPE (1962 [*477*]) and CARPENTER (1962, 1963 [*478*] to [*480a*]).

Rocket observations of Lyman α emission in the night sky enabled JOHNSON (1959 [*481*]) to estimate the distribution of protons to 6 earth radii. Owing to this gas within the cavity, the magnetic field produced at the front surface cannot be propagated with the speed of light. DESSLER (1958 [*482*]) suggested that it is propagated by hydromagnetic waves. WILSON and SUGIURA (1961 [*483*]), from a detailed analysis of rapid-run magnetic records for a number of sudden commencements, showed that complicated features of the sudden commncemeent can be explained by superposed differently polarized waves. Using the ray path theory, FRANCIS, GREEN and DESSLER (1959 [*484*]) and DESSLER, FRANCIS and PARKER (1960 [*485*]) attempted to examine the mode of propagation of hydromagnetic waves through the magnetosphere; however, in the magnetosphere the wave velocity may change drastically, as they indicated, within one wave length; therefore such an approach is subject to difficulties.

88. Neutral points and the aurora. As the initial formulation indicated, the solar plasma flow here considered does not provide any solar particles injected into the magnetosphere; they are sharply reflected back or scattered sideways. There are, however, two neutral points, one in each hemisphere, at the front of the plasma. It has been thought that "horns" of solar gas may flow through and near these neutral points towards the earth. Possibly this may occur because the solar plasma particles have thermal random motions, as well as general flow. LEBEAU and SCHLICH (1962 [*486*]) found some indication of this from their analysis of high latitude magnetic data. However, it seems unlikely that the neutral points play an important role in the theory of the aurora.

C. The ring current.

89. The magnetic field produced by the ring current. The magnetic effects of trapped particles in the magnetosphere have been extensively studied by DESSLER and PARKER (1959 [*487*]), AKASOFU and CHAPMAN (1961 [*153*]), APEL and SINGER

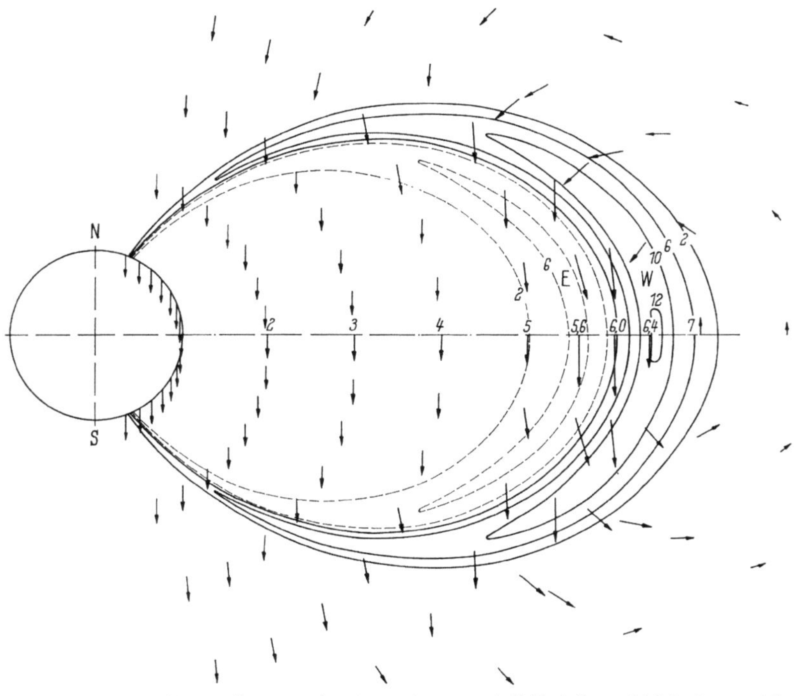

Fig. 112. The field vectors of F, the first approximation to the magnetic field of the model belt, in a meridian plane. and isolines of the equivalent current intensity in the belt (full lines indicate westward, broken lines eastward, current) The vector scale of force, and the unit in which the current intensity is expressed, are proportional to the energy density nE (keV/cm) at the center line of the belt, at 6 earth radii from the earth's center. (AKASOFU, S.-I., J. A. CAIN and S. CHAPMAN, 1961 [*154*].)

(1961 [*488*]), and AKASOFU, CAIN and CHAPMAN (1961 [*154*], 1962 [*125*]). In a steady state, both the gyration and the drift motion produce a net current within the trapping region. To a first approximation, AKASOFU and CHAPMAN (1961 [*153*]) and AKASOFU, CAIN and CHAPMAN (1961 [*154*]) showed that the magnetic field produced by the ring current can be expressed in terms of the five parameters r_{e0}, g_1, g_2, α and $n_0 E$, in the simple case of a ring current belt consisting of monoenergetic particles of energy E, when the number density n in the equatorial plane is given by

$$n = n_0 e^{-g_1^2 z^2} \quad (z<0\text{: the inner part of the belt})$$

$$n = n_0 e^{-g_2^2 z^2} \quad (z>0\text{: the outer part of the belt})$$

where

$$z = (r - r_{e0})/a$$

and r_{e0} denotes the distance at which n attains its maximum value n_0 in the equatorial plane.

The pitch-angle distribution F is assumed to be the same throughout the belt and to have the form

$$F = A(\alpha) \sin^{\alpha+1}\vartheta$$

where ϑ denotes the pitch angle, α is a constant, and $A(\alpha)$ is a normalization factor.

Fig. 112 shows the distribution of the magnetic field, together with the current intensity distribution, for the belt characterized by $r_{e0}=6a$, $g_1=g_2=1.517$, $\alpha=-0.5$ and $n_0E=150$ (kev/cm³). The field is fairly uniform within a radius of about $2a$, and is almost parallel to the dipole axis, agreeing with the observed field during magnetic storms. Beyond this distance the field becomes notably non-uniform. It shows considerable curl in the region of most intense current.

90. The energy of the ring current. AKASOFU, CHAPMAN and VENKATESAN (1963 [*157*]) estimated the total kinetic energy of the ring current. For different values of the center line of the belt r_{e0}, the table gives approximate values of (i) the maximum field ${}_1F_s$ at the equator at the earth's surface produced by the particular ring current belt ($g_1=2.990$, $g_2=0.419$, $\alpha=2.0$), (ii) the corresponding $Dst(H)_1=\frac{3}{2}\,{}_1F_s$, (iii) the maximum particle energy density n_0E at $r_e=r_{e0}$, (iv) N/n_0, (v) the total particle energy $N\varepsilon\,(=n_0\times N\varepsilon/n_0)$, (vi) the energy per unit Dst decrease (of 100 γ), and (vii) the magnetic field energy estimated by CHAPMAN (1964 [*514*]).

Table 9. *Model belt*; $g_1=2.990$, $g_2=0.419$, $\alpha=2.0$: $(Z^Y = Z\times 10^Y)$

$r_{e0}\,(a)$	2.0	3.0	4.0	5.0	6.0
Maximum ${}_1F_s\,(\gamma)$	−270	−155	−100	−55	−30
Maximum $Dst(H)\,(\gamma)$	−405	−233	−150	−83	−45
Maximum n_0 (/cm³)	3500	1200	500	200	85
N/n_0	2.7^{28}	4.5^{28}	6.8^{28}	9.6^{28}	1.3^{29}
Total energy $N\varepsilon$ (ergs)	1.5^{23}	8.7^{22}	5.5^{22}	3.1^{22}	1.8^{22}
The energy per unit Dst (H) decrease (of 100 γ) (ergs)	3.8^{22}	3.7^{22}	3.7^{22}	3.7^{22}	3.9^{22}
Magnetic field energy (ergs)	4.4^{21}	5.2^{21}	5.2^{21}	3.1^{21}	1.6^{21}

91. Generation of the ring current. The direct injection of energetic solar particles has long been thought to be the source of the ring current particles. In order to have $Dst(H')$ of order $-150\,\gamma$ at the earth's surface, however, the center line of the belt should be at $r_{e0}\sim 4a$ or less. Direct penetration to such a depth from outside the magnetospheric boundary could be attained by some non-relativistic solar cosmic rays, but their flux is much too low to form an appreciable ring current.

Assuming that 30 Mev protons can penetrate to $r_{e0}=4.0a$ in the equatorial plane (although such a deep *direct* penetration does not seem to occur), the number density n_0 required for $Dst(H')\sim -150\,\gamma$ is 500 (kev/cm³)/3×10^4 (kev) = 0.017/cm³; the flux is then $n_0w = 0.017$ (/cm³) $\times\, 7.58\times10^9$ (cm/sec) $=1.36\times$

$10^8/\text{cm}^2$ sec. Satellite observations indicate that the flux of 30 Mev solar protons outside the magnetosphere is of order $10^3/\text{cm}^2$ sec at most. This is negligible compared with the required number. The flux of non-relativistic solar protons increases rapidly towards lower energies, but it does not seem possible for 1 Mev protons to penetrate as close as $r_e \sim 4a$ or less. Satellites do not show any sign of the formation of the belt by these protons. This difficulty was pointed out by DESSLER, HANSON and PARKER (1961 [*489*]). Instead of direct injection, they proposed that hydromagnetic heating can raise the energy of thermal protons in the magnetosphere to a few kev. They infer that wherever the solar plasma comes in contact with the geomagnetic field, hydromagnetic waves are generated at the interacting surface. They infer also that the ratio $\Delta B/B$ of the amplitude ΔB of the waves to the dipole field intensity B would be of order unity at about $r_e = 3a \sim 5a$ (see also PARKER 1962 [*490*]; pp. 89—91): the waves generated at the interacting boundary develop into shock waves with a sharp crest, resulting in a rapid dissipation of the wave energy, presumably by instabilities of anisotropic thermal motions of particles. KERN (1962 [*491*]) further elucidated this problem.

However, it is uncertain whether the amplitude of hydromagnetic waves can be as large as $250\gamma \sim 1000\gamma$ at $r_e = 3a \sim 5a$. So far, satellites have not encountered any such violent magnetic fluctuations. Further, if the earth's field is so much disturbed as is suggested, it would hardly be possible to have any systematic ring current to produce the main phase. The difficulty would be increased further for a large Dst change of more than $300\,\gamma$. The formation of the ring current belt which produces such a considerable change would require ΔB to be as large as $4000\,\gamma$ at $r_e = 2a$. Furthermore, the impact of the solar plasma on the interacting surface does not necessarily lead to the formation of the ring current. It seems that their proposed mechanism in its present stage involves as many difficulties as direct injection.

The variety of development of magnetic storms suggests intrinsic differences between the solar streams, far beyond what we would expect from a mere difference between their pressures. The nature of these intrinsic differences is at present unknown. A simple model would be to assume an energy source, whatever it may be, for the main phase and for the auroral activity, distributed differently for different solar streams. The existence of storms with a large and long initial phase, but with no appreciable main phase, suggests that some of the solar plasma would not include any appreciable amount of such energy, at least along the sun-earth line. The variety of development would then be interpreted in terms of differences between the distribution and the intensity of such an energy source in the solar plasma. Satellite observations of various physical quantities in the interplanetary space, as well as in the magnetosphere, are extremely important from this view point (see AKASOFU (1964 [*492*])).

D. The formulation of the problem.

The aurora offers one of the most difficult problems in geophysics. Many partial theories have been put forward to explain particular auroral characteristics, but all are inevitably far from complete. There is no doubt that the ultimate source of the auroral energy is the sun and that the energy should finally appear as the kinetic energy of auroral electrons and protons. A solar energy must be transported from the sun to the earth, and then must be introduced into the magnetosphere in a *specific way* in order to be seen as the aurora.

92. Fundamental requirements. Here the discussion is limited to the auroras that appear in the auroral region.

a) Geometrical. α) Auroral zone. The occurrence of the aurora has a sharp maximum around the $L=6$ line. The lines of force "anchoring" in the auroral zone cross the equator approximately at a point at the geocentric distance of 6 earth radii. During an appreciable main phase, the equatorial distance is expected to change greatly.

β) Night-time appearance. The auroral zone defined by Fritz and others is of course derived from night-time observations. Radar can detect day-time auroras. It is thus found that auroras may extend into the hours of daylight and begin before sunset. As far as the auroral zones are concerned, one most important feature is their occurrence chiefly at night.

Within the auroral cap there are two maxima of the occurrence frequency; one around midnight and one in the early morning. As one approaches the gm pole from the auroral zone, these maxima shift, the former towards the early evening side and the latter towards the morning side.

γ) East-west extension. One fundamental form of the aurora is an arc, extending 5000 km or more in the east-west direction. During quiet periods, some of them extend all along the auroral zone from the evening twilight zone to the morning twilight zone. Each such arc has a planetary scale. In the auroral cap, arcs tend to be alined along the sun-earth direction.

δ) Thickness. The thickness of an arc may be typically 250 ∼ 500 m. This suggests that the equatorial region traversed by auroral primary electrons must be fairly narrow. The distance between two lines of force, separated by 250 m in the north-south direction in the auroral zone, is only of order 6 km in the equatorial plane.

ε) The multiple nature of arcs. In most cases auroral displays consist of several arcs.

b) Dynamical features. Each arc moves northward or southward with various speeds, the highest being of order 1 km/sec. Arcs can be folded to form large loops.

It has been thought that the aurora on the evening side is quiet, but *always* active around the midnight sector and beyond, and that different types of aurora are seen during the course of the night because the earth rotates under this pattern. However, the IGY data indicate no such fixed pattern of the aurora. During quiet periods the auroras all along the auroral zone are quiet and have a fairly simple structure. When an auroral substorm starts around the midnight meridian, the changes are different at different local times. The life time of such a substorm is typically a few hours. In the region of intense activation, auroras lose their simple arc structure, and the aurora in the evening sector *moves* westward and eastwards in the morning sector (see Akasofu (1964 [*515*])).

c) Relation with magnetic storm fields. The break-up phase is associated with polar magnetic substorms. Again, there is no fixed pattern of the current system under which the earth rotates once a day. However, once a polar magnetic substorm appears, its current system closely resembles the DS current system. The life time is of order a few hours, like the auroral substorm.

The intensity of the solar plasma flow indicated by the positive $Dst(H)$ does not necessarily lead to the generation of polar magnetic and auroral substorms.

The ring current tends to shift the belt of auroral activity equatorward.

It is not correct to assume that the aurora is continuously active during magnetic storms. Even during the greatest magnetic storms, the aurora can remain relatively quiet, particularly between magnetic and auroral substorms.

d) Nature of auroral particles and their injection. Major parts of the auroral energy are brought by electrons of energy 100 kev or less. Rocket and satellite observations seem to indicate that the energy spectrum is extremely steep, so that there are many more electrons of lower energy (<10 kev) than of higher (~ 100 kev).

During the active phase of the aurora, the injection appears to fluctuate rapidly, the time scale being of order 0.1 sec.

These primary electrons appear to come from the outer boundary of the outer radiation belt. There is an indication that an increase of the electron flux is associated with an enhancement of electrons with smaller pitch-angles, but not with an increase over the whole pitch-angle range.

e) Energy supply. There is an definite indication that the growth of the DR field and DP field is related. The DR field is associated with an increase in the particle kinetic energy density in the ring current belt, and the DP substorms are associated with enhanced injection of energetic particles into the auroral ionosphere. As discussed in Sects. 87 and 90, the mechanism of the supply of their energy from the solar plasma into the magnetosphere is little understood. Because the enhancement of the solar plasma pressure does not necessarily lead to the enhancement of the supply, one may speculate that there is some additional property associated with the solar plasma flow, which is at present unknown.

93. Breakdown of the invariances. Corresponding to the three component motions of trapped particles, namely the gyration, oscillation and drift motion, there exist three adiabatic invariants (Sects. 45—47). Because they change slowly they may be taken as constants of the motion to a high degree of accuracy. Therefore, unless their invariance breaks down, the particles in the inner allowed region remain trapped almost indefinitely.

It has been argued that the enhanced injection of auroral particles must be associated with the breakdown of the invariance. For the first invariant, the magnetic moment, this occurs when a particle passes through a region where the magnetic field is so weak that the guiding center approximation fails or when the magnetic field varies so rapidly that

$$(1/B)\,(d B/d t) \sim e B/m c\,.$$

Some earlier studies are summarized by CHANDRASEKHAR (1959 [*493*]).

In the magnetospheric region, hydromagnetic waves cannot satisfy the above equation. However, because trapped particles oscillate along the lines of force, they experience a Doppler shifted frequency of hydromagnetic wave. Resonance coupling occurs if

$$2\pi w_s \sim (e B/m)\,\lambda$$

where λ denotes the wave length of the hydromagnetic waves (DRAGT 1961 [*494*]; WENTZEL 1961 [*495*], 1962 [*496*]). This process is efficient for protons but not for auroral electrons. For auroral electrons, various VLF emissions appear to be more important, namely the trans-resonant process (cf. PARKER 1961 [*497*]; KELLOGG 1963 [*498*]). This possibility has not yet been explored in detail.

The breakdown of the second invariant was discussed by PARKER (1961 [*499*]). When hydromagnetic waves pass through the mirror point, its height will oscillate, so that the particles are reflected as from a moving mirror. This process tends to increase w_s, causing "dumping" of electrons into the dense atmosphere. However, this is a rather slow process; the life time of auroral electrons due to this loss mechanism is of order a few months under the most favorable conditions.

The breakdown of the third invariant leads to a rapid diffusion of electrons, inwards or outwards (PARKER 1960 [*500*], DAVIS and CHANG 1962 [*501*]). The importance of this effect on the auroral primary electrons has not yet been explored.

E. Present theories of the aurora.

94. The electric field model (ALFVÉN). ALFVÉN (1929 to date) has developed a theory of magnetic storms and auroras (ALFVÉN 1955 [*502*]). He assumes that the sun emits beams of an extremely rarified ionized gas, carrying "frozen in" magnetic fields of intensity $\boldsymbol{H}$. Because of the motion of the beam (of velocity $\boldsymbol{v}$), the magnetic field produces an electric field

$$\boldsymbol{E} = -\boldsymbol{v} \times \boldsymbol{B} = -\mu_0 \boldsymbol{v} \times \boldsymbol{H}.$$

The number density of the beam is so small that mutual electrostatic interaction is neglected, and the ionized particles execute circular gyrational motions and drift motions due to the electric field $\boldsymbol{E}$, the gradient of $\boldsymbol{H}$, and their inertia.

When the beam advances towards the earth, the current due to the inertia flows eastward, giving rise to an increase of the magnetic field at the earth's surface. He identified this increase with the storm initial phase, but this increase (as he recognized) lasts for only 10 minutes, rather than typically a few hours.

A further advance of the beam induces the current due to the gradient of the field, namely the current for the storm main phase. But the current intensity should not simply be $e\boldsymbol{u}_1$ ($\boldsymbol{u}_1$ = the drift velocity due to the gradient of the field; Sect. 47); a part of the gyrational motion cancels completely the current arising from this term (cf. PARKER 1957 [*296*], p. 928).

Therefore ALFVÉN's theory appears unsuccessful in this important aspect of magnetic storms. One may question whether the situation proposed by him exists at all around the earth.

95. The convective model of the magnetosphere (AXFORD, HINES, PIDDINGTON). AXFORD and HINES (1961 [*503*]) and PIDDINGTON (1962 [*505*], 1963 [*506*]) proposed that a frictional drag between the solar ion stream and the geomagnetic field tends to carry the material near the boundary surface to the geomagnetic tail. In a steady state there should be no pile-up of material at the tail, consequently a return flow must follow. Hence they suggest a large-scale convection within the magnetosphere (Fig. 113a) (see also AXFORD (1963 [*504*])).

If the geomagnetic field lines are frozen in the material, they are carried around by the convection. From the point of view of charged particles in the magnetosphere, this convective motion ($\boldsymbol{V}$) is accompanied by a polarization electric field (E), so that the charges of both signs can move together without the deflection produced by the Lorentz force (see COWLING 1954, [*507*], p. 543). The polarization field must be derivable from a potential φ such that

$$\operatorname{grad} \varphi = -\boldsymbol{E} = \boldsymbol{V} \times \boldsymbol{B}.$$

Thus the velocity vector V must lie in the equipotential surface.

Superposing such a convective motion upon the rotation of the magnetosphere, they obtained a pattern of the stream lines; in the equatorial plane, they consist of two large vortices, one on the morning side and the other on the evening side (Fig. 113a). The above equation is valid down to the ionospheric F level. The ionospheric motion can then be visualized by mapping the stream lines in the equator on to the F layer (Fig. 113b). In the E region, the equation is still valid for electrons, but for ions it can no longer be applied because of their frequent collision with neutral molecules. This results in a net current. At a height where ions are essentially immobile (~100 km), the direction of the arrow in Fig. 113b can be regarded as the direction of motion of ionospheric electrons (section 34).

The direction of ionospheric currents can thus be visualized by reversing the direction of the arrows. Such a current diagram closely resembles the *SD* current system obtained by CHAPMAN (1935 [*81*]). Further, because of a strong shear motion, any volume of auroral electrons will be elongated to form thin arcs. Thus the stream lines shown in Fig. 113b can be regarded as representing the alinement of auroral arcs.

This situation can be visualized by a simple analogy as follows: blowing across a cup of tea, it is possible to set up a convective motion in the cup; on pouring

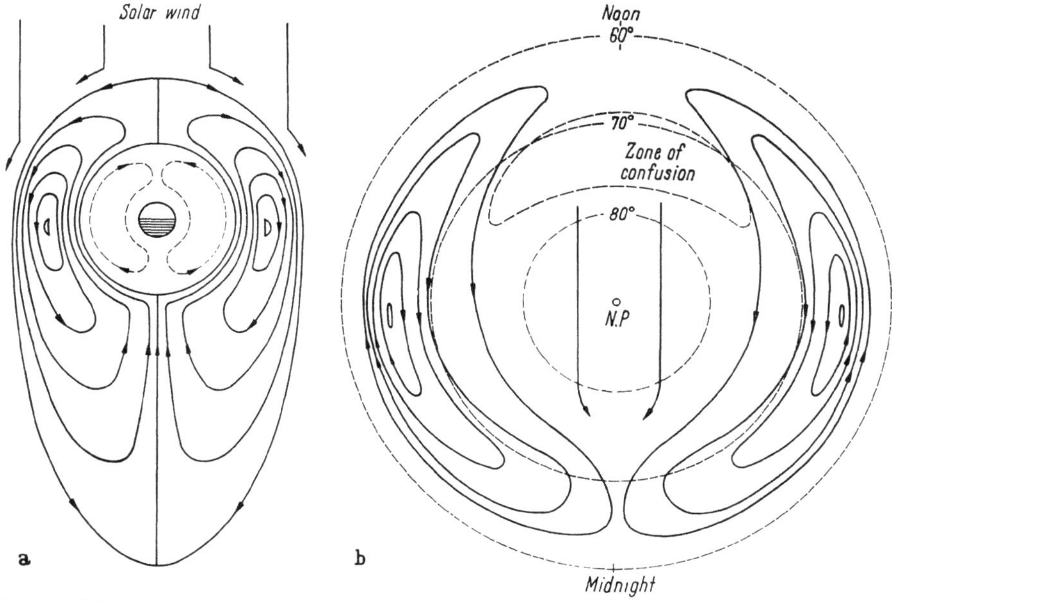

Fig. 113. (a) The proposed pattern of streamlines (or alternatively of the equipotentials of the electric field, at separations of 2 kilovolts) in the equatorial plane of the magnetosphere. The main circulation does not penetrate closer than about 4-1/4 earth radii; dotted lines within this region indicate a possible inner convection system. (AXFORD, W. I., and C. O. HINES, 1961 [*504*].) (b) The pattern of motion at ionospheric levels, obtained by mapping the streamlines of (a) down onto the northern hemisphere along the lines of force of the geomagnetic field. (AXFORD, W. I., and C. O. HINES, 1961 [*504*].)

a drop of milk in the tea at the blowing point; the drop will be quickly carried along by the convective motion, and will produce a pattern of the stream lines. In this analogy, a drop of milk can be regarded as a volume of auroral particles. At the opposite side of the blowing point (the night side), the streams of milk become branched, a part flowing towards the morning side and the rest toward the evening side.

It is not easy to see, however, in any mathematical deduction how and to what extent the gigantic convection takes place. Further, as seen in Sect. 30, there is no definite indication that arcs are alined north-south around the midnight meridian; no clear-cut case has shown any connection between the cap aurora and the auroral zone aurora. DAVIS' array of loops, which closely resemble the *SD* current lines, seems to be a statistical result of westward traveling folds.

It may further be added that an increase of the solar plasma pressure does not necessarily lead to *DP* activity (Sect. 35), although the convection model seems to imply their direct association. Although the convective model contains fascinating ideas, its validity is at present uncertain.

96. The charge separation model (CHAMBERLAIN, KERN). In a bounded plasma in a non-uniform magnetic field, positive and negative particles tend to

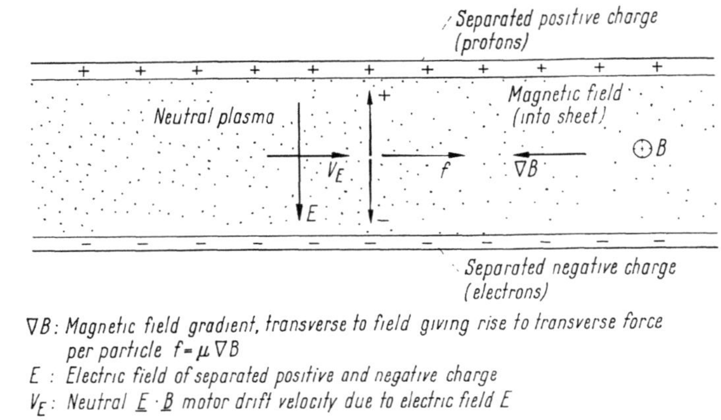

Fig. 114a. Charge separation in a bounded plasma due to transverse geomagnetic field gradient. (KERN, J.W., 1962 [*510*].

drift in opposite directions (Sect. 47); thus there appears an electric polarization field around the boundary of the plasma (Fig. 114a). Both CHAMBERLAIN (1961 [*508*]) and KERN (1962 [*510*]) proposed that a longitudinal gradient ($\nabla_\varphi B$) of the magnetic field will be produced by inhomogeneous plasma and also by the compressing effect of the solar plasma. The gradient is directed towards the midnight meridian on both the evening and morning sides (Fig. 114b). Charge separation is thus produced in the plasma; on the evening side positive charges drift inwards and negative charges outward; on the morning side positive charges outwards and negative charges inwards.

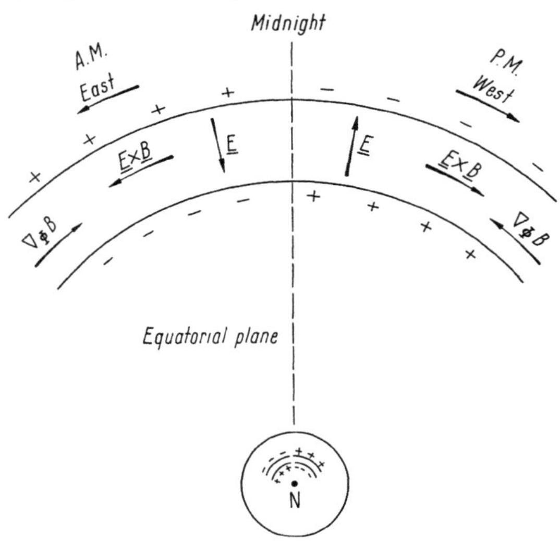

Fig. 114b. Charge separation in the equatorial plane.

On this assumption CHAMBERLAIN (1961 [*509*]) examined the motion of particles that drift to the boundaries where charge of one sign is accumulated. He found that such particles lose transverse kinetic energy, but are accelerated along the field lines, and a portion of them immediately spiral out along the geomagnetic field to the auroral ionosphere (see however COLE 1962 [*491*]).

Projecting the equatorial pattern in Fig. 114b on to the northern hemisphere, on the evening side the proton bombardment occurs to the south of the electron bombardment, and on the morning side the opposite occours. Because the boundary layer is expected to be thin, the bombarding area in the ionosphere is also narrow, suggesting a thin arc. The polarization electric field in the plasma produces an electric drift of the ionization. It is directed westward on the evening side and eastward on the morning side. He identified this motion with the east-west auroral motion (see also FEJER (1963 [*510a*])).

According to this model, however, there seems to be a discontinuity of an arc around the midnight meridian. Taking CHAMBERLAIN'S value of the thickness of the plasma shell (~1000 km) in the equatorial plane, the north-south discontinuity of an arc would be of order 50 km in the auroral zone. The analysis in Sect. 30 seems to show no such discontinuity (see also CHAMBERLAIN (1963 [*509*])).

97. The neutral point discharge model (HOYLE, DUNGEY). HOYLE (1949 [*511*], p. 102) and DUNGEY (1953 [*512*], 1958 [*10*]) assumed that some solar magnetic field, perpendicular to the equatorial plane, is carried away by the solar plasma, and that at times the earth's dipole field is immersed in such a rather uniform solar field. If so, whatever the direction of the interplanetary magnetic field may be, there will be two neutral points near the earth.

Dungey inferred that around the neutral points there appears a strong sheet current, and that lines of force from the neutral points lie on certain surfaces that intersect the earth in two closed curves. He suggested that these two lines are identified with the auroral zone.

It is not obvious how and to what extent such a model explains the aurora. He suggested further that by his model the SD current is reasonably explained.

98. The neutral line discharge model (AKASOFU, CHAPMAN, KENDALL). AKASOFU and CHAPMAN (1961 [*513*]) proposed that when the ring current belt is sufficiently enhanced, the earth's field near the center of the belt can be not merely reduced but reversed. The reversal occurs within a narrow strip of the equatorial plane, bounded by neutral lines along which the field intensity is zero. The lines are of two kinds, X and O. The magnetic lines of force cross on X lines and encircle O lines. The lines of force that pass through the X neutral line are connected with the earth. If the number density distribution is irregular and has more than one maximum, there may be more than one strip of reversed field, and corresponding pairs of neutral lines, X and O. Further, owing to the eastward current on the sunward side of the earth, at the surface of the hollow in the solar plasma, the reversal of the field is most likely to occur on the night side of the earth.

CHAPMAN and KENDALL (1962 [*460*]) showed that the pitch-angles of particles that pass near the X neutral line are scattered. There the guiding center approximation and the invariance of the dipole moments cease to be valid. Thus the pitch-angle distribution of the particles that pass near the X line changes, and ensures a continuous supply of particles with pitch-angle less than 3°, which can contribute to the thin ribbon-like auroral luminosity. When there is more than one X line, there will be more than one auroral arc.

If a new cloud of solar plasma is captured, the protons will tend to spread westward and the electrons eastward; thus an eastward electric field appears in a region containing X neutral lines. These provide an unimpeded channel for electric current flow. Elsewhere the magnetic field prevents current flow. The electric field will accelerate electrons, resulting in brighter auroras. AKASOFU and CHAPMAN (1961 [*513*]) suggested that the position, form, and movement of neutral lines determine the distribution, nature, and motion of the forms of the aurora.

The reversal of the direction of the earth's field within a limited area in the ring current belt was criticized by several workers; e.g., PARKER (1962 [*490*]) stated, "There is no reason to believe that overinflation can generate new lines of force with a reverse direction." It may be that a dip of the equatorial field intensity to a low value will suffice for the Akasofu-Chapman auroral theory.

References.

[*1*] MAIRAN, DE J. J. d'O.: Traité physique et historique de l'aurore boréale. Suite des Mém. Roy. Sci., p. 570. Paris. 1st ed. 1733; 2nd ed. 1754.
[*2*] FRITZ, H.: Das Polarlicht. 348 pp. Leipzig 1881.
[*3*] STÖRMER, C.: The Polar Aurora. 403 pp. Oxford: Oxford Univ. Press 1955.
[*4*] CHAMBERLAIN, JOSEPH W.: Physics of the Aurora and Airglow. 704 pp. New York: Academic Press 1961.
[*5*] CHAPMAN, S., and J. BARTELS: Geomagnetism. 1049 pp. Oxford: Oxford Univ. Press 1940; reprinted 1962.
[*6*] HARANG, L.: The Aurorae. 166pp. New York: John Wiley & Sons 1951.
[*7*] ARMSTRONG, E. B., and A. DALGARNO (Ed.): The Airglow and the Aurorae. 420 pp. London: Pergamon Press 1956.
[*8*] RATCLIFFE, J. A. (Ed.): Physics of the Upper Atmosphere. 586 pp. New York: Academic Press 1960.
[*9*] KUIPER, G. P. (Ed.): The Earth as a Planet. 751 pp. Chicago: Chicago Univ. Press 1954.
[*10*] DUNGEY, J. W.: Cosmic Electrodynamics. 183pp. Cambridge: Cambridge Univ. Press 1958.
[*11*] CHAPMAN, S.,: Aurora and Airglow. Annals of the IGY, vol. 4, part II, pp. 19—138. London: Pergamon Press 1957.
[*12*] AKASOFU, S.-I.: The dynamical morphology of the aurora polaris. Ann. IGY (1963, in press).
[*13*] ALFVÉN, H.: Cosmical Electrodynamics. 237 pp. Oxford: Oxford Univ. Press 1950.

Chapter I.

Sections 1—3.

[*14*] CHAPMAN, S.: Polar and tropical aurorae, and the iso-auroral diagram. Proc. Indian Acad. Sci. **37**, 175—188 (1953).
[*14a*] CHAPMAN, S.: Geomagnetic nomenclature. J. Geophys. Res. **68**, 1174 (1963).
[*15*] LOOMIS, E.: On the geographical distribution of auroras in the northern hemisphere. Amer. J. Sci. and Arts **30**, 89—94 (1860).
[*16*] VESTINE, E. H., I. LANGE, L. LAPORTE and W. E. SCOTT: The geomagnetic field, its description and analysis. Carnegie Institution of Washington, Publ. 580, Washington, D. C., 1947.
[*17*] VESTINE, E. H.: The geographic incidence of aurora and magnetic disturbance, northern hemisphere. Terr. Magn. **49**, 77—102 (1944).
[*18*] FELDSTEIN, Y. I., and E. K. SOLOMATINA: Some problems of the geographic distribution of aurorae in the northern hemisphere. Results of IGY Researches. Aurorae and Airglow No. 7, Sect. IV, IGY Program, 51—59. Publishing House Acad. Sci. USSR, Moscow (1961).
[*19*] HULTQVIST, B.: The geomagnetic field lines in higher approximation. Ark. Geofys. **3**, 63—77 (1958).
[*20*] HULTQVIST, B.: Auroral isochasms. Nature, Lond. **183**, 1478—1479 (1959).
[*21*] HULTQVIST, B.: Circular symmetry in the geomagnetic plane for auroral phenomena. Planet. Space Sci. **8**, 142—150 (1961).
[*22*] VESTINE, E. H., and W. L. SIBLEY: Lines of force of the geomagnetic field in space. Planet. Space Sci. **1**, 285—290 (1959).
[*23*] DAVIS, T. N.: The morphology of the auroral displays of 1957—1958, 1, Statistical analysis of Alaska data. J. Geophys. Res. **67**, 59—74 (1962). See also [*36*].
[*24*] DAVIES, F. T.: Observations of the aurora australis, Byrd Antarctic Expedition, 1929. Terr. Magn. **36**, 199—230 (1931).
[*25*] WHITE, F. W. G., and M. GEDDES: The Antarctic zone of maximum auroral frequency. Terr. Magn. **44**, 367—377 (1939).
[*26*] VESTINE, E. H., and E. J. SNYDER: The geographic incidence of aurora and magnetic disturbance, southern hemisphere. Terr. Magn. **50**, 105—124 (1945).

[27] GARTLEIN, C. W., B. NACK and G. SPRAGUE: Auroral observations — South pole — 1957, IGY WDC A, US Nat. Acad. Sci., IGY General Rep. No. 12, 38—56 (Washington 1960).
[28] FELDSTEIN, Y. I.: Geographical distribution of aurora and azimuth of auroral arcs. Investigations of Aurorae. Aurorae and Airglow No. 4, 61—78, Publ. House Acad. Sci. USSR, Moscow (1960).
[29] BOND, F. R., and F. JACKA: Distribution of auroras in the southern hemisphere. Austral. J. Phys. **13**, 611—612 (1960).
[30] SCHNEIDER, O.: Indices de actividad auroral. Contribution del Instituto Antarctico Argentino No. 56 (1961).

Section 4.

[31] QUENBY, J. J., and W. R. WEBBER: Cosmic ray cut-off rigidities and the earth's magnetic field. Phil. Mag. **8**, 90—113 (1959).
[32] VESTINE, E. H., and W. L. SIBLEY: The geomagnetic field in space, ring currents, and auroral isochasms. J. Geophys. Res. **65**, 1967—1980 (1960).
[33] McILWAIN, C. E.: Coordinates for mapping the distribution of magnetically trapped particles. J. Geophys. Res. **66**, 3681—3691 (1961).
[34] STONE, E. C.: The physical significance and application of L, B_0 and R_0 to geomagnetically trapped particles. J. Geophys. Res. **68**, 4157—4166 (1963).

Sections 5 and 6.

[35] FELDSTEIN, Y. I.: Magnetic disturbances and aurorae in the nearpolar region of the northern hemisphere. J. Phys. Soc. Japan **17**, Suppl. A—I, 225—226 (1962).
[36] DAVIS, T. N.: The morphology of the auroral displays of 1957—1958, 2, Detailed analysis of Alaska data and analysis of high-latitude data. J. Geophys. Res. **67**, 75—110 (1962).
[37] LASSEN, K.: Existence of an inner auroral zone. Nature, Lond. **184**, 1375—1377 (1959).
[38] NIKOLSKY, A. P.: On the position of the second auroral zone and the correlation of the morning aurorae with the magnetic disturbances. Results of IGY Researches. Aurorae and Airglow No. 7, 37—42, Publ. House Acad. Sci. USSR, Moscow (1961).
[39] DENHOLM, J. V.: Some auroral observations inside the southern auroral zone. J. Geophys. Res. **66**, 2105—2111 (1961).
[40] HATHERTON, T.: Geometry of the southern auroral zone and the evidence for the existence of an inner zone. Nature, Lond. **186**, 288—290 (1960).
[41] HATHERTON, T., and G. G. MIDWINTER: Observations of the aurora australis at New Zealand antarctic stations during IGY. J. Geophys. Res. **65**, 1401—1412 (1960).
[42] KHOROSHEVA, O. V.: Investigating planetary propagation of solar aurorae, Abstract, XIII General Assembly of IUGG, Berkeley, California, August 1963.
[43] ALFVÉN, H.: On the electric field theory of magnetic storms and aurora. Tellus **7**, 50—64 (1955).
[44] McINNES, B., and K. A. ROBERTSON: Latitude distribution and seasonal variation of aurora over the British Isles during 1957 and 1958. J. Atmosph. Terr. Phys. **19**, 115—125 (1960).
[45] ABBOTT, W. N.: The aurora of August 19, 1950, photographed in Greece. J. Atmosph. Terr. Phys. **1**, 343—344 (1951).
[46] AKASOFU, S.-I., and S. CHAPMAN: Large-scale auroral motions and polar magnetic disturbances-III, the aurora and magnetic storm of 11 February 1958. J. Atmosph. Terr. Phys. **24**, 785—796 (1962). For part I, see [*100*]; part II, see [*131*].
[47] SCHLOBOHM, J. C., R. L. LEADABRAND, R. B. DYCE, L. T. DOLPHIN and M. R. BERG: High-altitude 106.1-Mc/s radio echoes from auroral ionization detected at a geomagnetic latitude of 43°. J. Geophys. Res. **64**, 1191—1196 (1959).

Section 7.

[48] AKASOFU, S.-I., and S. CHAPMAN: The lower limit of latitude (US sector) of northern quiet auroral arcs, and its relation to Dst(H). J. Atmosph. Terr. Phys. **25**, 9—12 (1963).
[49] DAVIS, T. N., and D. S. KIMBALL: Incidence of auroras and their north-south motions in the northern auroral zone. Geophys. Inst. Univ. Alaska, UAG-R 100 (1960).
[50] MEINEL, A. B., B. J. NEGAARD and J. W. CHAMBERLAIN: A statistical analysis of low-latitude aurorae. J. Geophys. Res. **59**, 407—413 (1954).
[51] GARTLEIN, C. W., and R. K. MOORE: Southern extent of aurora borealis in North America. J. Geophys. Res. **56**, 85—96 (1951).
[52] DAVIES, F. T.: Visual auroral observations in Canada, 1943—1947. IATME (Int. Assoc. Terr. Magn. Electr.) Bull. No. 13. Transactions of Oslo meeting, 1948, pp. 255—273.

[53] PATON, J., and B. McINNES: Auroral Notes, Observatory **74**, 46—48, (1954).
[54] ELVEY, C. T., H. LEINBACH, J. HESSLER and J. NOXON: Preliminary studies of the distribution of auroras in Alaska. Trans. Amer. Geophys. Union **36**, 390—394 (1955).
[55] EVANS, S., and G. M. THOMAS: The southern auroral zone in geomagnetic longitude sector 20° E. J. Geophys. Res. **64**, 1381—1388 (1959).
[56] SHERET, M. A., and G. M. THOMAS: Auroral observations at Halley Bay, Antarctica, during 1959. Nature, Lond. **189**, No. 4767, 826 (1961).

Chapter II.

Sections 8—17.

[57] ELVEY, C. T.: Problems in auroral morphology. Proc. Nat. Acad. Sci. **43**, 63—75 (1957).
[58] AKASOFU, S.-I.: The dynamic morphology of the aurora polaris. J. Geophys. Res. **68**, 1667—1673 (1963).
[58a] KHOROSHEVA, O. V.: The space and time-distribution of auroras and their relationship with high-latitude geomagnetic disturbances. Geomagnetism and Aeronomy **1**, 615—621 (1961).
[59] GARTLEIN, C. W.: The visual auroral program. Trans. Amer. Geophys. Union **39**, 596—601 (1958).
[60] VEGARD, L.; see BATES, D. R., Chap. 6 Aurorae: Their characters and spectra p. 218. Phys. of the Upper Atmosph. 1960 [8].
[61] AKASOFU, S.-I.: Thickness of an active auroral curtain. J. Atmosph. Terr. Phys. **21**, 287—288 (1961).
[62] McEWEN, D. J., and R. MONTALBETTI: Parallactic measurements on aurorae over Churchill, Canada. Canad. J. Phys. **36**, 1593—1600 (1958).
[63] STÖRMER, C.: Statistics of heights of various auroral forms from southern Norway. Terr. Magn. **53**, 251—264 (1948).
[64] CURRIE, B. W.: Summary of some auroral height measurements and observations at Chesterfield, Canada. Terr. Magn. **39**, 293—297 (1934).
[65] REES, M. H., and C. S. DEEHR: The aurora of 27 November 1959 at College, Alaska including observations of a high altitude red arc. Planet. Space Sci. **8**, 49—58 (1961).
[66] VEGARD, L.: Altitude effects in the red part of the auroral spectrum and the two types of red aurorae. Nature, Lond. **141**, 200 (1938).
[67] CLARK, K. C., and A. E. BELON: Spectroscopic observations of the great aurora of February 10, 1958, Part I. Abnormal vibration of N_2^+. J. Atmosph. Terr. Phys. **16**, 205—219 (1959).
[68] DAHLSTROM, C. E., and D. M. HUNTEN: O_2^+ and H in the auroral spectrum. Phys. Rev. **84**, 378—379 (1951).
[69] HARANG, L., and W. BAUER: Über einen Nordlichtbogen in weniger als 80 km Höhe über der Erde. Gerlands Beitr. Geophys. **37**, 109—115 (1932).
[70] VEGARD, L., and O. KROGNESS: The position in space of the aurora polaris from observations made at the Haldde Observatory, 1913—1914. Geofys. Publ. I, No. 1, 1—172 (Oslo 1920).
[71] ABBOTT, W. N.: Displacements of the radiant point during the auroral disturbance of September 22, 1957. Canad. J. Phys. **36**, 643—648 (1958).
[72] GADSDEN, M., and C. G. LOUGHNAN: Observations of non-field-aligned auroral rays. J. Atmosph. Terr. Phys. **18**, 332—334 (1960).
[73] WEBSTER, H. F.: Structure in magnetically confined electron beams. J. Appl. Phys. **28**, 1388—1397 (1957).
[74] GOULD, R. W.: The dynamics of electron beams 78—118, Plasma dynamics (ed. by F. H. CLAUSER). Reading, Mass.: Addison-Wesley Publ. Co. 1960.
[75] AKASOFU, S.-I.: The auroral rays. J. Atmosph. Terr. Phys. **25**, 163—165 (1963).
[76] CURRIE, B. W., and C. K. JONES: Directional and diurnal characteristics of auroras at some places in Canada. Terr. Magn. **46**, 269—278 (1941).
[77] JENSEN, R. E., and B. W. CURRIE: Orientations of auroral displays in west-central Canada. J. Geophys. Res. **58**, 201—210 (1953).
[78] DAVIS, T. N.: The morphology of the polar aurora. J. Geophys. Res. **65**, 3497—3500 (1960).
[79] WEILL, G M.: Aspects de l'aurore observée à la base Dumont d'Urville en Terre Adelie. C. R. Acad. Sci., Paris **246**, 2925—2927 (1958).
[80] COLE, K. D.: Hydromagnetic radiation of the sun and its effect at the earth. J. Phys. Soc. Japan **17**, A—II. Internat. Conference on Cosmic Rays and the Earth Storm, Part II, p. 6, 1962.

[81] CHAPMAN, S.: The electric current-systems of magnetic storms. Terr. Magn. **40**, 349—370 (1935).
[82] VESTINE, E. H., and W. L. SIBLEY: Remarks on auroral isochasms. J. Geophys. Res. **64**, 1338—1339 (1959).
[83] HULTQVIST, B.: On the orientation of auroral arcs. J. Atmosph. Terr. Phys. **24**, 17—30 (1962).
[84] JACKA, F., and J. PATON: IQSY Instruction Manual No. 3, Aurora, Issued by IQSY Secretariat, 6 Cornwall Terrace, London, NW 1, England 1963.
[85] HURUHATA, M.: Aurora and airglow observations on February 11, 1958. Rep. Ionosph. Res., Japan **12**, 40—41 (1958).
[86] MANRING, E. R., and H. B. PETTIT: Photometric observations of the 5577 A and 6300 A emissions made during the aurora of February 10—11, 1958. J. Geophys. Res. **64**, 149—153 (1959).
[87] KRASOVSKII, V. I.: Results of investigations of aurorae and night airglow during the IGY and IGC. Planet. Space Sci **8**, 125—141 (1961).

Chapter III.

Sections 18—20.

[88] HEPPNER, J. P.: Time sequences and spatial relations in auroral activity during magnetic bays at College, Alaska. J. Geophys. Res. **59**, 329—338 (1954).
[89] FULLER, V. R.: A report of work on the aurora borealis for the years 1932—1934. Terr. Magn. **40**, 269—275 (1935).
[90] LASSEN, K.: Local aurorae in the morning hours at Godhavn, Greenland. Det Danske Meteorologiske Institut, Comm. Mag. No. 24, Charlottenlund, 1959.
[91] MALVILLE, J. M.: Antarctic auroral observations, Ellsworth Station, 1957. J. Geophys. Res. **64**, 1389—1393 (1959).
[91a] MOROZUMI: State Univ. of Iowa (Dept. of Physics and Astronomy).
[92] NIKOL'SKI, A. P.: Dual laws of the course of magnetic disturbance and the nature of mean regular variations. Terr. Magn. **52**, 147—173 (1947).
[93] MEEK, J. H.: The location and shape of the auroral zone. J. Atmosph. Terr. Phys. **6**, 313—321 (1955).
[94] THOMAS, L., and W. R. PIGGOTT: Some aspects of the incidence of polar blackout during the IGY, URSI Monographs. Some ionospheric results obtained during the International Geophysical Year, ed. W. J. G. Beynon, 61—71, Amsterdam, Elsevier Pub. 1960.
[95] AGY, V.: Polar blackout occurrence pattern, Polar atmospheric symposium, part II, Ionospheric section, pp. 129—134. J. Atmosph. Terr. Phys., Special Suppl. London: Pergamon Press 1957.
[96] AGY, V.: The types of blackout, their time variations, and the mechanisms producing them. J. Phys. Soc. Japan **17**, Suppl. A—I. Internat. Conference on Cosmic Rays and the Earth Storm, part I, pp. 93—97, 1962.
[97] DAVIS, T. N., and R. N. DE WITT: Twenty-four hour observations of aurora at the southern auroral zone. J. Geophys. Res. **68**, 6237—6241 (1963).

Sections 21—27.

[98] CHAPMAN, S.: Auroral observations in India and Pakistan. Bull. Nat. Inst. Sci. India Nr. 9, 180—192 (1957).
[99] KIM, J. S., and B. W. CURRIE: Horizontal movements of aurora. Canad. J. Phys. **36**, 160—170 (1958). See also [105].
[100] AKASOFU, S.-I.: Large-scale auroral motions and polar magnetic disturbances. I. J. Atmosph. Terr. Phys. **19**, 10—25 (1960). For part II, see [131]; part III [46].
[101] MEINEL, A. B., and D. H. SCHULTE: A note on auroral motions. Astrophys. J. **117**, 454—455 (1953).
[102] BLESS, R. C., C. W. GARTLEIN and D. S. KIMBALL: East-west motions in the aurora. Astrophys. J. **122**, 205—206 (1955).
[103] MEEK, J. H.: East-west motion of aurorae. Astrophys. J. **120**, 602—603 (1954).
[104] EVANS, S.: Systematic movements of aurorae at Halley Bay. Proc. Roy. Soc. Lond. A **256**, 234—240 (1960).
[105] KIM, J. A., and B. W. CURRIE: Further observations of the horizontal movements of aurora. Canad. J. Phys. **38**, 1366—1376 (1960). See also [99].
[106] BHATTACHARYYA, B. K.: Correlation studies of radio aurora, magnetic and earth-current disturbances. Canad. J. Phys. **38**, 624—637 (1960).
[107] BULLOUGH, K., T. W. DAVIDSON, T. R. KAISER and C. D. WATKINS: Radio reflections from aurorae-III. The association with geomagnetic phenomena. J. Atmosph. Terr. Phys. **11**, 237—254 (1957). See also [139].

[108] Nichols, B.: Drift motions of auroral ionization. J. Atmosph. Terr. Phys. **11**, 292—293 (1957).
[109] Lyon, G. F., and A. Kavadas: Horizontal motions in radar echoes from aurora. Canad. J. Phys. **36**, 1661—1671 (1958).
[110] Brown, R. R.: West-east motion of an auroral-zone X-ray event. J. Geophys. Res. **67**, 31—35 (1962).
[111] Stoffregen, W.: The east-west drift of auroral forms determined from all-sky camera films. J. Atmosph. Terr. Phys. **21**, 257—260 (1961).
[112] Omholt, A.: Velocities of very active auroral rays. Planet. Space Sci. **9**, 285—286 (1962).
[113] Davis, T. N.: Negative correlation between polar-cap visual aurora and magnetic activity. J. Geophys. Res. **68**, 4447—4453 (1963).
[114] De Witt, R. N.: The occurrence of aurora in geomagnetically conjugate areas. J. Geophys. Res. **67**, 1347—1352 (1962).
[115] Anderson, K. A., C. D. Anger, R. R. Brown and D. S. Evans: Simultaneous electron precipitation in the northern and southern auroral zones. J. Geophys. Res. **67**, 4076—4078 (1962)
[116] Nagata, T., and S. Kokobun: On the earth storms, 4. Polar magnetic storms with special reference to relation between geomagnetic disturbances in the northern and southern auroral zones. Rep. Ionosphere Res. Japan **14**, 273—290 (1960).
[117] Wescott, E.: Magnetic activity during periods of auroras at geomagnetically conjugate points. J. Geophys. Res. **67**, 1353—1355 (1962).
[118] Hook, J. L.: Some observations of ionospheric absorption at geomagnetic conjugate stations in the auroral zone. J. Geophys. Res. **67**, 115—122 (1962).
[119] Leinbach, H., and R. P. Basler: Ionospheric absorption of cosmic radio noise at magnetically conjugate auroral zone stations J. Geophys. Res. **68**, 3315—3382 (1963).
[120] Wilson, C. R.: Hydromagnetic interpretation of sudden commencements of geomagnetic storms. Ph. D. Thesis, Univ. of Alaska (1963).

Chapter IV.

Sections 32 and 33.

[121] Lovering, J.: On the periodicity of the aurora borealis. Mem. Amer. Acad. of Arts and Sciences, **10**, (1868—1871) (includes catalogues of auroras from B. C. 502—1869) (1870).
[122] Newton, H. W., and A. S. Milsom: The distribution of great and small geomagnetic storms in the sunspot cycle. J. Geophys. Res. **59**, 203—214 (1954).
[123] Parker, E. N.: Dynamics of the interplanetary gas and magnetic fields. Astrophys. J. **128**, 664—676 (1958); see also Interaction of the solar wind with the geomagnetic field. Phys. Fluids **1**, 171—187 (1958).
[124] Neugebauer, M., and C. W. Snyder: Preliminary results from Mariner II. Solar Plasma Experiment. Science **138**, 1095—1097 (1962).
[124a] Bonetti, A., H. S. Bridge, A. J. Lazarus, B. Rossi and F. Scherb: Explorer 10 plasma measurements. J. Geophys. Res. **68**, 4017—4063 (1963).
[125] Akasofu, S.-I., J. C. Cain and S. Chapman: The magnetic field of the quiet-time proton belt. J. Geophys. Res. **67**, 2645—2647 (1962).
[126] Coleman jr., P. J., C. P. Sonett and L. Davis jr.: On the interplanetary magnetic storm: Pioneer V. J. Geophys. Res. **66**, 2043—2046 (1961).
[127] Chapman, S., and V. C. A. Ferraro: A new theory of magnetic storms. Terr. Magn. **36**, 77—97, 171—185 (1931); **37**, 147—156, 421—429 (1932).
[128] Bartels, J.: Collection of geomagnetic planetary indices *Kp* and derived daily indices *Ap* and *Cp* for the years 1932 to 1961. IAGA Bull. No. 18 (1962).
[129] Bartels, J.: Discussion of time-variations of geomagnetic activity, indices *Kp* and *Ap*, 1932—1961. Ann. Géophys. **19**, 1—20 (1963).
[130] Bartels, J., and S. Chapman: A 21-years series of geomagnetic activity data, represented in connection with the appearance of aurora in lower latitudes. Beitr. Int. Geophys. Jahr Nr. 1 (Abhandl. Akad. Wiss. Göttingen, math.-phys. Kl.), 6—37 (1958).

Sections 34—37.

[131] Akasofu, S.-I.: Large-scale auroral motions and polar magnetic disturbances-II. The changing distribution of the aurora during large magnetic storms. J. Atmosph. Terr. Phys. **24**, 723—724 (1962). For part I, see [100]; part II [46].
[132] Birkeland, K.: The Norwegian aurora polaris expedition, 1902—1903, vol. 1, 801 pp. Aschehoug, Christiania, Norway 1908 and 1913.

[133] CHAPMAN, S.: The equatorial electrojet as detected from the abnormal electric current distribution above Huancayo, Peru, and elsewhere. Arch. Met. Geophys., Wien **4**, 368—390 (1951).

[134] ONWUMECHILLI, C. A.: A study of the equatorial electrojet. II. A model electrojet that fits *H* observations. J. Atmosph. Terr. Phys. **13**, 235—257 (1959).

[135] BAKER, W. G., and D. F. MARTYN: Electric currents in the ionosphere. I. The conductivity. Phil. Trans. Roy. Soc. Lond. A **246**, 281—294 (1953).

[136] MAEDA, H.: The vertical distribution of electrical conductivity in the upper atmosphere. J. Geomag. Geoelectr. **5**, 94—104 (1953).

[137] FEJER, J. A.: Semidiurnal currents and electron drifts in the ionosphere. J. Atmosph. Terr. Phys. **4**, 184—203 (1953).

[138] KAISER, T. R., see [6], p. 156, 1955.

[139] BULLOUGH, K., and T. R. KAISER: Radio reflections from aurorae. J. Atmosph. Terr. Phys. **5**, 189—200 (1954); **6**, 198—214 (1955). See also [107].

[140] DAGG, M.: The correlation of radio-star-scintillation phenomena with geomagnetic disturbances and the mechanism of motion of the ionospheric irregularities in the *F* region. J. Atmosph. Terr. Phys. **10**, 194—203 (1957).

[141] NICHOLS, B.: Auroral ionization and magnetic disturbances. Proc. Inst. Radio Engrs. **47**, 245—254 (1959).

[142] CHAMBERLAIN, J. W., J. KERN and E. H. VESTINE: Some consequences of local acceleration of auroral primaries. J. Geophys. Res. **65**, 2535—2537 (1960).

[143] MARTYN, F. D.: Electric currents in the ionosphere. III. Ionization drift due to winds and electric fields. Phil. Trans. Roy. Soc. Lond. A **246**, 306—320 (1954).

[144] VESTINE, E. H.: Disturbance field of magnetic storms. Trans. Wash. Assem. 1939, Publ. IATME (= Int. Assoc. Terr. Magn. Electr.) Bull. No. 11, 360—381 (1940).

[145] SILSBEE, H. C., and E. H. VESTINE: Geomagnetic bays, their frequency and current-systems. Terr. Magn. **47**, 195—208 (1942).

[146] FUKUSHIMA, N.: Polar magnetic storms and geomagnetic bays. J. Fac. Sci. Tokyo Univ. **8**, 293—412 (1953).

[147] WHITHAM, K., E. I. LOOMER and E. R. NIBLETT: The latitudinal distribution of magnetic activity in Canada. J. Geophys. Res. **65**, 3961—3974 (1960).

[148] BHATTACHARYYA, B. K.: Dipole-field type magnetic disturbances and auroral activities. Canad. J. Phys. **39**, 350—366 (1961).

[149] AKASOFU, S.-I., and S. CHAPMAN: The enhancement of the equatorial electrojet during polar magnetic storms. J. Geophys. Res. **68**, 2375—2382 (1963).

[150] GARTLEIN, C. W., and G. SPRAGUE: The aurora and the local magnetic field. Nat. Acad. Sci. IGY World Data Center A, General Rep. No. 12, 57—67 (1960).

[151] DAVIS, T. N.: Negative correlation between polar-cap visual aurora and magnetic activity. J. Geophys. Res. **68**, 4447—4453 (1963).

[152] AKASOFU, S.-I., and D. S. KIMBALL: Polar cap auroras (in preparation).

[153] AKASOFU, S.-I., and S. CHAPMAN: The ring current, geomagnetic disturbance, and the Van Allen radiation belts. J. Geophys. Res. **66**, 1321—1350 (1961).

[154] AKASOFU, S.-I., J. C. CAIN and S. CHAPMAN: The magnetic field of a model radiation belt, numerically computed. J. Geophys. Res. **66**, 4013, 4026 (1961).

[155] AKASOFU, S.-I.: The main phase of magnetic storms and the ring current. Space Sci. Rev. **2**, 91—135 (1963).

[156] KNAPP, D. G.: Some features of magnetic storms in high latitudes. J. Geophys. Res. **66**, 2053—2085 (1961).

[157] AKASOFU, S.-I., S. CHAPMAN and D. VENKATESAN: The development of the main phase of great magnetic storms. J. Geophys. Res. **68**, 3345—3350 (1963).

[158] AKASOFU, S.-I., and S. CHAPMAN: The development of the main phase of magnetic storms. J. Geophys. Res. **68**, 125—129 (1963); **69**, 1025 (1964).

[159] AKASOFU, S.-I., and S. CHAPMAN: Magnetic Storms: The simultaneous development of the main phase (DR) and of polar magnetic substorms (DP). J. Geophys. Res. **68**, 3155—3158 (1963).

[160] BARTELS, J., and N. FUKUSHIMA: A *Q*-index for magnetic activity in quarter-hourly intervals. Abh. Akad. Wiss. Göttingen, math.-phys. Kl., Sonderheft Nr. 2, 36 pp. (1956).

[161] BARTELS, J.: The technique of scaling *Q*-indices. Ann. of IGY **4**, 220—222 (1957).

[162] CAMPBELL, W. H.: Magnetic micropulsations and the pulsating aurora. J. Geophys. Res. **65**, 784—789 (1960).

[163] CAMPBELL, W. H., and M. H. REES: A study of auroral coruscations. J. Geophys. Res. **66**, 41—55 (1961).

[164] CAMPBELL, W. H., and H. LEINBACH: Ionospheric absorption at times of auroral and magnetic pulsations. J. Geophys. Res. **66**, 25—34 (1961).

[164a] TROITSKAYA, V. A.: Pulsation of the earth's electromagnetic field with periods of 1 to 5 seconds and their connection with phenomena in the high atmosphere. J. Geophys. Res. **66**, 5—18 (1961).
[165] FLOCK, W. L., A. E. BELON and R. R. HEACOCK: Geomagnetic agitation and overhead aurora. Proc. Int. Conf. Ionosphere, ed. A. C. STICKLAND, London, Institute of Phys. Soc. (1963).
[166] HEACOCK, R. R.: Notes on pearl-type micropulsations. J. Geophys. Res. **68**, 589—591 (1963).
[167] ELLIS, G. R. A.: Low frequency electromagnetic radiation associated with magnetic disturbances. Planet. Space Sci. **1**, 253—258 (1959).

Chapter V.

Sections 38—40.

[168] APPLETON, E. V., R. NAISMITH and L. J. INGRAM: British radio observations during the second international polar year, 1932/33. Phil. Trans. Roy. Soc. Lond. A **236**, 191—259 (1937).
[169] HARANG, L.: Änderungen der Ionisation der höchsten Atmosphärenschichten während der Nordlichter und erdmagnetischen Störungen. Gerlands Beitr. Geophys. **46**, 438—454 (1936).
[170] MEEK, J. H.: Correlation of magnetic, auroral, and ionospheric variations at Saskatoon. J. Geophys. Res. **58**, 445—456 (1953); **59**, 87—92 (1954).
[171] HEPPNER, J. P., E. C. BYRNE and A. E. BELON: The association of absorption and E_s ionization with aurora at high latitudes. J. Geophys. Res. **57**, 121—134 (1952).
[172] HARANG, L., and W. STOFFREGEN: Scattered reflections of radio waves from a height of more than 1000 km. Nature, Lond. **142**, 832 (1938).
[173] BOOKER, H. G.: Radar studies of the aurora, Chap. 8 in [8].
[174] MAEHLUM, B.: The sporadic *E* auroral zone. Geofysiske Publikasjoner **23**, No. 1 Oslo (1962).
[175] MONTALBETTI, R., and D. J. MCEWEN: Hydrogen emissions and sporadic *E* layer behaviour. J. Phys. Soc. Japan **17**, Suppl. A—I, 212—215 (1962).
[176] OMHOLT, A., W. STOFFREGEN and H. DERBLOM: Hydrogen lines in auroral glow. J. Atmosph. Terr. Phys. **24**, 203—211 (1962).
[177] MEEK, J. H., and A. G. MCNAMARA: Magnetic disturbances, sporadic *E* and radio echoes associated with the aurora. Canad. J. Phys. **32**, 326—329 (1954).
[178] POTAPOV, B. P., Z. RAPPOPORT and T. B. BARSUK: Study of the absorption of radio waves in the auroral zone. Results of IGY Researches. Aurorae and Airglow No. 2—3, 42 (Moscow 1960).
[179] BATES, H. F.: The slant *Es* echo — a high-frequency auroral echo. J. Geophys. Res. **66**, 447—454 (1961).
[180] PIGGOTT, W. R.: Abnormalities in the ionosphere at high latitudes. Nature, Lond. **171**, 124—125 (1953).
[181] AGY, V.: Geographic and temporal distribution of polar blackouts. J. Geophys. Res. **59**, 499—512 (1954).
[182] AGY, V.: The location of the auroral absorption zone. J. Geophys. Res. **59**, 267—272 (1954).
[183] PIGGOTT, W. R., and L. THOMAS: Variations in the incidence of polar blackout with magnetic activity; Some ionospheric results obtained during IGY, ed. W. J. G. BEYNON, 72—77. Amsterdam: Elsevier Pub. Co. 1960.
[184] AGY, V.: Spiral patterns in geophysics. J. Atmosph. Terr. Phys. **19**, 136—140 (1957).
[185] STOFFREGEN, W., H. DERBLOM and A. OMHOLT: Some characteristics of the *D*-region ionization during auroral activity. J. Geophys. Res. **65**, 1699—1704 (1960).
[186] LITTLE, C. G.: The measurement of ionospheric absorption using extra-terrestrial radio waves. Ann. of IGY, **3**, part II, 207 London: Pergamon Press 1957.
[187] LITTLE, C. G., and H. LEINBACH: Some measurements of high-latitude ionospheric absorption using extra terrestrial radio waves. Proc. Inst. Radio Engrs. **46**, 334—348 (1958).
[188] BASLER, R. P.: Radio wave absorption in the auroral ionosphere. J. Geophys. Res. **68**, 4665—4681 (1963).
[189] HARTZ, T. R., L. E. MONTBRIAND and E. L. VOGAN: A study of auroral absorption at 30 Mc/s. Canad. J. Phys. **41**, 581—595 (1963).
[190] HOLT, C., B. LANDMARK and F. LIED: Analysis of riometer observations obtained during polar radio blackouts. J. Atmosph. Terr. Phys. **23**, 229—243 (1961).
[191] ZIAUDDIN, S., and P. A. FORSYTH: Three-frequency measurements of auroral absorptions. J. Geophys. Res. **66**, 2315—2319 (1961).

[192] PARTHASARATHY, R., G. M. LERFALD and C. G. LITTLE: Derivation of electron-density profiles in the lower ionosphere using radio absorption measurements at multiple frequencies. J. Geophys. Res. **68**, 3581—3588 (1963).

[193] CHAPMAN, S., and C. G. LITTLE: The nondeviative absorption of high-frequency radio waves in auroral latitudes. J. Atmosph. Terr. Phys. **10**, 20—31 (1957).

[194] ANSARI, Z. A.: The spatial and temporal variations in high latitude cosmic noise absorption and their relation to luminous aurora. Ph. D. Thesis, Univ. of Alaska (1963).

[195] JELLY, D. H., and A. G. MATHEWS and C. COLLINS: Study of polar cap and auroral absorption. J. Atmosph. Terr. Phys. **23**, 206—215 (1961).

[196] LEINBACH, H., and G. C. REID: Ionization of the upper atmosphere by low-energy charged particles from a solar flare. Phys. Rev. Letters **2**, 61—63 (1959).

[197] HAKURA, Y., and T. GOH: Pre-SC polar cap ionospheric blackouts and Type IV solar radio out-burst. J. Radio Res. Lab. Tokyo **6**, 635—650 (1959).

[198] BAILEY, D. K.: Abnormal ionization in the lower ionosphere associated with cosmic-ray flux enhancements. Proc. Inst. Radio Engrs. **47**, 255—266 (1959).

[199] ANDERSON, K. A.: Ionizing radiation associated with solar radio noise storm. Phys. Rev. Letters **1**, 335—337 (1958).

[200] ROTHWELL, P., and C. E. MCILWAIN: Satellite observations of solar cosmic rays. Nature, Lond. **184**, 138—140 (1959).

[201] VAN ALLEN, J. A., and W. C. LIN: Outer radiation belt and solar proton observations with Explorer VII during March-April 1960. J. Geophys. Res. **65**, 2998—3003 (1960).

[202] BRYANT, D. A., T. L. CLINE, U. D. DESAI and F. B. MCDONALD: Solar cosmic rays following the flare of September 28, 1961. J. Geophys. Res. **67**, 4983—5000 (1962).

[203] BROWN, R. R., and R. G. D'ARCY: Observations of solar flare radiation at high latitudes during the period July 10—17, 1959. Phys. Rev. Letters **3**, 390—392 (1959).

[204] WINCKLER, J. R., and P. D. BHAVSAR: Low energy solar cosmic rays and the geomagnetic storm of May 12, 1959. J. Geophys. Res. **65**, 2637—2655 (1960).

[205] WINCKLER, J. R., P. D. BHAVSAR and L. PETERSON: The time variations of solar cosmic rays during July 1959, at Minneapolis. J. Geophys. Res. **66**, 995—1022 (1961).

[206] ANDERSON, K. A., R. ARNOLDY, R. HOFFMAN, L. PETERSON and J. R. WINCKLER: Observations of low-energy solar cosmic rays from the flare of 22 August 1958. J. Geophys. Res. **64**, 1133—1147 (1959).

[207] PFOTZER, G., A. EHMERT and E. KEPPLER: Time pattern of ionizing radiation in balloon altitudes in high latitudes. Mitt. Max-Planck-Inst. f. Aeronomie No. 9, parts A and B (1962).

[208] OGILVIE, K. W., D. A. BRYANT and L. R. DAVIS: Rocket observations of solar protons during the November 1960 events, I. J. Geophys. Res. **67**, 929—937 (1962).

[209] DAVIS, L. R., and K. W. OGILVIE: Rocket observations of solar protons during the November 1960 events, 2. J. Geophys. Res. **67**, 1711 (1962).

[210] LIN, W. C., and J. A. VAN ALLEN: Solar proton events observed with Explorer 7. Nuovo Cimento, in press (1964).

[211] MAEHLUM, B., and B. J. O'BRIEN: Solar cosmic rays of July 1961 and their ionospheric effects. J. Geophys. Res. **67**, 3269—3280 (1962).

[212] PIEPER, G. F., A. J. ZMUDA, C. O. BOSTROM and B. J. O'BRIEN: Solar protons and magnetic storms in July 1961. J. Geophys. Res. **67**, 4959—4981 (1962).

[213] VAN ALLEN, J. A.: Injun I — Explorer XII observations of solar cosmic rays, September 28 to October 4, 1961, Paper presented at the 43rd AGU Meeting, April 1962.

[214] FREIER, P. S., E. P. NEY and J. R. WINCKLER: Balloon observation of solar cosmic rays on March 25, 1958. J. Geophys. Res. **64**, 685—688 (1959).

[215] NEY, E. P., J. R. WINCKLER and P. S. FREIER: Protons from the sun on May 12, 1959. Phys. Rev. Letters **3**, 183 (1959).

[216] TREIMAN, S. B.: Effect of equatorial ring current on cosmic ray intensity. Phys. Rev. **89**, 130—133 (1953).

[217] RAY, E. C.: Effects of a ring current on cosmic ray intensity. Phys. Rev. **101**, 1142—1148 (1956).

[218] ROTHWELL, P.: Magnetic cutoff rigidities of charged particles in the earth's field at times of magnetic storms. J. Geophys. Res. **64**, 2026—2028 (1959).

[219] OBAYASHI, T., and Y. HAKURA: Propagation of solar cosmic rays through interplanetary magnetic field. J. Geophys. Res. **65**, 3143—3148 (1960).

[220] KELLOGG, P. J., and J. R. WINCKLER: Cosmic ray evidence for a ring current. J. Geophys. Res. **66**, 3991—4001 (1961).

[221] AKASOFU, S.-I., and W. C. LIN: The magnetic moment of model ring current belts and the cut-off rigidity of solar protons. J. Geophys. Res. **68**, 913—911 (1963).

[222] AKASOFU, S.-I., J. A. VAN ALLEN and W. C. LIN: The anomalous entry of non-relativistic solar protons into the geomagnetic field. J. Geophys. Res. **68**, 5327—5338 (1963).
[223] WEBBER, W. R.: The motion of low-rigidity cosmic rays in the earth's magnetic field and effects of external fields. J. Geophys. Res. **68**, 3065—3085 (1963).

Sections 41—43.

[224] BULLOUGH, K., and T. R. KAISER: Radio reflections from aurorae. J. Atmosph. Terr. Phys. **5**, 189—200 (1954). See also [107], [139].
[225] KAISER, T. R.: The geometry of auroral ionization. J. Gephys. Res. **62**, 297—298 (1957).
[226] UNWIN, R. S.: The geometry of auroral ionization. J. Geophys. Res. **63**, 501—506 (1958).
[227] SEED, T. J.: V. H. F. observations on the aurora australis. J. Geophys. Res. **63**, 517—526 (1958).
[228] POGORELOV, V. I.: A short review of the results of radar observations at Roschino Station. Spectral, Electrophotometric and Radar Researches of Aurorae and Airglow, No. 2—3, 32—35, Publ. House Acad. Sci. USSR, Moscow 1960.
[229] CHAPMAN, S.: The geometry of radio echoes from aurorae. J. Atmosph. Terr. Phys. **3**, 1—29 (1952).
[230] MOORE, R. K.: A VHF propagation phenomenon associated with aurora. J. Geophys. Res. **56**, 97—106 (1951).
[231] BOOKER, H. G., C. W. GARTLEIN and B. NICHOLS: Interpretations of radio reflections from the aurora. J. Geophys. Res. **60**, 1—22 (1955).
[232] BOOKER, H. G.: A theory of scattering by nonisotropic irregularities with application to radar reflections from the aurora. J. Atmosph. Terr. Phys. **8**, 204—221 (1956).
[233] BARKER, D., H. K. SUTCLIFFE and C. D. WATKINS: Some radar observations of meteors and aurorae at 300 and 500 Mc/s using a large radio telescope. II. J. Atmosph. Terr. Phys. **24**, 599—607 (1962).
[234] BOWLES, K. L.: Doppler shifted radio echoes from aurorae. J. Geophys. Res. **59**, 553—555 (1954).
[235] DYCE, R. B.: Auroral echoes observed north of the auroral zone on 51.9 Mc/sec. J. Geophys. Res. **60**, 317—323 (1955).
[236] OWREN, L.: High-latitude radio aurora. Ionospheric Radio, ed. W. J. G. BEYNON, 160—176. Amsterdam: Elsevier Pub. Co. 1961.
[237] HERLOFSON, N.: Interpretation of radio echoes from polar auroras. Nature, Lond. **160**, 867—868 (1947).
[238] FRICKER, S. J., R. P. INGALLS, M. L. STONE and C. S. WANG: UHF radar observations of aurora. J. Geophys. Res. **62**, 527—546 (1957).
[239] CURRIE, B. W., P. A. FORSYTH and F. E. VAWTER: Radio reflections from aurora. J. Geophys. Res. **58**, 179—200 (1953).
[240] HARANG, L., and B. LANDMARK: Radio echoes observed during aurorae and terrestrial-magnetic storms using 35 Mc/s and 74 Mc/s waves simultaneously. J. Atmosph. Terr. Phys. **4**, 322—338 (1954).
[241] GADSDEN, M.: Studies of the upper atmosphere from Invercargill, New Zealand. Part III. Radar echoes and visual aurora. Ann. Géophys. **15**, 403—411 (1959).
[242] KAISER, T. R., and K. BULLOUGH: Radio echoes from aurorae. Ann. Géophys. **11**, 279—283 (1955).
[243] HELLGREN, G., and J. MEOS: Localization of aurorae with 10 m high power radar technique, using a rotating antenna. Tellus **4**, 249—261 (1952).
[244] BULLOUGH, K.: Radio-echo observations of the aurora in Terre Adelie. II. Ann. Géophys. **18**, 1—17 (1962).
[245] FORSYTH, P. A.: On the geometry of radio reflections from aurora. Canad. J. Phys. **38**, 593—603 (1960).
[246] LYON, G. F.: The association of visible auroral forms with radar echoes. Canad. J. Phys. **38**, 385—389 (1960).
[247] MOORCRAFT, D. R.: Models of auroral ionization, Part 1: Auroral ionization models and their radio-reflection characteristics. Canad. J. Phys. **39**, 677—695 (1961).
[248] KELLY, P. E., D. R. HANSEN and P. A. FORSYTH: The azimuthal distribution of ultra-high frequency radar echoes from aurora. Canad. J. Phys. **39**, 1535—1543 (1961).
[249] LYON, G. F., and P. A. FORSYTH: Radio-auroral reflection mechanisms. Canad. J. Phys. **40**, 749—760 (1962).
[250] KAISER, T. R.: Radio investigation of aurorae and related phenomena, in [7], 1955.
[251] VERSHININ, YE. F.: Allowance for refraction in the interpretation of radio reflections from auroras. Geomagnetism and Aeronomy **2**, 243—245 (1962).
[252] ASPINALL, A., and G. S. HAWKINS: Radio echo reflections from the aurora borealis. J. Brit. Ast. Assoc. **60**, 130—135 (1950).

[253] UNWIN, R. S.: Studies of the upper atmosphere from Invercargill, New Zealand, 1, Characteristics of auroral radar echoes at 55 Mc/s. Ann. Géophys. **15**, 377—394 (1959).
[254] LEONARD, R. S.: Distribution of radar auroras over Alaska. J. Geophys. Res. **67**, 939—952 (1962).
[255] BULLOUGH, K., and T. R. KAISER: Radio reflections from aurorae. II. J. Atmosph. Terr. Phys. **6**, 198—214 (1955).
[256] BYERFELD, Y. G.: Some peculiarities in the diurnal course of radio reflections from aurorae according to observations at different stations and the character of daily reflections. Results of IGY Researches. Aurorae and Airglow No. 2—3, 15 (Moscow 1960).
[257] GRATCHEV, A. I.: Some results of radar of aurorae near the zone of their maximum recurrence. Spectral, Electrophotometric and Radar Researches of Aurorae and Airglow, No. 2—3, 15—18, Publ. House Acad. Sci. USSR, Moscow (1960).
[258] YARIN, V. I.: Diurnal and seasonal variations of radar reflections from aurorae. Spectral, Electrophotometrical and Radar Researches of Aurorae and Airglow, No. 2—3, 24—27. Publ. House Acad. Sci. USSR, Moscow (1960).
[259] YARIN, V. I.: Radar reflections from aurorae and their connection with magnetic disturbances. Spectral, Electrophotometrical and Radar Researches of Aurorae and Airglow, No. 5, 56—59. Publ. House Acad. Sci. USSR, Moscow (1961).
[260] HUNSUCKER, R. D., and L. OWREN: Auroral sporadic-E ionization. J. Res. Nat. Bur. Stand. **66**D, 581—592 (1962).
[261] LANGE-HESSE, G.: German aurora observations 1957—1962: Part A: ... by means of VHF Radio Waves; part B: Visual Aurora. Abh. Akad. Wiss. Göttingen, math.-phys. Kl. Beitr. Int. Geophys. Jahr. H. 10 (1963).
[262] LANGE-HESSE, G.: VHF-Long-Distance-Propagation in middle Europe by aurora-backscatter. Arch. elektr. Übertragung **16**, 251—261 (1962).
[263] LANGE-HESSE, G.: Seasonal influences on VHF-aurora-backscatter in Middle-Europe. Z. Geophys. **29**, 35—44 (1963).
[263a] LANGE-HESSE, G.: VHF-Bistatic-Aurora-Communications as a function of geomagnetic activity and magnetic latitude. To be published in AGARDOgraph on Radio communication in and to Actic region, Eighth Techn. Meeting of Avionics Panel of AGARD. London: Pergamon Press 1963.
[264] COHEN, R., and K. L. BOWLES: The association of plane-wave electron density irregularities with the equatorial electrojet. J. Geophys. Res. **68**, 2503—2556 (1963).
[265] BOWLES, K. L., B. B. BALSLEY and R. COHEN: Field-aligned E-region irregularities identified with acoustic plasma waves. J. Geophys. Res. **68**, 2485—2502 (1963).
[266] COHEN, R., K. L. BOWLES and W. CALVERT: On the nature of equatorial slant sporadic E. J. Geophys. Res. **67**, 965—972 (1962).
[267] DOVGER, V. I.: Radio-wave back-scattering by aurorae. Spectral, Electrophotometrical and Radar Researches of Auroras and Airglow, No. 6, 7—11. Publ. House Acad. Sci. USSR, Moscow (1961).
[268] PRESNELL, R. I., R. L. LEADABRAND, A. M. PETERSON, R. B. DYCE, J. C. SCHLOBOHM and M. R. BERG: VHF and UHF radar observations of the aurora at College, Alaska. J. Geophys. Res. **64**, 1179—1190 (1959).
[269] LEADABRAND, R. L., R. I. PRESNELL, M. R. BERG and R. B. DYCE: Doppler investigations of the radar aurora at 400 Mc. J. Geophys. Res. **64**, 1197—1203 (1959).
[270] LEADABRAND, R. L.: A note on the disposition of daytime ionization in space. J. Geophys. Res. **66**, 421—428 (1961).
[271] DYCE, R. B., L. T. DOLPHIN, R. L. LEADABRAND and R. A. LONG: Aurora-like radar echoes observed from 17° latitude. J. Geophys. Res. **64**, 1815—1818 (1959).
[272] PETERSON, A. M., and R. L. LEADABRAND: Long-range radio echoes from auroral ionization. J. Geophys. Res. **59**, 306—309 (1954).
[273] NAKATA, Y.: Auroral echoes in the ionograms obtained in the minauroral region. Rep. Ionosphere Res., Japan **12**, 1—5 (1958).
[273a] HEWISH, A.: The diffraction of galactic radio waves as a method of investigating the irregular structure of the ionosphere. Proc. Roy. Soc. Lond. A **214**, 494—514 (1952).
[274] LITTLE, C. G., and A. MAXWELL: Scintillations of radio stars during aurorae and magnetic storms. J. Atmosph. Terr. Phys. **2**, 356—360 (1952).
[275] COSTAIN, C. H.: Radio star scintillations close to the auroral zone. M. A. Thesis, University of Saskatchewan, Canada, 1955.
[276] SCHLOBOHM, J. C., R. L. LEADABRAND, R. B. DYCE, L. T. DOLPHIN and M. R. BERG: High-altitude 106.1 Mc radio echoes from auroral ionization detected at a geomagnetic latitude of 43°. J. Geophys. Res. **64**, 1191—1196 (1959).
[277] HARTZ, T. R.: Radio star scintillations and ionospheric disturbances. Canad. J. Phys. **37**, 1137—1152 (1959).

[278] BRIGGS, B. H.: A preliminary report on horizontal drifts in the F region during the IGY. Some ionospheric results obtained during the International Geophysical Year, ed. W. J. G. BEYNON, 297—303. Amsterdam: Elsevier Pub. Co. 1960.
[279] LITTLE, C. G., G. C. REID, E. STILTNER and R. P. MERRITT: An experimental investigation of the scintillation of radio stars observed at frequencies of 223 and 456 Mc/s from a location close the auroral zone. J. Geophys. Res. **67**, 1763—1784 (1962).
[280] BENSON, R. F.: Effect of line-of-sight aurora on radio star scintillations. J. Geophys. Res. **65**, 1981—1985 (1960).
[281] FORSYTH, P. A., and K. V. PAULSON: On the geometry of radio reflections from aurora. Canad. J. Phys. **39**, 502—509 (1961).
[282] MOORCROFT, D. R., and P. A. FORSYTH: On the relation between radio star scintillations and auroral and magnetic activity. J. Geophys. Res. **68**, 117—124 (1963).
[283] BRIGGS, B. H.: The correlation of radio star scintillations with geomagnetic disturbances. Geophys. J. Roy. Astr. Soc. **5**, 306—371 (1961).

Chapter VI.

Sections 44—47.

[284] MEREDITH, L. H., M. B. GOTTLIEB and J. A. VAN ALLEN: Direct detection of soft radiation above 50 kilometers in the auroral zone. Phys. Rev. **97**, 201—205 (1955).
[285] VAN ALLEN, J. A.: Direct detection of auroral radiation with rocket equipment. Proc. Nat. Acad. Sci., Wash. **43**, 57—62 (1957).
[286] VERNOV, S. N., A. E. CHUDAKOV, E. V. GORCHAKOV, J. L. LOGACHEV and P. V. VAKULOV: Study of the cosmic-ray soft component by the 3rd Soviet earth satellite. Planet. Space Sci. **1**, 86—93 (1959).
[287] VAN ALLEN, J. A.: The geomagnetically-trapped corpuscular radiation. J. Geophys. Res. **64**, 1683—1689 (1959).
[288] O'BRIEN, B. J., C. D. LAUGHLIN, J. A. VAN ALLEN and L. A. FRANK: Measurements of the intensity and spectrum of electrons at 100-km altitude and high latitudes. J. Geophys. Res. **67**, 1209—1225 (1962).
[289] O'BRIEN, B. J.: Direct observations of dumping of electrons at 1000-km altitude and high latitudes. J. Geophys. Res. **67**, 1227—1233 (1962).
[290] VAN ALLEN, J. A.: Van Allen Radiation Belt. A review of space research. National Academy of Sciences-National Research Council, Publication **1079**, Chap. 7, 15 (1962).
[291] ANDERSON, K. A.: Soft radiation events at high altitude during the magnetic storm of August 29/30, 1957. Phys. Rev. **111**, 1397—1405 (1957).
[292] WINCKLER, J. R., L. PETERSON, R. ARNOLDY and R. HOFFMAN: X-rays from visible aurorae at Minneapolis. Phys. Rev. **110**, 1221—1231 (1958).
[293] WINCKLER, J. R.: Atmospheric phenomena, energetic electrons, and the geomagnetic field, J. National Bureau of Standards, Radio Propagation **66** D, 127—143 (1962).
[294] ALFVÉN, H.: On the motion of a charged particle in a magnetic field. Ark. Mat., Astronom. Fys. **27** A, No. 22 (1940).
[295] WENTWORTH, R. C., W. M. MACDONALD and S. F. SINGER: Lifetimes of trapped radiation belt particles determined by Coulomb scattering. Phys. Fluids **2**, 499—509 (1959).
[296] PARKER, E. N.: Newtonian development of the dynamical properties of ionized gases of low density. Phys. Rev. **107**, 924—933 (1957).
[297] NORTHROP, T. G., and E. TELLER: Stability of adiabatic motion of charged particles in the earth's field. Phys. Rev. **117**, 215—225 (1960).
[298] KULSRUD, R. M.: Adiabatic invariant of the harmonic oscillator. Phys. Rev. **106**, 205—207 (1957).
[299] NORTHROP, T. G.: Adiabatic charged-particle motion. Rev. Geophys. **1**, 283—304 (1963).

Sections 48—51.

[300] ANDERSON, K. A.: Balloon observations in the auroral zone. I. J. Geophys. Res. **65**, 551—564 (1960).
[301] ANDERSON, K. A., and D. C. ENEMARK: Balloon observations of X-rays in the auroral zone. II. J. Geophys. Res. **65**, 3521—3539 (1960).
[302] BROWN, R. R.: Balloon observations of auroral-zone X-rays. J. Geophys. Res. **66**, 1379—1388 (1961).
[303] BROWN, R. R.: West-east motion of an auroral-zone X-ray event. J. Geophys. Res. **67**, 31—35 (1962).
[304] BARCUS, J. R., and R. R. BROWN: Electron precipitation accompanying ionospheric current systems in the auroral zone. J. Geophys. Res. **67**, 2673—2680 (1962).

[305] BROWN, R. R., and W. H. CAMPBELL: An auroral-zone electron precipitation event and its relationship to a magnetic bay. J. Geophys. Res. **67**, 1357—1366 (1962).
[306] EVANS, D. S.: A pulsating auroral-zone X-ray event in the 100-second period range. J. Geophys. Res. **68**, 395—400 (1963).
[307] PFOTZER, G., A. EHMERT, H. ERBE, E. KEPPLER, B. HULTQVIST and J. ORTNER: A contribution to the morphology of X-ray bursts in the auroral zone. J. Geophys. Res. **67**, 575—585 (1962).
[308] KEPPLER, E., A. EHMERT, G. PFOTZER and J. ORTNER: Sudden increase of radiation intensity coinciding with a geomagnetic storm sudden commencement. J. Geophys. Res. **67**, 5343—5346 (1962).
[309] KREMSER, G.: Mitt. Max Planck-Inst. Aeronomie (to be published 1963).
[310] WINCKLER, J. R., L. PETERSON, R. HOFFMAN and R. ARNOLDY: Auroral X-rays, cosmic rays, and related phenomena during the storm of February 10/11, 1958. J. Geophys. Res. **64**, 597—610 (1959).
[311] WINCKLER, J. R.: Balloon study of high-altitude radiations during the International Geophysical Year. J. Geophys. Res. **66**, 1331—1359 (1960).
[312] MAY, R. A.: A study of auroral X-rays at Minneapolis between 23 August 1959 and 1 August 1960. Tech. Rep., CR-36, Cosmic Ray Program, School of Phys., Univ. of Minnesota, 1961.
[313] BHAVSAR, P. D.: Scintillation counter observations of auroral X-rays during the geomagnetic storm of May 12, 1959. J. Geophys. Res. **65**, 679—692 (1961).
[314] HAMLIN, D. A., R. KARPLUS, R. C. VIK and K. M. WATSON: Mirror and azimuthal drift frequencies for geomagnetically trapped particles. J. Geophys. Res. **66**, 1—4 (1961).
[315] CAMPBELL, W. H., and S. MATSUSHITA: Auroral-zone geomagnetic micropulsations with periods of 5 to 30 seconds. J. Geophys. Res. **67**, 555—573 (1962).
[316] WINCKLER, J. R., P. D. BHAVSAR and K. A. ANDERSON: A study of the precipitation of energetic electrons from the geomagnetic field during magnetic storms. J. Geophys. Res. **67**, 3717—3736 (1962).
[317] O'BRIEN, B. J.: Lifetimes of outer-zone electrons and their precipitation into the atmosphere. J. Geophys. Res. **67**, 3687—3706 (1962).
[318] BROWN, R. R., T. R. HARTZ, B. LANDMARK, H. LEINBACH and J. ORTNER: Large-scale electron bombardment of the atmosphere at the sudden commencement of geomagnetic storm. J. Geophys. Res. **66**, 1035—1041 (1961).
[319] ANGER, C. D., J. R. BARCUS, R. R. BROWN and D. S. EVANS: Long-period pulsations in electron precipitation associated with hydromagnetic waves in the auroral zone. J. Geophys. Res. **68**, 3306—3310 (1963).
[320] ANDERSON, K. A., C. D. ANGER, R. R. BROWN and D. S. EVANS: Simultaneous electron precipitation in the northern and southern auroral zones. J. Geophys. Res. **67**, 4076—4077 (1962).

Sections 52 and 53.

[321] MEREDITH, L. H., L. R. DAVIS, J. P. HEPPNER and O. E. BERG: Rocket auroral investigations. In: Experimental results of the US Rocket program for the IGY to 1 July 1958. (J. HANESSIAN jr. and I. GUTTMACKER, eds.), pp. 169—178. National Academy of Sciences, IGY Rocket Report Series No. 1 (1958).
[322] McILWAIN, C. E.: Direct measurement of particles producing visible aurora. J. Geophys. Res. **65**, 2727—2748 (1960).
[323] McDIARMID, I. B., D. C. ROSE and E. BUDZINSKI: Direct measurement of charged particles associated with auroral zone radio absorption. Canad. J. Phys. **39**, 1888—1900 (1961).
[324] CHAMBERLAIN, J. W.: The excitation of hydrogen in aurorae. Astrophys. J. **120**, 360—366 (1954).
[325] MALVILLE, J. M.: Type B aurorae in the Antarctic. J. Atmosph. Terr. Phys. **16**, 59—66 (1959).

Sections 54—56.

[326] ARNOLDY, R. L., R. A. HOFFMAN and J. R. WINCKLER: Observations of the Van Allen radiation regions during August and September 1959, part 1. J. Geophys. Res. **65**, 1361—1376 (1960).
[327] ARNOLDY, R. L., R. A. HOFFMAN, J. R. WINCKLER and S.-I. AKASOFU: Observations of the Van Allen radiation region during August and September 1959. J. Geophys. Res. **67**, 3673—3686 (1962).
[328] FARLEY, T. A., and A. ROSEN: Charged-particle variations in the outer Van Allen zone during a geomagnetic storm. J. Geophys. Res. **65**, 3494—3496 (1960).

[329] FAN, C. Y., P. MEYER and J. A. SIMPSON: Dynamics and structure of the outer radiation belt. J. Geophys. Res. **66**, 2607—2640 (1961).
[330] FORBUSH, S. E., D. VENKATESAN and C. E. MCILWAIN: Intensity variations in outer Van Allen radiation belt. J. Geophys. Res. **66**, 2275—2287 (1961).
[331] O'BRIEN, B. J., J. A. VAN ALLEN, F. E. ROACH and C. W. GARTLEIN: Correlation of an aurora arc and a subvisible monochromatic 6300 A arc with outer-zone radiation on November 28, 1959. J. Geophys. Res. **65**, 2759—2766 (1960).
[332] O'BRIEN, B. J.: Private communication (1962).
[333] O'BRIEN, B. J.: High-latitude geophysical studies with satellite Injun III, Part III, State Univ. Iowa (Dept. of Phys. Astron.) SUI 63—31 (1963).
[334] O'BRIEN, B. J.: Lifetimes of outer-zone electrons and their precipitation into the atmosphere. J. Geophys. Res. **67**, 3687—3706 (1962).
[335] O'BRIEN, B. J., and H. TAYLOR: High-latitude geophysical studies with satellite Injun III, Part IV, State Univ. of Iowa (Dept. of Phys. Astron.) SUI 63—32 (1963).
[336] MAEHLUM, B., and B. J. O'BRIEN: Study of energetic electrons and their relationship to auroral absorption of radio waves. J. Geophys. Res. **68**, 997—1010 (1963).
[337] AKASOFU, S.-I.: Deformation of magnetic shells during magnetic storms. J. Geophys. Res. **68**, 4437—4445 (1963).
[338] KRASOVSKII, V., I. S. SCHKLOVSKII, Y. I. GALPERIN, E. M. SVETLITSKII, Y. M. KUSHNIR and G. A. BORDOVSKII: The detection of electrons with energies of approximately 10 kev in the upper atmosphere. Planetary and Space Sci. **9**, 27—40 (1962).
[339] O'BRIEN, B. J.: Review of studies of trapped radiation with satellite-borne apparatus. Space Sci. Rev. **1**, 415—484 (1962).
[340] EVANS, J. E., and A. E. BELON: IG Bulletin, No. 77, 1073 (1963).

Chapter VII.

Sections 57—60.

[341] CHAMBERLAIN, J. W., and A. B. MEINEL: Emission spectra of twilight, night sky and aurorae. In: The Earth as a Planet (ed. by G. P. KUIPER). Chicago: Chicago Univ. Press 1954.
[342] MCLENNAN, J. C.: The aurora and its spectrum (Bakerian Lecture). Proc. Roy. Soc. Lond. A **120**, 327—357 (1928).
[343] HARANG, L.: Height distribution of the red auroral line in polar aurorae. Geofys. Publ. **20**, No. 5, 1—9 (Oslo 1957).
[344] JORJGO, N. V.: Variation of relative brightness of the λ 6300 and 5577 Å emissions in different forms of aurorae. Spectral, Electrophotometric and Radar Researches of Aurorae and Airglow, No. 2—3, 45—48. Publ. House Acad. Sci. USSR, Moscow (1960).
[345] BARBIER, D., and D. R. WILLIAMS: Observations of the aurora borealis. J. Geophys. Res. **55**, 401—414 (1950).
[346] PETRIE, W., and R. SMALL: Auroral spectrum in the wavelength range 3300—8900 A. Astrophys. J. **116**, 433—441 (1952).
[347] OLIVER, N. J., S. J. WOLNIK, J. C. SCANLON and J. W. CHAMBERLAIN: Some remarks on spectra of low-intensity auroras. J. Opt. Soc. Amer. **43**, 710 (1953).
[348] WALLACE, L.: An analysis of a spectrogram of the aurora of February 11, 1958, in the wavelength range 3710—4420 A. J. Atmosph. Terr. Phys. **17**, 46—56 (1959).
[349] BELON, A. E., and K. C. CLARK: Spectroscopic observations of the great aurora of February 10, 1958. II. Unusual atomic features. J. Atmosph. Terr. Phys. **16**, 220—227 (1959).
[350] OMHOLT, A.: The red and near-infrared auroral spectrum. J. Atmosph. Terr. Phys. **10**, 320—331 (1957).
[351] DUFAY, M.: Le spectre d'une aurore de basse latitude dans le visible et le proche infrarouge. C. R. Acad. Sci., Paris **245**, 1648—1650 (1957).
[352] DUFAY, M.: Étude photoélectrique du spectre du ciel nocturne dans le proche infrarouge. Ann. Géophys. **15**, 134—152 (1959).
[353] VEGARD, L., and G. KVIFTE: An auroral spectrogram and the results derived from it. Geofys. Publik. **18**, No. 3, 1—23 (Oslo 1951).
[354] MEINEL, A. B.: The near-infrared spectrum of the night sky and aurora. Publ. Astronom. Soc. Pacific **60**, 373—378 (1948).
[355] PETRIE, W.: The near infrared spectrum of the polar aurora. Phys. Rev. **77**, 720 (1950).
[356] HARANG, L.: A study of the auroral OI feature at 8446 A. Planet. Space Sci. **9**, 383—387 (1962).
[357] SHEFOV, N. N.: The twilight OI 8446 A emission. Results of IGY Researches. Aurorae and Airglow No. 9, 55—58. Publ. House Acad. Sci. USSR, Moscow (1962).

[358] MEINEL, A. B.: O_2 emission bands in the infrared spectrum of the night sky. Astrophys. J. **112**, 464—468 (1950).
[359] CHAMBERLAIN, J. W., C. Y. FAN and A. B. MEINEL: A new O_2 band in the infrared auroral spectrum. Astrophys. J. **120**, 560—562 (1954).
[360] VEGARD, L.: Nouvelles recherches sur le spectre des aurores boréales. C. R. Acad. Sci., Paris **230**, 1884—1886 (1950); An auroral spectrogram obtained at Oslo on February 23. Nature, Lond. **65**, 1012—1013 (1950): Nouveaux résultats importants dans l'étude de spectre des auroraes boréales et la physique de l'ionosphére. Ann. Géophys. **6**, 157—163 (1950).
[361] GARTLEIN, C. W., and D. F. SHERMAN: Identification of O_2^+ bands and forbidden nitrogen in auroral spectra. Mém. Soc. Roy. Sci. Liège (4), **12**, Pts. 1—2, 187—190 (1952).

Sections 61—63.

[362] YARIN, V. I.: [N I] λ 5200 Å emission according to the observations in Yakutsk, Spectral, Electrophotometrical and Radar Researches of Aurorae and Airglow, No. 9, 53—54. Publ. House Acad. Sci. USSR, Moscow (1961).
[363] BERNARD, R.: Presence of the nitrogen forbidden line $2P-4S$ in the auroral spectrum. Phys. Rev. **55**, 511 (1939).
[364] WEISSBERG, O. L.: Auroral spectro and electro photometry of the emission N_2, N_2^+, O I and N II. Spectral, Electrophotometrical and Radar Researches of Aurorae and Airglow, No. 8, 43—49. Publ. House Acad. Sci. USSR, Moscow (1962).
[365] DUFAY, J., and TCHENG MAO LIN: Recherches spectrophotométriques sur la lumière du ciel nocturne dans la région visible. I. Ann. Géophys. **2**, 189—230 (1946).
[366] GÖTZ, F. W. P.: Zum Nordlichtspektrum des 17. April 1947. Experientia **3**, 185 (1947).
[367] MIRONOV, A. V.: On the N I 5200 Å emission. Spectral, Electrophotometrical and Radar Researches of Aurorae and Airglow, No. 2—3, 66—67. Publ. House Acad. Sci. USSR, Moscow (1960).
[368] BAGARIATSKII, B. A., and N. I. FEDOROVA: Auroral spectrum in the region 9000—10500 A (translated title). Dokl. Akad. Nauk S.S.S.R. **103**, 1009—1011 (1955).
[369] BAGARIATSKII, B. A., and N. I. FEDOROVA: Spectra of the aurora and nightglow in the infrared region up to 11800 A. In: The Airglow and the Aurorae [7], pp. 174—177. London: Pergamon Press 1956.
[370] HARRISON, W. A., and A. VALLANCE JONES: Measurements of the absolute intensity of the aurora and night airglow in the 0.9—2.0 μ region. J. Atmosph. Terr. Phys. **11**, 192—199 (1957).
[371] PETRIE, W.: Forbidden line of N II in the aurora. Phys. Rev. **87**, 1002 (1952).
[372] YEVLASHIN, L. S.: The outstanding aurora of February 11, 1958. Geomagnetism and Aeronomy **2**, 62—66 (1962).
[373] MIRONOV, A. V., V. S. PROKUDINA and N. N. SHEFOV: Auroral observations on 10—11 February 1958, Spectral, Electrophotometrical and Radar Researches of Aurorae and Airglow, No. 1, 20—24. Publ. House Acad. Sci. USSR, Moscow (1961).
[374] BAGARIATSKII, B. A.: Several data on the distribution of energy in the infrared spectrum of aurora borealis (translated title). Izv. Akad. Nauk S.S.S.R., Ser. Geofiz. No. 4, 540—542 (1957).
[375] PETRIE, W., and R. SMALL: The intensities of ultraviolet features of the auroral spectrum. J. Geophys. Res. **57**, 51—57 (1952).
[376] FAN, C. Y.: Emission spectra excited by electronic and ionic impact. Phys. Rev. **103**, 1740—1745 (1956).
[377] CHAPMAN, S.: An outline of a theory of magnetic storms. Proc. Roy. Soc. Lond. A **97**, 61—83 (1918).
[378] MEINEL, A. B.: The auroral spectrum from 6200 to 8900 A. Astrophys. J. **113**, 583—588 (1951).
[379] BAGARIATSKII, B. A., and N. I. FEDOROVA: Spectra of the aurora and nightglow in the infrared region up to 11800 A. In: The Airglow and the Aurorae [7], pp. 174—177. London: Pergamon Press 1956.
[380] HARRISON, A. W., and A. VALLANCE JONES: Observations of Meinel N_2^+ bands near 1.5 μ in the auroral spectrum. J. Atmosph. Terr. Phys. **13**, 291—294 (1959).
[381] NOXON, J. F., J. F. HARRISON and A. VALLANCE JONES: The infrared spectrum of the night airglow 1.4 μ to 4.0 μ. J. Atmosph. Terr. Phys. **16**, 246—251 (1960).

Sections 64—66.

[382] GARTLEIN, C. W.: Aurora spectra showing broad hydrogen lines. Trans. Amer. Geophys. Union **31**, 18—20 (1950).

[383] MEINEL, A. B.: Doppler-shifted auroral hydrogen emission. Astrophys. J. **113**, 50—54 (1951).
[384] FEDOROVA, N. I.: Observations of (O—O) 1 PG N_2 bands at 10420 Å and (O—O) bands of the Meinel negative system of N_2^+ 11109 Å. Geomagnetism and Aeronomy **1**, 622—626 (1961).
[385] FAN, C. Y., and D. H. SCHULTE: Variations in the auroral spectrum. Astrophys. J. **120**, 563—565 (1954).
[386] ROMICK, G. J., and C. T. ELVEY: Variations in the intensity of the hydrogen emission line H_β during auroral activity. J. Atmosph. Terr. Phys. **12**, 283—287 (1958).
[387] FAN, C. Y.: Time variation of the intensity of auroral hydrogen emission and the magnetic disturbance. Astrophys. J. **128**, 420—427 (1958).
[388] GALPERIN, Y. I.: Hydrogen emission and two types of auroral spectra. Planet. Space Sci. **1**, 57—62 (1959).
[389] REES, M. H., A. E. BELON and G. J. ROMICK: The systematic behavior of hydrogen emission in the aurora. I. Planet. Space Sci. **5**, 87—91 (1961).
[390] OMHOLT, A., W. STOFFREGEN and H. DERBLOM: Hydrogen lines in auroral glow. J. Atmosph. Terr. Phys. **24**, 203—209 (1962).
[391] STOFFREGEN, W., and H. DERBLOM: Auroral hydrogen emission related to charge separation in the magnetosphere. Planet. Space Sci. **9**, 711—716 (1962).
[392] MONTBRIAND, L. E., and A. VALLANCE JONES: Studies of auroral hydrogen emissions in west-central Canada, 1. Time and geographical variations. Canad. J. Phys. **40**, 1401—1410 (1962).
[393] GALPERIN, Y. I.: Proton bombardment in aurora. Planet. Space Sci. **10**, 187—193 (1963).
[394] MONTALBETTI, R., and A. VALLANCE JONES: H_α emissions during aurorae over west-central Canada. J. Atmosph. Terr. Phys. **11**, 43—51 (1957).
[395] MALVILLE, J. M.: Narrow hydrogen emission in the aurora. Planet. Space Sci. **2**, 130—132 (1959).
[396] INGRAM, M. F.: The nightglow spectrum, Part II. H_α emission in the night sky. Monthly Not. Roy. Astronom. Soc. London **124**, 523—532 (1962).
[397] SHEFOV, N. N.: On the nature of helium emission 10830 A in aurorae. Spectral, Electrophotometrical and Radar Researches of Aurorae and Airglow, No. 5, 47—48. Publ. House Acad. Sci. USSR, Moscow (1961).
[398] SHEFOV, N. N.: The helium emission in the upper atmosphere. Results of IGY Researches. Aurorae and Airglow No. 8, 50—62. Publ. House Acad. Sci. USSR, Moscow (1962).
[399] SHEFOV, N. N.: Helium in the upper atmosphere. Planet. Space Sci. **10**, 73—77 (1963).
[400] MALVILLE, J. M.: Excitation of helium in the aurora. J. Atmosph. Terr. Phys. **21**, 54—64 (1961).

Sections 67—69.

[401] STEWART, D. T.: Electron excitation functions of the first negative bands of N_2^+. Proc. Phys. Soc. Lond. A **69**, 437—440 (1956).
[402] OMHOLT, A.: Studies on the excitation of aurora borealis. II. The forbidden oxygen lines. Geofys. Publik. **21**, No. 1 (1959).
[403] CHAMBERLAIN, J. W., and C. SAGAN: The origin of nitrogen ionization in the upper atmosphere. Planet. Space Sci. **2**, 157—164 (1960).
[404] HUNTEN, D. M.: The production of N_2^+ in the atmosphere. Planet. Space Sci. **10**, 37—46 (1963).
[405] DALGARNO, A.: Charged particles in the upper atmosphere. Ann. Geophys. **17**, 1—30 (1962).
[406] LYTLE, E. A., and D. M. HUNTEN: Dawn enhancement of auroral N_2^+ emission. Canad. J. Phys. **38**, 477—481 (1960).
[407] TATE, T., and P. T. SMITH: The efficiencies of ionization and ionzation. Potentials of various gases under electron impact. Phys. Rev. **39**, 270—277 (1932).
[408] FITE, W. L., and R. T. BRACKMANN: Ionization of atomic oxygen on electron impact. Phys. Rev. **113**, 815—816 (1959).
[409] SEATON, M. J.: The calculation of cross-sections for excitation of forbidden atomic lines by electron impact. In: The Airglow and The Aurorae [7], pp. 289—301. London: Pergamon Press 1956.
[410] ASHBURN, E. V.: Photometry of the aurora. J. Geophys. Res. **60**, 202—212 (1955).
[411] MURCRAY, W. B.: Some properties of the luminous aurora as measured by a photoelectric photometer. J. Geophys. Res. **64**, 955—959 (1959).
[412] JORGIO, N. V.: Properties of auroral coruscations. Results of IGY Researches. Aurora and Airglow No. 8, 17—20. Publ. House Acad. Sci. USSR, Moscow (1962).

Sections 70—77.

[413] YEVLASHIN, L. S.: Type-A red auroras in high latitudes. Geomagnetism and Aeronomy 1, 472—474 (1961).

[414] REES, M. H.: Excitation of high altitude red auroral arcs. Planet. Space Sci. 8, 59—67 (1961).

[415] VEGARD, L., and E. TÖNSBERG: The temperature of the auroral region determined from band spectra. Geofys. Publik. 11, 1—9 (1937).

[416] STÖRMER, C.: Remarkable aurora-forms from southern Norway. III-IX. Geophys. Publik. 13, No. 7, 1—82 (Oslo 1942).

[417] BATES, D. R.: The intensity distribution in the nitrogen band systems emitted from the earth's upper atmosphere. Proc. Roy. Soc. Lond. A 196, 217—250 (1949).

[418] SHUISKAYA, F. K.: Determination of relative rates of population of the various vibrational levels of 1 NGN_2^+ and rotational temperatures in aurorae. Spectral, Electrophotometrical and Radar Researches of Aurorae and Airglow, No. 5, 49—52. Publ. House Acad. Sci. USSR, Moscow (1961).

[419] SEATON, M. J.: Excitation processes in the aurora and airglow. III. Low-latitude aurorae, in [7], pp. 225—243. London: Pergamon Press 1956.

[420] HIKOSAKA, T.: On the reat enhancement of the line [O I] 6300 in the aurora at Niigata on February 11, 1958. Rep. Ionosphere Res. Japan 12, 469 (1958).

[421] HIKOSAKA, T., and K. YANO: Low latitude red aurora and low energy protons in Van Allen belt. J. Phys. Soc. Japan 17, Suppl. A-I, 233—236 (1962).

[422] BARBIER, D.: L'activité aurorale aux basses latitudes. Ann. Géophys. 14, 334—355 (1958).

[423] ROACH, F. E., and E. MAROVICH: A monochromatic low-latitude aurora. J. Res. Nat. Bur. Stand. 63D, 297—301 (1959).

[424] BARBIER, D.: L'arc auroral stable. Ann. Géophys. 16, 544—549 (1960).

[425] ROACH, F. E., and J. R. ROACH: Stable 6300 A auroral arcs in mid-latitude. Planet. Space Sci. 11 (1963).

[426] ROACH, F. E., D. BARBIER and R. A. DUNCAN: Observation of a 6300 A arc in France, the United States and Australia. Ann. Géophys. 18, 390—391 (1962).

[427] REES, M. H., and S.-I. AKASOFU: On the association between subvisual red arcs and the Dst (H) decrease. Planet. Space Sci. 11, 105—107 (1963).

[428] TOHMATSU, T., and F. E. ROACH: The morphology of mid-latitude 6300 angstrom arcs. J. Geophys. Res. 67, 1817—1821 (1962).

[429] KING, G. A. M., and F. E. ROACH: Relationship between red auroral arcs and ionospheric recombination. J. Res. Nat. Bur. Stand. 65D, 129—135 (1961).

[430] MEGILL, L. R., M. H. REES and L. K. DROPPLEMAN: Electric fields in the ionosphere and the excitation of the red lines of atomic oxygen. Planet. Space Sci. 11, 45—56 (1963).

[431] BARBIER, D., G. WEILL et M. FAFIOTTE: Les arcs émis par la raie rouge du ciel nocturne en Afrique. C. R. Acad. Sci. Paris, 252, 3102—3103 (1961).

[432] BARBIER, D.: L'emission de la raie rouge du ciel nocturne et les propriétés de la couche F de l'ionosphère. C. R. Acad. Sci., Paris 252, 3315/16 (1961).

[433] BARBIER, D.: Etude de la couche F d'aprés l'émission de la raie rouge du ciel nocturne. Planet. Space Sci. 10, 29—35 (1963).

[434] SANDFORD, B. P.: Polar-glow aurora in polar cap absorption events. J. Atmosph. Terr. Phys. 24, 155—172 (1962).

[435] SANDFORD, B. P.: Optical studies of particle bombardment in polar cap absorption events. Planet. Space Sci. 10, 195—213 (1963).

[436] MONTALBETTI, R., and D. J. MCEWEN: Hydrogen emissions during the period November 9—16. Canad. J. Phys. 39, 617—627 (1961).

[437] STEIGER, W. R., and S. MATSUSHITA: Photographs of the high-altitude nuclear explosion "Teak". J. Geophys. Res. 65, 545—550 (1960).

[438] CULLINGTON, A. L.: A man-made or artificial aurora. Nature Lond. 182, 1356—1366 (1958).

[439] MALVILLE, J. M.: Artificial auroras resulting from the 1958 Johnston Island nuclear explosions. J. Geophys. Res. 64, 2267—2270 (1959).

[440] NEWMAN, P.: Optical, electromagnetic and satellite observations of high-altitude nuclear detonations, part I. J. Geophys. Res. 64, 923—932 (1959).

[441] ODENCRANTZ, F. K., P. SAINT-AMAND and J. G. MOORE: Zenith airglow observations during the high-altitude nuclear explosion of July 9, 1962. J. Geophys. Res. 67, 4091—4092 (1962).

[442] NEFF, S. H.: Photometric observations of an artifical aurora. J. Geophys. Res. 68, 587—588 (1963).

[442a] CARMAN, E. H., B. C. GIBSONWILDE, B. P. KILFOYLE and W. M. COLEMAN: Behavior of zenith O I 6300 Å airglow at Townsville during high-altitude nuclear explosions. J. Geophys. Res. 68, 2855—2857 (1963).

Sections 78—81.

[443] WARK, D. Q.: Doppler widths of the atomic oxygen lines in the airglow. Astrophys. J. 131, 491—501 (1960).
[444] NILSON, J. A., and G. G. SHEPHERD: Upper atmospheric temperatures from doppler line widths-I. Some preliminary measurements on O I 5577 A in aurora. Planet. Space Sci. 5, 299—306 (1961).
[445] TURGEON, E. C., and G. C. SHEPHERD: Upper atmospheric temperatures from Doppler line widths, part 2, Planet. Space Sci. 9, 295—304 (1962).
[446] MULYARCHIK, T. M., and P. V. SHCHEGLOV: Temperature and corpuscular heating in the auroral zone. Planet. Space Sci. 10, 215—218 (1963).
[447] ROSSELAND, S., and G. STEENSHOLD: Det Norsk Vidensk. Akad. Avk. 1, No. 5 (1933).
[448] HUNTEN, D. M., and G. G. SHEPHERD: Vibrational temperatures from second positive bands. J. Atmosph. Terr. Phys. 6, 64—66 (1955).
[449] JOHANSON, A. E., and A. VALLANCE JONES: Variations with height and form of N_2^+ rotational temperatures in aurora. Canad. J. Phys. 40, 24—32 (1962).
[450] SHUISKAYA, F. K.: Determination of relative rates of population of the various vibrational levels of ING N_2^+ and rotational temperatures in aurorae. Spectral, Electrophotometrical and Radar Researches of Aurorae and Airglow, No. 5, 49—52. Publ. House Acad. Sci. USSR, Moscow (1961).
[451] YARIN, V. I.: Continuous emission and the Herzberg O_2 bands in the night airglow. Spectral, Electrophotometrical and Radar Researches of Aurorae and Airglow, No. 5, 35—38. Publ. House Acad. Sci. USSR, Moscow (1961).

Chapter VIII.

Sections 82—85.

[452] BLOCK, L.: Model experiments on aurorae and magnetic storms. Tellus 7, 65—86 (1955).
[453] BRÜCHE, E.: Some new theoretical and experimental results on the aurora polaris. Terr. Mag. 36, 41—52 (1931).
[454] BENNETT, W. H.: The Störmertron. Ann. Géophys. 14, 206—207 (1958).
[455] SCHUSTER, A.: On the origin of magnetic storms. Proc. Roy. Soc. Lond. 85, 44—50 (1911).
[456] LINDEMANN, F. A.: Note on the theory of magnetic storms. Phil. Mag. 38, 669—684 (1919).
[457] CHAPMAN, S., and V. C. A. FERRARO: A new theory of magnetic storms. Terr. Magn. 36, 77—97, 171—186 (1931); 37, 147—156, 421—429 (1932).
[458] CHAPMAN, S., and V. C. A. FERRARO: The theory of the first phase of a geomagnetic storm. Terr. Magn. 45, 245—268 (1940).
[459] CHAPMAN, S.: Idealized problems of plasma dynamics relating to geomagnetic storms. Rev. Mod. Phys. 32, 919—933 (1960).
[460] CHAPMAN, S., and P. C. KENDALL: An idealized problem of plasma dynamics that bears on geomagnetic storm theory; oblique projection. J. Atmosph. Terr. Phys. 22, 142—156 (1961).
[461] LANGMUIR, I.: The interaction of electron and positive ion space charges in cathode sheaths. Phys. Rev. 33, 954—989 (1929).
[462] FERRARO, V. C. A.: On the theory of the first phase of a geomagnetic storm. J. Geophys. Res. 57, 14—49 (1952).
[463] MARTYN, D. F.: The theory of magnetic storms and auroras. Nature, Lond. 167, 92—94 (1951).

Sections 86—88.

[464] SINGER, S. F.: A new model of magnetic storms and aurorae. Trans. Amer. Geophys. Union 38, 175—190 (1957).
[465] GOLD, T.: Motions in the magnetosphere of the earth. J. Geophys. Res. 64, 1219—1224 (1959).
[466] ZHIGULEV, V. N., and E. A. ROMISHEVSKII: Concerning the interaction of currents flowing in a conducting medium with the earth's magnetic field. Soviet Phys. Doklady 4, 859—862 (1960).

[467] DUNGEY, J. W.: The steady state of the Chapman-Ferraro problem in two dimensions. J. Geophys. Res. **66**, 1043—1048 (1961).

[468] HURLEY, J.: Interaction of a streaming plasma with the magnetic field of a two-dimensional dipole. Phys. Fluids **4**, 854—859 (1961).

[469] SPREITER, J. R., and B. R. BRIGGS: On the choice of condition to apply at the boundary of the geomagnetic field in the steady-state Chapman-Ferraro problem. J. Geophys. Res. **67**, 2983—2985 (1962); see also Theoretical determination of the form of the boundary of the solar corpuscular stream produced by interaction with the magnetic dipole field of the earth. J. Geophys. Res. **67**, 37—51 (1962).

[470] SPREITER, J. R., and B. J. HYETT: The effect of a uniform external pressure on the boundary of the geomagnetic field in a steady solar wind. J. Geophys. Res. **68**, 1631—1642 (1963).

[470a] MIDGLEY, J. E., and L. DAVIS, jr.: Calculation by a moment technique of the perturbation of the geomagnetic field by the solar wind. J. Geophys. Res. **68**, 5111—5123 (1963).

[470b] HONES jr., E. W.: Motions of charged particles trapped in the earth's magnetosphere. J. Geophys. Res. **68**, 1209—1219 (1963).

[471] FERRARO, V. C. A.: An approximate method of estimating the size and shape of the stationary hollow carved out in a neutral ionized stream of corpuscles impinging on the geomagnetic field. J. Geophys. Res. **65**, 3951—3953 (1960).

[472] MEAD, G. D.: Deformation of the geomagnetic field by the solar wind. J. Geophys. Res. **69**, 1181—1197 (1964).

[473] AXFORD, W. I.: The interaction between the solar wind and the earth's magnetosphere. J. Geophys. Res. **67**, 3791—3796 (1962).

[474] KELLOGG, P. J.: Flow of plasma around the earth. J. Geophys. Res. **67**, 3805—3811 (1962).

[474a] FRANK, L.A., J. A. VAN ALLEN and E. MACAGNO: Charged-particle observations in the earth's outer magnetosphere. J. Geophys. Res. **68**, 3543—3554 (1963).

[474b] CAHILL, L. J., and P. G. AMAZEEN: The boundary of the geomagnetic field. J. Geophys. Res. **68**, 1871—1884 (1963).

[474c] FREEMAN, J. W., J. A. VAN ALLEN and L. J. CAHILL: Explorer 12 observations of the magnetospheric boundary and the associated solar plasma on September 13, 1961. J. Geophys. Res. **68**, 2121—2130 (1963).

[475] STOREY, L. R. O.: An investigation of whistling atmospherics. Phil. Trans. Roy. Soc. Lond. A **246**, 113—141 (1953).

[476] ALLCOCK, G. McK.: The electron density distribution in the outer ionosphere derived from whistler data. J. Atmosph. Terr. Phys. **14**, 185—199 (1959).

[477] POPE, J. H.: A correction of the exospheric electron density estimate using the nose whistlers of March 19, 1959. J. Geophys. Res. **67**, 412 (1962).

[478] CARPENTER, D. L.: New experimental evidence of the effect of magnetic storms on the magnetosphere. J. Geophys. Res. **67**, 135—146 (1962).

[479] CARPENTER, D. L.: Electron-density variations in the magnetosphere deduced from whistler data. J. Geophys. Res. **67**, 3345—3360 (1962).

[480] CARPENTER, D. L.: Whistler evidence of a "knee" in the magnetospheric ionization density profile. J. Geophys. Res. **68**, 1675—1682 (1963).

[480a] CARPENTER, D. L.: Whistler measurements of electron density and magnetic field strength in the remote magnetosphere. J. Geophys. Res. **68**, 3727—3730 (1963).

[481] JOHNSON, F. S.: The structure of the outer atmosphere including the ion distribution above the F 2 maximum. Lockheed Tech. Rep., April 1959.

[482] DESSLER, A. J.: The propagation velocity of world-wide sudden commencements of magnetic storms. J. Geophys. Res. **63**, 405—408 (1958).

[483] WILSON, C. R., and M. SUGIURA: Hydromagnetic interpretation of sudden commencements of magnetic storms. J. Geophys. Res. **66**, 4097—4111 (1961).

[484] FRANCIS, W. E., M. I. GREEN and A. J. DESSLER: Hydromagnetic propagation for sudden commencements of magnetic storms. J. Geophys. Res. **64**, 1643—1645 (1959).

[485] DESSLER, A. J., W. E. FRANCIS and E. N. PARKER: Geomagnetic storm sudden-commencement rise times. J. Geophys. Res. **65**, 2715—2719 (1960).

[486] LEBEAU, A., et R. SCHLICH: Sur une proprieté de l'activité magnétique diurne dans les régions de haute latitude (Stations Charcot et Dumont d'Urville). C. R. Acad. Sci., Paris **254**, 3014—3016 (1962).

Sections 89—91.

[487] DESSLER, A. J., and E. N. PARKER: Hydromagnetic theory of geomagnetic storms. J. Geophys. Res. **64**, 2239—2252 (1959).

[488] APEL, J. R., and S. F. SINGER: Geomagnetic field perturbations due to trapped particles. Ballistic Missile and Aerospace Technology **3**, 247—264 (1961). New York: Academic Press.
[489] DESSLER, A. J., W. B. HANSON and E. N. PARKER: Formation of the geomagnetic storm Main-phase ring current. J. Geophys. Res. **66**, 3631—3637 (1961).
[490] PARKER, E. N.: Dynamics of the geomagnetic storm. Space Sci. Rev. **1**, 62—99 (1962).
[491] COLE, K. D.: On CHAMBERLEIN'S theory of auroral bombardment. Astrophys. J. **136**, 677—678 (1962).
[492] AKASOFU, S.-I., and C. E. MCILWAIN: Energetic neutral hydrogen atoms as a source of the ring current particles (in preparation).

Sections 92 and 93.

[493] CHANDRASEKHAR, S.: Plasma Physics, pp. 1—213. Chicago: Chicage University Press 1959.
[494] DRAGT, A. J.: Effect of hydromagnetic waves on the lifetime of Van Allen radiation protons. J. Geophys. Res. **66**, 1641—1649 (1961).
[495] WENTZEL, D. G.: Hydromagnetic waves and the trapped radiation, Part 2: Displacements of the mirror points. J. Geophys. Res. **66**, 363—369 (1961).
[496] WENTZEL, D. G.: Hydromagnetic waves and the trapped radiation, Part 3: Effects on protons above the proton belt. J. Geophys. Res. **67**, 485—498 (1962).
[497] PARKER, E. N.: Transresonant electron acceleration. J. Geophys. Res. **66**, 2673—2676 (1961).
[498] KELLOGG, P. J.: Auroral x-rays, electron bombardment and trapped radiation. Planet. Space Sci. **10**, 165—178 (1963).
[499] PARKER, E. N.: Effect of hydromagnetic waves in a dipole field on the longitudinal invariant. J. Geophys. Res. **66**, 693—708 (1961).
[500] PARKER, E. N.: Geomagnetic fluctuations and the form of the outer zone of the Van Allen radiation belt. J. Geophys. Res. **65**, 3117—3130 (1960).
[501] DAVIS jr., L., and D. B. CHANG: On the effects of geomagnetic fluctuations on trapped particles. J. Geophys. Res. **67**, 2169—2179 (1962).

Sections 94—98.

[502] ALFVÉN, H.: On the electric field theory of magnetic storms and aurorae. Tellus **7**, 50—64 (1955).
[503] AXFORD, W. I., and C. O. HINES: A unifying theory of high-latitude geophysical phenomena and geomagnetic storms. Canad. J. Phys. **39**, 1433—1464 (1961).
[504] AXFORD, W. I.: Rotation of the magnetosphere. J. Geophys. Res. **68**, 5883 (1963).
[505] PIDDINGTON, J. H.: A hydromagnetic theory of geomagnetic storms. Geophys. J. **7**, 183—193 (1962).
[506] PIDDINGTON, J. H.: A hydromagnetic theory of geomagnetic storms and auroras. Planet. Space Sci. **9**, 947—960 (1962).
[507] COWLING, T. G.: Solar electrodynamics (Chap. 8), The sun, ed. G. P. KUIPER. Chicago: Chicago Univ. Press 1954.
[508] CHAMBERLAIN, J. W.: Theory of auroral bombardment. Astrophys. J. **134**, 401—424 (1961).
[509] CHAMBERLAIN, J. W.: Plasma instability as a mechanism for auroral bombardment. J. Geophys. Res. **68**, 5667—5674 (1963).
[510] KERN, J. W.: A charge separation mechanism for the production of polar auroras and electrojets. J. Geophys. Res. **67**, 2649—2665 (1962).
[510a] FEJER, J. A.: Theory of auroral electrojets. J. Geophys. Res. **68**, 2147—2158 (1963).
[511] HOYLE, F.: Some Recent Researches in Solar Physics. Cambridge: Cambridge University Press 1949.
[512] DUNGEY, J. W.: Conditions for the occurrence of electrical discharges in astrophysical systems. Phil. Mag. **44**, 725—738 (1953). See also [10].
[513] AKASOFU, S.-I., and S. CHAPMAN: A neutral line discharge theory of the aurora polaris. Phil. Trans. Roy. Soc. Lond. A **253**, 359—406 (1961).
[514] CHAPMAN, S.: The energy of magnetic storms. Geophys. J. Roy. Astron. Soc. 514—536 (1964).
[515] AKASOFU, S.-I.: The development of the auroral substorm. Planet. Space Sci. **12**, 273—282 (1964).

Time-Variations of Cosmic Rays.

By

Scott E. Forbush.

With 86 Figures.

First Part:

Results up to the International Geophysical Year.

A. Methods of observation.

1. Preliminary survey. Many investigators have undertaken systematic studies of time variations of cosmic-rays with the hope of finding some clue to their origin. The only definite evidence of this sort thus far obtained is that which indicates that the sun, on rare occasions emits charged particles in the cosmic-ray energy spectrum (below 10 BeV $= 10^{10}$ electron volts if protons).

Nevertheless, researches on time variations have established the existence of several world-wide variations in cosmic-ray intensity (in addition to variations, more local in character, of meteorological origin). These include: variations during some magnetic storms, 27-day quasiperiodic variations, variations with sunspot cycle, a diurnal variation, and a variation in the amplitude of the diurnal variation with time.

In addition, observations in balloons with ionization chambers and in jet planes with neutron detectors have established variations in the primary spectrum during part of the solar cycle; and rocket and satellite-experiments have provided additional evidence (see the Second Part). For all of the established variations evidence indicates the sun is directly or indirectly responsible.

One of the early attempts to systematically search for time variations was made by A. Corlin [*1*] in Northern Sweden, at various intervals between October 1929 and July 1933. In spite of careful observations and analyses the results failed to indicate significant variations. In the light of present knowledge the reasons for this failure now appear twofold; first the larger world-wide variations which are associated with some magnetic storms seldom occur during years near the minimum of sunspot activity (such as those years studied by Corlin); and second the smaller variations, especially in the polar regions, are likely to be completely obscured, in ionization-chambers, by the large variability of atmospheric effects (connected with weather-like changes in the atmospheric structure) which arise because of variations in the height at which μ-mesons are produced. The greater this height the greater is the number of μ-mesons which decay (into rapidly absorbed electrons and neutrinos) before they reach the ground, and consequently the smaller is the measured intensity at the ground. These (local) effects for example, at Godhavn, Greenland amount to several percent but at Huancayo, Peru they are much smaller, while the world wide changes at Huancayo are about the same in magnitude as at Godhavn.

Thus with ionization chambers, and with Geiger counters, the world-wide variations are best observed at stations at low latitudes. On the other hand the increases observed during some solar flares were never observed near the equator until February 1956; and the magnitude of these increases varies markedly among stations at different longitudes. Thus, until recently, fortune has played an important part in the location of stations for continuous observations.

2. Instruments: ionization chambers (COMPTON-BENNETT). *α) Stations.* Due to their simplicity, reliability, and facility of maintenance, ionization-chambers have proved especially valuable for continuous operation over long periods of time. The longest existing series of continuous observations with the same type of instrument, has been obtained with Carnegie Institution of Washington (CIW) model C Compton-Bennett ionization chambers [2]. Table 1 indicates the location and elevation of stations where these have been in operation since the dates indicated therein.

Table 1. *Location of CIW Compton-Bennett cosmic-ray meters, 12 cm Pb shield.*

Station	Latitude deg.	Longitude deg.	Geomagnetic latitude deg.	Elevation m	Barometric coefficient c/o (mm Hg)$^{-1}$	Operation began
Godhavn, Greenland	69.2 N	53.5 W	79.9 N	9	−0.18	October 1938
Cheltenham, Maryland	38.7 N	76.8 W	50.1 N	72	−0.18	March 1935
Climax, Colorado	39.4 N	106.2 W	48.1 N	3350		September 1953[1]
Teoloyucan, Mexico	19.2 N	99.2 W	29.7 N	2285	−0.345	February 1937[2]
Ciudad Universitaria, Mexico						September 1954
Huancayo, Peru	12.0 S	75.3 W	0.6 S	3350	−0.30	June 1936
Christchurch, New Zealand	43.5 S	172.6 E	48.6 S	8	−0.18	June 1936

[1] Previously operated intermittently at 3500 m.
[2] Not operated after 1945.

β) Description. The Compton-Bennett meter consists essentially of a spherical steel ionization chamber, volume 19.3 liters, filled with highly purified argon to 50 atmospheres pressure. Pure argon at high pressure gives larger ionization currents than any other gas; at 50 atmospheres the ionization is about 67 times that in air at normal pressure and temperature. The standard deviation of statistical errors arising from random arrival-times of individual rays is inversely proportional to the area of cross-section of the chamber. At Washington, from analysis of simultaneous hourly records from two Model-C meters, the standard deviation of statistical fluctuations in one hour's record from either of the instruments was found to be 0.7% of the total ionization due to cosmic-rays. Ions produced mainly by cosmic-rays in the argon of the main chamber are driven to the collector at the center of the sphere by a potential applied to the inner wall of the sphere. This is the main ionization current. A balancing current is supplied by ionization produced in a small auxiliary chamber, inside the sphere, by β-rays from metallic uranium (see Fig. 1). Turning the micrometer rod to which the uranium is attached permits varying the rate of entry of β-rays into the balance chamber as a consequence of changes in the amount of shielding. The balance current can thus be made about equal, and opposite in sign through choice of sweep field, to the average ionization current produced in the main chamber by cosmic rays. This permits ample sensitivity for recording changes in ionization due to cosmic rays on a photographic trace 60 mm wide. This method of balancing also compensates for changes in ionization which might result from changes in pressure or temperature of the gas. This compensation arises from the fact that the ionization current due to β-rays in the balance chamber is affected by changes in pressure and temperature in the same way as that due to cosmic rays in the main chamber.

γ) *Performance.* Direct tests involving two meters showed that the meters are not affected by temperature. Tests also show that at the higher pressures,

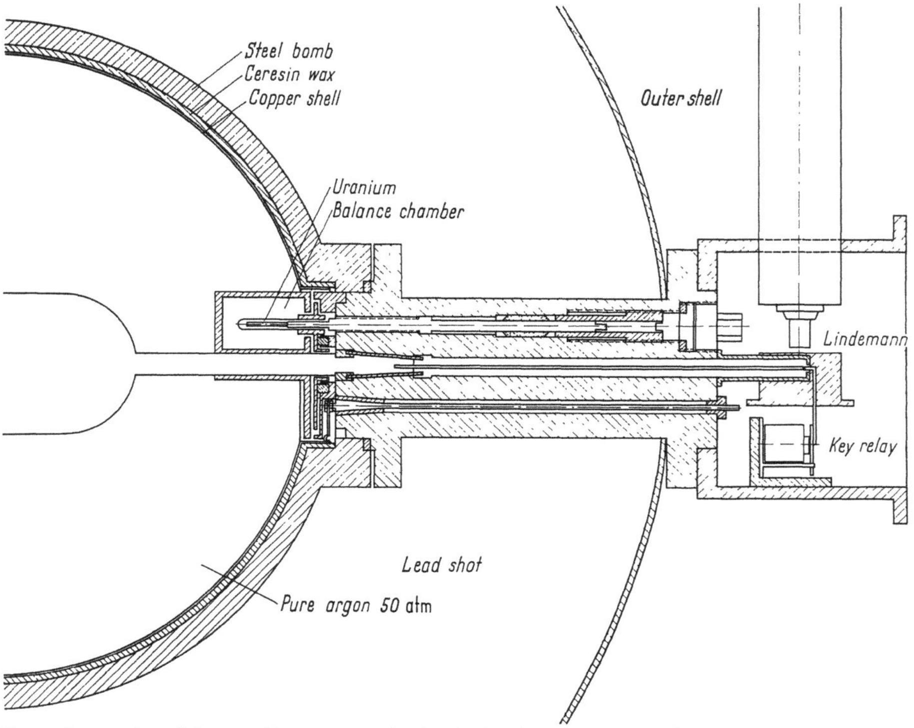

Fig. 1. Cross section of Compton-Bennett meter showing details of central electrode, β chamber and micrometer for adjusting position of uranium.

the balance is almost completely unaffected by changes in the pressure to which the chamber is filled. For this to be so, there must be negligible ionization in the main chamber due to α- and β-radiation from the walls, since this ionization will not increase with pressure as does that due to cosmic rays. Also, the dimension of the balance chamber must be so small that there is negligible absorption in the argon of the β-rays from the uranium. Shielding of the uranium with a thimble of aluminium 0.2 mm thick causes absorption of the α-particles and soft β-rays before they reach the balance chamber. There remain β-rays with a range of about 6 cm in argon at 50 atmospheres and some γ-rays. The β-rays traverse the balance chamber but are stopped within the 3 mm brass walls of the balance chamber. The γ-rays, which are not completely absorbed by the walls of the balance chamber, produce a small residual ionization in the main chamber which is proportional to the pressure and appears as a part of the zero correction.

It is known that the ionization in a closed chamber, filled with air at high pressure, increases with temperature because of less initial recombination of ions at the higher temperatures. In pure argon the initial recombination is small, and whatever small effect there may be is compensated by the balance arrangement. Changes in ionization current due to changes in the applied potential are also compensated by the balance chamber, since both chambers show the same degree of saturation at the voltages (250 V) used.

Batteries supply the sweeping potentials for the main chamber and the balance chamber. The batteries are connected across a 2.6 MΩ potentiometer, the variable center of which is grounded. In this way induced charges due to battery fluctu-

Fig. 2. Compton-Bennett meter at Cheltenham Magnetic Observatory.

ations are eliminated, provided the insulation of the batteries, potentiometer, and ionization-chamber walls is high compared to 2.6 MΩ. Elimination of induced charges on the collecting electrode from changes in the battery voltage also requires that the capacity coefficients C_1 and C_2 of the two chambers and the resistances r_1 and r_2 of the two arms of the potentiometer satisfy the relation: $r_1 C_1 = r_2 C_2$. By adjustment of the position of the balance chamber, C_1 and C_2 can be made very nearly equal. Balance is so effected that suddenly putting 500 V across the potentiometer induces a negligible deflection of the electrometer.

δ) Recording. The difference between the ionization currents in the two chambers is measured with a standard Lindemann electrometer housed in an airtight, dried chamber. At hourly intervals, the central electrode is grounded for about three minutes. The collecting electrode thus remains insulated or floating for about 57 min of each hour. Once every four hours, for about one minute, a relay connects the central electrode to a known potential. Thus the electrometer sensitivity is automatically recorded. Electrometer sensitivities of the order of one quarter of the maximum sensitivity attainable insure stability.

The shadow of the electrometer needle is projected through a compound microscope onto a strip of bromide paper 60 mm wide, moving 2 cm per hour for normal operation. A roll of paper 18 m in length is adequate for a month of continuous record. Visual observations can be made at any time from an image of the electrometer needle which is projected onto a ground glass scale.

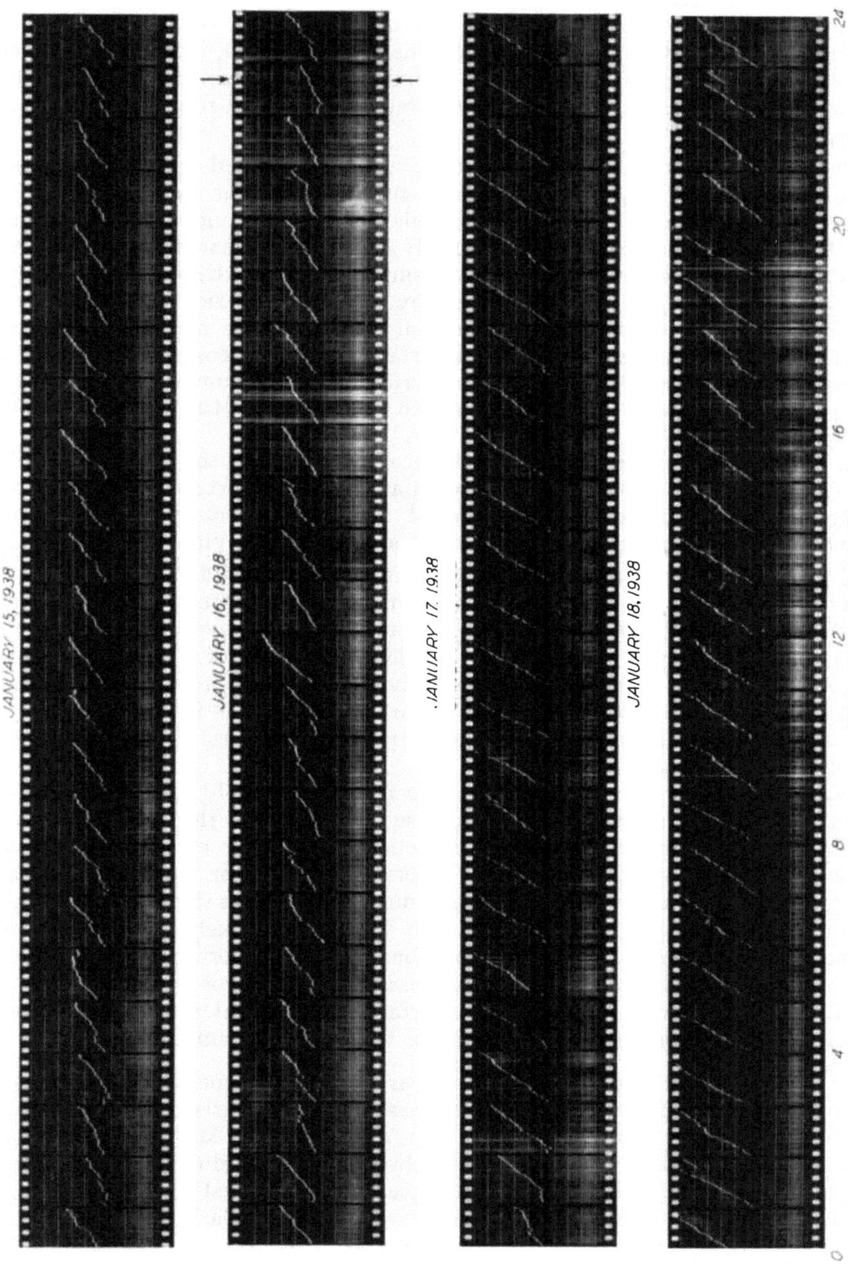

Fig. 3. Cosmic-ray records, Huancayo, Peru, showing effect of magnetic storm which began 22.7h GMT, January 16, 1938.

In addition to the cosmic-ray ionization, the barometric pressure from an aneroid barometer is usually recorded. Notes may be written on the back of the bromide paper in dim light without affecting the records. Fig. 2 shows the meter at Cheltenham and Fig. 3 shows cosmic-ray records from Huancayo for four days. The dark vertical lines are hour-marks. The upward slope of the electrometer trace in Fig. 3 indicates that the balance current exceeds the cosmic ray ionization current. The meter is shielded by lead shot of uniform size which is equivalent to a layer of about 10.7 cm of solid lead. The total shielding, including the steel walls of the chamber, is equivalent to about 12 cm of lead, which practically effects complete shielding from any local radiations unless radioactive sources are brought near.

ε) Evaluation. The absolute ionization, i_a, is determined from the relation $i_a=(i_1+i_2)/2$ in which i_1 is the ionization current when the same potential is applied to the main chamber and to the balance chamber, and i_2 is the current when these potentials are of opposite sign. If i_a and i_b are the ionization currents, respectively, in the main chamber (due to cosmic rays) and in the balance chamber, then $i_1=i_a+i_b$ and $i_2=i_a-i_b$. However, any residual ionization in the chamber contributes to i_a. To determine the residual ionization, the meters were taken into a coal mine 110 meters below the surface where the cosmic-ray intensity was estimated to be 0.2% of that at the surface. With the meter fully shielded, $i_a=(i_1+i_2)/2$ and $i_b=(i_1-i_2)/2$ were measured for different settings of the uranium micrometer.

In Fig. 3 an example of a burst at Huancayo is shown in the interval 2200 to 2300 GMT January 18, 1938. These records also show the decrease in ionization during the magnetic storm which started 2207 GMT January 16, 1938, and attained an intensity expressed as 8+ by the geomagnetic planetary index K_p.

ζ) Other types. Several other types of ionization chambers have been in operation for shorter periods of time. One of the more important of these is a 500 liter pressure chamber, shielded by 10 cm Fe, operated by the Physikalische Institut, under the direction of A. SITTKUS [3] of the Universität of Freiburg. Results from this meter have been published quarterly since January, 1950 in the Sonnen-Zirkular of the Fraunhofer-Institut. The larger size of this instrument results in a smaller standard deviation of statistical fluctuations of hourly values, arising from sampling.

However, as will be seen later, the uncertainties in daily means of cosmic-ray intensity from ionization chambers arise not only from the sample size but also from atmospheric effects which influence the number of μ-mesons which decay before reaching the meter. In general, even for the Compton-Bennett meter (except at Huancayo), the latter are much greater than the former. For this reason the reliability of daily means is, in general, not likely to be improved by increasing the size of the ionization chamber unless corrections for the μ-meson effects are applied. On the other hand, for studies of the variability of the diurnal variation the statistical uncertainty of diurnal variations derived from larger ionization chambers will be less than from the smaller ones.

3. Geiger counter—telescopes. One particular experimental arrangement of Geiger counters of especial interest in measuring time variations is that used by DOLBEAR [4] and ELLIOTT and earlier by ALFVÉN [5] and MALMFORS in Stockholm to measure variations in different directions. Recordings were obtained with the axes of the telescopes in a N-S plane, and inclined 45° on both sides of the vertical. The axes of the telescopes could be similarly arranged in an E-W plane. Each telescope comprised three trays of ten counters each, with

25 cm spacing between trays. Counters 4 cm in diameter with an effective length of 40 cm were used. Triple coincidences between counters in the three trays were registered on electromechanical counters which were photographed every 15 minutes together with a clock and an aneroid barometer to provide time and barometric pressure. Two arrays were used, one with 35 cm of Pb in the train and the other with no absorber. With the axes of the arrays inclined 45° to the vertical the counting rates were about 7000 and 10000 coincidences per hour respectively for the arrangement with and without absorber. Such an arrangement has the advantage that atmospheric effects should be the same for both telescopes so that differences between the two are due to real effects dependent on direction.

4. Neutron monitors. It has been shown by SIMPSON [6], [7], [8] and others that the low energy nucleonic component produced within the atmosphere by the low energy portion of the primary particle spectrum, exhibits the largest

Table 2. *Location and elevation of continuously recording neutron monitors.*

Station	Geomagnetic latitude deg.	Elevation Feet
Chicago, Illinois	52 N	600
Climax, Colorado	48 N	11000
Sacramento Peak, New Mexico	42 N	9800
Mexico City, Mexico	29 N	7600
Huancayo, Peru	0.5 S	11000

of the known geomagnetic latitude variations. At atmospheric depth 312 g cm^{-2} the variation in neutron intensity between the equator and 50° N was found to be about three times greater than that obtained in an ionization chamber. At sea level the ratio is even larger. A detailed analysis of neutron variations with latitude leads [7] to the conclusion that the ratio of the cross section for processes leading to neutron production to that for meson production increases rapidly with decreasing energy of primary particles. This means that neutron detectors are much more sensitive to changes in the low energy part of the primary spectrum than are ionization chambers which are mainly sensitive to the μ-meson component. For most of the time-variations the percentage changes from neutron counters are between 2.5 and 5 times greater than those from ionization chambers. The percentage increases in intensity associated with solar flares are of the order of twenty times greater than in ionization chambers [9]. Moreover, the neutron intensity is not affected by changes in the height of, say, the 100 mb pressure level in the atmosphere which can alter the meson intensity several per cent at the ground.

Several neutron piles have been constructed and put into operation by SIMPSON. These use proportional counters filled with enriched boron-10 trifluoride. The counters are surrounded by a lead-paraffin pile. Most of the neutrons are locally produced in the lead and slowed down to thermal energies by the paraffin, then captured by the boron-10. Table 2 indicates the location and elevation where neutron monitors are in continuous operation under the direction of JOHN A. SIMPSON, University of Chicago. Further description of the apparatus is given in reference [6]. Also, complete details of the construction, maintenance, and operation of the neutron monitors are given in reference [6a]. These standardized neutron monitors have been adopted for observations during the International Geophysical Year (July 1957 to December 1958).

B. Atmospheric effects.

I. Barometer effects.

5. Ionization chambers. The barometric effect on cosmic-ray intensity as measured in ionization chambers has been known since 1928 when it was discovered by MYSSOWSKY and TUWIN. The barometric coefficient is generally determined through the correlation between daily means of ionization and barometric pressure, an example of which, for 106 days of data from Cheltenham is shown in Fig. 4. The correlation coefficient for this sample is — 0.84. The slope of the regression line which minimizes the sum of the squares of the differences between the observed values of ionization and those calculated from a linear relation between pressure and ionization is simply the product of the correlation coefficient and the ratio of the standard deviation of ionization values to the standard deviation of the pressure values.

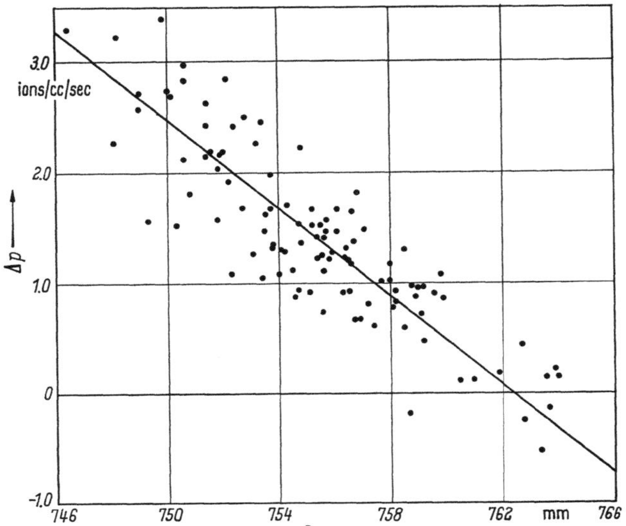

Fig. 4. Correlation between daily means (at Cheltenham, Maryland) of barometric pressure and departures from balance (Δ_p) 106 days, June 1 — September 30, 1936.

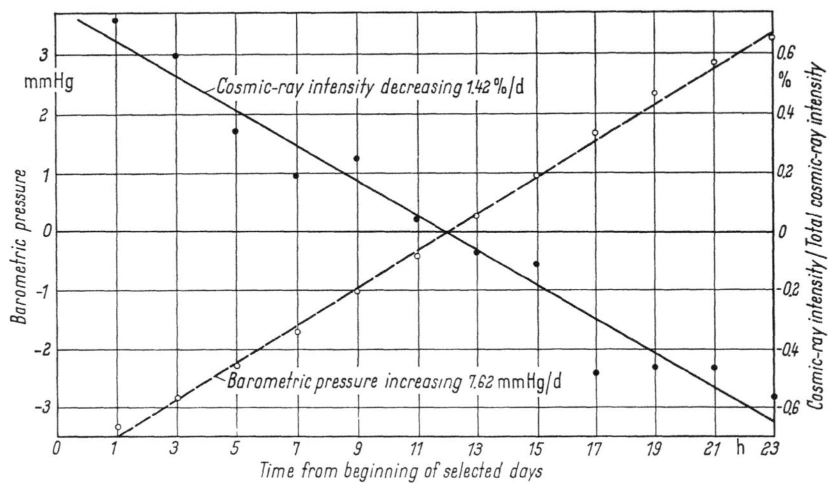

Fig. 5. Mean departures from average of bihourly values of cosmic-ray intensity and barometric pressure, derived from 15 selected 24-hour intervals with increasing pressure. October 7 to November 18, 1938, Godhavn, Greenland.

This procedure suffers the disadvantage that the daily means of ionization are often altered by world-wide changes, or by the passage of meteorological fronts over stations in temperate or polar latitudes. These effects can introduce appreciable errors into the barometric coefficient when it is calculated as described above. A better procedure, which mitigates the consequences of changes in ionization which are not due to pressure, is to select several intervals of 24 (or 48) hours length during each of which the barometric pressure is increasing (or decreasing), and to average, over all the selected intervals, the barometric pressure and ionization for each of the twelve bihourly divisions in each of the selected 24-hour intervals. A sample of results obtained at Godhavn (Greenland) is shown in Figs. 5 and 6 for 15 selected intervals with increasing pressure and in Figs. 7 and 8 for 15 selected intervals with decreasing pressure. Figs. 5 and 7 indicate that on the average the bihourly means of pressure and of ionization were changing linearly during the selected intervals. Figs. 6 and 8 indicate the correlation between corresponding averages of bihourly means of barometric pressure and ionization for the two sets of selected intervals. With this procedure the agreement between results from different

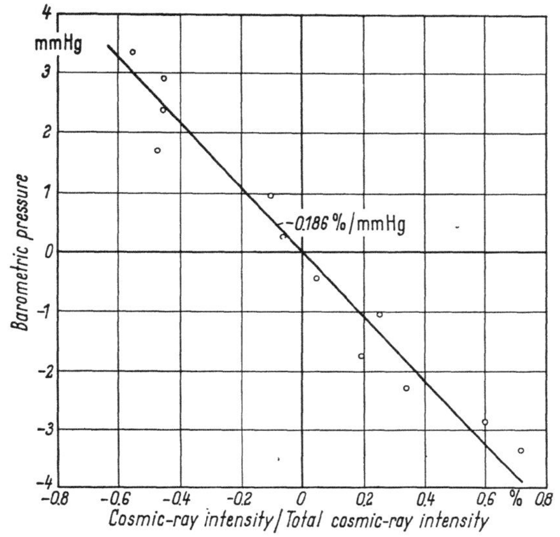

Fig. 6. Correlation between mean departures, from average, of bi hourly values of cosmic-ray intensity and barometric pressure, derived from 15 selected 24-hour intervals with increasing pressure. October 7 to November 18, 1938, Godhavn, Greenland. Correlation coefficient $r = +0.982$.

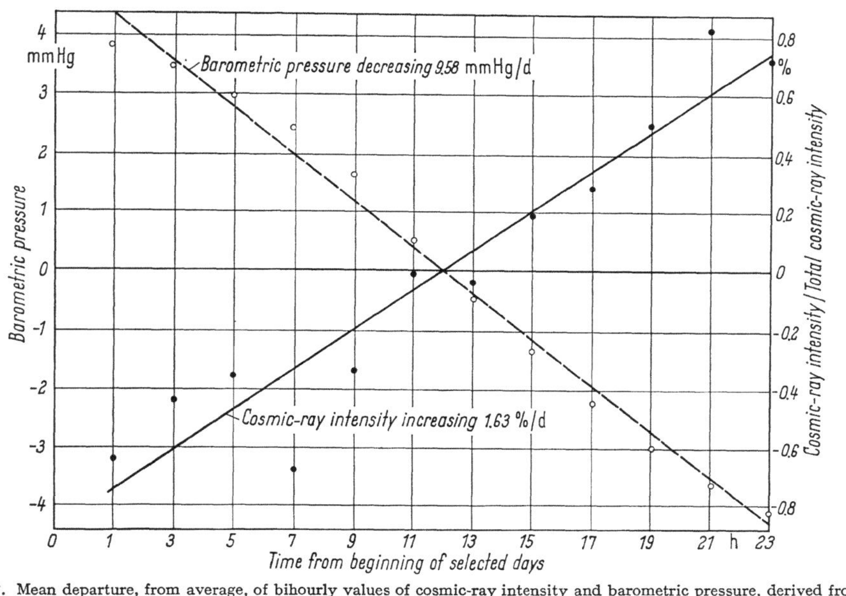

Fig. 7. Mean departure, from average, of bihourly values of cosmic-ray intensity and barometric pressure, derived from 15 selected 24-hour intervals with decreasing pressure. October 7 to November 18, 1938, Godhavn, Greenland.

samples is generally quite close—and no systematic differences were found between results for increasing and decreasing pressure.

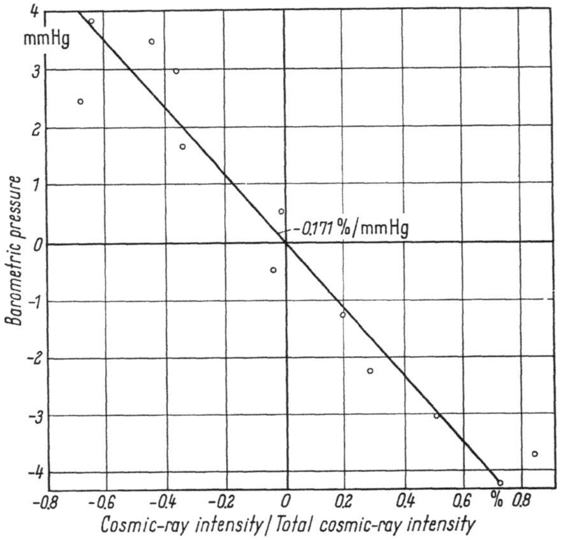

Fig. 8. Correlation between mean departures from average, of bihourly values of cosmic-ray intensity and barometric pressure from 15 selected 24-hour intervals with decreasing pressure. October 7 to November 18, 1938. Godhavn, Greenland. Correlation coefficient $r = +0.966$.

6. Special case: Huancayo. At Huancayo the daily mean barometric pressure changes only a very few mm of Hg during a month while the cosmic-ray ionization may ordinarily change a few percent. On the other hand the 12-hourly and 24-hourly waves in barometric pressure each have amplitudes of the order of one mm Hg at Huancayo. To mitigate the effect of the real changes in cosmic-ray intensity on the reliability of the determination of barometric coefficient, daily means were used only for selected intervals. The majority of intervals selected comprised four days; no interval was shorter than four days and few were as long as eight or nine days. The intervals selected were characterized by monotonic increases (or decreases) of the daily means of barometric pressure. For each interval, the departures, from the average of the interval, of each daily

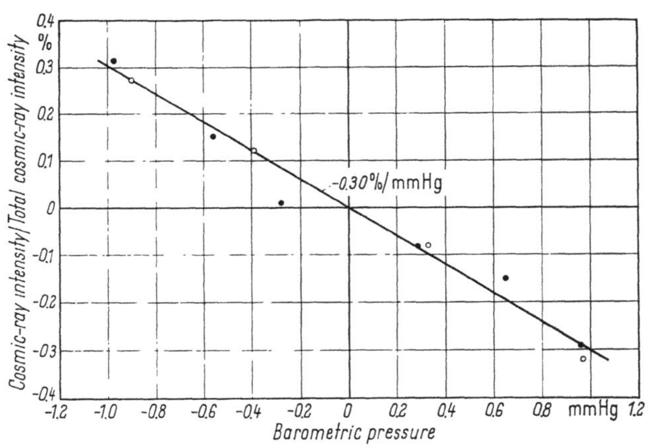

o = Average of 61 departures from mean of 61 selected intervals from March 1938 to March 1940.
• = Average of 30 departures from mean of 30 selected intervals from June 1936 to September 1937.

Fig. 9. Barometric coefficient for cosmic-ray intensity at Huancayo Observatory from correlation between average departures in cosmic ray- intensity and in barometric pressure from means for selected days. Days selected from weekly intervals with greatest changes in pressure. Regression line indicated, $r = 0.99$.

mean pressure and ionization were obtained. The departures in pressure were ranked (by magnitude and sign) and averages of all departures in each rank-interval were derived for pressure and for ionization. These are plotted in Fig. 9.

Each circled point is based on the average of 61 departures and each of the remaining points on the average of about 30 departures. The indicated slope of the line (-0.30%/mm Hg) was determined by least squares giving all weight to the pressure departures. A similar determination, in which the ranking was based on the ionization departures according to size, resulted in a barometric coefficient of -0.68%/mm Hg when all weight was given to the ionization departures. The large difference in the two barometric coefficients is due to the rather low ($r = -0.70$) correlation between pressure and ionization departures for single days. The slope of the regression line which gives all weight to barometric pressure is that shown in Fig. 9. It is the coefficient actually used for Huancayo.

At Huancayo the phase of the 12-hour sine-wave in ionization is practically 180° from that in barometric pressure, which has an amplitude of about 1 mm Hg. The ratio of the amplitudes of these two waves gave an apparent barometric coefficient in good agreement with that shown in Fig. 9. The barometric coefficients indicated in Table 1 are those used in correcting the long series of observations published by the Carnegie Institution of Washington [*10*] for the period 1936 to 1959.

7. Counter telescopes. During 1951 and 1952 Dawton [*11*] and Elliott carried out experiments at Manchester to investigate time variations of the hard and soft component of cosmic radiation. Three counter trays each 40×40 cm were placed vertically above each other with a separation of 25 cm between trays. Between the two lower trays they placed a slab of lead 10 cm thick. Two-fold coincidences (C_{12}) between the upper two trays and between the bottom two trays (C_{23}) were recorded. The coincidence rates, C_{12} and C_{23}, thus respectively measured the intensities of the soft and hard components. Following Duperier's procedure [*12*] they computed the partial correlation coefficients between the intensities and atmsopheric pressure, between the intensities and the height of the 100 mb atmospheric pressure level, and between the intensities and the temperature at the 100 mb level. From the correlation between the intensity, C_{23}, of the hard component and barometric pressure a pressure coefficient of -0.17%/mm Hg was obtained. This is in good agreement with the values in Table 1 for Compton-Bennett meters at Godhavn, Cheltenham, and Christchurch with 12 cm Pb shielding.

8. Neutron monitors. As already indicated in Sect. 5, the intensities from neutron monitors are not affected by μ-meson decay effects. The only atmospheric effect of consequence is the barometric effect which is large, for which Simpson [*6*] and coworkers find the value 0.94%/mm Hg. They also show that their neutron barometric coefficient is essentially independent of latitude and of altitude for atmospheric depths greater than 600 g cm^{-2}. In order to obtain useful information from their very high counting rates, they take special precautions to obtain accurate values of barometric pressure to insure that the statistical reliability of the data is not limited by uncertainties in the barometric corrections.

II. μ-Meson decay effects and seasonal variations.

9. μ-Meson decay effects. Several investigators [*13*], [*14*] using ionization chambers have indicated negative correlations between the cosmic-ray intensity and temperature at the ground level. Blackett [*15*] was the first to correctly explain these effects as due to the instability of μ-mesons. He pointed out that the pressure level in the atmosphere where most of the μ-mesons were formed, now known to be in the region of the 100 mb pressure level, would be higher as the atmospheric temperature increased, and consequently from the greater height more μ-mesons would decay before reaching the ground thus resulting in a decrease of ionization. The correlations between ground temperature and cosmic-ray ionization were due entirely to the fact that both exhibit a seasonal variation. This was shown [*16*] from the fact that the variability of daily means of ground temperature within periods of a month or so exhibited no statistically significant correlation with the cosmic-ray ionization. Using the arrangement described in Sect. 7 DAWTON and ELLIOTT found from the partial correlations between: intensity and barometric pressure, intensity and height of the 100 mb level, and intensity and temperature at the 100 mb level, that an increase of 1 km in the height of the 100 mb level resulted in a decrease of about 4.0% ± 0.4% in the intensity of the hard component under 10 cm Pb. This effect of changing height of the 100 mb level is thus large enough to introduce serious errors of purely meteorological origin into the results at some stations. The consequences of this effect will be pointed out later in discussing the comparison of world-wide changes at several stations.

10. Seasonal variations. With the arrangement described above, in Sect. 7, DAWTON and ELLIOTT found (1950—1951) a seasonal wave of amplitude about 1.5%, with maximum near February. Their results also show a seasonal variation, of amplitude about 300 m, in the height of the 100 mb level at Manchester, with maximum about February. With the decay coefficient of −4.0% per km, discussed in Sect. 9, an amplitude of about 1.2% in the hard component would be inferred from the height changes; this differs little from the observed amplitude of 1.5%, and the phases are consistent. It should be noted that DAWTON and ELLIOTT derived the decay coefficient, not through the correlation between monthly means of height of the 100 mb layer and intensity, but from the correlation between the departures of the daily means of these from their monthly averages. Thus it seems quite certain that most of the seasonal wave is due to the seasonal variation in height of the 100 mb layer.

In an analysis [*17*] of cosmic-ray data (ionization chambers) from several stations no seasonal wave was found for Huancayo. However, it was shown that if the seasonal waves at other stations were deducted from the data the residual variations were quite similar at all stations. In the harmonic dials of Fig. 10 the seasonal wave for each of several years is shown for Huancayo, and for Godhavn, Cheltenham, and Christchurch after deducting the monthly means for Huancayo from the monthly means at these stations. This procedure removes the world-wide variations [*17*]. For Cheltenham the seasonal wave is quite consistent for the several years indicated, and the amplitude[1], about 1.7%, of the average wave and its phase are in fair agreement with that obtained by DAWTON and ELLIOTT at Manchester. At Godhavn there is considerable variability among the twelve month waves for different years.

It is doubtful if the average seasonal wave for Huancayo is significant in view of the large variability which is probably due to the erratic nature of world-wide changes. If the monthly mean heights, averaged for ten years, of the 100 mb pressure level at Washington are plotted against the corresponding monthly mean values of ionization averaged for the same ten years, for Cheltenham minus Huancayo, the points do not fall on a straight line. The points deviate from a line in a manner suggesting that the differences, Cheltenham minus Huancayo, contain a seasonal wave not in phase with the seasonal wave in the height of the 100 mb layer at Cheltenham. This Lissajou characteristic disappears

[1] Amplitude denotes here, and throughout the chapter, the factor c in the sine-wave $c \sin(n t + \varepsilon)$; the difference of maximum and minimum is $2c$.

if a wave with amplitude about 0.3% and maximum in October (as in Fig. 10) is first deducted from Huancayo. Thus it seems likely that there is a small seasonal wave at Huancayo.

No data are yet available from neutron meters for long periods to determine if significant seasonal variations are present in neutron intensities. In any case

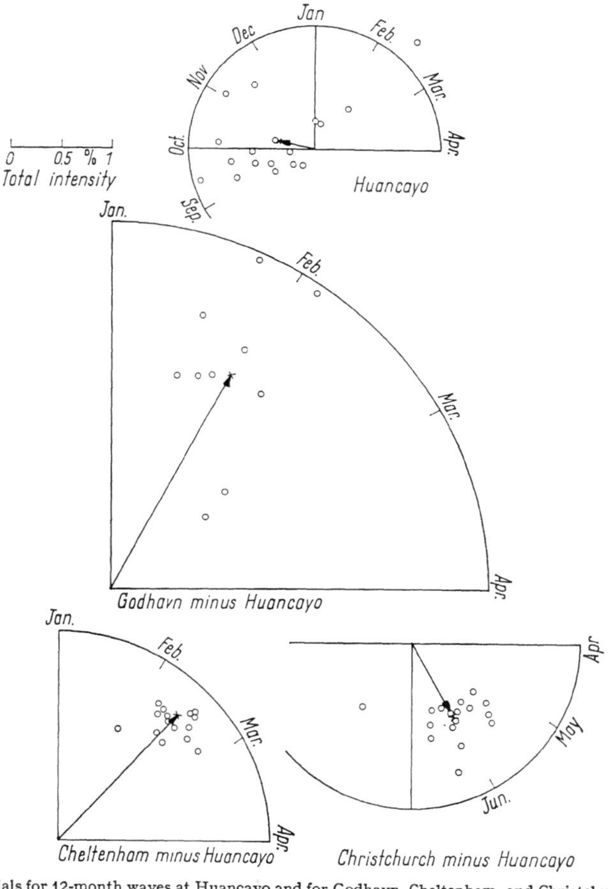

Fig. 10. Harmonic dials for 12-month waves at Huancayo and for Godhavn, Cheltenham, and Christchurch after deducting 12-month waves at Huancayo.

the neutron data are free of the μ-meson decay effects which contribute much to the variability of ionization-chamber data especially in temperate and polar latitudes. However, evidence will later be given indicating that variations in μ-meson decay effects contribute little to the variability of the daily means of cosmic-ray intensity measured at Huancayo with an ionization-chamber.

C. Diurnal variations.

I. Solar diurnal variations.

11. Methodology. *α) Point clouds in harmonic dials.* Many early investigations of time variations of cosmic-ray intensity have published results indicating diurnal variations. Most of these results have been in the form of averages with inadequate estimates for the reliability or the statistical reality of those averages. Those fields of geophysics which involve time variations and their variability demand the application of modern statistical procedures. Indeed in many cases these methods provide the only rational basis for deciding what is significant. No one has done more to foster the use of such techniques and to make clear the fruitfulness and power of such procedures than J. BARTELS [18], [19]. It is not the purpose here to attempt any complete descriptions of his contributions but only to illustrate by means of examples their applicability to studies involving the diurnal variation of cosmic-ray intensity.

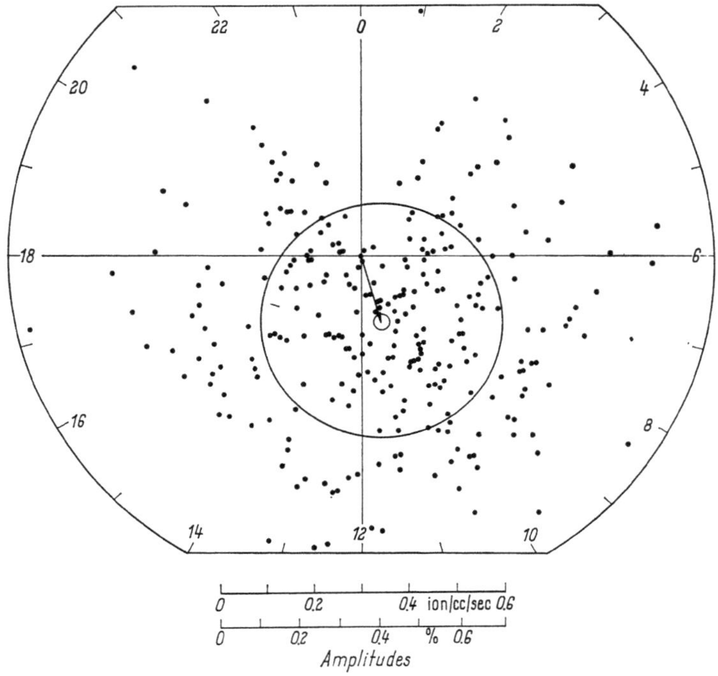

Fig. 11. 24-hour harmonic dial, apparent cosmic-ray intensity, 273 single days during April 20, 1935 to October 27, 1936, Cheltenham, Maryland (times of maximum in 75° West Meridian Mean Hours).

The harmonic dials in Figs. 11 and 12 show respectively the 24-hour and 12-hour waves for each day in a sample of 273 days of cosmic-ray ionization at Cheltenham corrected for barometric pressure. The vectors indicate the average 24-hour and 12-hour waves. The question to be answered is whether the average wave is statistically significant. The coordinates for the points in Figs. 11 and 12 are respectively a_1, b_1 and a_2, b_2 which derive from the harmonic analysis (corrected for non-cyclic change) of the 12 bihourly means of cosmic-ray ionization for each day. Each point in the dials is the end point of a vector, the length of which is the amplitude of the wave. The vector points to the time of the wave-maximum marked on the peripheral dial. The two-dimensional Gaussian frequency-distribution which best fits the cloud is in general elliptical [18]. The axes P_1, and P_2 of the probable ellipse, that is the ellipse which contains half the points inside it are given by:

$$P_1, P_2 = 0.833\left\{(\sigma_a^2+\sigma_b^2) \mp \left[(\sigma_a^2-\sigma_b^2)^2+4r^2\sigma_a^2\sigma_b^2\right]^{\frac{1}{2}}\right\}^{\frac{1}{2}}$$

in which σ_a^2 and σ_b^2 are the sample variances of a and b, and r is the coefficient of correlation between a and b. In case of circular symmetry (which applies to Figs. 11 and 12) $P_1 = P_2 = 0.833\,M$, with $M^2 = \sigma_a^2 + \sigma_b^2$. For the dial in Fig. 11: $\sigma_{a_1} = 0.245$, $\sigma_{b_1} = 0.258$, $M_1 = 0.357$, and $C_1 = \sqrt{a_1^2 + b_1^2} = 0.169$, all in units of 1% of the total cosmic-ray ionization. The larger circle in Fig. 11 is the so-called probable-error circle for single days, its radius is $0.833\,M = 0.298\%$, and the number of points inside it is 138 or about half the total (273). If the deviations, from the average, of points for successive days are independent then the radius of the probable-error circle for the mean of 273 days should be $0.298/\sqrt{273} = 0.0180\%$. The mean amplitude $C_1 = 0.169$ is thus about 9.4 times the radius of its probable error circle. Thus for samples of 273 days from a population with $C_1 = 0$ the probability [*19*] that the average vector has a length $\geqq C_1 = 0.169$ is $(\frac{1}{2})^{9.4^2} = 10^{-26}$, and the hypothesis $C_1 = 0$ is rejected. For the dial in Fig. 12 the corresponding probability is about 3×10^{-4} so that the hypothesis that the population value $C_2 = 0$ would also be rejected.

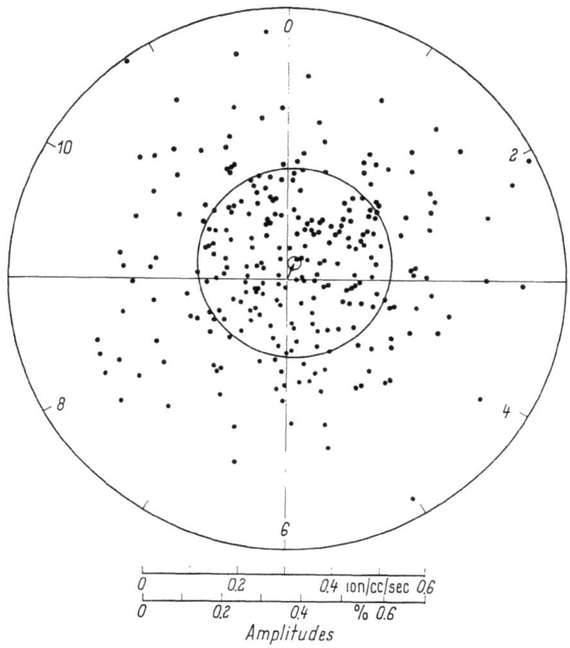

Fig. 12. 12-hour harmonic dial, apparent cosmic-ray intensity, 273 days during April 20, 1935 to October 27, 1936, Cheltenham, Maryland (times of maximum in 75° West Meridian Mean Hours).

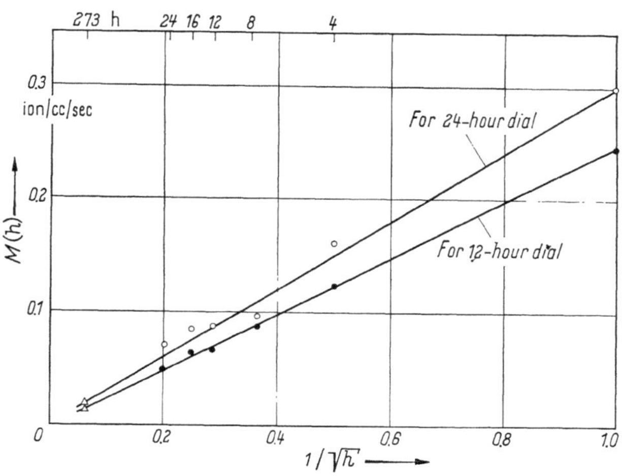

Fig. 13. Test for independence of points for successive days in 24-hour and 12-hour harmonic dials, April 20, 1935 to October 27, 1936, Cheltenham, Md.

β) Random walk. A slightly different approach is through the application of random walk theory discussed in a famous paper by Bartels [*19*] which leads to the same conclusion, namely that both waves are statistically significant. In either case the probabilities for reality depend upon the assumption of statistical independence of points in the dial derived from successive days. Fig. 13 tests this independence for both dials by indicating the dependence [*19*] on h, of the two dimensional standard deviation, $M(h)$, derived from means of h successive days, when values of $h=1$, 4, 8, 12, 16, and 24 were used.

The lines in Fig. 13 indicate how $M(h)$ would decrease for independence. Thus in this sample there is no indication of lack of independence. In Fig. 13 $M(h)$ is given in units of ions $\mathrm{cm^{-3}\,sec^{-1}}$ in which units the total intensity was 84. It may be noted in Fig. 13 that $M_1=0.300$, and $M_2=0.246$ for the 24-hour and 12-hour waves respectively. From an analysis of the differences between hourly means from two Compton-Bennett meters operating simultaneously at Cheltenham, a value of 0.7% was found for the standard deviation of statistical fluctuations in hourly mean values from either instrument. It has been shown [*19*] that the two-dimensional standard deviation M, in harmonic dials derived from harmonic analysis of sets of r random ordinates with standard deviation ξ is given by: $M=2\xi/\sqrt{r}$. For Cheltenham putting $\xi=0.7\%$ and $r=24$, $M=0.286\%$ or $M=0.240$ ions $\mathrm{cm^{-3}\,sec^{-1}}$. This means that the values of M derived from the data in the harmonic dials of Figs. 11 and 12 can not be less than the value $M=0.240$ ions $\mathrm{cm^{-3}\,sec^{-1}}$. In these units the values $M_1=0.300$ and $M_2=0.246$ were obtained for the 24-hour and 12-hour clouds respectively. Thus practically all the scatter in the 12-hour dial can be ascribed to the statistical fluctuations of hourly means.

The scatter in the 24-hour dial can be conceived as due to two independent causes, one of which is that arising from statistical fluctuations in hourly values and the other from variations, from other causes, in the 24-hour wave from day to day. The standard deviation for the first cause was 0.286% and for the second the value 0.214%. If for example the standard deviation of statistical fluctuations in hourly values were only 0.15%, as it might well be in a sufficiently large ionization chamber, then from this cause alone the two-dimensional standard deviation for points in the harmonic dial would be: $2\times 0.15/\sqrt{24}=0.061\%$. Suppose the variability of the wave itself from day to day, — a physical phenomenon which would be present even if the standard deviation of hourly values were negligible (i.e. an infinitely large chamber or counting rate), — were characterized by a two-dimensional standard deviation of say 0.214%, as above. The total standard deviation, from these two independent causes, for the points in the harmonic dial would be $\{0.214^2+0.061^2\}^{\frac{1}{2}}$ or about 0.222%. In this case it is quite evident that the mere specification of the statistical uncertainty of hourly values provides a practically useless lower limit for estimating the reliability of individual or average waves in harmonic dials. Nevertheless, this unfortunate practice continues.

12. 24-hour variation from counter telescopes. It was long realized that the small 24-hour wave in cosmic-ray intensity measured by ionization chambers and counters might be due entirely to atmospheric effects. Partly for this reason ALFVÉN [*20*] and MALMFORS in Stockholm, KOLHÖRSTER [*21*] in Berlin and more recently DOLBEAR [*22*] and ELLIOTT in Manchester using counter telescopes of the type described in Sect. 7 have measured the diurnal variation in different directions: With the axes of the two telescopes in the North-South plane and inclined 45° on opposite sides of the zenith. ELLIOTT and DOLBEAR [*23*] made measurements over a period of a year, and confirmed the earlier results of ALFVÉN and MALMFORS indicating the diurnal variation was significantly different for the North and South pointing telescopes. Since the particles recorded by the two sets of telescopes (N and S) pass through the same amount of atmosphere, it was concluded that at least part of the variation was due to causes outside the atmosphere. The difference between the variations in the two directions was interpreted as evidence for lack of isotropy of the primary rays.

This conclusion was later confirmed from the difference in diurnal variation obtained by ELLIOTT and DOLBEAR [*24*] with telescopes inclined in an East-West plane, both without absorber and with 35 cm of lead. They suggested that the anisotropy of the primary radiation appeared to be due entirely to solar influences, which as suggested by ALFVÉN [*25*] might arise from polarization effects in the so ar streams responsible for magnetic disturbances. They dichotomized data for 360 days according to the sum of the $8K$ indices (per day) of magnetic activity. For the "quiet" group this sum was $\leqq 17$, and for the other "disturbed" group it was $\geqq 18$; this division was selected to obtain the same number of days in each group. It was found that the amplitude of the average diurnal variation was about twice as large (about 0.2%), in the South minus North diurnal variation, on the geomagnetically "disturbed" days as on the "quiet" days.

13. World-wide variation in annual mean 24-hour wave from ionization chambers and counter telescopes. THAMBYAHPILLAI and ELLIOTT [*26*] were the first to point out that the local time of maximum for the yearly averages of the 24-hour wave, exhibited a large systematic variation over the period of 20 years or so for which data from various stations were available. Fig. 14 extends the results to include data for 1953 and 1954 for Huancayo and Cheltenham and Christchurch. It is evident that the variations indicated in Fig. 14 are undoubtedly real considering the general agreement between the results from different stations for the same year. The results from all the stations [except those for 1932, 1933, and 1934, from Hafelekar] are based on data which were corrected only for variations of barometric pressure. Those from Hafelekar were, in addition, unfortunately corrected for temperature. THAMBYAHPALLAI and ELLIOTT noted, from the results through 1952, the possibility of a 22 year wave in the phase. SARABHAI and KANE [*27*] claim, from observational data published by the Carnegie Institution that there are world-wide variations not only in the phase but also in the amplitude of the 24-hour diurnal variation.

14. Diurnal variation from neutron meters, its variability and comparison with ionization chamber results. FONGER [*28*] and SIMPSON [*29*], FIROR, and TREIMAN have compared the average diurnal variation obtained with neutron monitors at Climax with that from the large ionization chamber at Freiburg. The results are shown in Fig. 15 in which the observed values at Freiburg have been multiplied by 5. The authors estimated the standard deviations of random statistical fluctuations for each point, in Fig. 15, on the curve for Climax to be about 0.08% and that for points on the Freiburg curve to be about 0.023%. However, it seems quite improbable that the random statistical fluctuations in hourly values are the only cause for variability in the differences of corresponding hourly values on individual days at the two stations. For example it is quite possible that uncertainties in the corrections for μ-meson decay effects may introduce a variability into the diurnal variation at Freiburg, which should not affect the neutron data. In addition there were world-wide variations of appreciable magnitude during the period when the data for Fig. 15 were obtained. These world-wide variations will be simultaneous at all stations—thus introducing differences between the *local* time variations at the two stations. Thus (see also Sect. 11) the standard deviations for points in Fig. 15 may be grossly underestimated. In view of this, the differences between the two curves in Fig. 15 may not be significant. The similarity of the curves is direct experimental indication that the diurnal variation from pressure corrected data from ionization chambers is not seriously affected by other atmospheric effects.

Fig. 16a indicates the diurnal variation (d.v.) in neutron intensity [29] at Climax and at Huancayo averaged for 10 selected days in 1952 with *small* d.v.

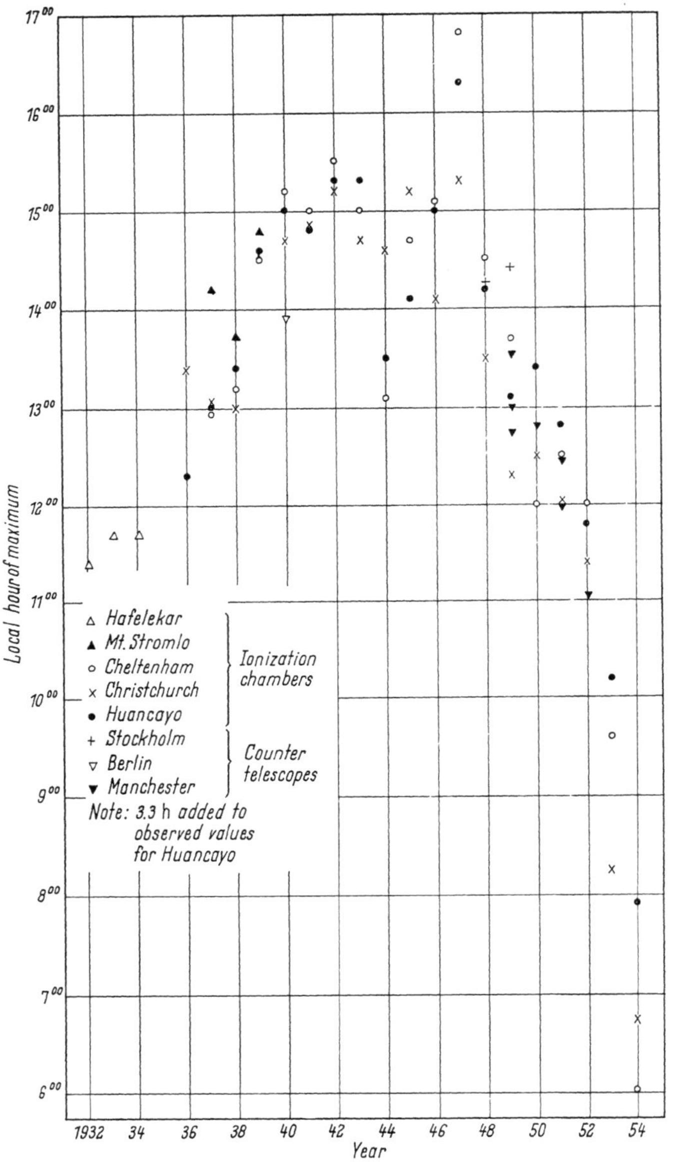

Fig. 14. Variation of local time of maximum of sketch 24-hour wave in cosmic-ray intensity.

at Climax while Fig. 16b indicates the averages for 10 selected days in 1952 with *large* d.v. at Climax. The results indicate *world-wide* variations in the amplitude of the diurnal variation on local solar time. SITTKUS [30] has shown, from data registered by the large ionization chamber at Freiburg, that days with unusually large diurnal variation, 1% or so, frequently occur on several successive days and that these occurrences exhibit a 27-day recurrence tendency. A comparison of these results and the data of EHMERT [31] in Weissenau with those of SEKIDO [32] in Tokyo indicated the world-wide occurrence of days with large diurnal variations *on local time*.

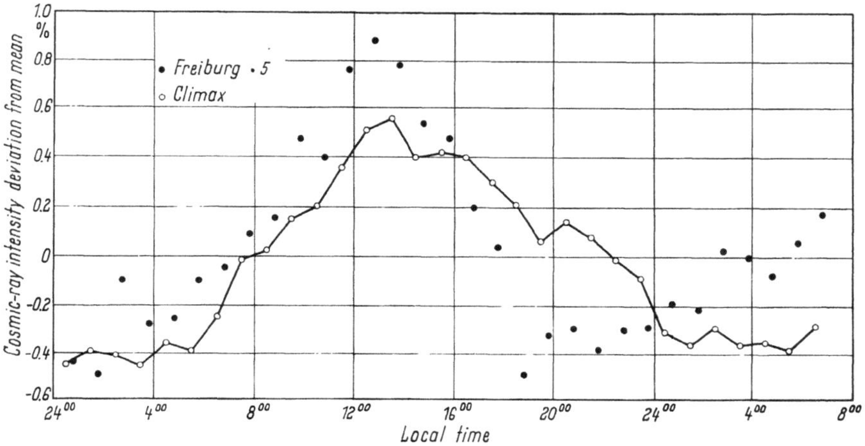

Fig. 15. Mean hourly values of cosmic-ray intensity averaged for 74 days in the interval July 14 to October 17, 1951 from neutron data at Climax and from ionization data at Freiburg (FONGER).

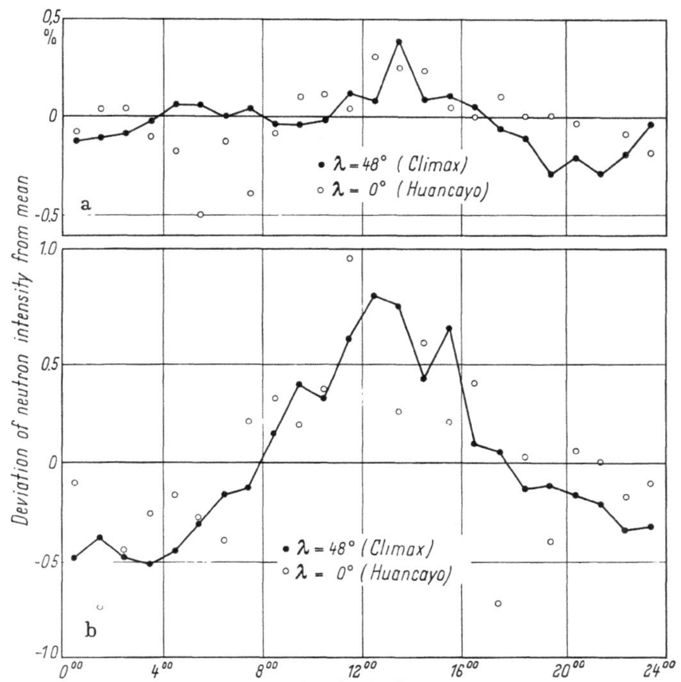

Fig. 16a and b. Diurnal variation (d.v.) in neutron intensity at Climax and Huancayo. a Average of 10 days with small d.v. at Climax. b For average of 10 days with large d.v. at Climax (FIROR, FONGER and SIMPSON).

15. Magnetic activity and the diurnal variation. In Sect. 12 it was indicated that the bihourly differences of cosmic-ray intensity from Northward and Southward pointing telescopes exhibited a diurnal variation which was greater on geomagnetically disturbed days than on quieter days. SEKIDO and YOSHIDA [*32*] concluded, from data with vertical counter telescopes, that the amplitude of the diurnal variation increased during magnetic storms and that the local time of maximum of the 24-hour wave came earlier. Also ELLIOTT and DOLBEAR [*33*] concluded from two magnetic storms (with associated decreases in cosmic-ray intensity) that the amplitude of the diurnal variation was much greater than normal. Finally YOSHIDA and KONDO [*34*] claim a 27-day recurrence tendency in the amplitude of the diurnal variation at Huancayo based on results for the period 1936—1940.

During these magnetic storms which are accompanied by large decreases in cosmic-ray intensity there are simultaneous world-wide variations (on universal time) within each storm. These variations may well have serious effects on the results for the diurnal variation on local time. Thus it would appear that it is difficult to establish the effects on the local time variation of cosmic-ray intensity during magnetic storms.

II. Sidereal diurnal variation.

16. Basis for sidereal variation. COMPTON and GETTING [*35*] first pointed out that if cosmic-rays originated outside our galaxy then a sidereal diurnal variation in intensity should arise as a consequence of the linear velocity of the earth due to the rotation of our galaxy and to the fact that the earth is far from the center of the galaxy. On the other hand no sidereal variation would be expected if the cosmic-rays originated in our own galaxy. The theory of the effect of galactic rotation was developed by COMPTON and GETTING without taking account of the deflection of charged particles by the earth's magnetic field. From rough estimates they concluded that the effect of deflections in the earth's magnetic field would decrease the amplitude of the sidereal diurnal variation to about one fifth of that computed in the absence of the earth's field.

VALLARTA, GRAEF, and KUSAKA [*36*] extended the theory of the galactic rotation effect to take account of the geomagnetic deflection of charged particles. Their investigation was confined to particles moving in the plane of the geomagnetic equator for which situation the equations of motion are integrable. For trajectories not in the geomagnetic equator and for non-equatorial latitudes they pointed out that the calculations are more difficult since the equations of motion are no longer integrable. They showed that the time of maximum for the 24-hour wave for all positive primaries depended rather critically, for particles in the geomagnetic equatorial plane, upon the energy distribution of primaries. With the number of primaries, all positive, varying inversely as the cube of their energy the maximum of the 24-hour wave was calculated to occur at about 13 hours sidereal time and with amplitude 0.17%. With the number of primaries, all positive, decreasing exponentially with energy, the calculated maximum occurred at 18 hours sidereal time with amplitude 0.24%. This comparison of observed and theoretical sidereal variations was only valid for observations made at the geomagnetic equator and then with rigor only for counter telescopes to insure that the particles counted arrive in or near the geomagnetic equatorial plane.

17. Search for a sidereal variation in the observations. Several investigators have reported 24-hour sidereal waves in cosmic-ray intensity. Most of these results have presented only the average 24-hour sidereal wave, with no determination of its variability, without which a reliable test for the statistical reality of the average wave is impossible. Using data for 595 days from a Compton-Bennett meter at Cheltenham FORBUSH [37] found a 24-hour sidereal wave with amplitude 0.03% and with maximum near 22 hours sidereal time.

To test whether this average amplitude was too large to be ascribed to chance he used the two dimensional standard deviation derived from the scatter of points for individual days in the harmonic dial for the 24-hour solar diurnal variation in a sample of 273 days [38]. On this basis the average amplitude, 0.03%, of the sidereal wave at Cheltenham was small enough to be ascribed to chance and was, therefore, not regarded as real.

ELLIOTT and DOLBEAR [39] derived the 24-hour sidereal diurnal wave from the south minus north variation from counter telescopes at Manchester as described in Sect. 3. They found a 24-hour sidereal variation with amplitude about 0.03% and maximum near 8 hours sidereal time. This amplitude is about three times its two-dimensional standard deviation as they derived it, which would give a probability of about 10^{-4} of obtaining, by chance variations, an amplitude $\geqq 0.03\%$. However, their standard deviation was estimated on the basis of counting rate and as indicated in the last paragraph of Sect. 11 this is certainly an underestimate of the actual variability since it takes no account of the variability of the solar diurnal variation itself. ELLIOTT and DOLBEAR also derived from nine years of data the 24-hour sidereal waves for Huancayo, and for the average of Christchurch and Cheltenham. They obtained in both cases an amplitude of about 0.02%. The sidereal times of maxima for Huancayo and for the average of Cheltenham and Christchurch were respectively 4 hours and 6 hours.

Fig. 17 is a 24-hour sidereal harmonic dial based on data from Huancayo for the interval 1937—1954. Each point in the dial was obtained by deducting the yearly mean solar 24-hour wave from the average solar wave for one of the indicated bimonthly intervals, and then transforming the coordinates of this difference vector to a 24-hour sidereal dial by appropriate rotation of axes. The standard deviations of the ordinates and abscissae of the points, from their means shown by the end point of the average vector, were found to be respectively 0.047 and 0.044%. Since these two standard deviations are not significantly different the cloud of points has circular symmetry. The standard deviation of the distances between individual points and the end point of the average vector is thus the geometric mean of these two standard deviations or 0.064%. This corresponds to BARTELS' expectancy for single vector deviations from the mean [19]. The expectancy for the averages of 108 statistically independent vector deviations (corresponding to the 108 points in Fig. 17) is then $0.064/\sqrt{108}$ or about 0.0062%. The length of the average vector is 0.0155% so that $\varkappa = 0.0155/0.0062 \approx 2.5$ (see Sect. 11 and Ref. [18]) and $e^{-\varkappa^2} = e^{-6.25} \approx 0.002$. Thus the probability is about one in five hundred of obtaining by chance a vector as large or larger than the average actually obtained in the sample of 108 in Fig. 17. The standard deviations derived for averages of three chronologically successive vectors was found no larger than would be expected if these were independent, so the above probability is not underestimated from this cause. The large circle in Fig. 17 is the so-called probable error circle; it contains approximately half the points. The small circle is the probable error circle for the mean.

It is thus apparent that at Huancayo the 24-hour sidereal wave is small in amplitude, about 0.015%; furthermore its phase is about opposite to that calculated by VALLARTA and GRAEF on the basis of positive primaries confined to orbits in the plane of the geomagnetic equator. Finally, it must be emphasized that small systematic variations during the year in the phase and amplitude of the solar 24-hour wave can give rise to an apparent sidereal variation (see for example Fig. 14). *Thus from the 24-hour sidereal variation the evidence is certainly not convincing that any detectable fraction of cosmic-rays arrive at the earth from outside our galaxy.*

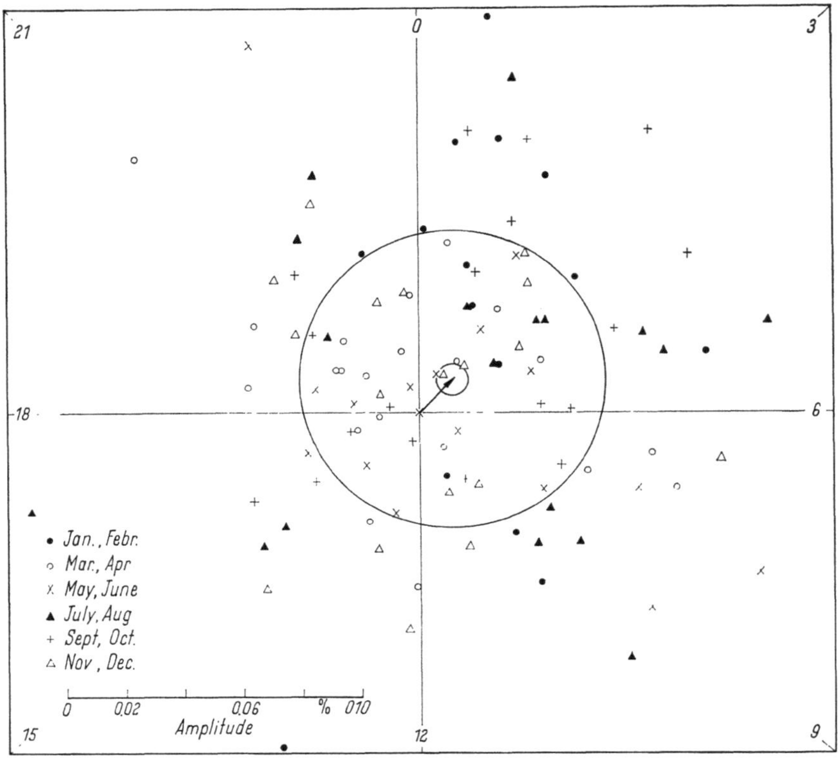

Fig. 17. Sidereal harmonic dial from data at Huancayo 1937—1954.

D. World-wide variations.

I. Variations associated with magnetic storms.

18. Some varieties of individual magnetic storm-effects. Fig. 3 (p. 163) is a reproduction of cosmic-ray records from Huancayo for the period January 15 to 18, 1938, and shows the decrease in cosmic-ray intensity during the magnetic storm which began at 22.7 hours G.M.T., January 16, 1938. Fig. 18 shows the bihourly means of cosmic-ray intensity averaged for three stations (all near the same longitude) and the daily mean horizontal magnetic intensity at Huancayo, Peru, for the period January 15 to 29, 1938. Fig. 19 indicates the correlation between the daily means of horizontal intensity at Huancayo and the daily means of cosmic-ray intensity for each of the three stations for which the average bihourly values were shown in Fig. 18. It is obvious in Fig. 19 that the ratio of the change in cosmic-ray intensity to that of horizontal magnetic intensity is more than twice as large for the storm beginning January 16 as for the storm following the sudden commencement on January 22. Fig. 20 compares the daily means of cosmic-ray intensity at three stations with those in horizontal intensity at Huancayo for the period January 11 to 31, 1938. Fig. 21 shows a similar comparison for the period April 16 to May 10, 1937.

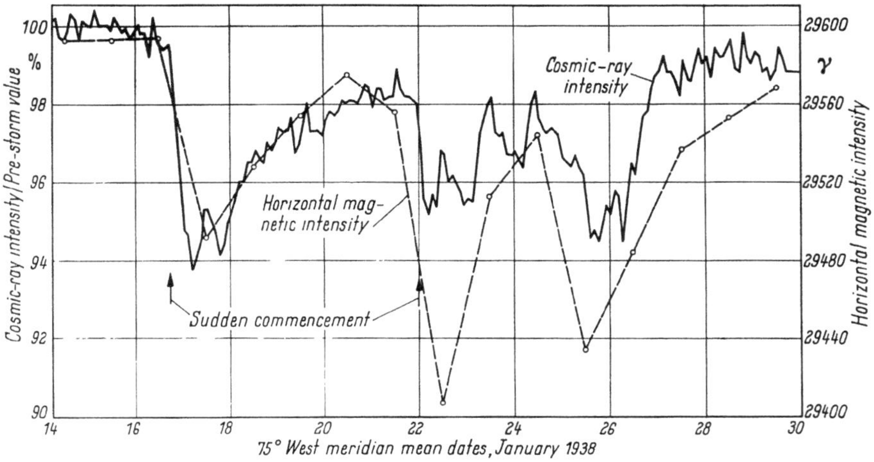

Fig. 18. Magnetic storm effects on bihourly mean cosmic-ray intensity averaged for Boston, United States, Cheltenham United States and Huancayo, Peru, and on daily mean magnetic horizontal intensity at Huancayo, Peru.

That not all magnetic storms are accompanied by decreases in cosmic-ray intensity is evident from Fig. 22 which shows no decrease in cosmic-ray intensity for a storm in which the daily mean horizontal magnetic intensity was depressed, on August 22, 1937, about 120 γ below normal (the geomagnetic planetary index Kp reached 8o).

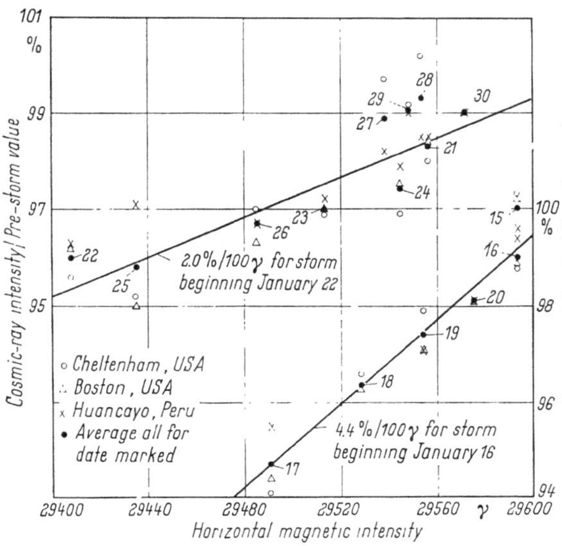

Fig. 19. Correlation between daily means of horizontal magnetic intensity at Huancayo, Peru and of cosmic-ray intensity at Boston, United States, Cheltenham, United States, and Huancayo, Peru, January 15—30, 1938.

Finally, Fig. 23 a and b shows the daily means in cosmic-ray intensity at Cheltenham and Huancayo together with the daily mean horizontal magnetic intensity for the period February 1 to 13, 1946. In this case the major decrease in cosmic-ray intensity preceded the major decrease in horizontal magnetic intensity by some three days. It may be of interest to indicate that a magnetic sudden commencement occurred at Huancayo and Watheroo toward the end of February 2 (75° WMT) but it is conjectural whether this was in any way connected with the mechanism responsible for the decrease in cosmic-ray intensity (see also Sect. 22)

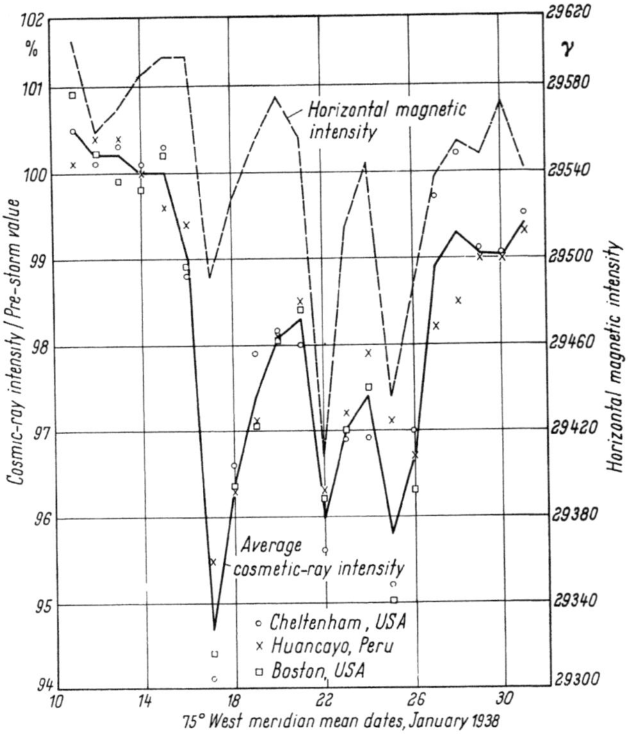

Fig. 20. Magnetic storm effects on daily mean cosmic-ray intensity at Boston, United States, Cheltenham, United States, and Huancayo, Peru, and Christchurch, New Zealand, and on magnetic horizontal intensity at Huancayo, Peru.

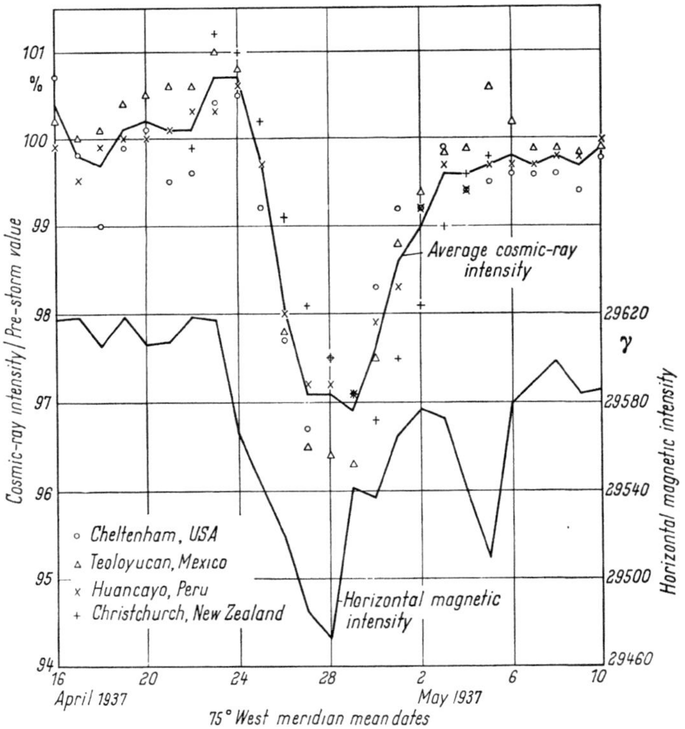

Fig. 21. Magnetic storm effects on daily mean cosmic-ray intensity at Cheltenham, United States, Teoloyucan, Mexico, Huancayo, Peru and Christchurch, New Zealand, and on magnetic horizontal intensity at Huancayo, Peru.

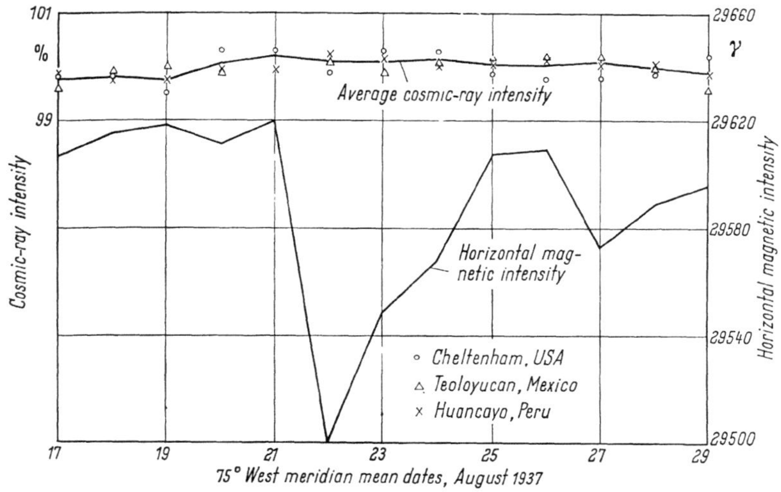

Fig. 22. Daily means horizontal magnetic intensity at Huancayo, Peru, and cosmic-ray intensity at Cheltenham, United States, Teoloyucan, Mexico, and Huancayo, Peru, showing no change in cosmic-ray intensity during magnetic storm beginning August 21, 1937.

Fig. 23a and b. Daily means cosmic-ray intensity at Cheltenham and at Huancayo (a), and daily means magnetic horizontal intensity at Huancayo (b), February 1–13, 1946.

Additional examples of storms with and without associated decreases in cosmic-ray intensity may be seen in Fig. 28 in which are plotted the daily means of cosmic-ray intensity at Huancayo for the period 1937—1953 together with all the available daily means of horizontal magnetic intensity at Huancayo. Daily means for selected years for other stations are also included in the figure.

19. Cosmic-ray effects and the ring current. The decreases in daily mean horizontal intensity at Huancayo during the magnetic storms of January 16, 1938 and of August 21, 1937, shown in Fig. 20 and 22 respectively, were of about equal magnitude although the associated cosmic-ray effects were radically different. If the mechanism responsible for magnetic storms were in any way connected with that responsible for cosmic-ray decreases, this would require that in some respect the storm of January 16, 1938, should differ from the storm of August 21, 1937. An attempt [*40*] was therefore made to estimate for these two storms, the radius of the hypothetical ring current, concentric with the earth, and in the plane of the geomagnetic equator which would give rise to the main storm field changes. The magnetic potential of the storm field was assumed to be represented by a series of zonal harmonics, of odd degree since the field is known to be symmetrical with respect to the magnetic equator. Only two terms were used. These two terms in the expression for the external potential suffice to determine the ratio, R/a, of the radius, R, of the assumed ring current source, to the radius, a, of the earth. Data for both storms were obtained from several magnetic observatories and were corrected for the effect of ionospheric auroral zone currents. The results indicated that if the source were such a ring current then for neither of the two storms could R/a have been much less than two. However, for either storm R/a could have been indefinitely larger than two. This uncertainty in R/a arises from the fact that it is not possible to eliminate with certainty the magnetic effects of the S_D currents which flow in the ionosphere.

In any case when cosmic-ray decreases occur during magnetic storms, they are world-wide, having been observed not only near the equator, but also near the magnetic poles, at stations like Godhavn, geomagnetic latitude 80° N, and Thule, geomagnetic latitude 88° N. If cosmic-rays were excluded from reaching the earth by the magnetic field of a ring current it would certainly not be expected that they would be excluded from regions so near the geomagnetic pole as Godhavn and Thule. Theoretical calculations of the effects of the magnetic field of a ring current on the trajectories of cosmic-ray particles have given results which do not appear to explain the observed cosmic-ray effects [*41*]. The results of an investigation by Treiman [*42*] indicate that an increase in cosmic-ray intensity should arise from a decrease in the magnetic field at the equator if this decrease is due to a spherical current sheet concentric with the earth.

II. Geomagnetic activity effects.

20. Cosmic-ray intensity for magnetically quiet and disturbed days. Fig. 24a and b indicate, respectively, for Huancayo and Cheltenham, and for Huancayo and Godhavn, the correlation between the average difference of cosmic-ray intensity for the five magnetically disturbed days of each month less that for the five quiet days. It is evident from Fig. 24a and b that the frequency of positive values of the differences for Huancayo is only about one-fifth that for negative values, which indicates definitely that the cosmic-ray intensity tends to be less for the five magnetically disturbed days than for the five quiet days. It is also evident from these figures that the correlation between the differences for Huancayo and Godhavn is less than for Huancayo and Cheltenham. This, as will be shown later, is probably due to greater variations in vertical air-mass distribution at Godhavn as compared with those at Cheltenham. Fig. 25 indicates the relation

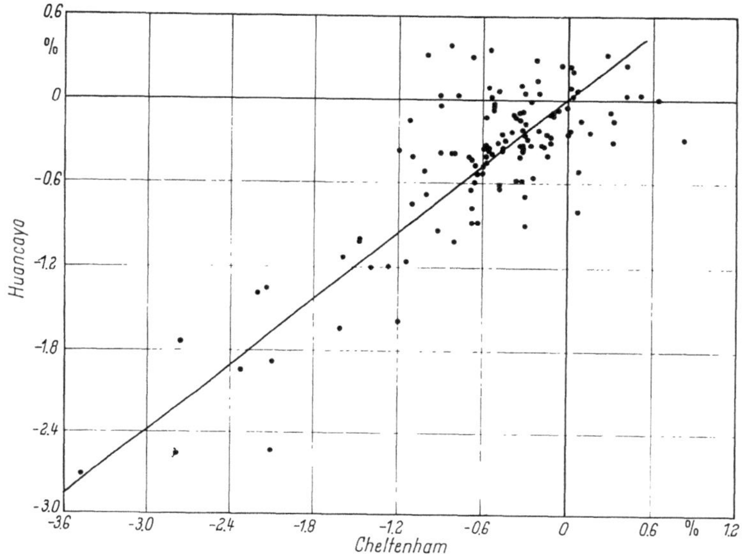

Fig. 24a. Average difference cosmic-ray intensity for five disturbed days less that for five quiet days in each month, April 1937 to December 1947, at Cheltenham and Huancayo.

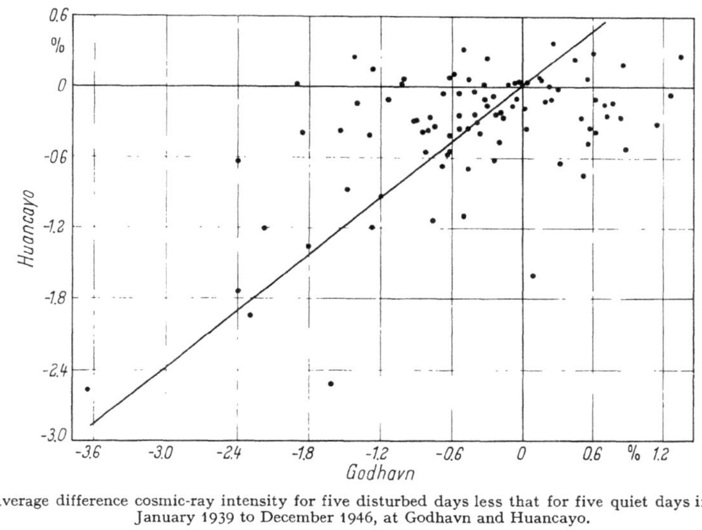

Fig. 24b. Average difference cosmic-ray intensity for five disturbed days less that for five quiet days in each month, January 1939 to December 1946, at Godhavn and Huancayo.

between the differences, disturbed minus quiet days, for cosmic-ray intensity and magnetic horizontal intensity at Huancayo. The differences are always negative for the horizontal intensity and preponderantly negative for cosmic-ray intensity at Huancayo.

The correlation coefficient for data of Fig. 25 would obviously be low. This is expected from the fact that the ratio between changes in cosmic-ray intensity to those in horizontal intensity is known to vary from one storm to another. Fig. 26 indicates the variation in the annual means, for disturbed minus quiet days, in cosmic-ray intensity at three stations and in horizontal intensity, H, at Huancayo (values for H were unavailable after 1947). It will be noted that the annual mean difference, disturbed minus quiet days, is always negative for cosmic-ray intensity at all three stations.

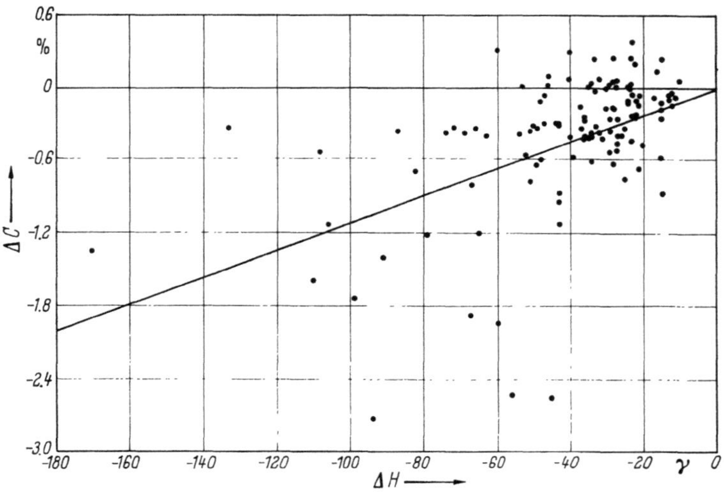

Fig. 25. Average difference for cosmic-ray intensity (ΔC) and for horizontal mangetic field (ΔH) for five disturbed days less that for five quiet days in each month, April 1937 to December 1946 at Huancayo.

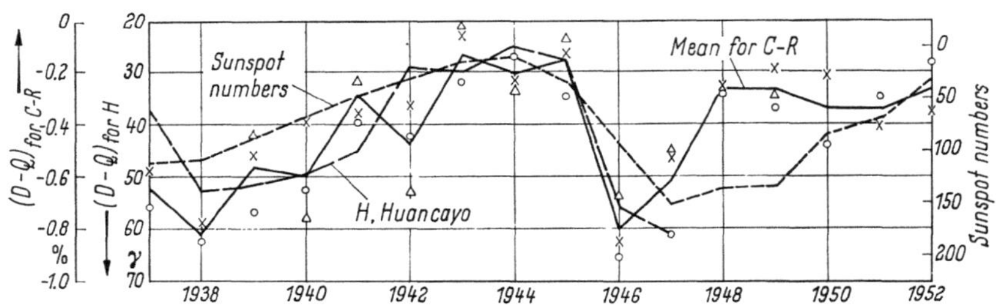

Fig. 26. Annual means for differences (D-Q) for magnetically disturbed less quiet days (5 each per month) for cosmic-ray intensity (C-R) and for horizontal magnetic intensity, H. Legend for (C-R): o = Chettenham, × = Huancayo, ▵ = Godhavn.

These facts together suggest that the mechanism responsible for the decrease in cosmic-ray intensity is connected with that responsible for mangetic storms. Alfvén [*43*] has proposed that the cosmic-ray decreases arise from deflection of cosmic-ray particles by magnetic fields carried away from the sun in the conducting streams which also give rise to magnetic storms.

21. Variability of daily means at Huancayo. Fig. 27 indicates the standard deviation of daily means from monthly means at Huancayo derived from pooling the variance of daily means from monthly means for each year 1937—1952. The curves show that standard deviations of departures from the monthly means are roughly four times smaller near sunspot minimum than near sunspot maximum. They are only slightly less when the five magnetically disturbed days of each month are excluded. This is probably due to the fact that in most months the variation arises principally from 27-day quasiperiodic variations. For 1944, the standard deviation of daily means from monthly means is about 0.21% (excluding the five magnetically disturbed days). Since this figure included the variability of the world-wide component, it is an upper limit for the combined effects of statistical fluctuation in the records and those from variations of μ-meson decay due to changes in vertical distribution of air-mass. It is thus evident that the latter effects are quite small at Huancayo, and that the daily means (relative to the mean of the month) at Huancayo are reliable to within at most 0.2% (that is, their standard deviation $s \leqq 0.2\%$).

From a previous investigation [*44*] it was found that the world-wide changes at Teoloyucan, Mexico, were about 1/0.63 times those at Huancayo, Peru. For

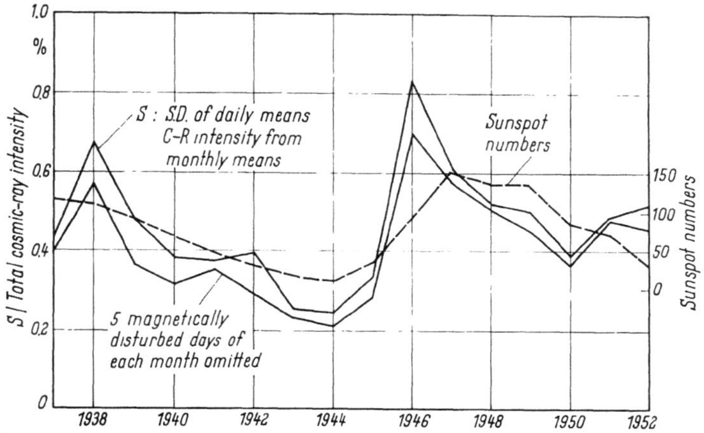

Fig. 27. Annual means: sunspot numbers and variability cosmic-ray intensity at Huancayo.

1937, daily means, with seasonal wave removed, were available from Teoloyucan for all months except January and November. These daily means for Teoloyucan were multiplied by 0.63 to reduce them to Huancayo. The difference between the daily mean at Huancayo and the reduced daily mean at Teoloyucan was found for each day of the ten months. The standard deviation of single differences about their average for the month was found to be 0.24% from pooling the ten samples of one month each. Assuming equal variance for statistical fluctuations at both stations, the standard deviation for the statistical fluctuations in single daily means is only 0.17%, which is slightly less than the figure of 0.2 derived from the fluctuations of daily means from the monthly means for Huancayo in 1944. The standard deviation for hourly values at Huancayo is about 0.6% (from differences between the same pair of hours on a number of quiet days) which gives for 24 independent hours a standard deviation of 0.12%. The standard deviation of the ten monthly mean differences about their average is about three times greater than would be expected from the fluctuations in the differences in daily means. This may indicate some small systematic change which would arise if the seasonal variations at Teoloyucan deviated from a pure 12-month wave or if there were a small seasonal wave at Huancayo. It will be shown later that variations arising from non-world-wide changes are much greater at the other stations than at Huancayo and Teoloyucan and that the data from Huancayo and Teoloyucan provide more reliable measures of the world-wide component than do those from the other stations. The absence of any large seasonal variation at Huancayo is further indication that the vertical distribution of air-mass there must vary little with season.

The curves for the standard deviation of daily means, in Fig. 27, are similar to that derived by MEYER [45] and SIMPSON from the variation of the amplitude of the 27-day variation, over the period 1936—1954. These were based on published ionization chamber data for Cheltenham, Christchurch, and Huancayo,

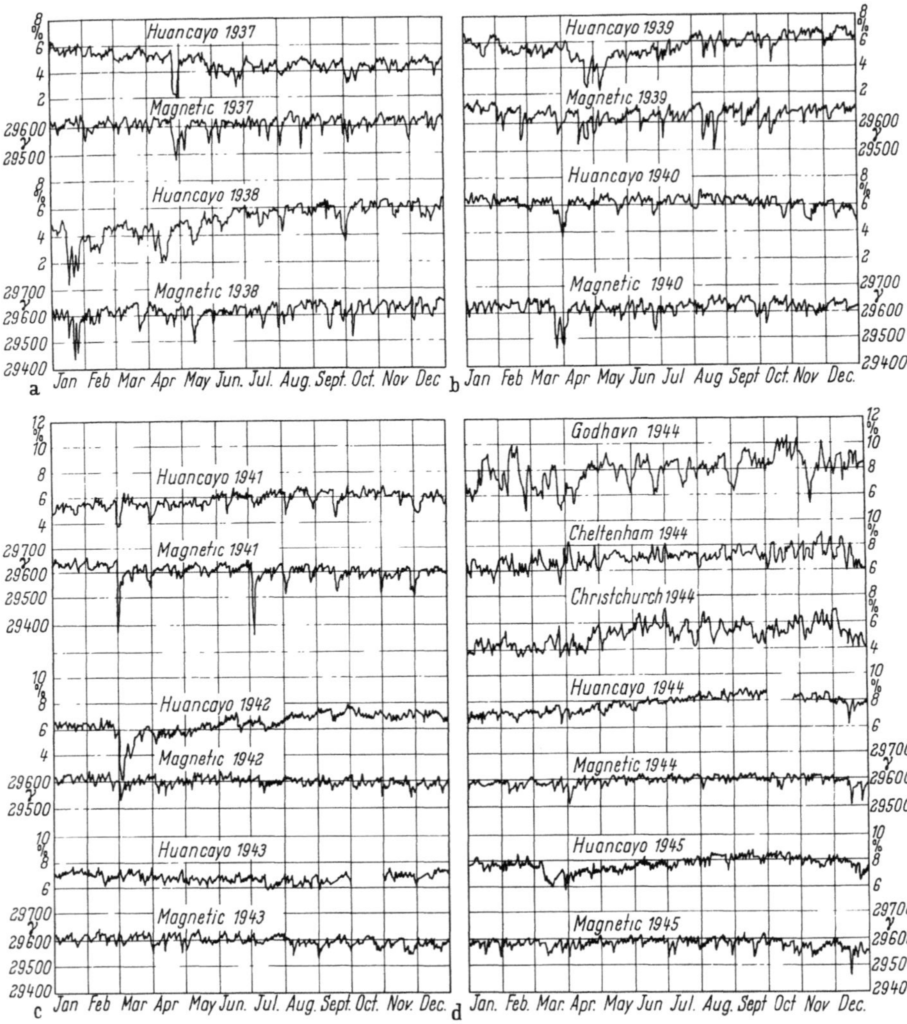

Fig. 28a—f. Daily means cosmic-ray intensity for: Huancayo (75° WMT), Cheltenham (75° WMT), Godhavn (45° WMT), and Christchurch (172.5° EMT) and daily mean horizontal magnetic component at Huancayo (75° WMT).

for the period 1936—1946, and on neutron pile data at Climax and Chicago for the period 1951—1953. From the data at two stations, A and B, they derived the standard deviation of what was termed the tracking component of the variation. For a measure of the amplitude of the 27-day variation at station (A) they used the standard deviation, σ_A, of δ_A with $\delta_A = [I_A(t) - I_A(t-T)]/\bar{I}_A$ in which $I_A(t)$, and $I_A(t+T)$ are respectively the daily mean intensities on day t and $(t+T)$, and $\bar{I}_A$ is the average intensity. They used $T = 14$ days since as a consequence of the 27 day recurrence tendency σ_A was found [46] to have a maximum with $T=14$ days. Similarly σ_B was determined for station B. For corresponding values of t the correlation coefficient r_{AB} between δ_A, and δ_B, was determined using data for six month periods. The standard deviation of the tracking component was taken as $|r_{AB}|\,\sigma_A$ which, except for the absolute value of r_{AB}, derives from the well-known relation for the standard deviation of the linearly dependent variations in A and B. This procedure has the advantage that it shows the 27 day

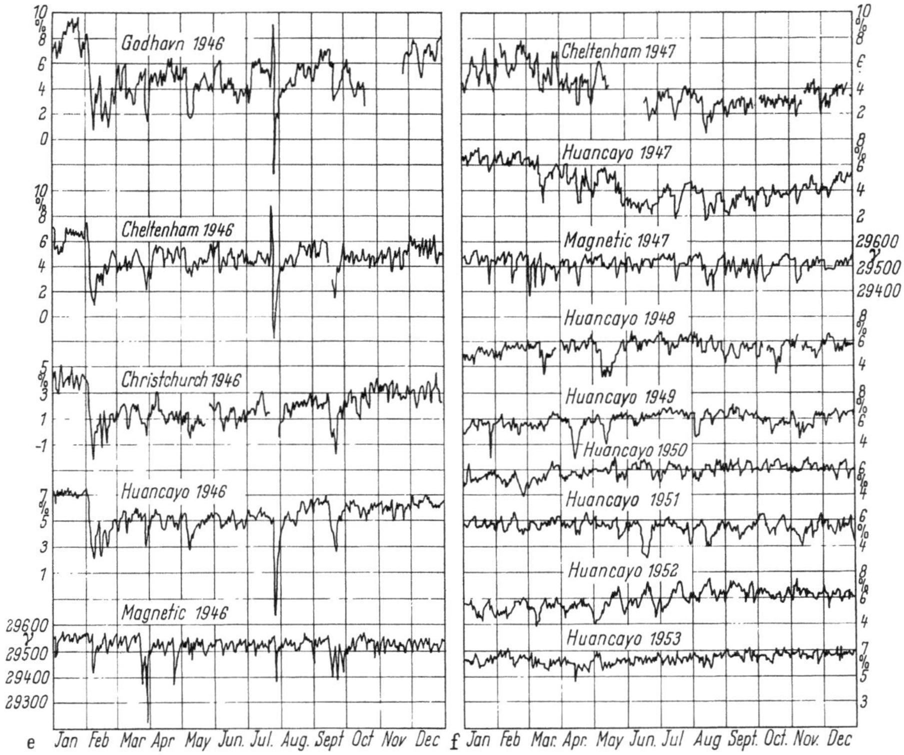

Fig. 28 e and f.

variation is correlated between the different stations and is thus a world-wide phenomenon. The world-wide nature of the 27-day variation had also been demonstrated in a different manner which will be discussed in the section on 27-day variations.

22. Variations in daily mean cosmic-ray intensity at Huancayo, 1937—1953 compared with variations in the earth's magnetic field and with cosmic-ray variations at other stations for selected years. Fig. 28 is a graph of daily means of cosmic-ray intensity at Huancayo 1937—1953, together with daily means for Godhavn, Cheltenham, and Christchurch for selected years. These daily means have been corrected for bursts, barometric pressure and seasonal variation, also for a linear drift the magnitude and nature of which will be discussed in connection with the sunspot-cycle variation in cosmic-ray intensity. Inspection of the graphs of cosmic-ray intensity for Huancayo in Fig. 28 indicates a marked difference in the variability means in different years, which is particularly evident if the curves for 1944 are compared with those for 1946 and 1947; this variability was shown quantitatively in Fig. 27.

During 1946 and 1947, there were large variations at Huancayo, which in general follow those at Cheltenham, Godhavn (1946 only), and Christchurch (1946 only). There is, of course, the large increase at Godhavn, and Cheltenham on July 25, 1946, which occurred during a large solar flare on that date, and which is absent at Huancayo, and at Christchurch where the meter was out of operation for eight days starting July 23.

On the other hand, a comparison of the graphs for the four stations for the sunspot minimum year 1944 shows that the variability of the daily means is decidedly less at Huancayo than at the other three stations, and that the variability is greatest at Godhavn. Moreover, the major variations at Godhavn, Cheltenham, and Christchurch during 1944 were seen (by overlaying the original curves) to be essentially uncorrelated. At Cheltenham, and to some extent at Godhavn, the larger variations in 1944 (which were absent at Huancayo) occurred more often in winter than in summer. At Godhavn and at Cheltenham, it was found that the large variations in 1944 generally occurred during periods when the barometer was changing rapidly. These large variations are thus probably due to changes of the vertical air-mass distribution accompanying the movement of a front over the station and the consequent effects arising from meson decay. Although smaller variations occurring at Huancayo are often obscured at the other stations by this meteorological effect, it will be shown that the *averages* of a sample of such variations are very nearly the same at all four stations.

Fig. 28 also shows the daily mean values of the horizontal magnetic component (H) at Huancayo, 1937—1947, from which it can be seen whether decreases in H, which occur during magnetic storms, are accompanied by decreases in cosmic-ray intensity. From these graphs, a tabulation showed 48 cases (1937 to 1947) when from one day to the next a decrease, in H, of 75 γ or more occurred. In 36 of these cases, the change in cosmic-ray intensity at Huancayo was negative, although in only 22 cases was the decrease in cosmic-ray intensity greater than 0.4%.

The graphs were also used to tabulate the dates between which the daily means of cosmic-ray intensity at Huancayo decreased continuously (successive days with no change were included) for a total decrease of 1.0% or more. There were 92 such intervals from 1937 to 1947. The change, ΔH, in daily mean horizontal magnetic intensity at Huancayo from the first to the last day of each of the above-selected intervals was also tabulated; in 71 (out of 92) of the intervals, ΔH was negative. Examination of magnetograms for Huancayo (Peru) and Watheroo (Australia) indicated magnetic disturbance in most of the 21 cases for which ΔH was either zero or positive. It thus seems evident that during most of the periods when the cosmic-ray intensity at Huancayo is decreasing there is evidence for magnetic disturbance, which suggests that the cause of the cosmic-ray decreases is quite probably connected with the mechanism giving rise to magnetic disturbance.

Finally, in this connection, attention should be called to the graphs of daily means for cosmic-ray intensity and magnetic horizontal intensity, H, for February 1946 in Fig. 28. Between February 3 and 6, 1946, the five per cent decrease in cosmic-ray intensity at Huancayo was accompanied by only a small decrease in H, while the large decrease in H after February 6 was accompanied by only a small further decrease in cosmic-ray intensity. This can be better seen in Fig. 23 which was discussed in Sect. 18. There was at Huancayo and Watheroo a marked magnetic sudden commencement at $08^{h}42^{m}$ (75° West Meridian Time) a few hours before the start of the decrease in cosmic-ray intensity. Attention should also be called to the fact that after Februray 6, 1946, both the cosmic-ray intensity and H at Huancayo remained low during the rest of the year. While it seems clear that most of the decreases in cosmic-ray intensity occur during periods of magnetic disturbance, no measurable characteristic of magnetic disturbance has yet been found which is quantitatively well correlated with changes in cosmic-ray intensity.

Attention has been called in Fig. 28 to the fact that the variability of daily means at Huancayo, obvious in 1944, is in general less than that for the other stations (see Fig. 28d). It was also indicated that these variations at Godhavn, Cheltenham, and Christchurch were generally uncorrelated and could well obscure small variations which can be reliably seen at Huancayo on account of the lack of appreciable μ-meson decay effects arising from variations in the height of the 100 mb pressure level.

To determine whether small variations at Huancayo can be traced in the averages of several effects at the other stations, Fig. 28 was used. From these plots of daily means for Huancayo ten intervals of 20 days length were selected with each interval exhibiting a variation similar to that shown for Huancayo

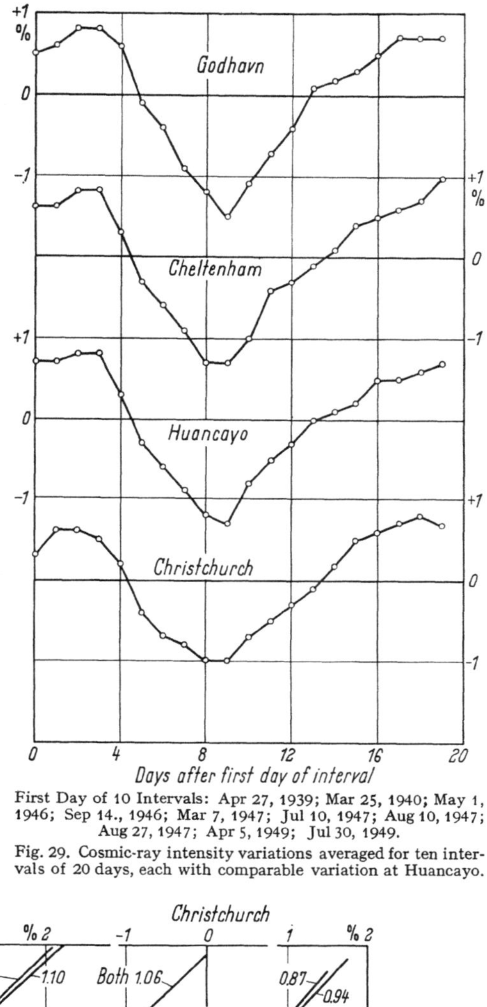

First Day of 10 Intervals: Apr 27, 1939; Mar 25, 1940; May 1, 1946; Sep 14., 1946; Mar 7, 1947; Jul 10, 1947; Aug 10, 1947; Aug 27, 1947; Apr 5, 1949; Jul 30, 1949.

Fig. 29. Cosmic-ray intensity variations averaged for ten intervals of 20 days, each with comparable variation at Huancayo.

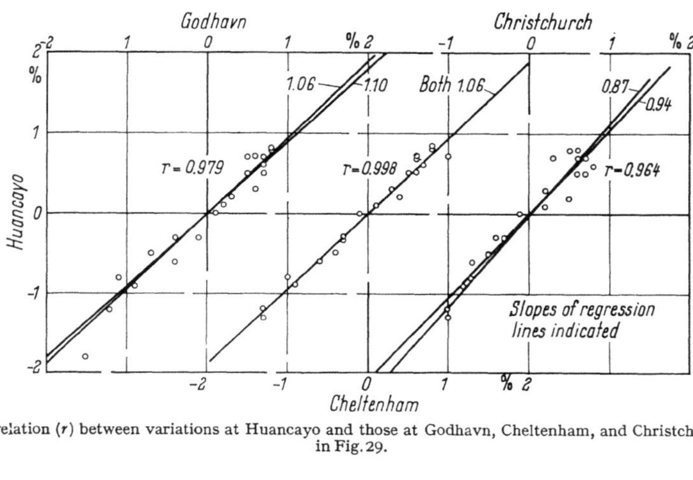

Fig. 30. Correlation (r) between variations at Huancayo and those at Godhavn, Cheltenham, and Christchurch from data in Fig. 29.

in Fig. 29. Fig. 29 indicates the variation averaged for the same ten intervals for each of the three stations. Fig. 30 indicates the correlation between the averaged variation for Huancayo and that for each of the other three stations. The correlation coefficients and slopes of the regression lines are also indicated. The smaller of the two slopes results from the assumption of no statistical error in the means for Huancayo. Except for Christchurch, the factors are in fair agreement with those derived in an earlier study of world-wide changes.

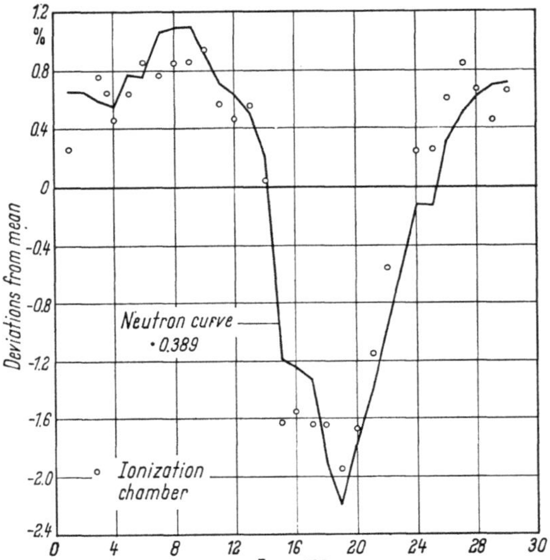

Fig. 31. Comparison neutron daily means (New Mexico) multiplied by 0.389, with those from ionization chamber at Huancayo. Standard deviation of differences is 0.25%.

23. Sample comparisons of variations of daily means from ionization chambers and neutron monitors. Fig. 31 is a comparison of the variation of daily means for June 1951 from the Compton-Bennett meter at Huancayo and those published by SIMPSON [47] from neutron counters at Sacramento Peak, New Mexico. The standard deviation (s.d.) of the differences between daily means from the Compton-Bennett meter at Huancayo and those from the neutron counters (multiplied by 0.389) at Sacramento Peak is about 0.25%. The series is

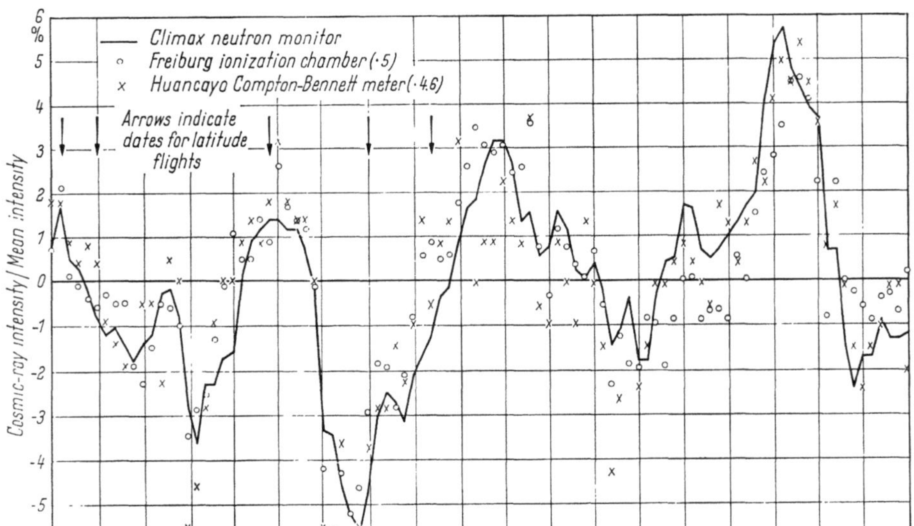

Fig. 32. Comparison of daily means from Huancayo and Freiburg ionization chambers with those from Climax neutron monitor, July—October, 1951.

too short to determine whether there are systematic changes in background in either instrument involved. The occurrence of any such changes would result in increasing the s.d. If the value of 0.17% is accepted (see Sect. 21) for the s.d. of single daily means from the monthly average at Huancayo, then from the value of 0.25% for the s.d. of differences between the Huancayo daily means and those for neutrons in June 1951, the s.d. of daily means for the latter is found to be about 0.19%, in the reduced neutron units. Or, the s.d. of the neutron daily means would be 0.19/0.389 = 0.49%. Since this figure is many times greater than would be expected in view of SIMPSON's high neutron counting rate, it is evident that during the period of this comparison one of the two instruments was subject to variations, either real or instrumental, which did not affect the other. Since the Instituto Nacional de la Investigacion Cientifica and the University of Mexico, Mexico, D.F., are now collaborating (since September 1, 1954) with the Department of Terrestrial Magnetism in the operation of a Compton-Bennett meter at the University of Mexico, Mexico, D.F., it will be possible in future to compare results from it with those from the neutron monitor which is in operation there by SIMPSON's group.

Fig. 32 shows a comparison published by FONGER [*48*] of daily means from the neutron monitor at Climax and from the large ionization chamber at Freiburg to which has been added the daily means for Huancayo. While the changes at the three stations are in generally good agreement, it may be noted that the ratio of changes at Climax to those at Huancayo is decidedly greater than that indicated for Fig. 31. This indicates that further comparisons are needed before it is certain whether or not the ratio is constant.

III. Variation with sunspot cycle.

24. Ionization chamber results. α) *Evidence in long series of observation.* When all available annual means, from Godhavn, Cheltenham, Huancayo, and Christchurch, of cosmic-ray ionization, corrected for bursts, and barometric pressure, were examined, a secular decrease was obvious in the results for Christchurch. Since there was no evidence in the results for Cheltenham of any significant secular change, other than the sunspot variation as shown in Fig. 33, no correction for drift was applied to the data for Cheltenham. By comparing results for the other stations with those for Cheltenham, the following linear changes were found: Christchurch, $-1.40\%\ \mathrm{yr}^{-1}$; Godhavn, $-0.25\%\ \mathrm{yr}^{-1}$; and Huancayo, $+0.40\%\ \mathrm{yr}^{-1}$. The annual means in Figs. 33 and 34 have been corrected for the above linear changes, which are assumed to be instrumental and probably arise from decay of radioactive contamination in the main chamber or in the balance chamber of the meters.

The agreement between the annual means of cosmic-ray intensity for the four stations, or their average, and that for annual mean sunspot numbers is evidence that the mechanism responsible for these changes in cosmic-ray intensity involves some phenomenon associated with solar activity. It is known that some magnetic storms are accompanied by large decreases in cosmic-ray intensity, and it was shown that most of the major decreases which occur during intervals of a few days are associated with magnetic storms or periods of magnetic disturbance. Thus, there arises the question of whether these decreases are mainly responsible for the variation of cosmic-ray intensity with sunspot numbers shown in Fig. 33. To answer this question, the variation of annual means of cosmic-ray intensity at Huancayo for all days (as used in Fig. 33) is compared in Fig. 34 with that for international magnetic quiet days and with that for international magnetic disturbed days. It is evident from Fig. 34 that the variation of annual means for all days, which in Fig. 33 follows the curve of sunspot numbers, is very little different from that for quiet days and not greatly different from that for disturbed days. Thus, *the main features of the variation of cosmic-ray intensity*

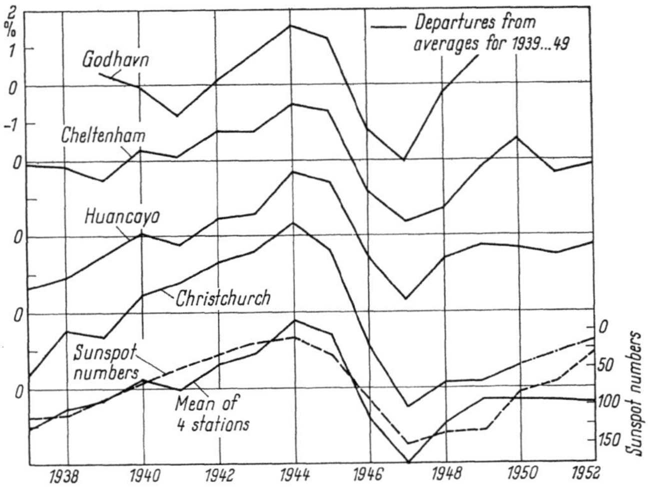

Fig. 33. Annual means cosmic-ray intensity at four stations.

with sunspot numbers persist for long periods (six months or more) and are not ascribable to transient decreases accompanying some magnetic storms.

Further evidence of effects that persist for long periods of time is indicated in Fig. 28d (p. 188), in which the curve for daily means of cosmic-ray intensity at Huancayo shows a gradual increase of about 1.5% from January 1944 to September 1944 during a period in which there was no very large transient decrease in cosmic-ray intensity and no great magnetic storm. It thus appears that the transient decreases in cosmic-ray intensity which occur during some magnetic storms and magnetically disturbed periods are superimposed upon a variation with sunspot cycle.

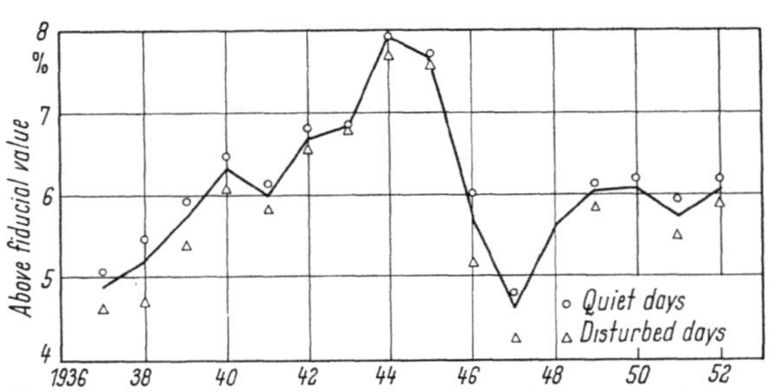

Fig. 34. Annual means cosmic-ray intensity at Huancayo for all days, international quiet days and international disturbed days.

β) Comparison with geomagnetic data. Fig. 35 shows the variation of the monthly means of cosmic-ray intensity for four stations after removing the seasonal wave and linear trend. Also shown is the monthly mean horizontal magnetic component at Huancayo corrected for a linear estimate of secular change. It is evident that the horizontal intensity and cosmic-ray intensity were both markedly lower throughout 1946 and 1947 than in 1944. This fact suggests the possibility that the same mechanism may be responsible for both effects. In this connection, it should be mentioned that VESTINE [*49*], from a long series of magnetic data from many observatories, found evidence for an 11-year variation in the horizontal component of the earth's field. In deriving the latitude distribution for the three geomagnetic components of the storm-time field, D_{st} (disturbed minus quiet days) he found that the eastward geomagnetic component of D_{st} was zero, on the average. However, he points out that the yearly average of the east component for all days is not only not zero but varies during the sunspot cycle, indicating that the cause of this variation (and probably also that for horizontal intensity on all days) may be distinct from that for D_{st}. Thus, this unexplained variation in the earth's field is possibly connected with the sunspot variation in cosmic-ray intensity.

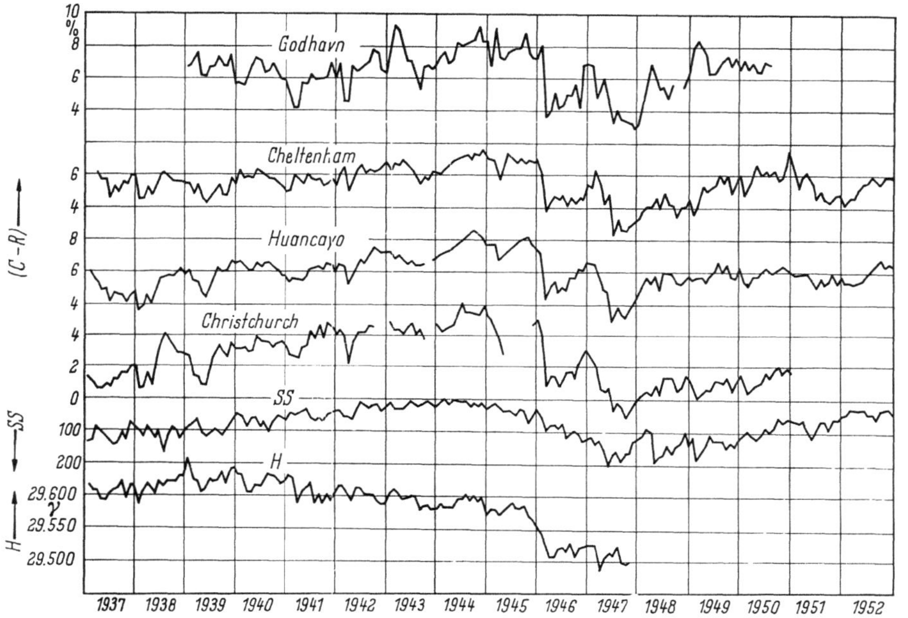

Fig. 35. Monthly means: cosmic-ray intensity (*C-R*) at four stations; sunspot numbers (*SS*) and magnetic horizontal intensity (*H*) at Huancayo.

The variation of monthly means is further compared in Fig. 36. To effect this comparison, the monthly means for Huancayo were categorized in six intervals of 1% (3% to 9%). Monthly means for each of the other stations were averaged for each of these six categories. These group means are reasonably well fitted by the straight lines shown in Fig. 36. These lines thus approximate the regression lines obtained by assuming the monthly means at Huancayo are free of statistical errors, which are presumed present only in the means for the other stations. The factor of 1.23 for the ratio of changes in Christchurch to those at Huancayo is roughly 20% greater than that found earlier from a shorter series of data. It is also greater than that obtained in the last paragraph of Sect. 22. For Cheltenham and Godhavn, the factors shown on Fig. 36 are more nearly consistent with those derived earlier and with those derived in Sect. 22.

γ) Comparison with the 100 mb level. The question arises whether variations with solar cycle in the height of the 100 mb pressure level might explain, through μ-meson decay effects, the sunspot variation in cosmic-ray ionization. The average seasonal wave in cosmic-ray intensity at Cheltenham has an amplitude of 1.45% and is opposite in phase to the average seasonal wave, amplitude 260 m, in the height of the 100 mb pressure level derived from U.S. Weather

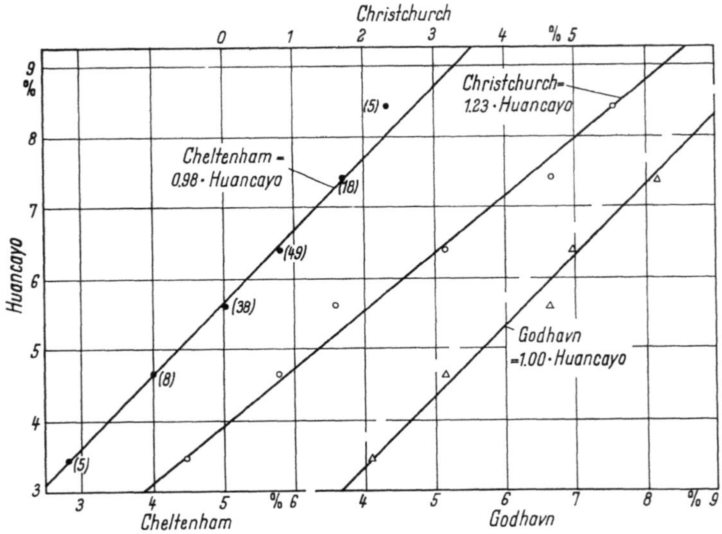

Fig. 36. Correlation between averages of groups of months for Huancayo and each of three other stations. Numbers in parenthesis indicate number of months in each group.

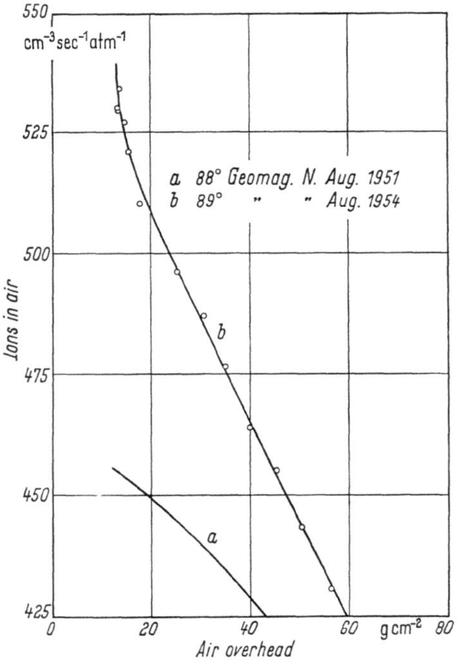

Fig. 37. On an expanded scale the ionization vs. air mass overhead is shown for the two years, 1951 and 1954, near the geomagnetic pole. The increased slope in 1954 at the lowest pressures is evidence for low energy particles (150 MeV if protons) present in 1954 that were absent in 1951 (NEHER and STERN).

Bureau radiosonde data at Washington for the period 1944—1953. The ratio of these amplitudes gives a 5.6% decrease in cosmic-ray intensity per kilometer increase in height of the 100 mb level. The range in annual mean heights of the 100 mb level (1944—1953) was only 75 m and the annual means exhibited no variation similar to the solar cycle. Thus the sunspot variation in cosmic-ray intensity does not arise from this cause.

25. The "knee" of the cosmic-ray latitude curve. At sea level the "knee" of the cosmic-ray latitude curve has been placed between geomagnetic latitudes 40 and 45°. The "knee" at sea level is doubtless due to atmospheric absorptions. However, at high altitudes divergent results have been obtained by different investigators. NEHER [*50*] points out that these discordant results may arise from the fact that the data were taken at different times. Fig. 37 shows NEHER's results for the ionization as a function of air mass overhead for high altitude (95 000 ft) flights made in the summers of 1951 and 1954 near the geomagnetic pole. The increase of ionization for a given decrease in air overhead is markedly greater for the 1954 curve than for the 1951 curve. This effect was found on each of the five 1954 flights which reached sufficient altitude.

Fig. 38 shows, for the indicated atmospheric dephts, the variation of ionization with latitude for 1951 [51] and 1954. It is evident that large changes in the radiation took place in the intervening three years at the northern latitudes at these high altitudes. While the change in 1951 from 58° to 68° was less than 1% at 20 g cm^{-2}, in 1954 a change of about 6% was found at the same depth, covering the same range of latitude. This difference is indication that low energy particles which were present in 1954 were absent in 1951. From geomagnetic theory it is found that 150 MeV protons can arrive vertically at geomagnetic latitudes north of 65°. The evidence from Fig. 38 is that particles (if protons) were present in 1954 in the primary radiation down to 150 MeV. The increase in the ionization from 68° to 90° shown in the curves of Fig. 38, amounting to 12 ions cm^{-2} sec^{-1} atm^{-1} of air at 20 g cm^{-2}, was the same in 1954 as in 1951. Since no new particles, admitted by the opening of the STÖRMER cones, could reach the instrument north of 68°, this increase is ascribed to the opening of the shadow cones [51] for both occasions. From Fig. 37 the continued increase in the slope of the 1954 curve at 15 g cm^{-2} probably indicates that particles were present which had ranges equal to and less than this value. These results indicate that there was no cutoff of the primary particles down to at least 150 MeV (assuming protons) in the summer of 1954 at northern latitudes. In contrast, the apparent cutoff for protons in 1951 was estimated at 800 MeV. These experiments provide additional evidence of an inverse relationship between solar activity and cosmic-ray intensity. During the summer of 1954 the sun was at its lowest ebb in 22 years and particles of low energy were arriving at the earth which at other times were excluded.

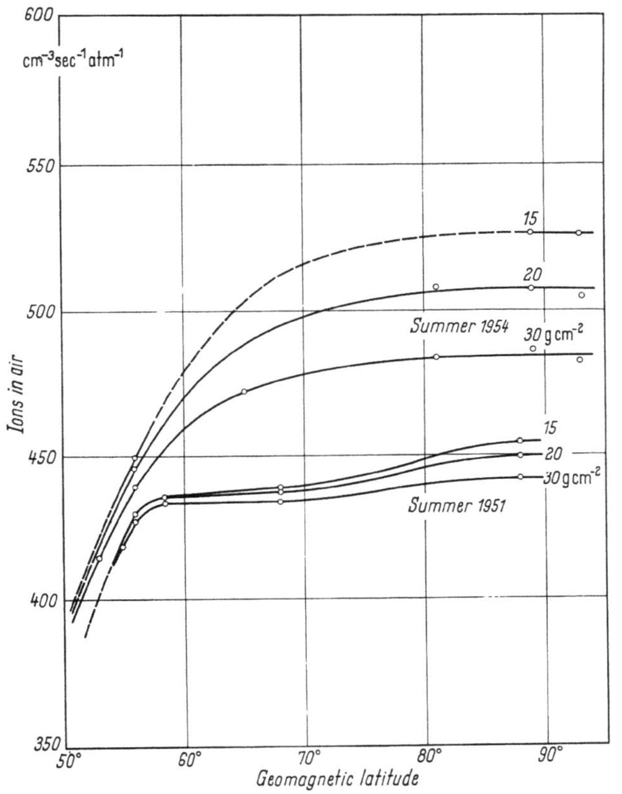

Fig. 38. The increase in ionization with increasing latitude in 1954 up to at least 68° N shows a lack of cut-off of primary particles down to at least 150 MeV, if protons (NEHER and STERN).

26. Changes in the low energy particle cut-off and primary spectrum. In a communication kindly sent in advance of publication the authors, PETER MEYER and J. A. SIMPSON, at the University of Chicago show from measurements with a nucleonic component detector, in B 47 jet aircraft at 310 g cm^{-2} atmospheric depth, that the low rigidity cut-off for particles in the primary cosmic-ray spectrum has decreased within the period 1948 through 1951. They determine that this decrease corresponds to a northward change, between 1948 and 1951, of 3° in the "knee" for the nucleonic component. This change is accompanied by a change in the primary spectrum for particle rigidities less than about 4 Bv and by an increase in the total primary intensity. They find that if the differential primary intensity at low rigidities is given by $j = C/(p/z)^2$ for 1948 then the spectrum for 1951 through 1954 is approximately $j = C'/(p/z)^{2.7}$, where p/z is proportional to the magnetic rigidity of particles with charge ze. The total change of intensity due to changes in spectrum and low rigidity cut-off is more than 13%. Thus the decrease in low rigidity cut-off discovered by NEHER (see preceding Sect. 25) to occur between 1951 and 1954 appears to be an extension of the decrease in low rigidity cut-off found by MEYER and SIMPSON to occur between 1948 and 1951. These authors conclude that neither known terrestrial magnetic fields, assumed geoelectric fields, nor a solar magnetic dipole (even if changing with time) could produce the observed effects. They suggest that solar system fields may be found which will prevent low energy particles present within the galaxy from entering the solar system near the earth's orbit.

IV. 27-day variations.

27. Waves of 27-days period in cosmic-ray intensity, magnetic activity and horizontal intensity. Figs. 39a and b are harmonic dials [*52*] for the departures of individual 27-day waves from the average 27-day wave for the daily American magnetic character figure (from 0.0 for very quiet to 2.0 for intense storm), and for cosmic-ray intensity at Huancayo, respectively. Figs. 39c and d were obtained, respectively, by rotating the vectors in a to vertical and by rotating vectors for corresponding intervals in b (of 27 days) through the same angle. Statistical tests show that the probability, P, of obtaining an average vector as large or larger than that in Fig. 39d in a sample of 34 vectors from a population in which the components of the vectors are independent and random, with standard deviations estimated from the 34 vectors in d is only about 2×10^{-6}. This indicates that the vectors in a as well as in b definitely tend to have similar phases. Furthermore, the phases in c and d of Fig. 39 indicate that the phases in a and b are opposite, that means that the maxima of the 27-day waves in cosmic-ray intensity tend to occur near the minima of the 27-day waves in magnetic activity, as measured by the American character-figure.

Fig. 40 indicates the results of a similar comparison between magnetic horizontal intensity and cosmic-ray intensity at Huancayo. For c of Fig. 40, the probability, P, is 7×10^{-5}, indicating correlation between the phases of the vectors in a and b of Fig. 40. Here it will be noted that the maxima of the waves in cosmic-ray intensity and those in horizontal intensity tend to be in phase, which is consistent with the results in Fig. 39, since low horizontal intensity occur at times of high magnetic activity.

28. World-wide and quasi persistent nature of 27-day waves. Using a procedure similar to that in Figs. 39 and 40 the 27-day waves were shown [*52*] to be

Fig. 39a–d. Harmonic dials for departures from average of 27-day waves in American character-figure (a) and in cosmic-ray intensity at Huancayo, Peru (b), computed from 34 rotations (intervals of 27 days) beginning June 13, 1936; harmonic dials (c) and (d) obtained respectively by rotating vectors in (a) to vertical and by rotating vectors for corresponding intervals in (b) through the same angle.

Fig. 40a–c. Harmonic dials for departures from average of 27-day waves in magnetic horizontal intensity at Huancayo, Peru (a), computed from 34 rotations (intervals of 27 days) beginning June 13, 1936; harmonic dials (b) and (c) obtained respectively by rotating vectors in (a) to vertical and by rotating vectors for corresponding interval in (b) of Fig. 39 through the same angle.

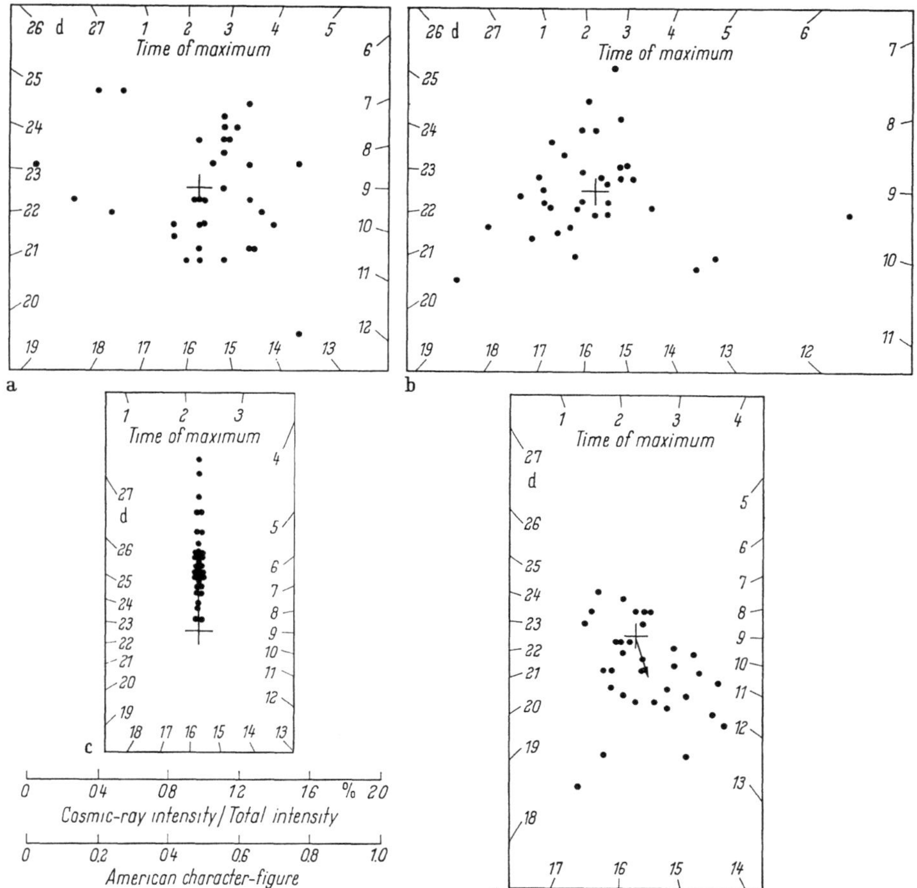

Fig. 39 a—d.

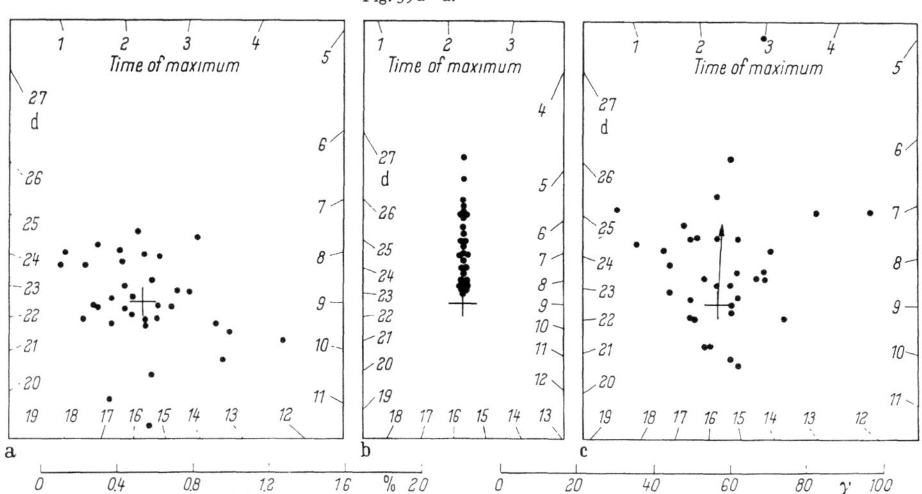

Fig. 40 a—c.

correlated at Cheltenham, Huancayo, and Christchurch in a sample of 34 intervals of 27-days. Figs. 41 and 42 indicate the 27-day waves at Cheltenham, for 158 rotations (intervals of 27 days) and at Huancayo for 169 rotations. These figures exhibit the range in amplitude (and phases) obtained from individual rotations. The large circles are the so-called probable error circles, inside which fall about

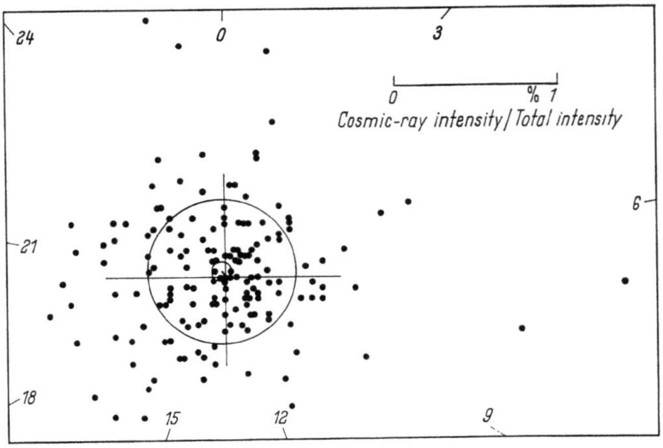

Fig. 41. 27-day harmonic dial for cosmic-ray intensity at Cheltenham, 158 rotations.

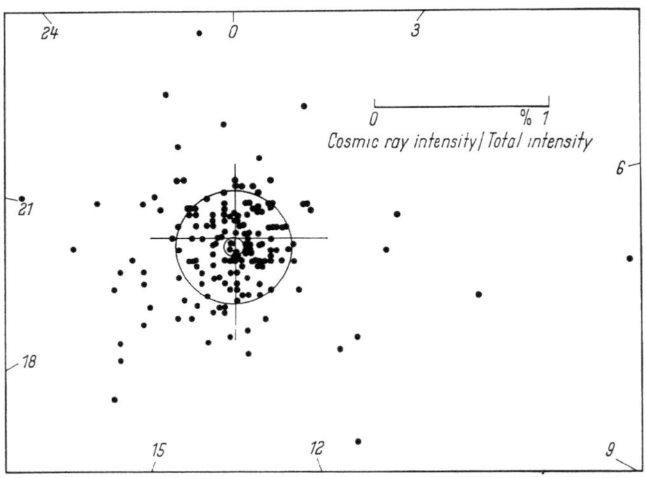

Fig. 42. 27-day harmonic dial for cosmic-ray intensity at Huancayo, 169 rotations.

half the points. Their radii for Figs. 41 and 42 are respectively about 0.45% and 0.35%. The larger radius for Cheltenham is doubtless due to the fact that at Cheltenham the variability in the 27-day waves due to μ-meson decay effects is greater than at Huancayo. The small circles in the figures are the so called probable error circles of the *average* 27-day vectors. In both cases the averages are so small that no statistically real persistent 27-day wave is indicated.

The correlation between the 27-day waves summed for 5 successive rotations, at Huancayo and Cheltenham is shown in Fig. 43 as derived from a total of 155 rotations. Incidentally a test made [52] for a *persistent* 27-day wave in the difference of 27-day waves, Cheltenham minus Huancayo, indicated nothing large enough to be regarded as statistically significant. Such a wave would only be expected from a sufficiently large solar magnetic dipole inclined to the sun's axis of rotation.

The quasi persistent nature of the 27-day waves is evident in Fig. 44 which shows the BARTELS' [53] characteristic diagram as derived from the data of Figs. 41 and 42. In a harmonic dial let the two-dimensional standard deviation of points about their average be M, then if the deviations for a successive chronological sequence of points are statistically independent, the standard deviation $M(h)$ for means of h successive points has the expected value $M(1)/\sqrt{h}$, or with $c(h) = M(h)\sqrt{h}$, $c(h)$ has the expectation $M(1)$. For pure persistence, or complete lack of independence, $c(h)$ increases linearly with $\sqrt{h}$. For quasi persistence $c(h)$ at first increases with $\sqrt{h}$ but for larger values of $\sqrt{h}$, approaches asymptotically a constant value, indicating that for sufficiently large h the means of h successive deviations are essentially independent. In Fig. 44 it is evident that the ordinates, which are proportional to $c(h)$ increase little for h greater than about 3 or 4. Thus the 27-day waves are quasi-persistent.

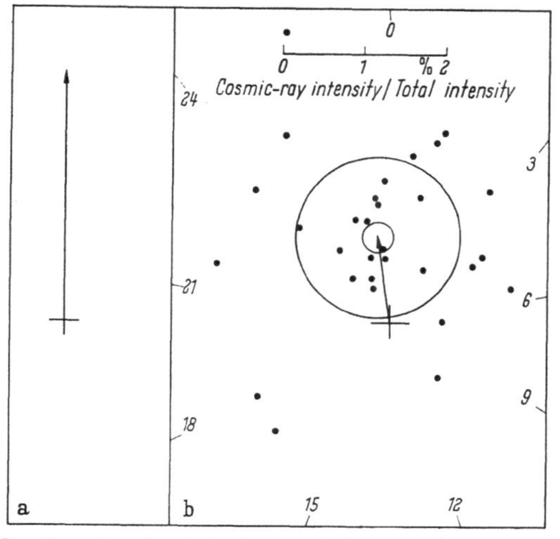

Fig. 43a and b. Correlation between 27-day waves in cosmic-ray intensity at Huancayo and Cheltenham; (a) average vector (sum for 5 successive rotations) for Huancayo after rotating each to vertical, (b) harmonic dial for Cheltenham after turning each vector (sum for 5 rotations) for corresponding interval through same angle as for (a).

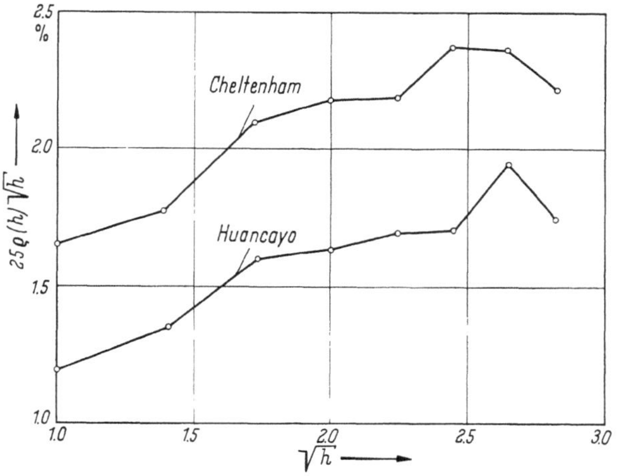

Fig. 44. Radius $\varrho(h)$ of probable error circle for means of h successive 27-day rotations as function of $\sqrt{h}$ indicating quasipersistence in 27-day waves.

29. 27-day changes in the nucleonic component and its latitude variation. Fig. 32 (p. 192) showed the variation in daily mean values from the neutron monitor at Climax. FONGER [48] determined the autocorrelation function, $r(T)$ for the Climax neutron curve of Fig. 32. From unity with $T=0$ (T is the lag in days), $r(T)$ decreases to a minimum of about -0.4 for T near 14 days. From $T > 14$ days $r(T)$ rises to a maximum of $+0.4$ for $T=28$ days after which it decreases to a second minimum of about -0.4 near $T=42$ days. While the series of data is rather short for such purposes $r(T)$ clearly indicates the 27-day recurrence tendency.

A most important contribution to the better understanding of the 27-day variation has been made by Simpson [54] who made an extensive series of measurements on the latitude variation of the nucleonic component at atmosphere depth of $312\ \mathrm{g\,cm^{-2}}$. The measurements were made in a type RF-80 jet aircraft with which data over the geomagnetic latitude, λ, interval 40° to 53° (where the change of intensity with latitude is large) could be obtained in a few hours. However, complete latitude curves were obtained over a fixed route between $\lambda = 40°$ and 65° N. Referring to Fig. 32 the vertical arrows show the dates on which one of the series of latitude flights was made. It will be seen that the latitude flights were planned to take place near maxima and minima on the curves of daily means from neutron ground station monitors. In this way the flights were made

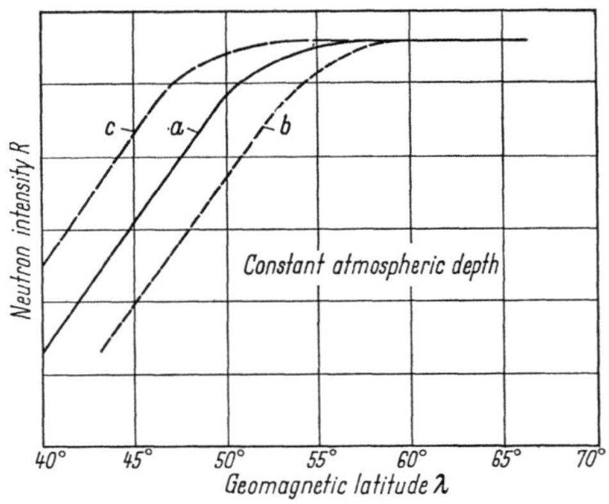

Fig. 45. The predicted behavior of neutron intensity as a function of latitude based upon the assumption that the primary cosmic-radiation intensity variation is produced by a *geomagnetic field variation* (Simpson).

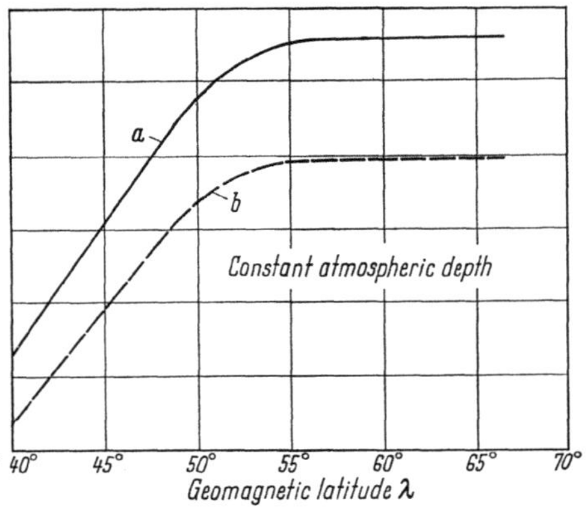

Fig. 46. The predicted behavior of neutron intensity as a function of latitude based upon the assumption that the cosmic radiation intensity variation is produced by a *change in the primary flux with time* (Simpson).

at times between which there were large intensity variations. The purpose of the experiments was to determine whether the recurring 27-day variations are the result of changes in primary particle flux or of variations of the geomagnetic field.

Fig. 45 shows the latitude variation to be expected if the 27-day variation is due to variations of the geomagnetic field. Simpson [55], Fonger and Treiman have obtained the relationship between the counting rate R of a neutron detector located at atmospheric depth x, geomagnetic latitude λ and the vertical, differential primary flux $j_Z(p/Z, t)$ of particles of momentum p and charge Z. To relate these functions they defined the specific yield of neutrons as a function $S_Z(p/Z, x)$ which is experimentally determined from the time averaged parameters j_Z and R to yield the neutron counting rate at depth x arising from a unit flux of vertically incident primary particles of charge Z and rigidity p/Z. Thus, they found [55]:

$$R_v(\lambda, x, t) = \sum_Z \int_{[p/Z]_\lambda}^{\infty} S_Z(p/Z, x)\, j_Z(p/Z, t)\, d(p/Z)$$

where $[p/Z]_\lambda$ is the cut off for vertical arrival at λ and $R_v(\lambda, x, t)$ is the counting rate due only to those primaries which arrive from the vertical direction per unit solid angle at time t. It was shown that R_v and the observed rate R are related in good approximation by the Gross transformation. A variation of R_v may thus be produced either by a variation of the lower limit of the integral which is determined by parameters of the geomagnetic field or by a variation of j_Z

in the integrand, namely, a variation of j_Z with time. Thus in Fig. 45, a is a typical latitude curve, curve b shows approximately how the latitude curve will appear for a variation $+[\delta p/Z]_\lambda$ and curve c that for a variation $-[\delta p/Z]_\lambda$. The integral counting rate, R_v, is unchanged by variations of $[p/Z]_\lambda$ above the knee of the curve.

On the other hand, if the fast neutron latitude curve at time t is represented as curve a, Fig. 46, then if at time t_1 a variation occurs to produce a fractional change of intensity $-\delta R/R$, which variation for simplicity was made independent of latitude, then at t_1 the neutron latitude curve will appear as curve b, Fig. 46. For the special case where the function $S_Z(p/Z, x)$ vanishes for finite value of $j(p/Z, t)$ the observed cutoff of the latitude curve will be determined by S rather than j, and again, if j undergoes a variation of intensity with time, the counting rate of the detector will change with time above and below the cutoff determined by S_Z. From the equation above for R_v the maximum value for the integral at time t is obtained by setting the lower limit equal to the minimum particle rigidity $[p/Z]_{\min}$ observed in the primary spectrum. Hence, for all values of $[p/Z]_\lambda < [p/Z]_{\min}$ the integral is a constant. Now if $j_Z(p/Z, t)$ undergoes a variation the observed counting rate will change for observations at latitudes corresponding to $[p/Z]_\lambda < [p/Z]_{\min}$ as well as for latitudes corresponding to $[p/Z]_\lambda \geqq [p/Z]_{\min}$. Thus, an observer within the atmosphere would measure a change in secondary particle intensity above and below the cutoff of the latitude curve.

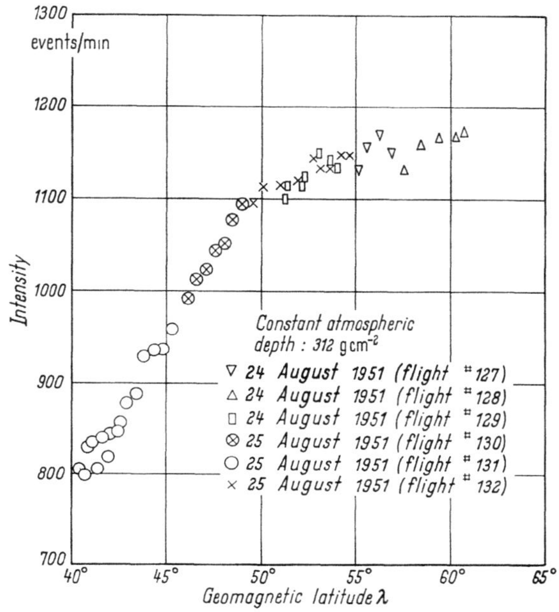

Fig. 47. The experimental neutron intensity data which are used to establish the latitude curve for 25 August 1951 (curve c in Fig. 48). The magnitudes of the standard deviations are given by the size of the flight indication symbols (SIMPSON).

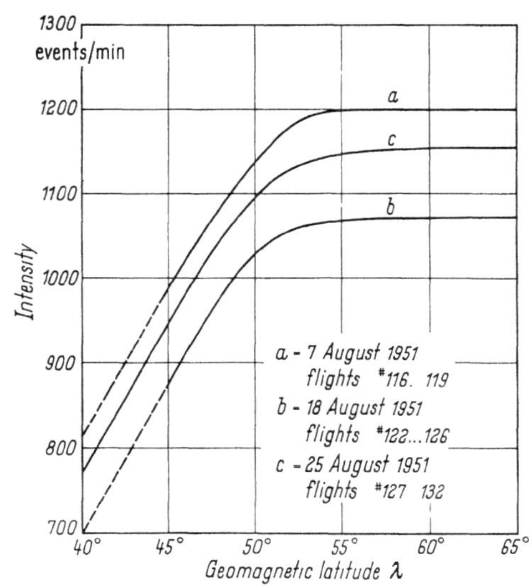

Fig. 48. The latitude curve c derived from Fig. 47 and similar curves (a) and (b). The extrapolated portions of the curves are dashed (SIMPSON).

Fig. 47 shows a typical set of latitude observations. As already indicated the data in the latitude interval 40° to 53°, where $dR/d\lambda$ is large, were obtained within a few hours. The additional data required to construct the latitude curve, such as that between $\lambda = 55°$ to 65°, are corrected for the intensity at the time of the flight between 40° and 55°. From other sets of data for August 7 and 18 similar to that in Fig. 47, the curves in Fig. 48 are derived. It is evident when Fig. 48 is compared with Figs. 45 and 46 that *the changes due to the 27-day variation are produced by changes of primary particle flux rather than by variations of geomagnetic field intensity.*

E. Solar flare effects.

I. Large increases of cosmic-ray intensity associated with solar flares.

30. Results from ionization-chambers. During some 17 years of continuous registration of cosmic-ray ionization with Compton-Bennett meters at several stations four unusual increases have been observed [*56*]. The sudden increase in cosmic-ray intensity which began at $10^{h}45^{m}$ GMT, November 19, 1949, was the largest yet recorded. The other three increases occurred on February 28, 1942, March 7, 1942 (observed also by EHMERT [*56a*] in Germany), and July 25, 1946. All were registered with Compton-Bennett ionization chambers completely shielded with 12-cm Pb. Three of the increases began during intense chromospheric eruptions or solar flares [*56*]. While no solar flare was actually observed during the increase of cosmic-ray intensity on March 7, 1942, a radio fadeout occurred very near the time the increase in cosmic-ray intensity began. The fadeout, which occurred only on the day-light side of the earth definitely indicates the occurrence of a solar flare. The terrestrial magnetic effect of such solar flares is an increase, on the daylight side of the earth, in the normal diurnal variation in the earth's field [*57*]. The known small diurnal variation [*58*] in cosmic-ray intensity excludes the possibility that the increases were due to changes in the earth's external magnetic field resulting from an augmentation of the magnetic diurnal variation. The evidence thus indicates [*56*] that the four increases in cosmic-ray intensity were probably due to charged particles accelerated by some mechanism [*59*] on the sun. Unless the particles responsible for the increases were charged, it would be difficult to explain either the simultaneous occurrence of the increases on both the daylight and dark hemispheres or the absence of the increases at the equator.

The sudden increases in cosmic-ray intensity on February 28 and March 7, 1942, are shown in Fig. 49, in which the curves are drawn through the bihourly means, after correcting these to constant barometric pressure. It is evident that neither increase occurred at Huancayo and that the increase on February 28 did not occur at Teoloyucan. The decrease in cosmic-ray intensity during the magnetic storm following the sudden commencement on March 1 is evident at all the stations.

The observations of the sudden increase of July 25, 1946 at Godhavn and Cheltenham is shown in Fig. 50. Again, no increase occurred at Huancayo, although the decrease during the subsequent magnetic storm is evident there. In Fig. 51 the increase observed [*60*] at geomagnetic latitude 88° N on July 25, 1946, with a Millikan-Neher electroscope is compared with that observed at Godhavn with the Compton-Bennett meter.

The sudden increase in cosmic-ray intensity at the time of the solar flare on November 19, 1949, is shown in Fig. 52. This is the first instance when an increase in cosmic-ray intensity accompanying a solar flare has been recorded at a mountain station and at sea-level stations. The increase, in percent of the total cosmic-ray ionization, is obviously very much greater at Climax than at Cheltenham. In fact, if the ordinates on the curve showing the increase at Cheltenham are multiplied by 4.2, the resulting points, shown in Fig. 53, lie on the curve for Climax. It may also be noted in Fig. 52 that the increase on November 19 was not followed

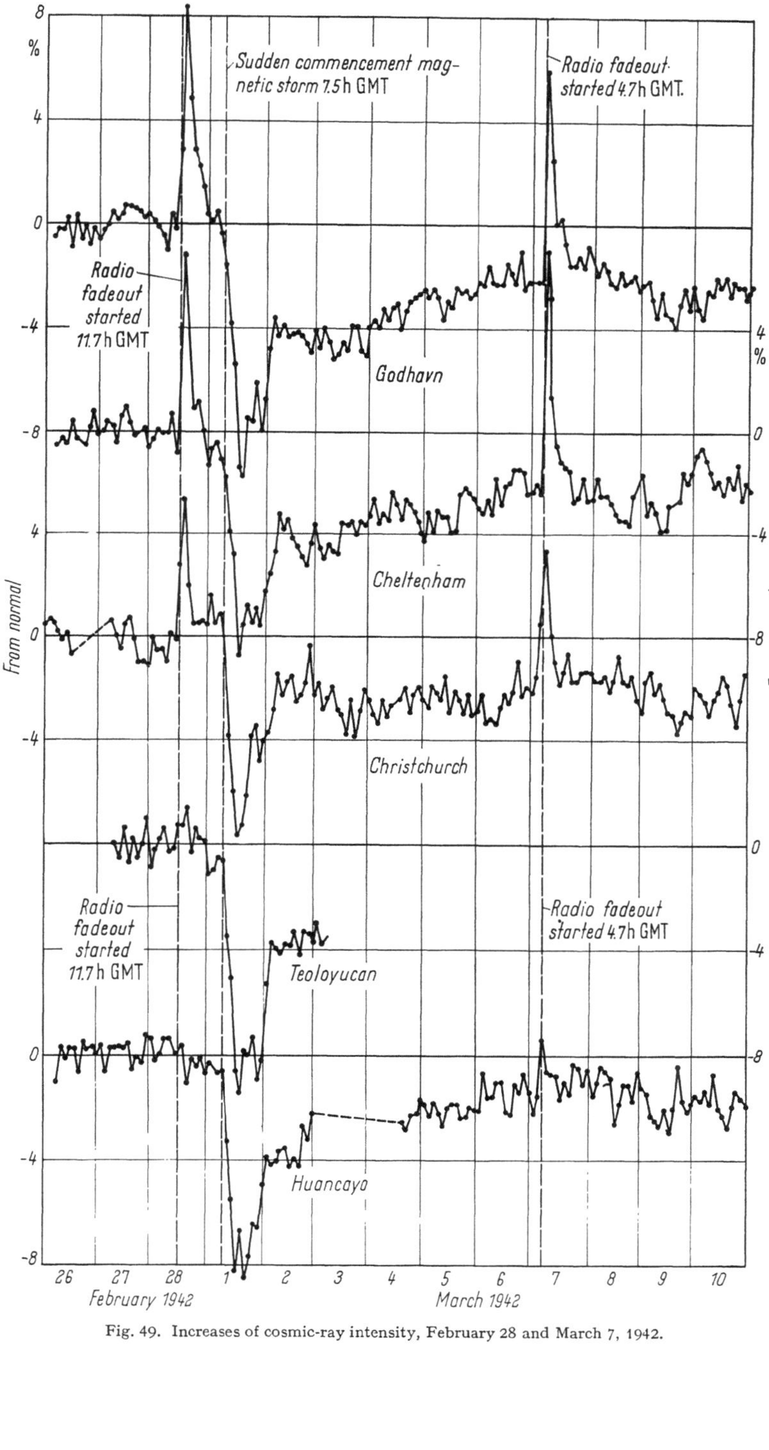

Fig. 49. Increases of cosmic-ray intensity, February 28 and March 7, 1942.

by a decrease in cosmic-ray intensity during the magnetic storm which began about 18^h GMT on November 19 (see also the discussion of events on that day in [60a]).

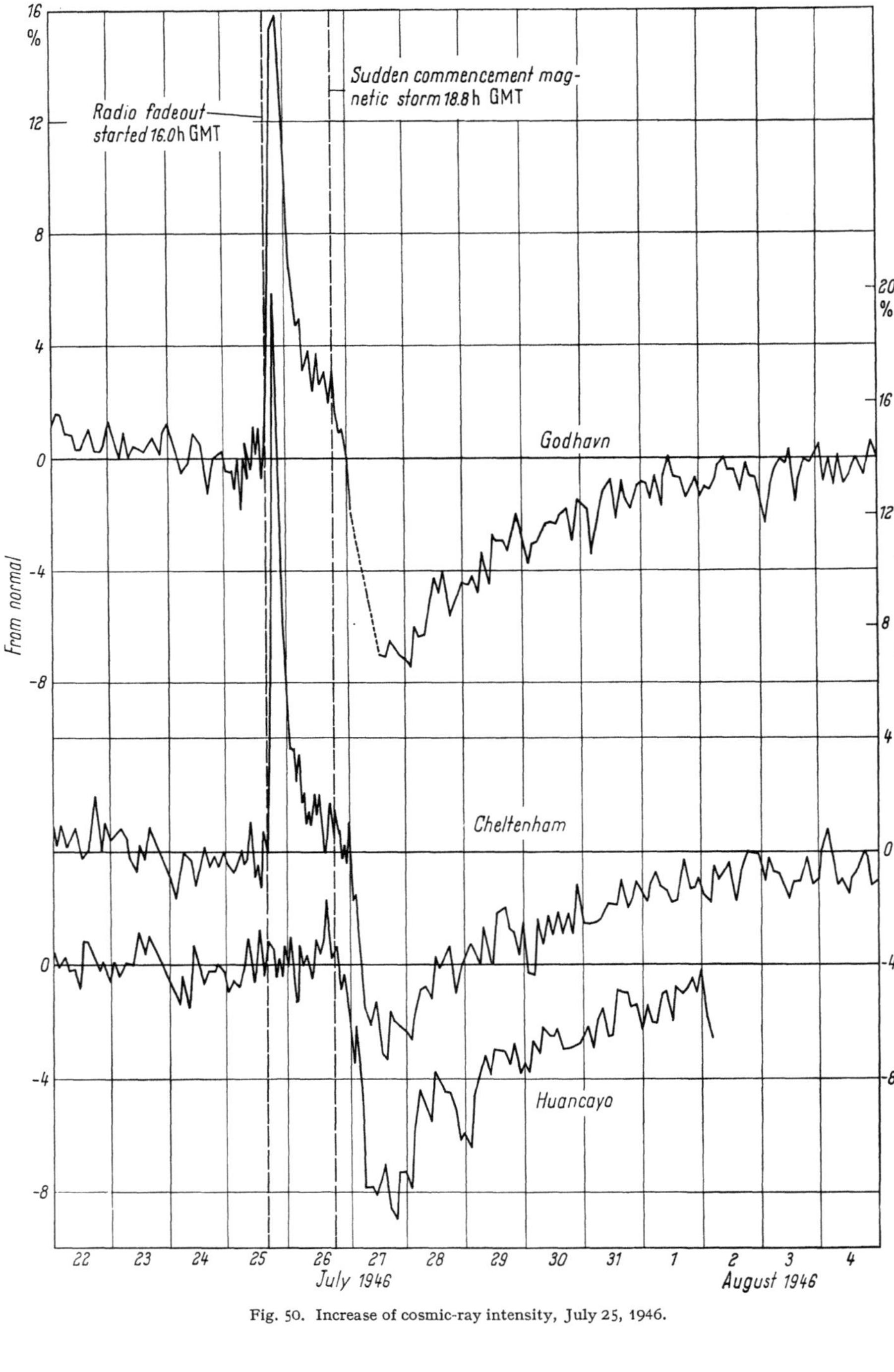

Fig. 50. Increase of cosmic-ray intensity, July 25, 1946.

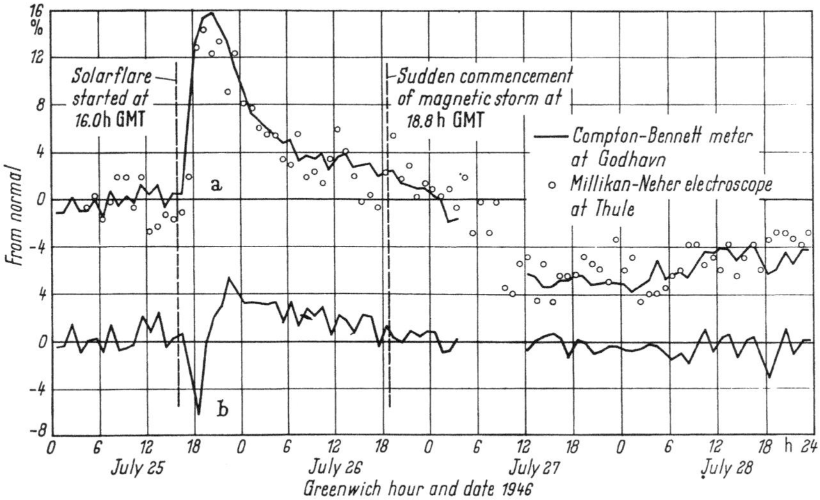

Fig. 51 a and b. Cosmic-ray ionization 1946, July 25—28. (a) Hourly means at Thule and at Godhavn, Greenland. (b) Hourly means at Godhavn less hourly means at Cheltenham.

In addition to complete shielding by 12-cm Pb, the meter at Climax, during the flare of November 19, 1949, was under a rectangular iron shield 4 ft long, 1 ft wide, and 16.5 cm thick. The absorption mean free path for nucleons of medium energy in iron is approximately 240 g cm^{-2}. Taking the dimensions of the shield into account and figuring the zenith angle distribution for a radiation exponentially absorbed with an absorption coefficient of about 145 g cm^{-2}, it is estimated that the increase at Climax on November 19, 1949, would have been 15% greater without the iron shield. Thus it is estimated [*61*] that the maximum of the increase on November 19 at Climax would have been about 207% without the iron shield, instead of the uncorrected 180% as shown in Figs. 52 and 53. This correction also results in a factor of 4.8 for the ratio of the percentage increase at Climax on

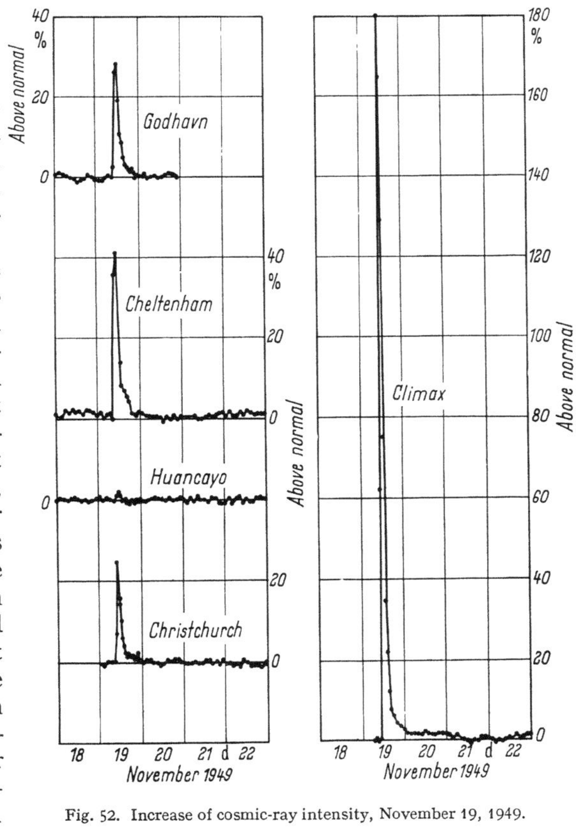

Fig. 52. Increase of cosmic-ray intensity, November 19, 1949.

November 19 relative to that at Cheltenham instead of 4.2 as is indicated in Fig. 53.

31. Flare increases due to nucleonic-component. Since the total ionization at Climax (under 12-cm Pb) is about 2.5 times that at Cheltenham, and since the percentage increase on November 19, 1949, was about 4.8 times that at Cheltenham, the actual magnitude of the increase on that date at Climax was about 12 times greater than at Cheltenham [61]. Since the difference in the atmospheric layer is equivalent to 340 g cm^{-2}, the radiation responsible for the increase during the flare has an absorption coefficient of about 137 g cm^{-2}. This is just about the rate at which the nucleonic component responsible for star production in photographic emulsions, increases with altitude. The increase in total ionization under 12-cm Pb by a factor of 2.5 from Cheltenham to Climax is mainly due to mesons. It is thus evident that the magnitude of the flare effect increases too rapidly with altitude to be ascribed to ordinary mesons. The latitude effect in chambers under 12-cm Pb, due principally to mesons, is small, whereas the flare effect exhibits a strong dependence on latitude (being zero at the equator). This also indicates that ordinary mesons contribute negligibly to the flare effect.

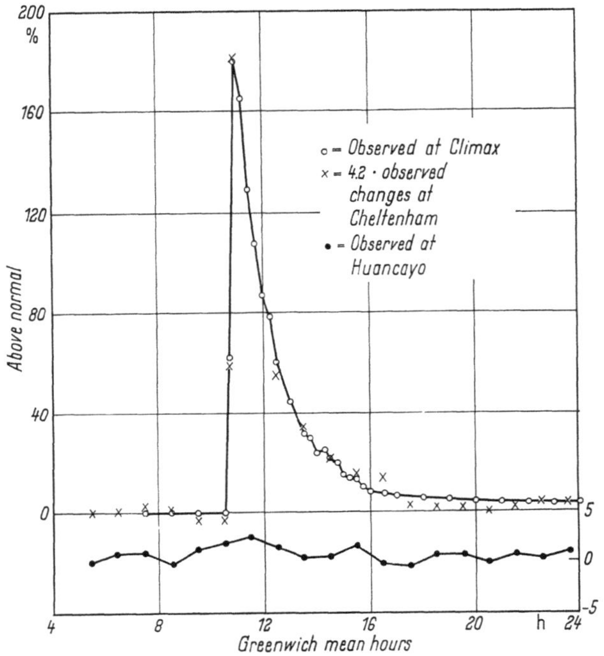

Fig. 53. Cosmic-ray intensity, November 19, 1949.

The results on the latitude variation of the proton and neutron intensity suggested that the cross section for nucleon production relative to that for meson production decreases rapidly with increasing energy of primary particles. This is in accord with the conclusion that the increase in intensity during the solar flare of November 19, 1949, was due principally to the nucleonic component generated in the atmosphere by relatively low energy primaries, and not to ordinary mesons. At Climax, under 12-cm Pb, probably not more than about 10% of the total ionization is normally due to local radiation originating from the nucleonic component. If this radiation is produced entirely by particles in the same band of energy as those responsible for the increase of 207% in ionization on November 19, 1949, then the number of primary particles, reaching there per unit time, in that band of energy, must have increased to at least 20 times the normal value. Three sets of triple coincidence counters and one set of four-fold coincidence counters, located above the meter at Climax and arranged to record air showers, were in continuous operation during the period of the increase in ionization on November 19, 1949. There was no evidence of any significant increase in the rate of air showers during this period.

The conclusion in the last paragraph that the increase of cosmic-ray intensity on November 19, 1949 was due to the nucleonic component generated in the atmosphere by relatively low energy primaries and that the increase in this component must have been at least twenty fold is confirmed by the results in Fig. 54. These observations were obtained by ADAMS [62] with a neutron counter in a large "pile" of high purity graphite. The hourly mean counting rate increased to a maximum 550% above normal. This was about 12 times the corresponding maximum hourly increase in ionization recorded at Cheltenham, and about 40 times the maximum increase recorded on three fold-coincidence meson telescopes pointing north and south, 45° to the horizontal, at Manchester [62]. The increase in neutron intensity at Manchester was nearly 70 times that observed by CLAY [63] in an ionization chamber at Amsterdam. Data for the solar

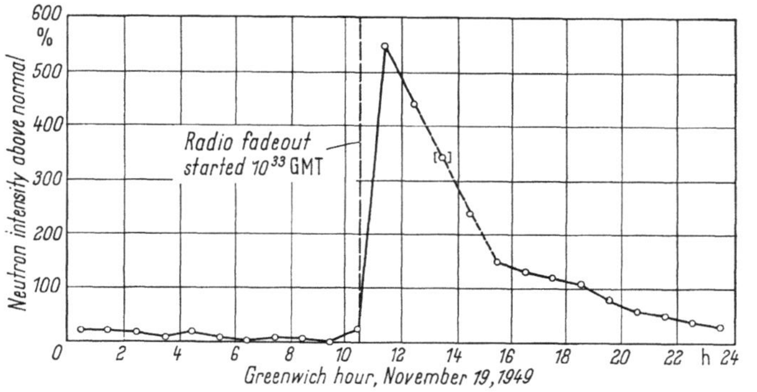

Fig. 54. Increase in neutron intensity at Manchester, November 19, 1949 (ADAMS).

flare increases which have been observed at many stations have been carefully summarized by SEKIDO and YOSHIDA [64] who from its analysis determined that the maximum of the flare-effect occurs between 3 and 8 hours local time. In two review articles ELLIOTT [65] and BIERMANN [66] have also summarized some of the observations of the four large flare effects.

II. Expected geographical distribution of solar-flare increases if these come from the sun.

32. Theoretical impact zones. When all the available observations of the four large solar-flare increases in cosmic-ray intensity are summarized, as it has for example been done by SEKIDO, and YOSHIDA [64], it is quite evident that the magnitude of the observed increase depends not only on latitude but also on the local time at which the increase occurred. EHMERT [67] by utilizing some of the trajectories of cosmic-ray particles integrated by STÖRMER showed that the distribution of the increase on February 28, 1942, observed in America but not in Germany, indicated positive particles arriving from the direction of the sun. SCHLÜTER [68] has integrated twenty trajectories of cosmic-ray particles using the method originated by STÖRMER, in which particles are assumed to be initially moving along a line parallel to the sun-earth line. He concludes from an examination of the impact points on the earth of these orbits that for positive particles from the sun there should be a sharp maximum near 0900 local time, and he regards the observation of one of the large increases in Germany in the afternoon as in disagreement with the theory. Using STÖRMER's Nullbahnen, i.e., orbits which would pass through the dipole if extended, as representative of all orbits striking the earth, SCHLÜTER discusses the seasonal change in the impact points on the earth for particles from the sun, as well as the existence of forbidden zones near the poles of the earth for solar particles. The observation of the increases at Godhavn, Greenland ($\lambda = 80°$) he also regards as in disagreement with the theory, since this station lies in such a forbidden region (see also Fig. 51).

FIROR [69] has estimated the geographical distribution of the solar flare increase in cosmic-ray intensity. He showed the existence of a background or latitude zone within which the magnitude of the increase would have no strong dependence on local time. He also made rough estimates of the relative intensities in the different zones, within which the magnitude of the increase depends strongly upon local time. In addition FIROR considered the effect of a finite source larger than the visible sun. To determine the latitude and longitude of the impact zones for particles arriving from a point source FIROR utilized the available data for about 80 orbits which arrive vertically thus insuring that the results will be of greatest importance for detectors deep in the atmosphere for which vertically incident particles are most effective. The trajectories used involved those calculated by STÖRMER [70], DWIGHT [71], and SCHLÜTER [68], together with those obtained from the model experiments of MALMFORS [72] and BRUNBERG [73]. To complete the picture required interpolation between the known orbits and in some cases extrapolation beyond the known orbits. In this way curves in Fig. 55 were obtained for the geomagnetic latitude of impact points on the earth for particles arriving vertically with each of several different magnetic rigidities between 1 and 10 BV, as a function of the geomagnetic latitude (latitude of asymptotic velocity vector) of the source. The lower limit of the latitude at which flare increases have been observed indicates that few of the particles involved had magnetic rigidities of more than 10 BV, corresponding to cutoff at geomagnetic latitude 30°.

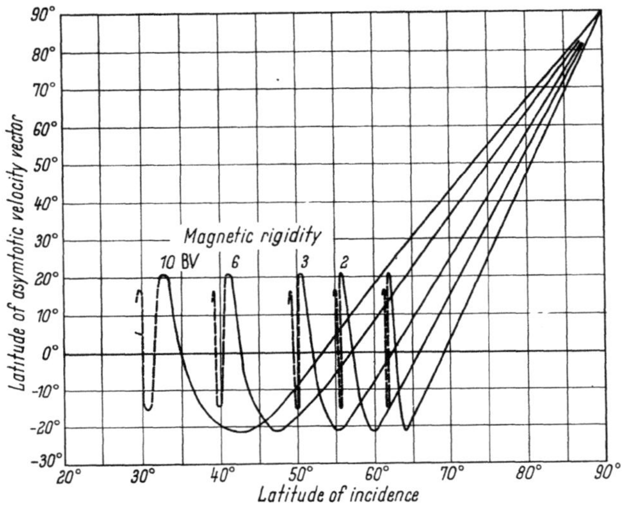

Fig. 55. Geomagnetic latitude of impact on the earth of particles arriving vertically with magnetic rigidities of 1—10 BV as a function of the geomagnetic latitude of the source (FIROR).

Curves obtained for the geomagnetic longitude of impact points on the earth, for magnetic rigidities between 1 and 10 BV, as a function of the source latitude are shown in Fig. 56. Here it will be noted that for rigidities between 1 and 10 BV the curves do not depend critically on rigidity and they form a narrow band about one hour wide. For particles from the sun these zones will be several hours wide. Fig. 56 which does not involve rigidity as a parameter greatly facilitates the construction of impact zones for various situations. From Fig. 55 the latitudes on the earth of possible impact points are readily obtained for a particular rigidity and for a particular source latitude. For the same latitude of the source the longitudes on the earth of the impact points are readily obtained from Fig. 56.

In Fig. 57 are plotted as smooth curves the impact points in the nothern hemisphere for particles leaving a point source far from the earth and in the plane

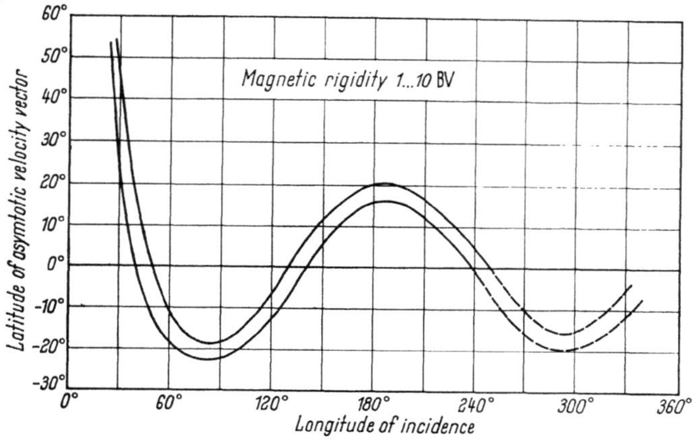

Fig. 56. Geomagnetic longitude of impact on the earth of particles with magnetic rigidities of 1—10 BV. Curves shown are the envelope of the individual curves for each rigidity (FIROR).

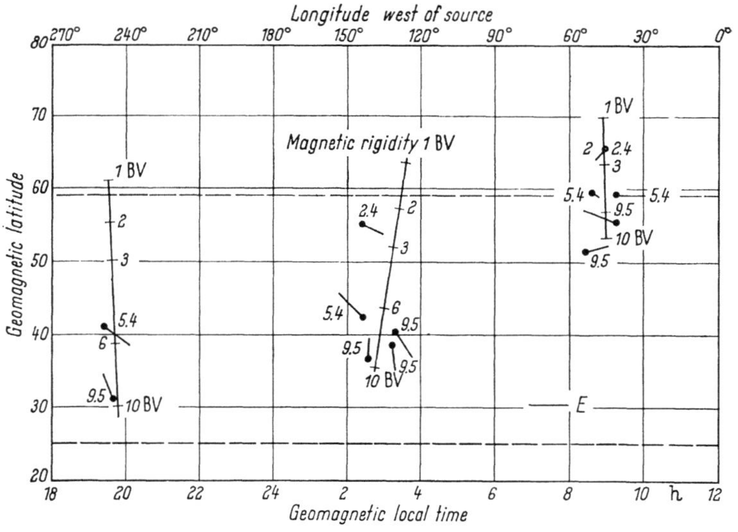

Fig. 57. Impact points on the earth for particles aprpoaching in the equatorial plane. The solid lines show the impact points for particles arriving vertically. The circles give the impact points for some orbits which arrive with large zenith angles. The line attached to each of the latter shows the azimuth and zenith angles of arrival. This line is the projection on the earth's surface of the velocity vector, length "E", at the time of arrival. The longitude scale in hours local time assumes the source of particles to be at noon local time. The dashed lines indicate the band of latitudes which is filled with impact points of 1 to 10 BV particles (FIROR).

of the geomagnetic equator, and arriving at the earth with vertical or near vertical incidence. Also shown are the impact points of some orbits integrated by SCHLÜTER which arrive with large zenith angles. The line attached to each SCHLÜTER orbit in Fig. 57 is a projection on the earth's surface of the velocity vector of the particle at the time of arrival. The length marked "E" in the figure gives the reference length of the velocity vector. Since the source is to be identified with the sun the lower scale is in hours, with the source at local noon.

In Fig. 57 the impact points fall into three groups. The group near 0900 local time, results from orbits that remain north of the equatorial plane. The group at 0400 results from orbits that pass once through the equatorial plane, and the group at 2000 comes from orbits passing twice through the equatorial plane, and so on. There are in fact for a particular source latitude, within limits, an infinite number of intersections with the impact curve of Fig. 56, since as STÖRMER [*74*], [*75*] pointed out this curve oscillates through the ordinate 0° for increasing longitudes. This results in groups of impact point, not shown in Fig. 57, which have latitude distributions similar to the group at 2000 hours. Fig. 57 was constructed for the source in the geomagnetic equatorial plane. Figs. 55 and 56 may be used to estimate impact points for the source not in the equatorial plane. Moving the source south of the equator causes the two morning impact groups to move closer together, in the nothern hemisphere, while the evening group occurs earlier. This is reversed if the source is moved north of the equator. In this way the effect of a finite particle source upon the impact points can be calculated. For a source 10° in diameter the lines of Fig. 56 become zones about an hour wide. The zone at 0900 has the least spread, while the zone at 2000, and presumably all the similar zones are the widest. Since each of the latter is at a greater longitude than the preceding one, this infinity of impact zones fills the whole range of local times between geomagnetic latitudes 25° to 60° for 1 to 10 BV particles. This results in three distinct zones one at 0900, and one at 0400, and the background zone. FIROR [*69*] finds that the expected magnitude of the increase for the above three zones, respectively, should be roughly in the ratio: 7:2:1 for a source 15° in dimater and in the ratio 7:3:1 for a source 30° in diameter.

The theoretical *results* thus far obtained indicate that if the sun, while near the geomagnetic equator, emits a pulse of particles, with rigidities up to 10 BV, the following effects should have been recorded by identical detectors:

1. Detectors located at latitudes less than 25° would receive no new particles.

2. Detectors between 25° and 35° would record an increase in counting rate. This increase would be due to the background zone and would have no strong longitude (local time) dependence.

3. Detectors above 35° would see an increase as in 2, but in addition those detectors at local times around 0400 would see an increase about three times as large as in 2.

4. Detectors above 50° would see an increase as in 2, but detectors near 0400 or 0900 would see additional increases, up to seven times as large as in 2. The highest latitude at which detectors would see increases, either the background type as in 2 or the strongly local time dependent increase as in 3 and 4, would depend on the lowest rigidity particles to which the detector was sensitive.

33. Comparison with observations. The observed increases were compared with the predicted distribution by constructing an impact zone diagram for the position of the sun at the time of the increase, and marking on the diagram the location of stations reporting the increase.

Figs. 58, 59, and 60 are the resulting diagrams respectively for the flare increases which occurred November 19, 1949, February 28, 1942 and March 7, 1942. The stations reporting one or more of the flare events are summarized in the following table which includes references to reports of the events:

In Figs. 58, 59, and 60 the density of the crosshatching indicates roughly the relative intensities to be expected in the different zones. The lack of symmetry

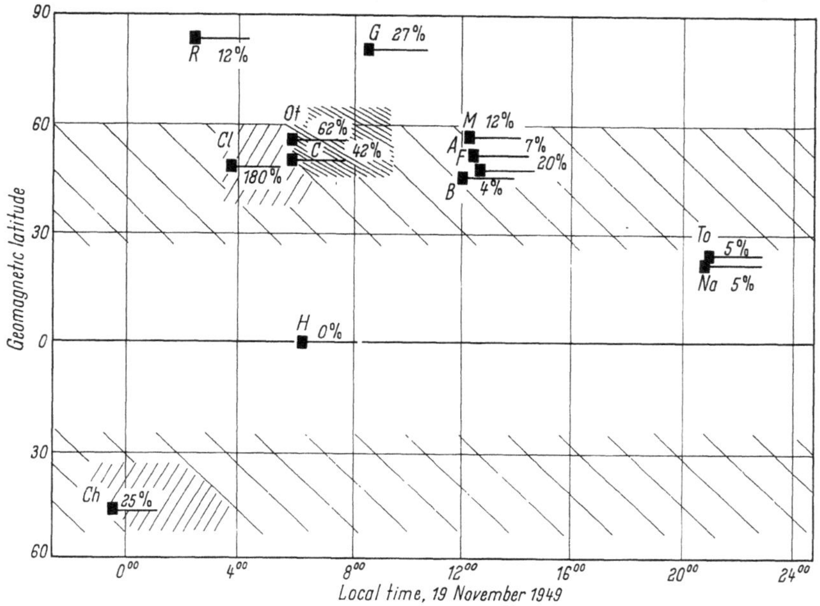

Fig. 58. Impact zones on the earth at the time of one of the large cosmic-ray increases—1100 UT on 19 November 1949. The cross-hatching indicates the positions of the zones with the density of crosshatching giving roughly the relative intensities predicted for the different zones. The cosmic-ray stations are indicated by solid squares; the lines attached to each station show its motion during the increase. Near each station is an identifying letter (see Table 3), p. 215 and the percentage increase observed (FIROR).

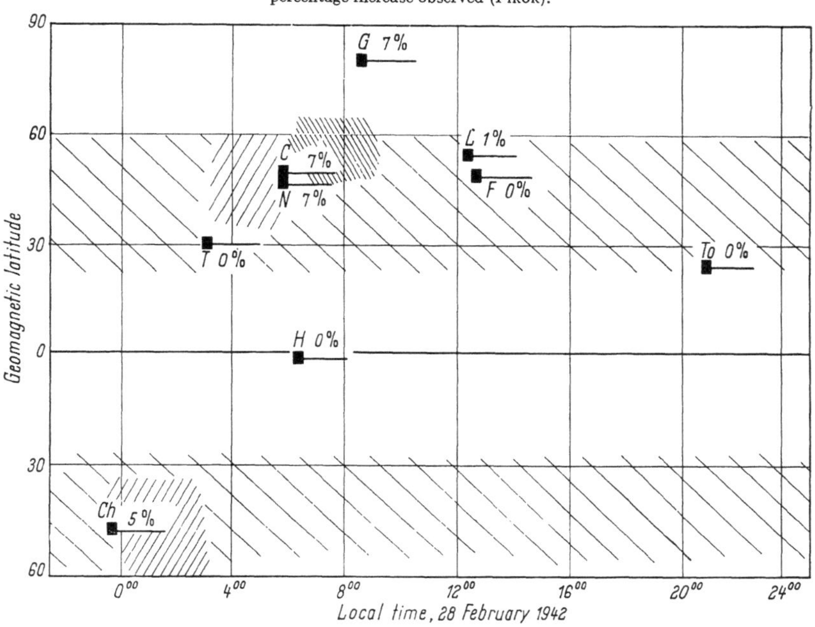

Fig. 59. Impact zones and observations at cosmic-ray stations at 1100 UT on 28 Febrauary 1942. Symbols same as in Fig. 58 (FIROR).

between the zones in the northern and southern hemispheres is an example of the seasonal effect mentioned earlier. The small squares are the positions of the cosmic-ray stations reporting the increase (or its absence), with the line attached to each indicating its motion during the increase. Near each station is marked the percentage increase observed using for the most part short time intervals (5 to 15 min). The zones were constructed assuming a source 15° in diameter.

In Fig. 58 there is evident a tendency for the stations in the morning impact zones to show the largest increase. In particular the station at Ottawa, in the 0900 zone, recorded an increase about five times as large as that at Manchester,

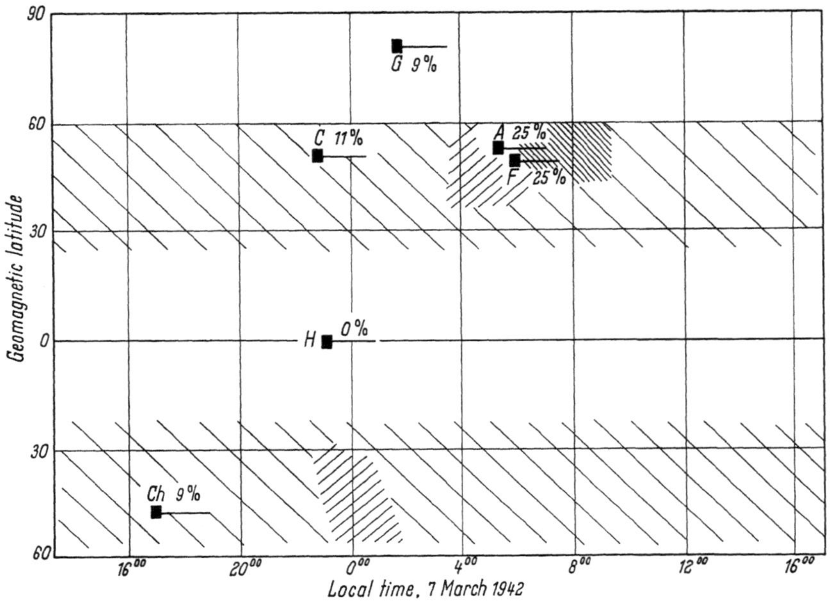

Fig. 60. Impact zones and observations at cosmic-ray stations at 0400 UT on 7 March, 1942. Symbols same as in Fig. 58 (FIROR).

located in the background zone. This is consistent with the ratio seven predicted above. The 0900 and 0400 zones may be compared by using Cheltenham and Christchurch, giving a ratio of about two, consistent with the predicted 7:3. The 20% increase at Freiburg seems large compared with the other European observations and is probably due to the fact that this figure is derived from a very short time interval at the peak intensity. Similarly, the 7% at Amsterdam is a one hour average and should be made somewhat larger for comparison with other stations. Two German stations, not shown for lack of room, gave increases of 15% for each of two similar counter telescopes and 17% for a shielded ion chamber. The very large increase at Climax can in part be attributed to the high altitude of that station. A reasonable altitude correction to sea level lowers the percentage increase to about the same value as at Cheltenham [61]. The high latitude stations, Resolute and Godhavn, recorded increases although they were not in any impact zone.

Fig. 59 shows the impact zones and results observed on February 28, 1942. In this case the increases marked in the figure are for one hour averages centered on the half hour. It is seen (with Godhavn again the possible exception) that the distribution of the increase is again consistent with positive particles arriving at the earth from the direction of the sun. All of the increases for this event began roughly an hour after the flare began. The ratio of the increases in and out of the morning impact zones is, in this case, seven or greater.

Table 3. *Observations of large cosmic-ray increases*

Station	Symbol used in Figures	λ (degrees)	Altitude (a) (meters)	References: 28 February 1942	7 March 1942	25 July 1946	19 November 1949
Amsterdam	*A*	54 N		b	b	c	d
Bagneres	*B*	46 N	550	—	—	—	e
Bargteheide		54 N		—	—	—	f
Cheltenham	*C*	50 N		g	g	g	g
Christchurch	*Ch*	48 S		g	g	—	g
Climax	*Cl*	48 N	3500	—	—	—	g
Darmstadt		50 N		—	—	—	f
Freiburg } Friedrichshafen }	*F*	48 N		h	h	—	f
Godhavn	*G*	80 N		g	g	g	g
Huancayo	*H*	1 S	3350	g	g	g	g
London	*L*	54 N		i	i	—	—
Manchester	*M*	57 N		—	—	j	k
Nagoya	*Na*	25 N		—	—	—	l
Norfolk	*N*	49 N		m	—	—	—
Ottawa	*Ot*	56 N		—	—	—	n
Resolute	*R*	83 N		—	—	—	n
Teloyucan	*T*	30 N	2285	g	—	—	—
Tokyo	*To*	25 N		o	o	—	p
Weissenau		49 N		—	—	—	f
Mt. Wilson		43 N	1800	—	—	q	—
Thule		88 N		—	—	r	—

a Altitude listed only for mountain stations.
b Clay, Jongen and Dijker: Proc. Kon. Ned. Akad. Wet. **52**, 923 (1949).
c Clay, J.: Proc. Kon. Ned. Akad. Wet. **52**, 899 (1949).
d Clay, J., and H. F. Jongen: Phys. Rev. **79**, 908 (1950).
e Dauvillier, A.: C. R. Acad. Sci., Paris **229**, 1096 (1949).
f Müller, R. et al.: J. Atmosph. Terr. Phys. **1**, 37 (1950).
g Forbush, S. E., M. Schein and T. B. Stinchcomb: Phys. Rev. **79**, 501 (1950).
h Ehmert, A.: Z. Naturforsch. **3**a, 264 (1948).
i Duprerier, A.: Proc. Phys. Soc. Lond. **57**, 468 (1945).
j Dolbear, D. W. N., and H. Elliot: Nature, Lond. **159**, 58 (1947).
k See references [*62*] and [*63*].
l Sekido, Kodama and Yagi: Rept. Ionos. Res. Japan **4**, 207 (1950).
m Berry, E. B., and V. F. Hess: Terr. Magn. Atm. Electr. **47**, 251 (1942).
n Rose, D. C.: Canad. J. Phys. **29**, 227 (1951).
o Nishimura, J.: J. Geomag. Geoelectr. **2**, 121 (1950). — Sekido, Yoshida and Kamiya: Rept. Ionos. Res. Japan **6**, 195 (1952).
p Miyazaki, Wada and Kondo: Rept. Ionos. Res. Japan **4**, 176 (1950).
q Neher, H. V., and W. C. Roesch: Rev. Mod. Phys. **20**, 350 (1948).
r Graham, J. W., and S. E. Forbush: Phys. Rev. **98**, 1348 (1955).

In Fig. 60 is shown the result, based on hourly averages, for the flare of March 7, 1942. Here also the largest increase occurred in the morning impact zone.

For the increase on July 25, 1946 no results were reported from a station in the morning impact zones at this time. The increase was seen by five stations (see Table above) with roughly the same amplitude of about 15%. No increase was seen at Huancayo. These observations again agree with the predicted distribution.

Thus from Figs. 58, 59, and 60, it is evident that the results for middle and low latitude stations are consistent with predictions based on the assumption that the flare increases resulted from particles with magnetic rigidities of less than 10 BV coming from a source about 15° in diameter, centered at the sun. However, the increases at high latitudes reported for Resolute, Godhavn, and Thule (see Fig. 51) are not consistent with the predictions. The explanation for the increases at these high latitudes has yet to be found.

34. Effects of small flares. Since the model described above gave reasonably consistent agreement with observations for large solar flare increases it was used by FIROR [*69*] as a basis for investigating the question of whether small increases might also be statistically detected during small solar flares. The model was used to determine the range in local time during which the flare effects at Climax would be expected to have the greatest amplitude. Data from the Climax neutron monitor were used since these are the most sensitive to the low energy particles from flares. The reported flaıes, with importance 1+ or more, were dichotomized so that the first group contained flares the starting time of which occurred when Climax was in a morning impact zone, which was considered four hours wide. The second group contained the flares which started when Climax was at least an hour away from the edge of a morning impact zone. The cosmic-ray data were averaged on "flare time" for each group. Some indication was obtained for an increase of 0.8% lasting an hour or so during the time of occurrence of 12 flares in 1951 in the first group. An increase of about 0.5% was similarly indicated for four flares of 1952 which were in the first group. In neither case was an increase evident in the second group. In neither case for the first group was the increase greatly in excess of the indicated standard deviations and the exact, manner in which these were obtained is not described.

FIROR [*69*] also compared the diurnal variation in neutron intensity at Climax on days with large flare indices with that for days with small flare indices. The results indicated, for days with the larger flare indices, a slightly greater average intensity during the period when Climax was in a morning impact zone. FIROR also found a correlation of 0.6 between the flare index and the size of the cosmic-ray increase from a sampl e of 16. Thus, the probability is about $^1/_{50}$ that the sample was from a population with zero correlation.

While these results suggest the possibility of small increases during small solar flares the effects are not so large relative to the variability as to be thoroughly established without more rigorous tests of their statistical reality.

F. Summary of results up to the Geophysical Year.

35. Observational procedures and meteorological effects. Experience has shown that ionization chambers have the merit of simplicity and reliability for registering changes in cosmic-ray intensity over long periods of time. This advantage is offset to a considerable extent by the influence of meteorological factors on the recorded intensity. In addition to barometric effects such measurements are affected by the height of the region of μ-meson production, where the barometric pressure is about 100 mb. Since this height varies seasonally, corresponding more or less regular seasonal variations are introduced into the measurements. Irregular variations in this height during the passage of meteorological fronts also affect the measurements. Unless meteorological data are regularly available to heights of 16 km or more, no reliable corrections for these latter effects are possible. Near the equatorial zone these effects are quite small. Geiger-counter apparatus is similarly affected. Neutron monitors, on the other hand, are affected only by barometric pressure and while the pressure coefficient is large, the advantage of freedom from other meteorological influence is great. In addition neutron monitors are more sensitive to changes in the lower energy portion of the primary cosmic-ray spectrum. While the time variations from neutron monitors are generally similar to those found from ionization chambers the magnitude of percentage changes is generally four of five times larger for the former.

36. Time variations. The solar 24-hour diurnal variation has been established Its amplitude undergoes world-wide changes. A large world-wide secular change in its phase is evident. There are indications that the amplitude of the diurnal variation changes with magnetic activity but the mechanism involved is not understood. The 24-hour sidereal variation is extremely small, and since an apparent sidereal variation could arise from other causes it is not regarded as significant.

Large decreases in cosmic-ray intensity occur during some magnetic storms but storms also occur without decreases in cosmic-ray intensity. Furthermore, there are some decreases in cosmic-ray intensity without marked magnetic activity. Nevertheless, in a statistically significant majority of cases, the cosmic-ray intensity is less on the five international magnetically disturbed days of each month than on the five quiet days. The variability of the daily means of cosmic-ray intensity is significantly less near sunspot minimum than near sunspot maximum. There is a world-wide sunspot cycle variation in cosmic-ray intensity with maximum near sunspot minimum. At high altitudes and near the geomagnetic pole, particles of lower energy are detected near sunspot minimum than at other times, and there is evidence that the "knee" of the latitude curve moves northward as the sunspot numbers approach the minimum.

A world-wide quasi-persistent (q.p.) 27-day variation in cosmic-ray intensity is established with minima near the times of maximum of the 27-day (q.p.) wave in magnetic character figure. These times of minimum are near those for the 27-day (q.p.) waves in the horizontal component of the earth's magnetic field at the equator. The character of the latitude variations, measured with neutron detectors in jet planes at times of maximum and minimum of the 27-day (q.p.) variations shows that these variations result from changes in primary particle flux.

Sudden increases in cosmic-ray intensity during a few chromospheric eruptions are caused by charged particles evidently accelerated by some solar mechanism. The large variation in observed intensity of these events with geographical position is generally consistent with calculation, if the particles (magnetic rigidity up to 10 BV) are assumed to come from the neighborhood of the sun. On the other hand, the increases observed near the geomagnetic poles during these events are not yet explained. Finally, there is some inconclusive evidence that increases in cosmic-ray intensity may occur during small solar-flares or chronospheric eruptions.

Thus most of the established time-variations in cosmic-rays are the direct or indirect result of solar activity, and a better comprehension of the mechanisms responsible will doubtless advance understanding of other phenomena related to solar activity. The results of observation and experiment in many related fields of geophysics executed in the International Geophysical Year (July 1957 to December 1958) have provided valuable material for interpreting the mechanisms which cause the time variations in cosmic-rays; this will be discussed in the Second Part.

Second Part:

Results obtained in the International Geophysical Year and afterwards.

37. Introduction. The first part reviewed in some detail much of what had been learned about temporal variations of cosmic-ray intensity until about the end of 1955 from the time of the discovery of cosmic radiation, which resulted from the historic balloon flight made by V. F. HESS [*76*] fifty years ago on August 17, 1912. Since 1955 research activities concerned with temporal variations and other geophysical aspects of cosmic radiation have increased enormously. This was due principally to the interest and opportunities generated by the *International Geophysical Year* (IGY), July 1957 to December 1958, and the *International Geophysical Cooperation Year* (1959). Many of these activities have continued and will continue during the forthcoming *International Quiet Sun Years* (IQSY), planned for the 24 months January 1964 to December 1965.

During the IGY cosmic ray measurements were carried out at 128 stations. The stations [*77*] included: 67 with neutron monitors; 31 ionization chambers; 67 with meson telescopes plus 18 with narrow angle telescopes; 13 using emulsions; 8 using balloons, aircraft, or rockets; and several stations using other instruments. The location of these stations is given in [*77*]. In addition, *artificial earth satellites* and *space probes* have since provided opportunities for measuring cosmic-ray intensity in new environments. Many of the IGY and IGC cosmic-ray data are available from the IGY world data centers where all observational results have been collected.

For the IGY the choice of a period near an expected maximum in the cycle of solar activity was especially fortunate, since during this interval sunspot numbers and magnetic activity reached the highest values yet recorded. During the IGY and since, the results obtained in each of several areas of investigation have provided results important to the understanding of phenomena in other related areas. It is now established that most, if not all, of the temporal changes of cosmic-ray intensity are the consequence of solar phenomena. *The solar cycle variation in cosmic-ray intensity and the so-called Forbush decreases, which generally are associated with magnetic storms, both arise from electromagnetic phenomena which originate on the sun and modulate or change the primary cosmic ray intensity.* Electromagnetic phenomena associated with *solar flares* also accelerate charged particles which occasionally increase the apparent cosmic ray intensity in ground level detectors. At high altitude and high latitude where effects from lower energy solar particles can be measured, such *solar flare events* are much more frequent. The number density and energy spectrum of these "solar cosmic-ray particles" has been measured during several of such events. The energy spectrum of primary cosmic-ray particles has been determined at different epochs of the solar cycle and also during Forbush decreases of intensity. These decreases have been observed in interplanetary space showing that they are not due to the geomagnetic storm field. Programs, for the study of cosmic-ray intensity variations and their causes, initiated during the IGY and subsequently, have led to the discovery of the *Van Allen geomagnetically trapped radiation belts*, and to the discovery of *strong bursts of X-rays coincident with visible zenith aurora.*

G. Solar cycle variations.

I. Results from ionization chambers and neutron monitors.

38. Comparison with sunspot numbers. The longest series of continuous registration of cosmic-ray intensity is that obtained from ionization chambers [*78*]. Fig. 61 shows the monthly means of cosmic-ray ionization at Huancayo,

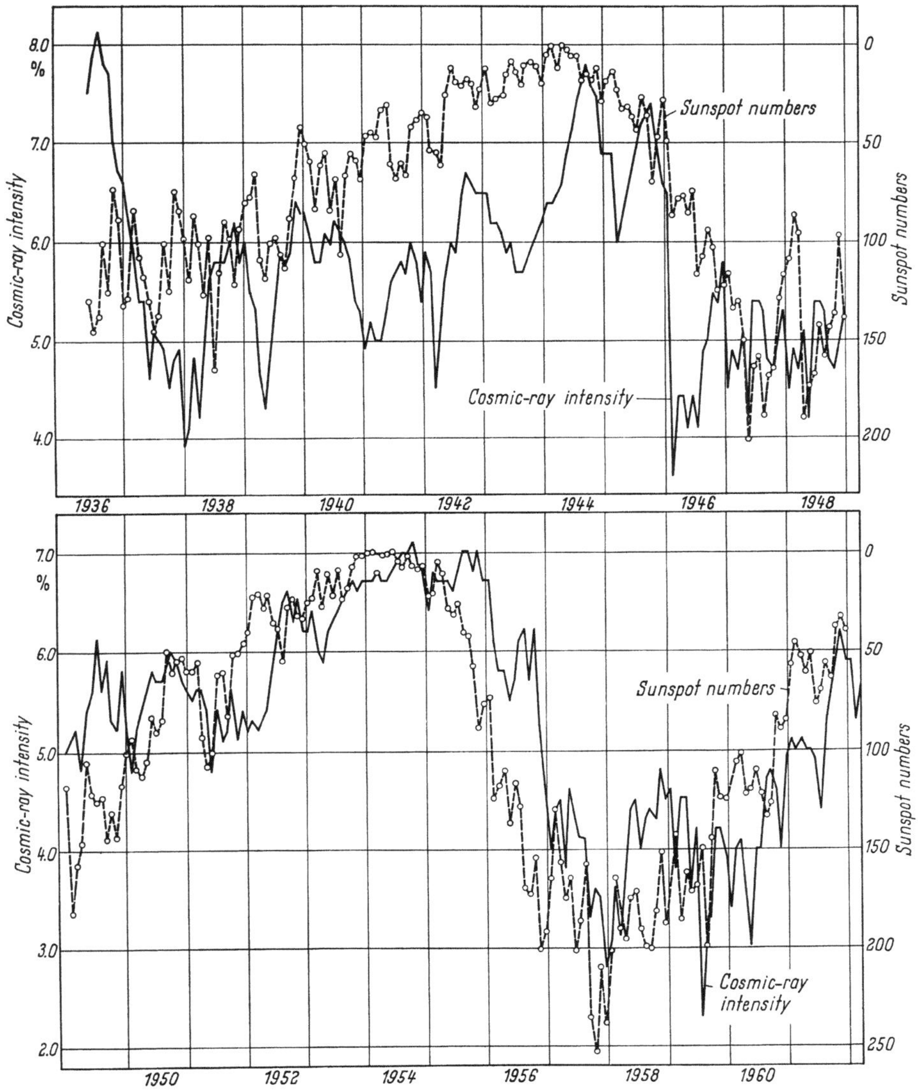

Fig. 61. Monthly means cosmic-ray intensity *C-R* and sunspot-numbers *SS*, plotted downward. 1936–1961.

and monthly mean sunspot numbers for the period from June 1936 to March 1962. For this interval the correlation coefficient between annual means of sunspot numbers, (*SS*), and cosmic-ray intensity, (*C-R*) is −0.85. Fig. 61 indicates, at least in 1955, that following sunspot minimum the increase in SS numbers continues for some months without an accompanying decrease in C-R at Huancayo.

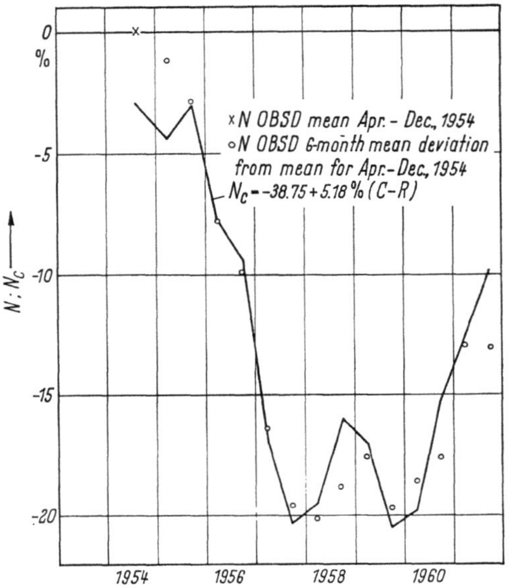

Fig. 62. Six-month means neutron intensity, N, observed at Ottawa (ROSE) and computed, N_c, from cosmic-ray C-R ionization at Huancayo. April 1954 to December 1961.

Fig. 62 shows the variation in 6-month means of C-R ionization at Huancayo and of neutron intensity at Ottawa. After 1954 the neutron intensity at Ottawa appears to start decreasing some months earlier than the ionization at Huancayo.

39. Variations for quiet and disturbed days. Fig. 63 shows the variation, of cosmic-ray ionization at Huancayo from 1937 to 1961 for international magnetically quiet and for international magnetically disturbed days (five each per month). The annual mean cosmic-ray intensity, (C-R), at Huancayo is practically always less for disturbed days although the solar cycle variation is similar for disturbed and quiet days. Also shown in Fig. 63 are the yearly means of the southward geomagnetic field of the equatorial ring current field, ERC, for 1939 to 1945 from the values

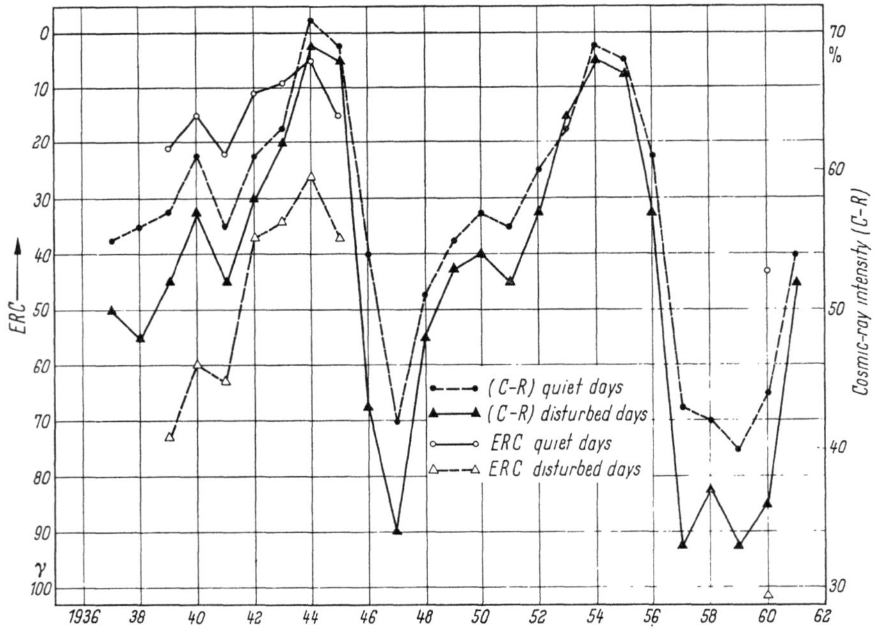

Fig. 63. Yearly means for magnetically quiet and disturbed days for C-R, 1937—1961 and for ERC 1939—1945 and 1960.

published by KERTZ [79] and for 1960, similarly derived. Fig. 63 indicates that relative to the amplitude of the solar cycle variation for quiet days the difference, disturbed minus quiet days, is several times larger for ERC than for cosmic-ray intensity. The large decrease in *C-R* from January to February 1946, in Fig. 61, is conspicuous for the fact that the monthly mean in February

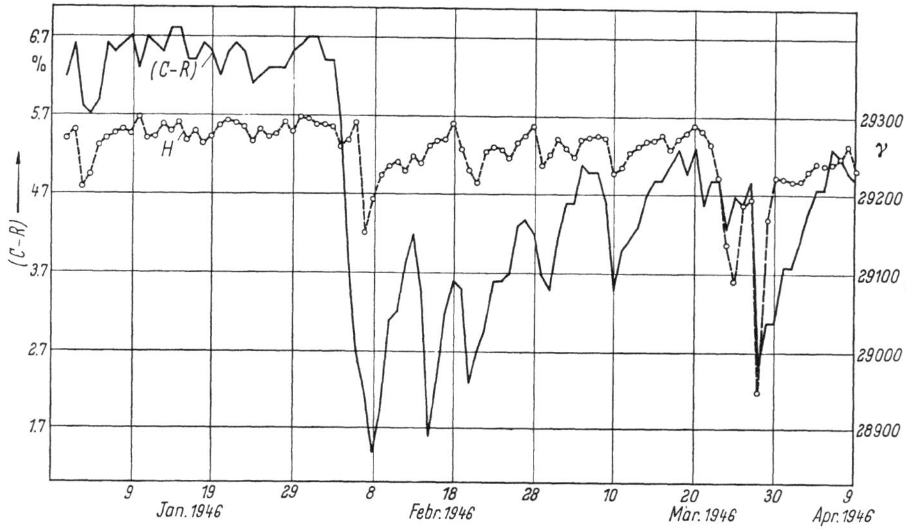

Fig. 64. Daily means (75° West Meridian Time) *C-R* and *H* at Huancayo. January 1 to April 9, 1946.

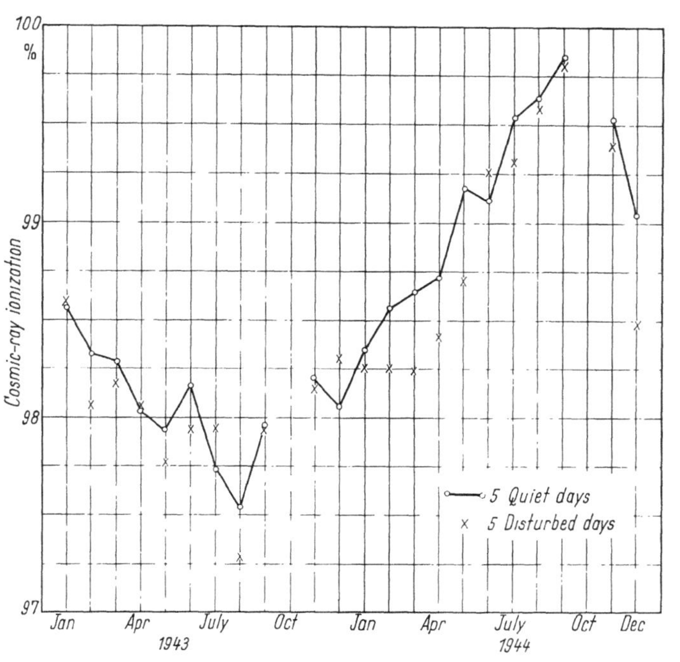

Fig. 65. Means cosmic-ray ionization at Huancayo for 5 quiet and for 5 diturbed days by months for 1943 and 1944.

was the lowest until 1957, and for the fact that a monthly mean as high as that for January 1946 did not occur thereafter until 1952.

The exceptional decrease in cosmic-ray intensity in Februray 1946 is shown in greater detail in Fig. 64 in which daily means are plotted together with those for horizontal magnetic field, H, at Huancayo Peru, for the period January 1 to April 9, 1946. In Fig. 64 it is evident that the conspicuous decrease in H from February 6 to 7 occurred a few days *after* the large decrease in C-R. It is not clear from Fig. 64, whether the low value of C-R prevailing for months after February 8, 1946 was due to a sequence of independent decreases (each occurring

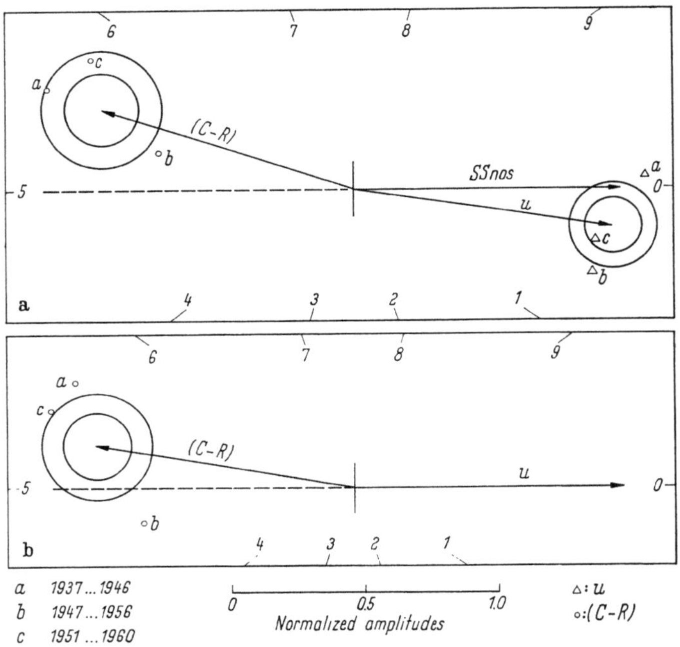

Fig. 66a and b. Harmonic dials 10-year waves. (a) u and C-R relative to sunspot numbers (SS). (b) C-R relative to u.

before the post perturbation recovery from preceding decreases) or whether the initial decrease in early February was the result of some large scale solar plasma emission which resulted in a depression of intensity, which even without the subsequent superimposed decreases having faster recoveries, would have continued for some months.

Fig. 65 shows the means of cosmic-ray intensity for the five quiet and for the five disturbed days in each month in 1943 and 1944. The steady rise of intensity from November 1943 to September 1944 appears definitely not be to the consequence of recovery from storm time decreases after August 1944 since the latter are quite small. The slow increase from November 1943 to September 1944, which continues in spite of whatever small stormtime decreases occurred during the interval, indicates a large scale mechanism the effect of which changes very slowly, with time constant of the order of a year.

40. Ten-year waves in cosmic-ray intensity, sunspot numbers, and magnetic activity. Fig. 66 shows harmonic dials for 10-year waves in cosmic-ray intensity (C-R), at Huancayo, SS numbers, and geomagnetic activity as measured by

Bartels [80] u-measure. In Fig. 66a times of maxima for the 10-year waves in (C-R) and u are relative to that for the weighted mean for SS numbers. In Fig. 66b the times of maxima for C-R vectors are relative to that for u. For the interval 1937—1960 the period of the solar cycle was very near 10 years. The vectors in Fig. 66 are weighted means. Points lettered a and b (see legend) were given weight one and those numbered c half weight. The weighted mean amplitude for each of the indicated vectors is normalized to unity on the scale shown. In addition, in Fig. 66, the amplitude for each of the three rotated vectors (end points lettered: a, b, and c) was corrected for amplitude deviations from the weighted mean amplitude for the vector rotated to zero time of maximum, using the approximately linear relation between amplitudes. The so-called probable error circles are indicated for the weighted averages and for single waves. The sizes of these error circles must be regarded as lower limits. For a "cloud" of only 2.5 points (1.5 degrees of freedom) the dispersion can not be reliably estimated and the statistically sound procedures developed by Bartels [81] can not be used, in the usual manner, with so few data. Thus, to obtain some estimate for the error circles for (C-R), for example in Fig. 66b, the regression of yearly means of C-R on yearly means of u was derived ($r = -0.86$). From this regression line yearly values (C-R) were computed from u and the standard deviation, σ, of yearly differences between the observed and computed values of (C-R) were derived.

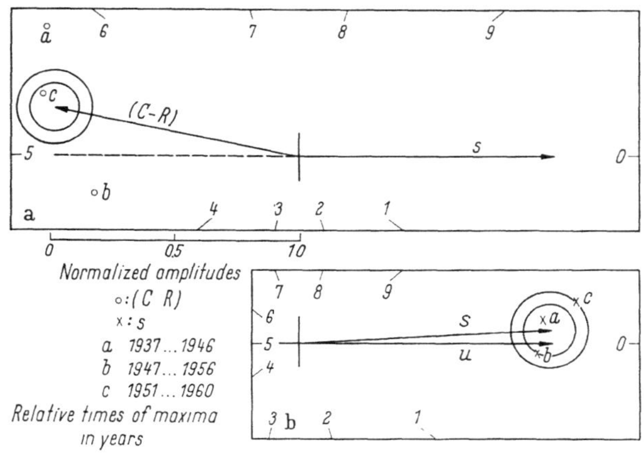

Fig. 67a and b. Harmonic dials 10-year waves. (a) C-R relative to s. (b) s relative to u.

Regarding these differences as random the estimated radius, ϱ_1, of the error circle [81] for single vectors was obtained from: $\varrho_1 = (0.833 \times 2\sigma/\sqrt{10})$ in which 10 is the number of ordinates used in the harmonic analysis. Similarly the radii of the error circles in Fig. 66a were obtained. If the above differences are not due mainly to random effects then the size of the circles is underestimated. Tests made indicated the differences were essentially random although these tests with such a limited number of differences are not very reliable. At any rate this procedure gives some indication of the uncertainty in the 10-year waves indicated in Fig. 66.

The results indicate that the average 10-year wave in u has its maximum about 3 months after the maximum in the 10-year wave for SS numbers. Similarly, the minimum in C-R occurs about 6 months after the maximum for SS numbers, or about 3 months after the maximum for u.

In a similar manner Fig. 67 shows the phase of the 10-year wave in cosmic-ray intensity at Huancayo, C-R, relative to s, and of s relative to u, for which s is for cosmic-ray intensity the yearly pooled standard deviation of daily means from monthly means [78]. Fig. 67a indicates that the 10-year wave in cosmic-ray intensity has its minimum about 4 months after the maximum for s. Fig. 67b indicates no significant difference in phase between the waves for *s* and *u*. Together these results also indicate that the 10-year wave in cosmic-ray intensity may be due in part at least to a mechanism with a much longer time constant than that for the Forbush-type decreases, the effects of which are principally measured by *s*, for which the 10-year wave is essentially in phase with *u*.

II. Results from high altitude observations.

41. Ionization chambers. Using carefully calibrated and standardized ionization chambers NEHER [*82*] has measured cosmic-ray ionization at high altitude near the geomagnetic pole over Thule, Greenland in each of most years in the period 1951 to 1961. In addition he made latitude surveys in four different epochs of the solar cycle. Fig. 68 shows the results obtained, under 15 g cm^{-2} of air, in four different years. Also shown in Fig. 68 are the latitudes at which protons coming in at 45° W and 45° E of the zenith, and having a range in air of 15 g cm^{-2}, are eliminated by the earth's magnetic field. Except for the earliest of these, flights were also made from a base station at Bismark, North Dakota, to monitor the radiation. Temporal changes observed from the base station flights were used to make appropriate corrections to the roving station.

The sharp "knee" which is a prominent feature of the curve for 1937, 1951, and 1958, does not appear in the curve for 1954. NEHER [*83*] concluded that the knee was also absent in 1955. He showed [*84*] that the latitude of the "knee"

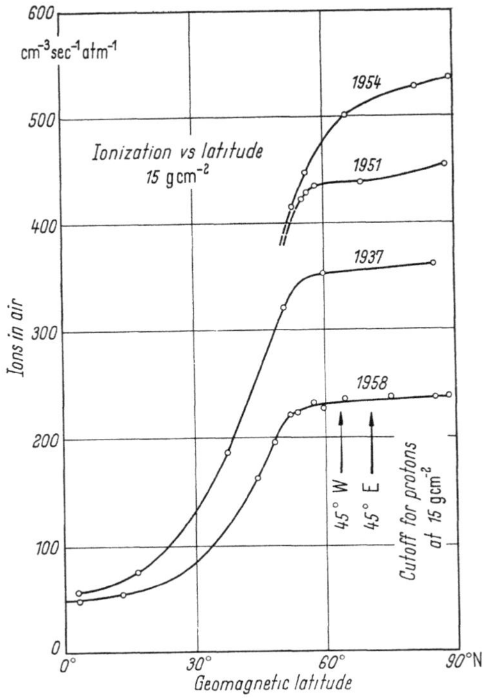

Fig. 68.

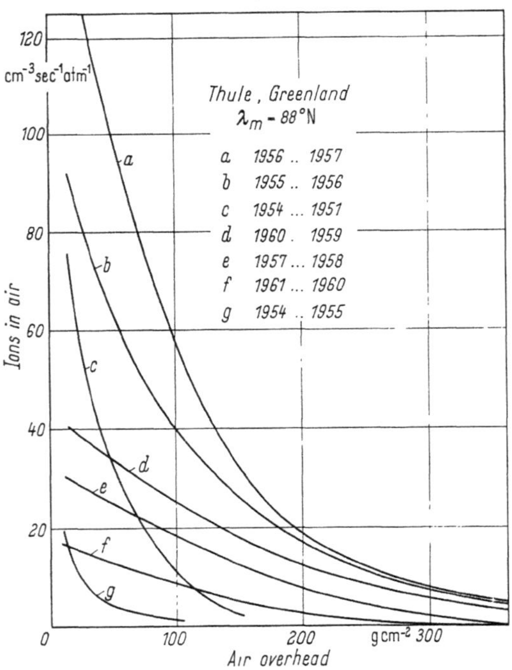

Fig. 69.

Fig. 68. Results of four latitude surveys. The knee moves only over about 2° of latitude, whereas the ionization changes by a factor of 2. In 1954 the knee apparently disappeared (after NEHER).

Fig. 69. These curves represent the way in which the radiation that changes from one year to another is absorbed in the atmosphere (after NEHER).

in 1951 and in 1958 differed by only 2° although in 1951 the ionization, at 15 g cm^{-2}, above the knee was about twice that in 1958. To study the absorption in the atmosphere of the particles responsible for the changes from one year to another NEHER plotted the differences for successive years as in Fig. 69. These data are taken from curves for the averages of the flights for any one year. From Fig. 69 it is evident that the decrease in the intensity is quite different from one curve to another. For example, the radiation responsible for the increase from 1951 to 1954, curve c, was quite absorbable compared with some other years. With the approach of sunspot maximum the radiation that was removed by the modulation mechanism became increasingly energetic.

Comparing the ratio of percentage changes in neutron intensity at Uppsala to those in the ionization chamber at Huancayo for several rapid decreases (Forbush effects) SANDSTRÖM [85] et al. concluded that this ratio probably decreased from sunspot minimum to sunspot maximum. This also would indicate that *the radiation removed by the solar cycle modulation mechanism becomes increasingly energetic near sunspot maximum.*

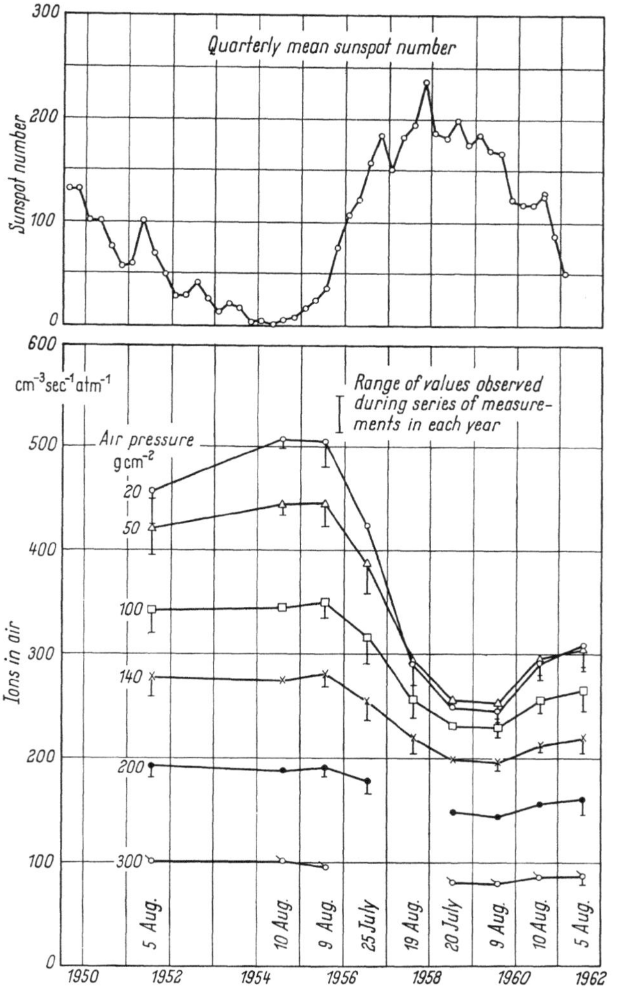

Fig. 70. Cosmic-ray ionization at Thule, Greenland at definite atmospheric depths for various years. Note the nearly anticorrelation with Zürich sunspot numbers (after NEHER).

Fig. 70 shows the ionization at different atmospheric depths obtained by NEHER [84] over Thule, Greenland. It is clear that the magnitude of the solarcycle variation (see sunspot numbers at the top of Fig. 70) increases very rapidly with diminishing air overhead.

During sunspot minimum in 1954 NEHER [86] concluded, from atmospheric absorption and rigidity requirements, that protons with energy $\leqq 150$ MeV were required to explain his results at high latitude and high altitude. The knee of the latitude variation, at least for protons down to 150 MeV, was completely absent in the summer of 1954.

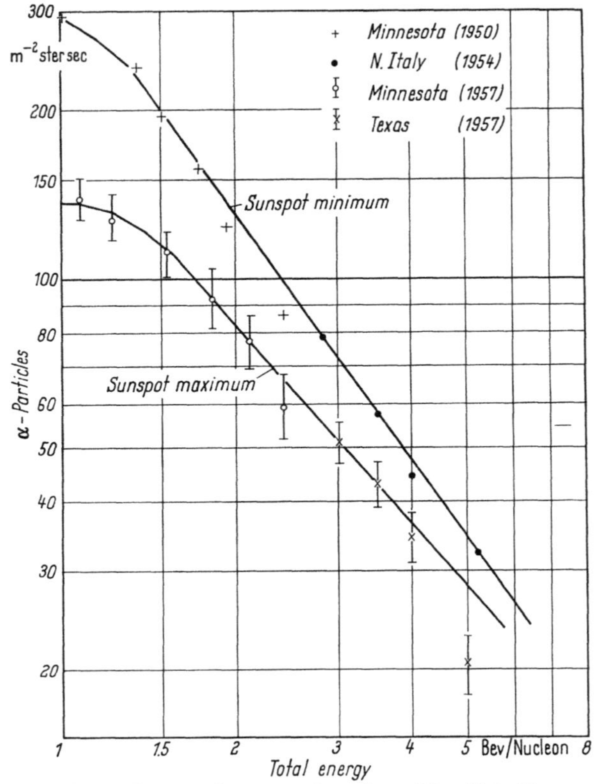

Fig. 71. The integral energy spectrum of α-particles obtained from scattering measurements (after NEY).

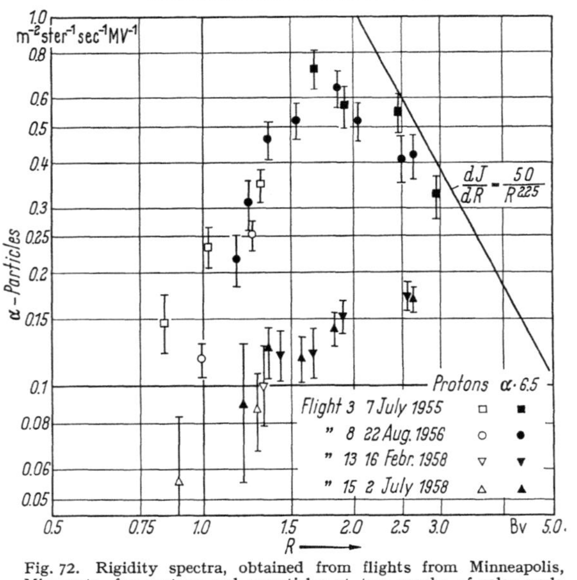

Fig. 72. Rigidity spectra, obtained from flights from Minneapolis, Minnesota, for protons and α-particles at two epochs of solar cycle after multiplying α-particle flux by 6.5 (after MCDONALD).

At 20 g cm^{-2} the ionization in 1958 and 1959 was only about 50% of that in 1954 and 1955. The corresponding change in the neutron intensity in Ottawa was about 20% and in the ionization chamber at Huancayo only about 4% (see Fig. 62). NEHER [84] et al. concluded that the largest negative correlation between sunspot numbers and ionization at 15 g cm^{-2} obtained when ionization was correlated with sunspot numbers observed from 9 to 12 months earlier.

The cosmic-ray group at the University of Minnesota [87], [88] has made numerous balloon flights with ion chambers since 1956. They found that the solar cycle variation of ionization at 10 g cm^{-2} was about three times the corresponding variation in neutron monitors near sea level.

42. Nuclear emulsions, other detectors and spectrum changes. NEHER used the results of his high altitude latitude surveys in 1954 to estimate the proton spectrum that would account for the increased ionization between latitudes 56° and 89°. MEYER and SIMPSON [89] used neutron monitors in high altitude latitude surveys with aircraft and from the results showed that the exponent in the power law spectrum, as well as the total intensity, were both greater near sunspot minimum in 1954 than in 1948 or 1956.

Changes in the low-energy part of the spectrum during the solar cycle are probably best derived from measurements on *primary α particles* in nuclear emulsion packages carried to high altitudes in balloons.

Although there are numerous experimental difficulties in separating the proton component from other singly charged components it is possible to discriminate between α particles and the background of secondaries. This procedure has been used by the group at the University of Minnesota [*90*] to determine the flux and energy distribution of primary α particles at different epochs of the solar cycle.

Fig. 71 shows their integral energy spectrum for α particles at two different epochs in the solar cycle in the energy range from 1—5 MeV/nucleon. The total flux of α particles near sunspot minimum is about twice that near sunspot maximum. The slope of the integral energy spectrum near sunspot maximum is definitely less than near sunspot minimum, which agrees with the results from neutron monitors [*89*]. Investigations of the differential energy spectrum of α particles near sunspot minimum and near sunspot maximum have shown that the solar cycle modulation of cosmic-ray intensity results in a decrease in flux of particles in the energy spectrum at least up to 30 BeV/nucleon.

The spectrum of both protons and α particles has been measured by McDonald [*91*], [*92*] using a combination of Čerenkov and scintillation counter detectors. Fig. 72 shows his results for the differential rigidity spectra for protons and for α particles at two epochs of the solar cycle. After multiplying the α particle flux by 6.5, which is the ratio of protons to α particles in the primary cosmic rays, the spectra for protons and α particles are alike near sunspot minimum and near sunspot maximum, although the flux of α particles and of protons, for rigidity $R<3.0$ Bv, is only about one quarter as great in 1958 as in 1955, and 1956. Since particles of the same rigidity are modulated in the same way a magnetic modulation mechanism is indicated to account for the changes observed. The *solar cycle variations* in cosmic-ray intensity are most likely *produced by the magnetic fields in plasma clouds ejected from the sun*. When the earth is inside such a cloud the cosmic-ray intensity is reduced due to deflection of primary trajectories in these magnetic fields which may extend throughout a large part of the planetary system.

H. Forbush decreases.

I. Results from continuous monitors.

43. Ionization chambers. Sect. 22 described the variations of daily means of cosmic-ray ionization from several stations during the period 1937—1953. Fig. 28a—f (p. 188—9) showed daily means of cosmic-ray intensity for Huancayo for the period 1937—1953 together with daily means of horizontal magnetic field H at Huancayo. Daily means of ionization from Godhavn and Christchurch were plotted for selected years, mainly to show that at these two stations there are large fluctuations of ionization which do not appear at Huancayo. These undoubtedly are due principally to variations in the height of the 100 mb pressure level where μ-mesons are predominantly produced. In several months near sunspot minima in 1944 and 1954, the standard deviation of daily means from monthly means was as low as 0.16%.

During the period January to September 1954, 53 intervals of three consecutive days were selected such that within each interval the daily mean barometric pressure was essentially the same. The variance of D was computed with: $D=[(M_1+M_3)/2-M_2]$ in which, $M_{1,2,3}$ are daily means of ionization, corrected for bursts (not for barometric pressure), for the three consecutive days of each interval. This value of D eliminates any linear change over the three day interval;

thus the values of D arise principally, if not entirely near sunspot minimum, from statistical fluctuations. Thus if s_1 designates the standard deviation of statistical fluctuations for means for single *days* then the standard deviation

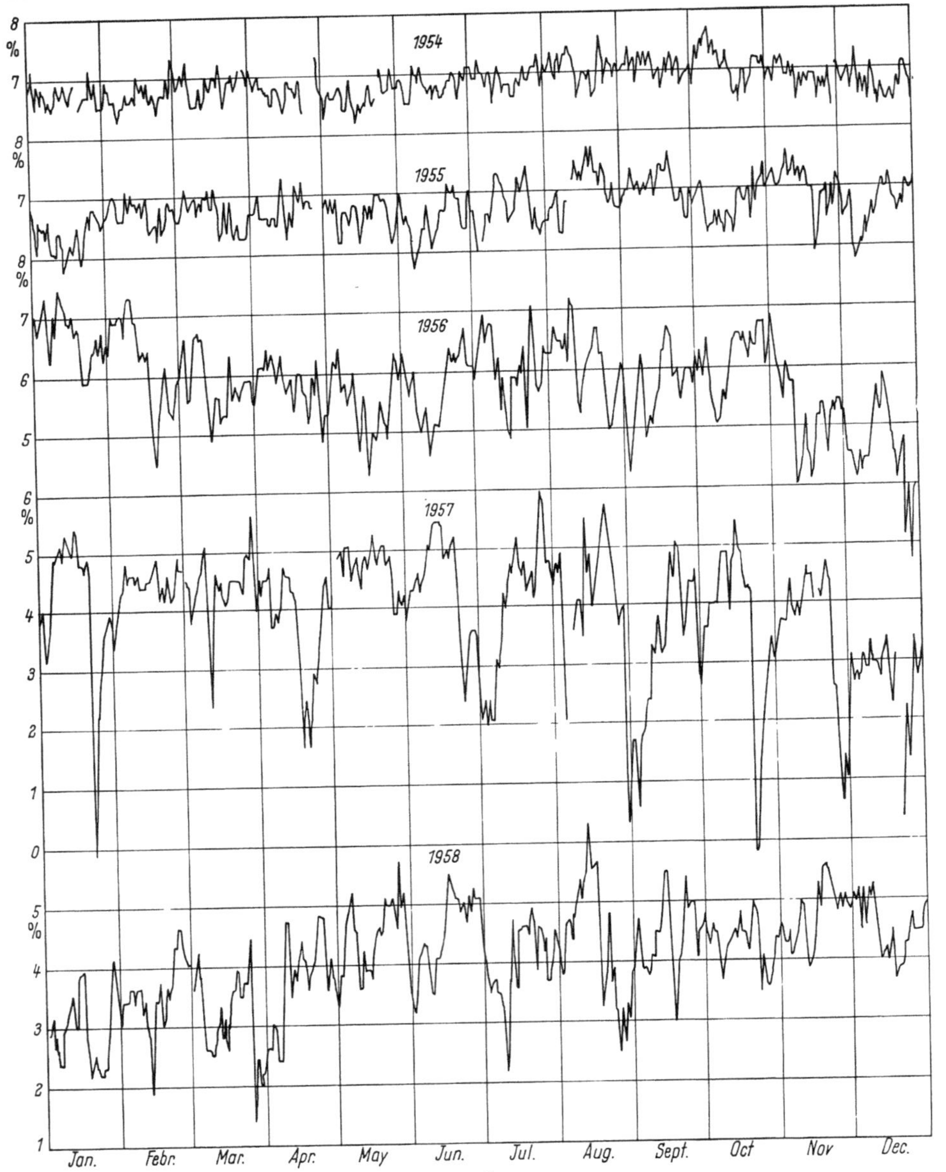

a

Fig. 73a and b. Daily means cosmic-ray intensity Huancayo (ordinates in percent from fiducial value). The daily means for 1954 refer to days between midnights of 75° West Meridian Time, those for 1955–1961 are for Greenwich days.

of D is $s_1\sqrt{3/2}$, from which s_1 is obtained. From the above 53 intervals the value 0.14% was obtained for s_1. In a similar manner the standard deviation for *hourly* means (uncorrected for pressure) was found to be 0.59%.

At Cheltenham the standard deviation for hourly means was also obtained experimentally from differences between simultaneous hourly values in two identical ionization chambers. The value obtained was 0.70%. The total ionization at Cheltenham is about 0.63 times that at Hunacayo so that the standard

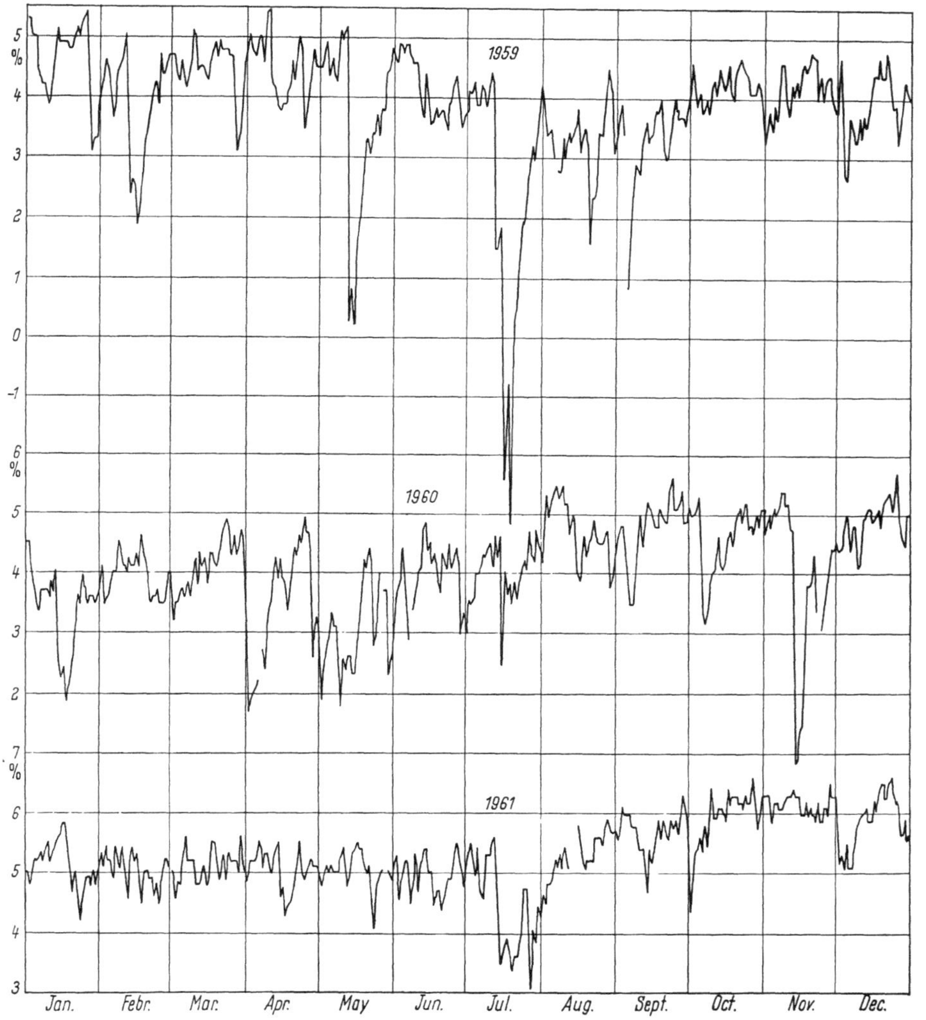

Fig. 73b.

deviation of statistical fluctuations at Huancayo would be expected to be smaller by the factor, $0.63^{\frac{1}{2}} = 0.80$, or about 0.56% which is essentially the same as the value, 0.59% obtained at Huancayo for hourly values, or about 0.12% for daily means. For 1954 the yearly pooled standard deviation of pressure corrected daily means from monthly means is 0.21%. Since any real changes of cosmic-ray intensity within months increase this standard deviation it is evident that, within months, real variations must be quite small in 1954. Moreover, changes due to meteorological effects, other than the small changes in barometric pressure, at Huancayo must be quite small also.

The above procedure is described in detail since it may be used as well for data from neutron counters to compare with the standard deviation, $N^{-\frac{1}{2}}$ in per cent, derived from the counting rate. Appreciable discrepancies between these two values have been used to reveal instrumental malfunction.

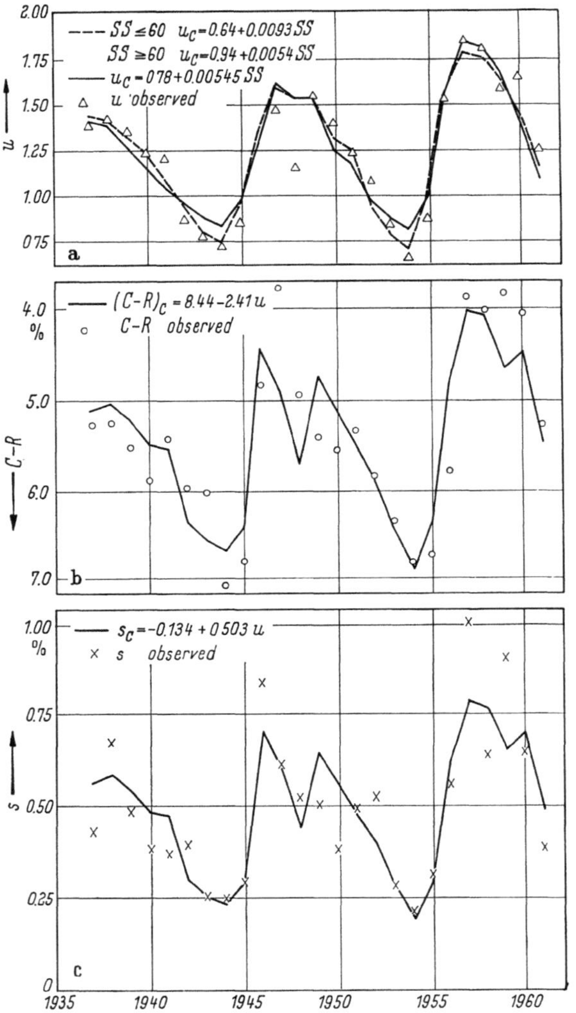

Fig. 74a—c. 1937—1961 comparison of yearly means, observed and computed. (a) u with u_c computed from SS Nos. (b) C-R with C-R computed from u. (c) s with s_c computed from u.

The corrections for instrumental drift at Huancayo and the possibility of a seasonal wave with amplitude about 0.27% have been discussed by FORBUSH [78]. Figs. 73a and b show the daily means of cosmic-ray intensity from the ionization chamber at Huancayo corrected for barometric pressure and instrumental drift for each day in the interval 1954—1961. In these figures the ordinate scale distance for 1% is twice that used in the First Part for Figs. 28a—f. Fluctuations of daily means are obviously least in 1954 and are much greater (about 5 times) in 1957. In 1961 fluctuations are considerably less than in 1960.

Fig. 74c compares the observed yearly pooled standard deviation, s (crosses), with that computed from the linear regression of s on yearly means of u, for which $r = +0.84$. The geomagnetic activity index u was computed from the interdiurnal variability of daily means of horizontal magnetic intensity as described by BARTELS [80]. Similarly, Fig. 74b compares observed annual means of cosmic-ray ionization, C-R, at Huancayo with values computed from the linear regression of annual means of $(C$-$R)$ on u for which $r = -0.86$. Fig. 74a compares observed annual means of u with those computed from the regression of u on sunspot numbers, SS. The solid line indicates values of u computed from a single regression line ($r = +0.90$) for annual means. The dashed curve shows u computed from two regression lines as indicated.

For the intense magnetic storm of July 15, 1959, Fig. 75 shows the bihourly means of cosmic-ray ionization, C-R, at Huancayo, and three hour means for the southward (positive downward) geomagnetic field, ERC, of the equatorial ring current computed by the method of KERTZ [79]. For many of the magnetic storms in the interval 1955—1961 the correspondence between cosmic-ray intensity and ERC may likely be less marked than for the July 1959 storms. In fact, the intense storm of August 1937 (see Fig. 22, p. 183) was not accompanied by any detectable decrease in cosmic-ray ionization. The July 1959 storms [107], however, suggest a large scale solar emission of plasma clouds.

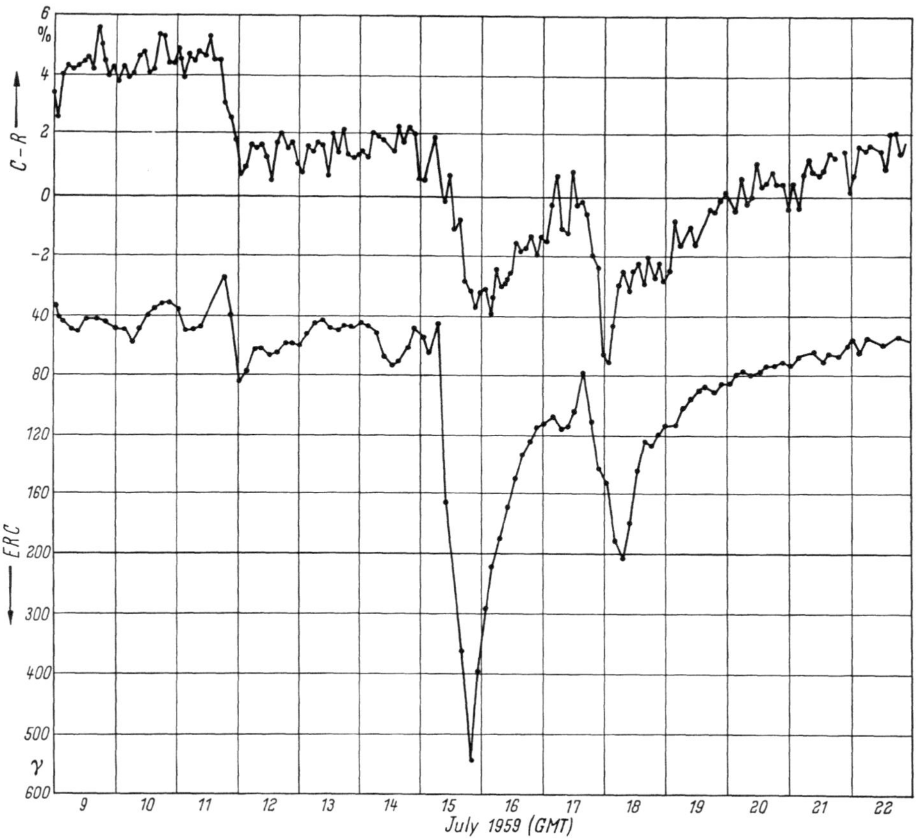

Fig. 75. Bihourly means cosmic-ray ionization, *C-R*, at Huancayo and three hour means of southward geomagnetic equatorial ring current field, ERC, July 9—22, 1959.

44. Neutron monitors. The numerous Forbush decreases during the period 1955—1961 derived from neutron monitors in general show changes somewhat similar, except for magnitude, to those shown in Fig. 73. For the IGY period July 1, 1957 to December 31, 1958, monthly means, daily means, and bihourly means, of pressure corrected neutron intensity at some 72 stations are shown separately graphed in publications of the IGY National Committee of Japan [93].

LOCKWOOD [94] has made an analysis of the large and rapid cosmic-ray Forbush decreases in cosmic radiation which occurred from 1954—1959. Data from the IGY network of neutron monitor stations were used together with other observations, to obtain the changes in the primary rigidity spectrum, onset times and the existence of anisotropies during the decreases. The decreases analyzed were preceded within 3 hours by a sudden commencement magnetic storm and most occurred from 6—36 hours after a polar cap absorption event [95] preceded by solar flares. Since the primary rigidity spectrum changes during the solar cycle, LOCKWOOD determined the rigidity spectrum *before* each decrease in order to obtain the changes in rigidity spectrum produced *during* the decreases.

In general, the observed decreases were found to diminish with increasing values of the primary vertical cutoff rigidity in the range 1—15 BV, although for the event of December 19, 1957, the decrease was found to be independent of rigidity in the energy region to which neutron monitors respond. In addition, among the several events analyzed, there were other significant differences in the slope of the curves for percentage decreases as a function of vertical cutoff rigidity. The decrease in intensity during July 1959 as shown in Fig. 75 represents an extreme case of modulation near the recent solar activity maximum. According to WEBBER [96] the integral intensity of all particles at the top of the atmosphere during the period of minimum cosmic-ray intensity in July 1959 was probably only about one-fifth that at sunspot minimum. He estimates that at 15 BV the intensity was at least 30% less than at sunspot minimum. WEBBER [96] also discusses other Forbush decreases in detail and gives curves showing the rigidity dependence for these events. He gives a critical review of each of the several models to account for the solar cycle variation and the Forbush decreases.

II. Results from satellites and space probes.

45. Satellites. That the mechanism to account for Forbush decreases does *not* depend on the geomagnetic storm time field (i.e., the equatorial ring current field) was demonstrated by the results obtained by SIMPSON [97] and colleagues at the University of Chicago. In the Explorer VI satellite they used a triple coincidence, proportional counter system which measured protons with energy greater than 75 MeV or electrons with energy greater than 13 MeV. The triple coincidence detector is not affected by bremsstrahlung from electrons trapped in the geomagnetic field. Fig. 76 compares the triple coincidence rate in the satellite, at distances $\geqq 35 \times 10^3$ km from the earth, with the counting rate of the neutron monitor at Climax, Colorado, during the period August 14 to August 22, 1959. For the period August 19 to 21 (excluding passages through the trapped radiation zone) they find a correlation coefficient of $+0.96$ between triple coincidence rates in the satellite and neutron intensity at Climax. The relative change in the satellite detector was about twice that in the neutron monitor. The relative intensity changes in the primary intensity outside the atmosphere

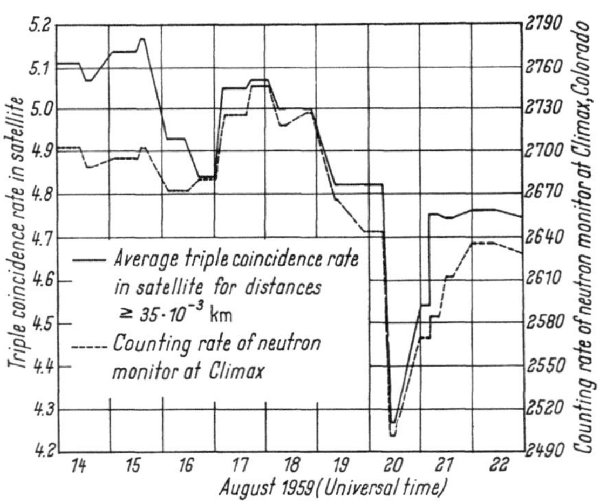

Fig. 76. Cosmic-radiation intensity as a function of time showing on August 20, 1959 a Forbush-type intensity decrease of $\sim 15\%$ within five hours. The nucleonic component monitor provides a measure of primary intensity changes at the earth. The light-dashed and solid lines are used to connect the periods for which satellite data are available (after FAN et al.).

were estimated to be about twice those in the neutron monitor. This estimate was based on a comparison of Forbush decreases in neutron monitors in high altitude balloons with those observed in the Climax neutron monitor for that period of the solar cycle. The authors thus conclude that the magnitude of the decrease observed out to distances of 7.5 earth radii were essentially the same as those estimated for the changes in primary radiation near the earth.

46. Space probe results. On the space probe Pioneer V SIMPSON [98] and colleagues at the University of Chicago had equipment identical with that used on Explorer VI which was described in the preceding paragraph. Fig. 77 shows their results obtained between March 27 and April 3, 1960, together with magnetometer results obtained by COLEMAN [99] et al. A decrease of 28% in the triple coincidence counter rate occurred on March 31, and April 1, 1960, during an intense magnetic storm. The authors estimate that the relative decrease of intensity in the Pioneer V detector, on April 1, at about 5×10^6 km from the earth was about 30% greater than the relative changes observed in the neutron monitor at Climax and extrapolated to the top of the atmosphere. They conclude that the larger decrease measured in Pioneer V was due to the removal of cosmic-ray primaries with magnetic rigidities below the cutoff (2.4 BV) for the Climax monitor but not below the threshold of the detector in Pioneer V. They found that this diminution of low energy primaries persisted for more than 30 days after April 1, although the higher energy particle flux had returned much earlier to the level observed before the March 31 storm (see also Fig. 73f). From analysis of neutron monitor data extending to the geomagnetic equator where the cutoff reaches 15 BV they found independent evidence that the intensity decrease in this case was more strongly dependent upon particle magnetic rigidity than for the event of August 19—20, 1959 observed in Explorer VI.

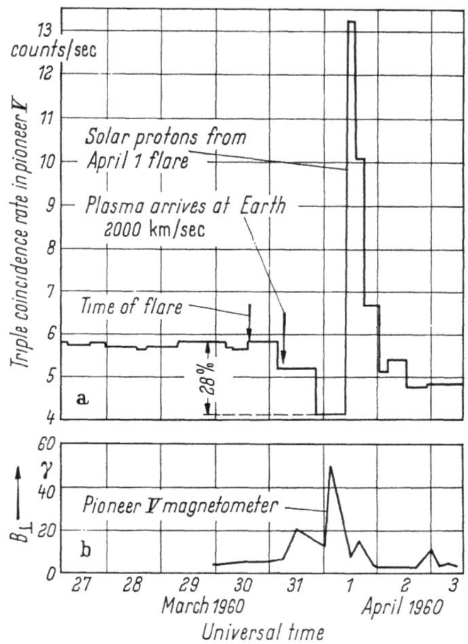

Fig. 77. Telemetered data from the space probe Pioneer V at distances $4-5.5\times10^6$ km from the earth. The time for the first arrival of solar protons April 1 was determined by the onset of enhanced ionization from protons at the polar cap. The magnetometer measurements in Pioneer V are published by COLEMAN et al. (after FAN et al.).

From the interval between the solar flare on March 30, 1960 and the sudden commencement of the magnetic storm the velocity of the solar plasma was estimated as 2×10^3 km · sec^{-1}. FAN [98] et al. estimate that about 28% of the particle flux detected by Pioneer V disappeared in less than 20 hours, with half of this decrease occurring in less than six hours. From these data and the computed radial velocity of 2000 km/sec for the advancing "front" of solar plasma, a penetration of about 0.3 astronomical units behind the front was required to reduce the intensity by 14% — half the full intensity decrease. For other events of this kind where the rate of decrease may be as high as 5—6% per hour the corresponding depth of penetration may be smaller by a factor of 3. This requires the rapid appearance of enhanced interplanetary magnetic field intensities.

The magnetometer results in Fig. 77 show magnetic field intensities 10 to 20 times those for the quiescent field. These results directly indicate the existence of magnetic fields frozen in conducting plasma ejected from solar flares. The authors conclude that *the cosmic-ray decrease is caused by the convective removal of galactic primaries by particle collisions with advancing large scale magnetic field irregularities such as a shock front.* From Fig. 78 it may be seen that by 0600 April 1 the advancing region which produced the full decrease of intensity had passed outward beyond Pioneer V and the earth.

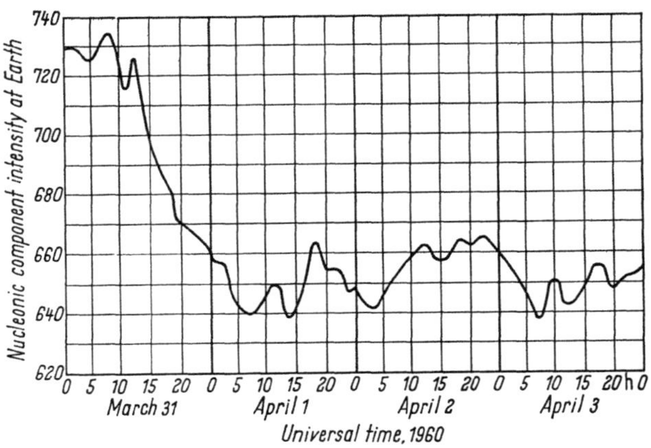

Fig. 78. The changes in galactic cosmic-ray intensity for particles above 2.4 BV magnetic rigidity. The neutron monitor is located at Climax, Colorado.

Fig. 77 shows that solar protons arrived at the Pioneer V detector about one hour after the April 1 flare. The solar particle increase is shown in Fig. 77 for protons >75 MeV (but less than 1 BeV since no increases were observed by neutron monitors at the earth). Maximum intensity was reached within 50 minutes. From this the authors conclude that the interplanetary magnetic field conditions behind the advancing front are either smooth and radial or weak, irregular fields ($B_{\mathrm{rms}} < 5 \times 10^{-6}$ Gauss), otherwise the low energy solar protons would not penetrate to the orbit of the earth in less than one hour.

I. Solar flare effects.

I. Results from ground level monitors.

47. Ionization chambers. In Sects. 30 and 31, the four increases in cosmic-ray intensity which had been observed in the period 1937—1955 were discussed and results shown in Figs. 49—54. The fifth and *largest increase* recorded in ionization chambers to date (December 1962) occurred on *February 23, 1956.* In the Cheltenham ionization chamber the maximum intensity was about 85% above the pre-flare value [*100*]. The increase began about 18 min after the solar flare was first observed at 0330 GMT February 23, 1956. From a large shielded ionization chamber at Derwood, Maryland, the increase was determined for one minute intervals during the first hour of the increase and for six minute intervals thereafter. Fig. 79 shows the curve from the Derwood meter and points from the ionization chambers at Godhavn, Cheltenham, Mexico, Huancayo, and Christchurch.

This is the only occasion, from 1937—1962, that any particles accelerated in a solar flare acquired sufficient energy, greater than 15 BeV for protons, to be *detected at the equator*: The neutron monitor intensity at Huancayo, as reported by SIMPSON and his colleagues [*101*] registered a maximum of 20% above the pre-flare value.

48. Neutron monitors: the event of February 23, 1956. The first observational evidence that the flux of particles responsible for solar flare effects is much greater for particles in the lower energy band to which neutron detectors are sensitive than in the band to which ionization chambers respond, was obtained during the solar flare of November 19, 1949 by FORBUSH et al. [61]. The solar flare increase of November 19, 1949 observed in an ionization chamber at Climax,

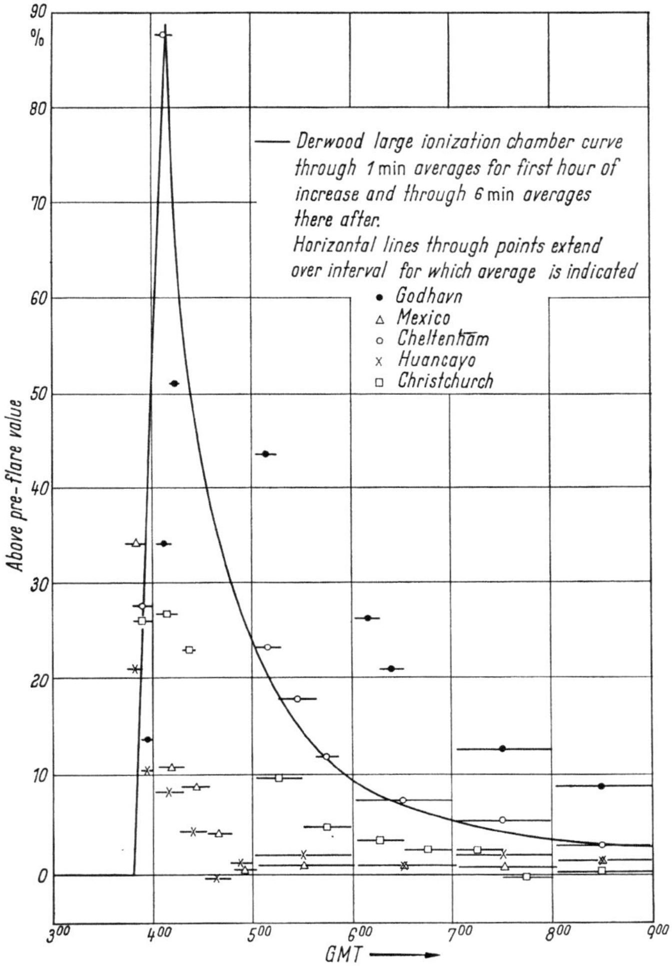

Fig. 79. Cosmic-ray increase during solar flare of February 23, 1956.

Colorado (altitude 3500 m) was nearly five times that observed in an ionization chamber at Cheltenham near sea level.

From this result, discussed in Sect. 31, it was predicted that for the flare of November 19, 1949, a neutron monitor should have shown an increase of the order of twenty times that in the ionization chamber at Cheltenham. This prediction was confirmed by the results obtained in a neutron pile by ADAMS at Manchester [62], see Sect. 31, Fig. 54, who observed an increase of 550% or 12 times that observed in the ionization chamber at Cheltenham.

The first solar flare increase to be observed in several neutron monitors was that on February 23, 1956. This increase was recorded by 72 instruments which included 17 ionization chambers (with 10—12 cm Pb shielding), 14 neutron monitors and 41 counter telescopes.

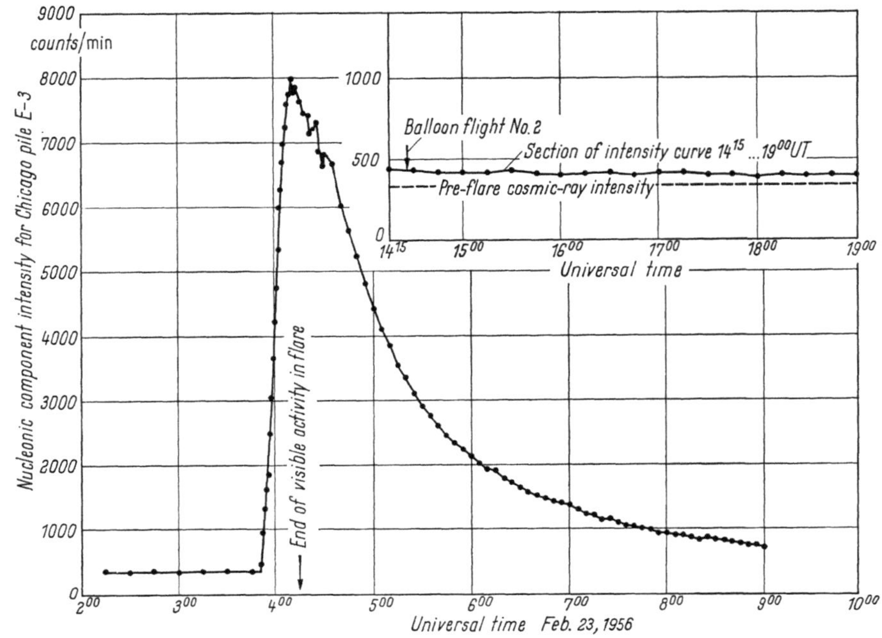

Fig. 80. The intensity increase of secondary neutrons generated in the atmosphere from primary solar-flare protons. These observations were obtained with a neutron monitor pile at Chicago, Ill. (after SIMPSON et al.).

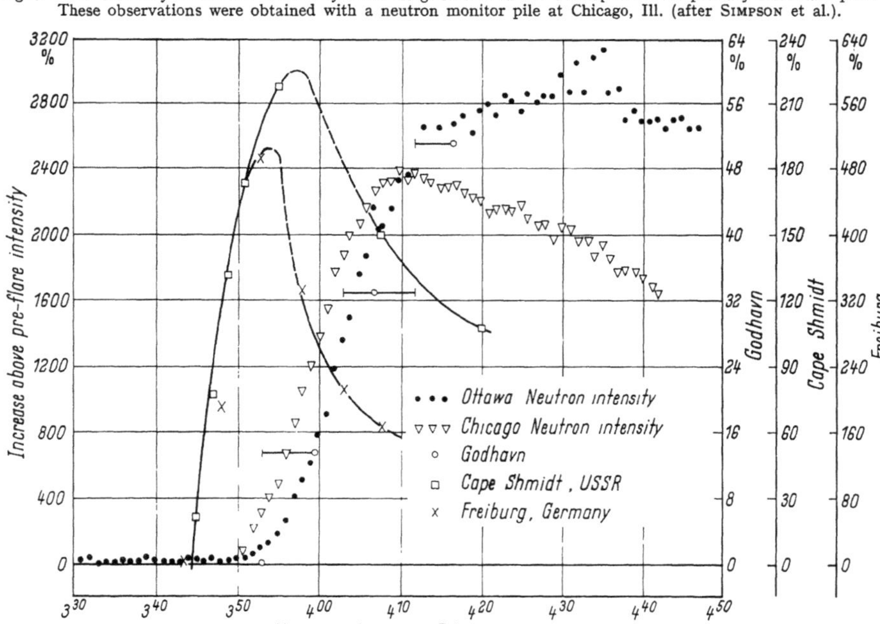

Fig. 81. Detectors located in different parts of the geomagnetic field respond to different energies of solar protons provided they come from a point source—the impact zone effect (after SIMPSON et al.).

Fig. 80 shows the increase observed in the neutron monitor at Chicago. From an analysis of data from this and several other stations the group at the University of Chicago discovered that one of the characteristics regarding the arrival of the first particles from this flare is that high-energy particles appear to arrive ahead of low-energy particles with a spread in arrival time of the order of 10—15 min for an energy-range of 10 BeV for protons. This dispersion effect is shown in Fig. 81 which shows examples of prompt and delayed onset times, taken from a world-wide distribution of cosmic-radiation intensity recorders. The primary flare particle spectrum derived by SIMPSON et al. [101] by taking account of the different cut off energies imposed by the earth's magnetic field at different locations is shown in Fig. 82. The spectra are rather similar at the three different times indicated although the intensities differed by an order of magnitude. The differential rigidity spectra of Fig. 82 follow approximately the power law N^{-7}. These investigators obtained independent evidence for the character of the spectrum of flare particles at the lower particle rigidities from neutron detectors flown in balloons from 9—12 hours after the start of the increase.

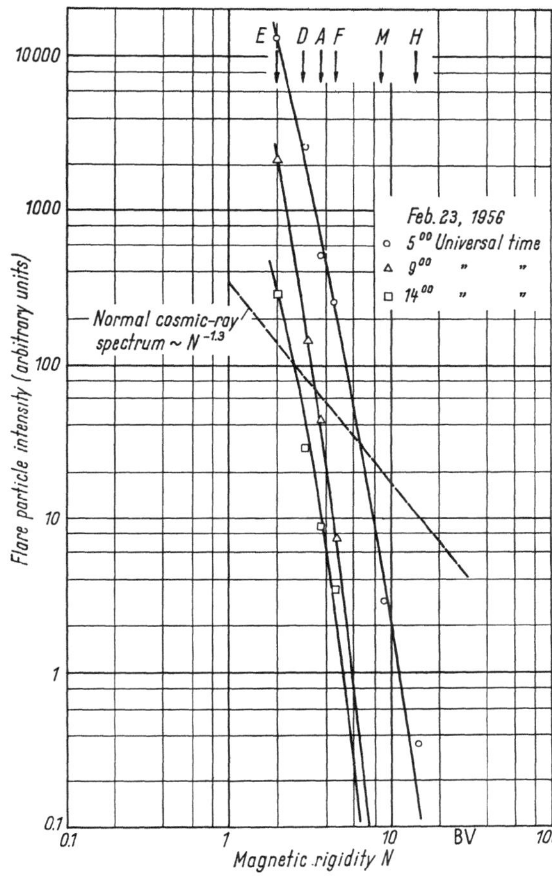

Fig. 82. The differential magnetic rigidity spectrum for the solar protons after particle storage or trapping in the solar system had taken over. Here the particle magnetic rigidity $N = p\,c/Z\,e$, where p is momentum, c is velocity of light, $Z\,e$ is the particle charge, and N is measured in volts. The scale is given in BV = billion volts = 10^9 V (after SIMPSON et al.).

The development of the February 23, 1956, event is described by SIMPSON [102] as follows. The apparent source is a relatively large area which includes the sun. Initially, the particles arriving are of the highest energies in the flare spectrum. Subsequently, the whole sky becomes "illuminated" with arriving particles extending to lower energies, and for these low-energy particles and late-arriving high-energy particles, there develops a remarkable isotropy in space near the earth. Following the onset of isotropy, the particle intensity gradually diminishes with the flare particle spectrum essentially unchanged over many hours, as shown in Fig. 82, while the particles escape from the magnetic fields which store them in the solar system. The intensity at the earth is observed to diminish to its pre-flare level within a period of 20 hours. These phenomena suggest three intervals of time in the development of the cosmic-ray flare:

1. Beginning with the initial release of high-energy particles and ending at the time when the solar particles reach a maximum intensity at the earth, the particles come from a limited source direction. This interval of time is of the order 10—30 minutes, depending upon the solar event.

2. A brief period of transition sets in when particles begin to arrive from directions other than the source—suggesting that particles arrive later as a result of scattering or passage through indirect magnetic channels in the interplanetary magnetic fields connecting the earth to the solar region.

3. At late times when isotropy has been established, all evidence for the release of energy in the solar flare region has vanished. But the influx of cosmic-ray particles at the earth continues for many hours. This fact and the fact that no source direction persists for even the highest energy particles strongly support the view that the sun accelerates particles only during a short interval of time and that these particles are trapped and stored in the interplanetary magnetic fields only to be lost subsequently from the solar system or arrive at the earth. The decay mode of the particles in the vicinity of the earth opens the possibility for determining the characteristics of the storage or trapping in magnetic fields. This kind of evidence for the storage of charged particles is at present the strongest evidence for the existence of interplanetary magnetic fields. From the arrival of solar protons at the equator over Huancayo, Peru, and their detection in balloons where the cutoff for protons is as low as 1—2 BV, it is clear that the energy range certainly exceeds 1—24 BV.

DORMAN [*103*] using all the available data for the event of February 23, 1956, has contributed an extensive analysis and discussion of the implications of the results, including possible acceleration mechanisms for the solar flare particles. Besides the diffusion model of SIMPSON et al. [*101*] to explain the persistence of the cosmic-ray increase after the disappearance of the flare, other models have been proposed. These are critically discussed by WEBBER [*96*]. A model involving the "magnetic tongue" from the sun is discussed by COCCONI et al. [*104*].

49. Other events. CARMICHAEL [*105*] and STELJES showed that the small solar flare increase on July 17, 1959, was observed only in neutron monitors at latitudes where the cutoff rigidity was less than about 1.1 BV. Fig. 83 shows the increase in neutron intensity at several stations plotted against QUENBY [*106*] and WEBBER cutoff rigidities. Only the point for Sulphur Mountain in Fig. 83 is quite off the curve and it is doubtless due to the high altitude of the Sulphur Mountain Station. This increase was only about 5% at Churchill and since it occurred during a large Forbush decrease (see Fig. 75), it was only detected from differences of intensities between neutron monitors at quite high latitudes and the neutron monitor at Uppsala.

The same data as in Fig. 83 when plotted against cutoff rigidities determined for the earth's *eccentric* dipole resulted [*105*] in points with very large deviations from any smooth curve that could be drawn through them. It is probable that so small an increase could not have been reliably ascribed to the solar flare without the availability of cutoff rigidities as determined by QUENBY and WEBBER's procedure [*106*]. This method takes account of the actual field of the earth in deriving cutoff rigidities for any points on the earth. Cutoff rigidities similarly determined accounted very well [*96*] for the cosmic-ray equator determined by SIMPSON in high altitude air-craft using neutron monitors.

The cosmic-ray events of July 1959 have been discussed by several investigators. These and other geophysical events of closely related interest were the subject of a symposium at Helsinki in 1960. The collected papers presented at this symposium have been published [*107*].

Three other large increases of cosmic-ray intensity have been recorded in neutron monitors. The first of these occurred on May 4, 1960, and the other two on November 12 and November 15, 1960 [*108a*]. The increase in neutron intensity is shown at Ottawa in Fig. 84, together with the cosmic-ray ionization at Huancayo which shows no increase but only the large decrease associated with the magnetic storm for which the geomagnetic equatorial ring current field, ERC, is shown. By an ingenious procedure, McCracken [*108*] has analyzed the neutron

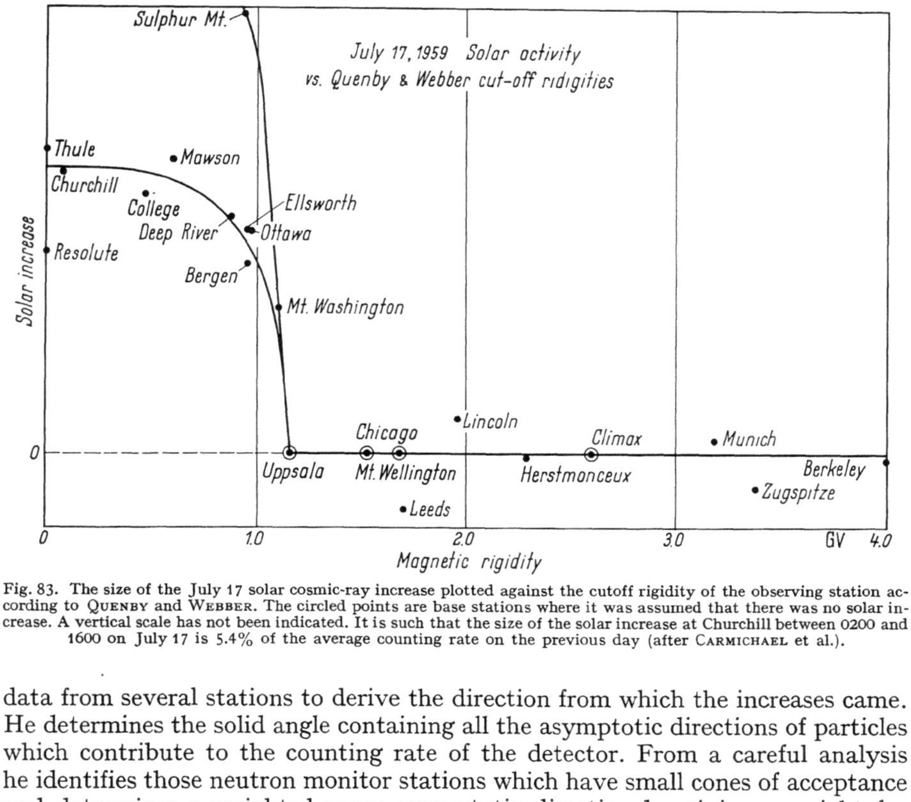

Fig. 83. The size of the July 17 solar cosmic-ray increase plotted against the cutoff rigidity of the observing station according to Quenby and Webber. The circled points are base stations where it was assumed that there was no solar increase. A vertical scale has not been indicated. It is such that the size of the solar increase at Churchill between 0200 and 1600 on July 17 is 5.4% of the average counting rate on the previous day (after Carmichael et al.).

data from several stations to derive the direction from which the increases came. He determines the solid angle containing all the asymptotic directions of particles which contribute to the counting rate of the detector. From a careful analysis he identifies those neutron monitor stations which have small cones of acceptance and determines a weighted mean asymptotic direction by giving a weight for each rigidity which is proportional to the fraction of the total counting rate which it contributes. He has also shown for certain stations in high latitude that standard neutron monitors at these stations will record identical percentage enhancements when an isotropic flux of solar particles is incident upon the earth.

Differences in the observed magnitude of events at these different stations then indicate anisotropy in the radiation. Moreover, when the radiation is anisotropic he determines whether there is some asymptotic direction about which the radiation fluxes are symmetrical. For the solar flare increase of May 4, 1960, he finds that the direction of maximum intensity was persistently inclined about 50° to the west of the sun. For the solar flare event of November 12, 1960, the direction of maximum intensity was also about 50° to the west of the sun during the first 5 hours of the increase, from 1400—1900 GMT. Near 1900 GMT an abrupt increase of intensity occurred soon after which the radiation was found to be isotropic. For the increase of November 15, 1960, the radiation became isotropic very rapidly after an initial phase of very marked anisotropy.

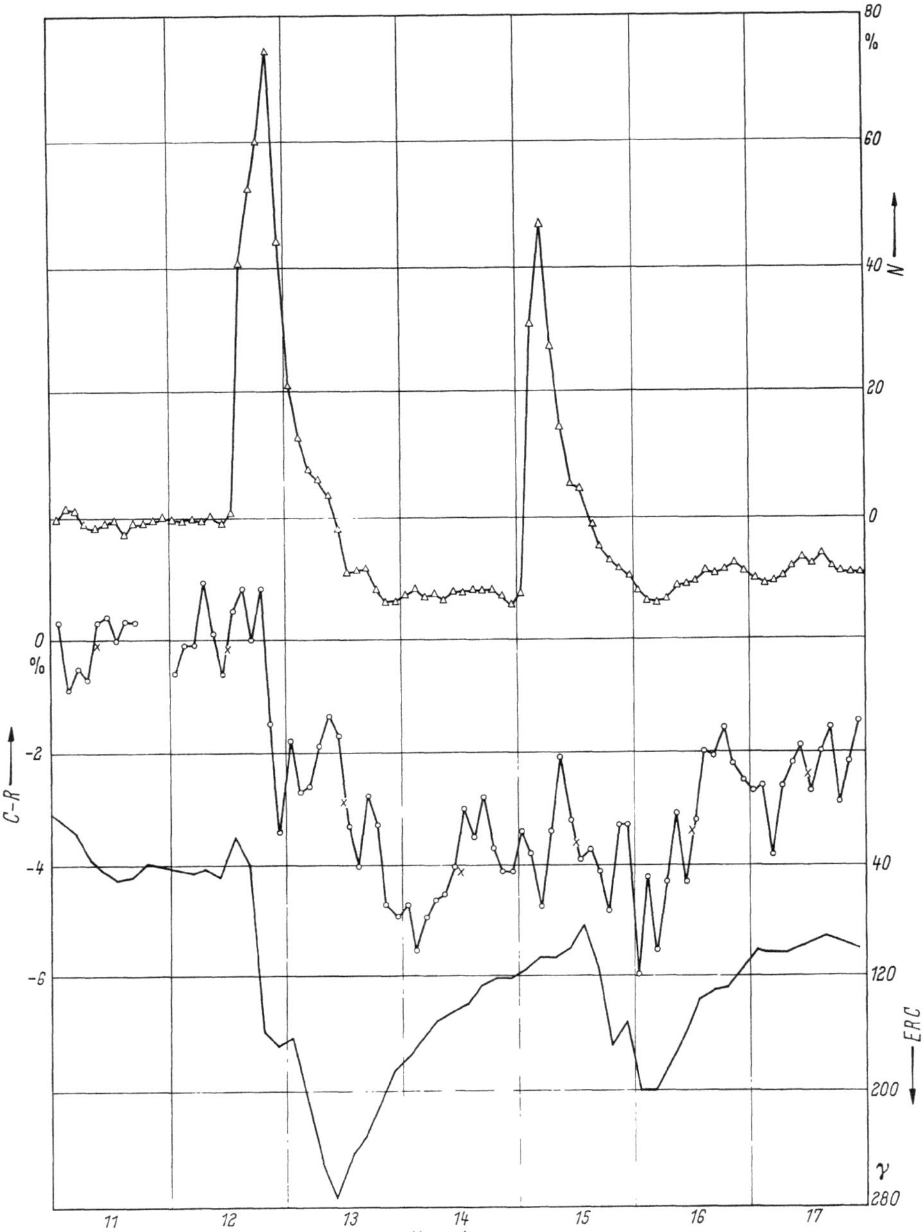

Fig. 84. Bihourly means (GMT) in per cent from prestorm values for neutron intensity, N, at Ottawa (after ROSE), cosmic-ray ionization, C-R, at Huancayo, and three hour means of southward geomagnetic equatorial ring current field, ERC, November 11—17, 1960. Note different ordinate scales for N and C-R.

These time variations of the anisotropy and their important significance in determining the magnetic regime, and changes in it, through which the solar cosmic rays must pass to reach the earth are discussed by McCracken [*108*].

II. Results from other methods of detection.

50. Riometers. The magnitudes of solar flare events observed in neutron monitors have been enormously greater than in ionization chambers, due to their much greater sensitivity to low energy primaries, down to rigidities of about 1 BV (for ground level instruments). Nevertheless, solar flare events are not often detected by ground level neutron monitors. The discovery in 1957 of solar proton beams of low energy, however, has shown that the neutron monitor usually does not indicate the arrival at the top of the atmosphere of solar cosmic rays. That incoming solar streams of heavy ions would produce, and did produce, in the February 1956 case, great ionization of the ionospheric D-layer and consequent absorption of radiowaves was first realized by Bailey [*109*].

For the period of the solar flare of February 23, 1956, Bailey [*109*] investigated oblique-incidence signal intensities and simultaneous observations of the background cosmic noise at very high frequency for a number of high latitude communication links employing the ionosphere scatter mode of propagation. During the flare and for some time afterward, all paths were in the dark hemisphere. Simultaneously with the arrival of solar cosmic rays he observed a sharp enhancement of signal intensity which he attributed to solar protons. In summarizing his conclusions Bailey [*109*] made the following statement:

"... it seems necessary to recognize a new class of signal-intensity enhancement for waves propagated by ionosphere scattering and a new kind of high-altitude absorption phenomenon. It would seem inappropriate to identify the observed absorption effects as merely a special case of the well-known polar blackout absorption. It is clear that an event such as that reported is rare. To the extent that such events are associated generally with outbursts of solar cosmic rays, they may on the basis of very meager statistics be expected to occur about once in four years. Actually smaller events may occur more frequently, but are likely to be associated with important flares. The particle velocities thought necessary to account for the absorption effects are of the order of a tenth of cosmic-ray particle velocities, and the associated particle energies are correspondingly lower. The sun is, therefore, more likely to eject absorption-producing particles than particles having cosmic-ray energies."

These predictions have been verified by subsequent events registered in high latitudes from *riometers*, which register on a frequency of about 27 megacycles the intensity of galactic radio emission. Following the solar flare of July 29, 1958, Leinbach and Reid [*110*] observed a large attenuation of cosmic radio noise which they ascribed to absorption resulting from increased ionization in the ionospheric D-region. The effect differed from cosmic noise absorption produced in the auroral zone in that absorption was observed at stations well inside the auroral zone and in addition, the absorption exhibited a latitude cutoff. They concluded that this absorption resulted from enhanced ionization produced near 60 km altitude by ions from the solar flare, as had been previously suggested by Bailey [*109*] and by Little and Leinbach [*111*]. Penetration to 60 km altitude requires proton energies of 20—30 MeV.

From signal strength recordings obtained since 1952 from several VHF high latitude scatter paths Bailey and Harrington [*112*] have detected 44 such events. They list occasions when the polar cap absorption extends much further southward during the main phase of magnetic storms. During some of these occasions Ney et al. [*113*] have also observed protons with energy well below that for which protons are normally excluded by the earth's field at the latitude of their balloon flights.

51. Balloons, satellites. From May 1957 through July 1959, REID and LEINBACH [*114*] observed 24 so-called Type III absorption or PCA (polar cap absorption) events on riometer records. Detailed observations of solar cosmic rays from balloon observations, by the Minnesota group and others, were made possible by taking advantage of the riometer indications of the arrival of solar particles, and launching balloons as soon thereafter as possible. The nature of the solar particles arriving at the earth was established by ANDERSON [*115*], who measured an increase in charged particle intensity above Churchill, Canada, and Fairbanks,

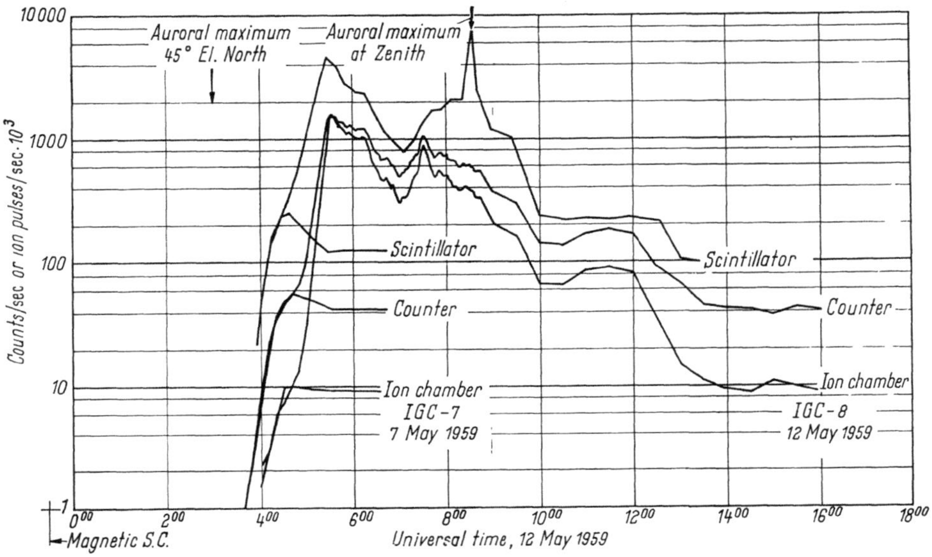

Fig. 85. Total counting rate of sodium iodide scintillator (50-kV threshold), single Geiger counter, and integrating ionization chamber on balloon flight IGC-8. For comparison, note same instruments on flight IGC-7, five days earlier. Solar particles are first detectable at 0430 UT at a balloon depth of 100 g/cm². The balloon reached ceiling at 0530 UT. Note large selective response of scintillator to auroral x-rays at auroral maximum. The solar particles temporarily increased during the aurora at 0730 UT (after NEY et al.).

Alaska, on August 22 and 23, 1957. They measured with ion chambers the variation in counting rate as a function of altitude following the solar flare and compared this counting rate with normal ones on days of no solar activity. The variation of counting rate with altitude allowed them to infer a proton energy spectrum and to set some limits on the flux of primary electrons. ANDERSON deduced a differential number energy spectrum $N(E)\,dE = KE^{-5}dE$ and found that the results were consistent with protons arriving in the energy range from 100—400 MeV. ROTHWELL and MCILWAIN [*116*], in experiments with Explorer IV, found large solar flare increases occurring at 0432 UT on August 16 and at 0005 UT on August 22. They measured the counting rates in shielded and unshielded Geiger counters during the period of the increases and concluded that the incoming beam must consist largely of protons. The first observations of incoming solar cosmic rays in emulsions were made by FREIER [*117*] et al. in the event which occurred on March 26, 1958.

Ionizing radiation recorded in balloon altitudes in high latitudes (in northern Sweden and in North America) have been discussed by PFOTZER et al. [*116a*].

One of the more spectacular events from which detailed information was obtained concerning solar cosmic rays was that of May 12, 1959, reported by

NEY et al. [113]. Fig. 85 shows the results of their observations derived from four balloon flights made on May 12, 1959. Preceding these observations a 3+ solar flare of 3 hours' duration occurred at 2000 GMT on May 10 and gave rise to a very strong solar noise storm. At 2340 GMT, May 11, there occurred a magnetic sudden commencement followed by a magnetic storm accompanied by a large decrease of cosmic-ray intensity. At Huancayo, as seen from Fig. 73b (p. 229), the daily mean ionization on May 12 was about 5% less than on May 11. The riometer at College, Alaska, indicated absorption starting about 0100 GMT on May 11. From the latter part of May 11, the riometer indicated absorption exceeding 17 db for nearly one day. In Fig. 85 the ion chamber curve shows that the ionization increased to 160 times that normally due to cosmic rays. From emulsion tracks NEY [113] et al. estimated the vertical flux of solar cosmic-ray particles at 1000 times the normal cosmic-ray flux.

Fig. 86 shows the proton rigidity spectrum obtained from measurements in emulsions for the May 12 event and also that for another event on March 26, 1958. The figure also shows the proton rigidity spectra near solar maxima and minima.

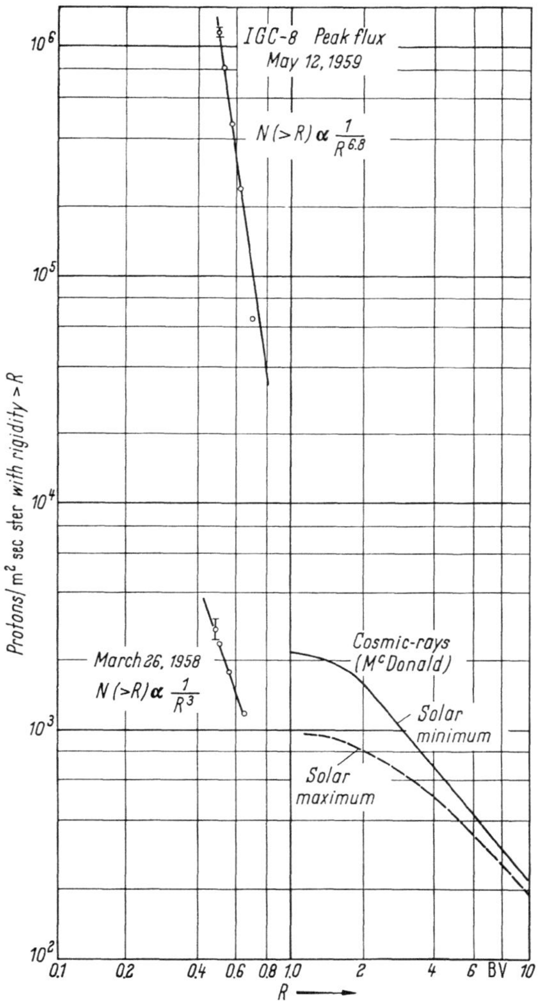

Fig. 86. Integral rigidity spectrum of protons observed during the solar flares of May 12, 1959 and March 26, 1958. The cosmic-ray proton spectrum during solar maximum and solar minimum is that measured by McDONALD (F. B. McDONALD and W. R. WEBBER to be published). F. B. McDONALD (to be published). (After NEY et al.)

52. Summary on solar proton events. From this and other similarly observed solar proton events NEY [118] summarizes the general features that have been established for solar proton events as follows:

a) The atmospheric effects are produced by protons primarily in the energy range from 30 MeV to something in excess of 500 MeV.

b) α particles appear to be present with approximately the same rigidity spectrum as the incoming protons.

c) The arrival of the proton beam heralded by the onset of cosmic noise absorption may occur within hours of the onset of the solar flare. The longest delay observed was the case of the storm of March 28, 1958, in which the arrival of the proton beam was coincident with the Forbush decrease and the magnetic storm. This occurred 20 hours after the flare. In contrast to this long delay, half of the events have delay times less than 5 hours.

d) The measurements at high latitudes of incoming particle fluxes confirm the suspicion that the cosmic noise absorption shown by the riometer is a good measure of the intensity of arrival of the incoming beam and that the incoming protons die off quasiexponentially with a time constant of the order of several days.

e) At latitudes as far south as Minneapolis, the events are very much more complicated. Even through particles may be arriving at high latitude, as shown in balloon measurements as well as riometer indications, the particles do not enter the atmosphere at Minnesota until the onset of the geomagnetic storm. The reason for this is that the normal cosmic ray cutoffs at Minnesota exclude the majority of the protons of the energies contained in the solar beam. It is, therefore, impossible for these particles to enter at this latitude without a perturbation of the earth's magnetic field. Although the cutoff energy for protons at Minnesota is of the order of 300 MeV, during the magnetic storm the solar particles with energies down to 50 MeV are able to enter. The time decay of the particle intensity at this low latitude, however, seems to be entirely governed by the behavior of the magnetic field as affected by the incoming solar stream. In the event in May 1959, the particle intensities persisted at their very high value for only several hours although the riometer showed that they continued to persist at high latitude for a number of days. Direct observations of the high latitude cosmic rays and the corresponding protons at Minnesota were made in the July 1959 event during which it became quite clear that the particle fluxes could arrive at Minnesota in essentially full intensity as long as the magnetic field was perturbed, but when the earth's magnetic field returned to normal the particles were not allowed to enter in Minnesota but were still arriving at the latitude of Churchill, Canada.

f) The incoming solar beams appear to be almost entirely positively charged. Certainly less than 10% of the particles in these beams are electrons. Since α particles of the same rigidity appear to be present, the question arises as to whether electrons accelerated with the positive particles can lose their energy in some specific way and therefore not be allowed to reach the earth. One possible explanation for the absence of the electrons would be that these electrons are trapped in the magnetic field of the solar corona and lose their energy by synchrotron radiation, thereby emitting radio waves. It has been established now by THOMPSON and MAXWELL [*119*] and by KUNDU and HADDOCK [*120*] that the radio emission from the sun at the time of the flare acceleration of the cosmic rays is of a particular kind which could, in fact, be attributed to synchrotron radiation of electrons.

g) The incoming proton beams, several hours after the beginning of the event, seem to be isotropic at the top of the atmosphere. This conclusion must be taken with some reservation, however, until direct measurements are made simultaneously at satellite altitudes and at balloon altitudes.

WEBBER [*96*] has provided a detailed comprehensive analysis and survey of individual solar flare increases including associated phenomena, general characteristics, and a discussion of specific models for the propagation and control of solar flare effects.

References.

See also Vol. XLVI, Cosmic Rays, of this Encyclopedia.

[1] CORLIN, A.: Ann. Obs. Lund Nr. 4 (1934).
[2] COMPTON, A. H., E. O. WOLLAN and R. D. BENNETT: Rev. Sci. Instr. **5**, 415 (1934).
[3] SITTKUS, A.: Z. Naturforsch. **1**, 204 (1946).
[4] DOLBEAR, D. W. N., and H. ELLIOTT: J. Atmosph. Terr. Phys. **1**, 215 (1951).
[5] ALFVÉN, H., and K. G. MALMFORS: Ark. Mat. Astronom. Fys. **29** A, No. 24 (1943).
[6] SIMPSON, J. A., W. FONGER and S. B. TRIEMAN: Phys. Rev. **90**, 934 (1953).
[6a] SIMPSON, J. A.: Cosmic radiation neutron monitor. Institute of Nuclear Studies, University of Chicago.
[7] SIMPSON, J. A.: Phys. Rev. **81**, 895 (1951).
[8] SIMPSON, J. A.: Phys. Rev. **83**, 1175 (1951).
[9] ADAMS, N., and H. J. BRADDICK: Phil. Mag. **41**, 501 (1950).
[10] Cosmic-ray results from Huancayo Magnetic Observatory, Peru, June 1936—December 1946. Including summaries from observatories at Cheltenham, Christchurch, and Godhavn through 1946. Researches of the Department of Terrestrial Magnetism, Vol. XIV, Carnegie Institution of Washington Publication 175, Washington, D.C. (1948). See Vols: XX (1956) and XXI (1961) for subsequent results through 1959.
[11] DAWTON, D. I., and H. ELLIOTT: J. Atmosph. Terr. Phys. **3**, 295 (1953).
[12] DUPERIER, A.: Proc. Phys. Soc. Lond. A **62**, 684 (1949).
[13] HESS, V. F., H. GRAZIADEI u. R. STEINMAUER: Sitzgsber. Akad. Wiss. Wien **144**, 53 (1935).
[14] COMPTON, A. H., and R. N. TURNER: Phys. Rev. **52**, 799 (1937).
[15] BLACKETT, P. M. S.: Phys. Rev. **54**, 973 (1938).
[16] FORBUSH, S. E.: Terr. Magn. **42**, 1 (1937).
[17] FORBUSH, S. E.: Phys. Rev. **54**, 975 (1938).
[18] BARTELS, J.: Terr. Magn. **37**, 291—302 (1932).
[19] BARTELS, J.: Terr. Magn. **40**, 1—60 (1935). — The contents of these papers are partly reproduced in Chap. 16 and 19 of CHAPMAN, S., and J. BARTELS: Geomagnetism. p. 543—605. Oxford Univ. Press 1940 (reprinted 1951).
[20] ALFVÉN, H., and K. G. MALMFORS: Ark. Mat. Astronom. Fys. **29** A, No. 24 (1943).
[21] KOLHÖRSTER, W.: Phys. Z. **42**, 55 (1941).
[22] DOLBEAR, D. W. N., and H. ELLIOTT: J. Atmosph. Terr. Phys. **1**, 215 (1953).
[23] ELLIOTT, H., and D. W. N. DOLBEAR: Proc. Phys. Soc. Lond. **63** A, 137 (1950).
[24] ELLIOTT, H., and D. W. N. DOLBEAR: J. Atmosph. Terr. Phys. **1**, 205 (1951).
[25] ALFVÉN, H.: Phys. Rev. **75**, 1732 (1949).
[26] THAMBYAHPILLAI, T., and H. ELLIOTT: Nature, Lond. **171**, 918 (1953).
[27] SARABHAI, V., and R. P. KANE: Phys. Rev. **90**, 204 (1953).
[28] FONGER, W.: Phys. Rev. **91**, 351 (1953).
[29] SIMPSON, J. A., J. FIROR, W. FONGER et S. B. TREIMAN: Recueil des travaux de l'observatoire du Pic-du-midi, Série: Rayons cosmiques No. 1, Congr. Internat. sur le rayonnement cosmique, **4** (1953).
[30] SITTKUS, A.: Recueil des travaux de l'observatoire du Pic-du-midi, Série: Rayons cosmiques No. 1, Congr. Internat. sur le rayonnement cosmique, **11** (1953).
[31] EHMERT, A., u. A. SITTKUS: Z. Naturforsch. **6** a, 618 (1951).
[32] SEKIDO, Y., and S. YOSHIDA: Rep. Ionosph. Res. Japan **4**, 37 (1950).
[33] ELLIOTT, H., and D. W. N. DOLBEAR: J. Atmosph. Terr. Phys. **1**, 205—214 (1951).
[34] YOSHIDA, S., and I. KONDO: J. Geomag. Geoelectr. **6**, 15 (1954).
[35] COMPTON, A. H., and I. A. GETTING: Phys. Rev. **47**, 817 (1935).
[36] VALLARTA, M. S., C. GRAEF and S. KUSAKA: Phys. Rev. **55**, 1—5 (1939).
[37] FORBUSH, S. E.: Phys. Rev. **52**, 1254 (1937).
[38] FORBUSH, S. E.: Terr. Magn. **42**, 1 (1937).
[39] ELLIOTT, H., and D. W. N. DOLBEAR: J. Atmosph. Terr. Phys. **1**, 205 (1951).
[40] FORBUSH, S. E.: Terr. Magn. **43**, 203 (1938).
[41] HAYAKAWA, S., J. NISHIMURA, T. NAGATA and M. SUGIURA: J. Sci. Res. Inst. Japan **44**, 121 (1950).
[42] TREIMAN, S. B.: Phys. Rev. **89**, 130 (1953).
[43] ALFVÉN, H.: Cosmical electrodynamics. Oxford: Clarendon Press 1950.
[44] FORBUSH, S. E.: Phys. Rev. **54**, 978 (1938).
[45] MEYER, P., and J. A. SIMPSON: Phys. Rev. **96**, 1085 (1954).
[46] FONGER, W. H.: Phys. Rev. **91**, 351 (1953).
[47] SIMPSON, J. A.: Phys. Rev. **94**, 426 (1954).
[48] FONGER, W. H.: Phys. Rev. **91**, 351 (1953).

[49] Vestine, E. H., L. Laporte, I. Lange and W. E. Scott: The geomagnetic field, its description and analysis. Washington, D.C., Carnegie Inst. Pub. 580 (1947).
[50] Neher, H. V., and E. A. Stern: Phys. Rev. **98**, 845 (1955).
[51] Neher, H. V., V. Z. Peterson and F. A. Stern: Phys. Rev. **90**, 655 (1953).
[52] Forbush, S. E.: Trans. Washington Meeting. Int. Union Geod. Geophys., Assoc. Terr. Magn. Electr., Bull. No. 11, 438 (1940).
[53] Bartels, J.: Terr. Magn. **37**, 1 (1935).
[54] Simpson, J. A.: Phys. Rev. **94**, 426 (1954).
[55] Simpson, J. A., W. H. Fonger and S. B. Treiman: Phys. Rev. **90**, 934 (1953).
[56] Forbush, S. E.: Phys. Rev. **70**, 771 (1946).
[56a] Ehmert, A.: Z. Naturforsch. **3**a, 264 (1948).
[57] McNish, A. G.: Terr. Magn. **42**, 109 (1937).
[58] Forbush, S. E.: Terr. Magn. **42**, 1 (1937).
[59] Forbush, S. E., P. S. Gill and M. S. Vallarta: Rev. Mod. Phys. **21**, 44 (1949).
[60] Graham, J. W., and S. E. Forbush: Phys. Rev. **98**, 1348 (1955).
[60a] Müller, R., et al.: J. Atmosph. Terr. Phys. **1**, 37 (1950).
[61] Forbush, S. E., M. Schein and T. B. Stinchcomb: Phys. Rev. **79**, 501 (1950).
[62] Adams, N.: Phil. Mag. **41**, 503 (1950).
[63] Clay, J., and H. F. Jongen: Phys. Rev. **79**, 908 (1950).
[64] Sekido, Y., and S. Yoshida: Rep. Ionosph. Res. Japan **7**, 147 (1953).
[65] Elliott, H.: Progress in cosmic ray physics, Chap. VIII. Amsterdam: North Holland Publishing Company 1952.
[66] Biermann, L.: In W. Heisenberg (ed.), Kosmische Strahlung. Berlin-Göttingen-Heidelberg: Springer 1953.
[67] Ehmert, A.: Z. Naturforsch. **3**a, 264 (1948).
[68] Schlüter, A.: Z. Naturforsch. **6**a, 613 (1951).
[69] Firor, J.: Phys. Rev. **94**, 1017 (1954).
[70] Störmer, C.: Astrophys. Norv. **1**, 1 (1934).
[71] Dwight, K.: Phys. Rev. **78**, 40 (1950).
[72] Malmfors, K. G.: Ark. Mat. Astronom. Fys., Ser. A **32**, No. 8 (1945).
[73] Brunberg, E.: J. Geophys. Res. **58**, 272 (1953).
[74] Störmer, C.: Astrophys. Norv. **1**, 115 (1934).
[75] Störmer, C.: Terr. Magn. and Electr. **22**, 23 (1917).
[76] Hess, V. F.: Sitzgsber. Wien. Akad. IIa, 2001-32 **121** (1912).
[77] Annals of the International Geophysical Year, Vol. VIII, pp. 209—214. London: Pergamon Press 1959.
[78] Forbush, S. E.: J. Geophys. Res. **63**, 651 (1958).
[79] Kertz, Walter: Ein neues Maß für die Feldstärke des erdmagnetischen äquatorialen Ringstroms. Abh. der Akad. der Wiss. in Göttingen, Math.-physik. Kl. Beiträge zum Internat. Geophysikalischen Jahr, H. 2. Göttingen: Vandenhoeck and Ruprecht 1958.
[80] Bartels, J.: Terrestrial magnetic activity and its relation to solar phenomena. Terr. Magn. **37**, 1 (1932).
[81] Bartels, J.: Random fluctuations, persistence and quasipersistence in geophysical and cosmical periodicities. Terr. Magn. **40**, 1 (1935).
[82] Neher, H. V., and H. R. Anderson: J. Geophys. Res. **67**, 1309 (1962).
[83] Neher, H. V.: Phys. Rev. **107**, 588 (1957).
[84] Neher, H. V.: J. Geophys. Res. **66**, 4007 (1961).
[85] Sandström, A. E., and S. E. Forbush: J. Geophys. Res. **63**, 876 (1958).
[86] Neher, H. V.: Phys. Rev. **103**, 228 (1956).
[87] Ney, E. P., J. R. Winckler and P. S. Freier: Phys. Rev. Letters **3**, 183 (1959).
[88] Winckler, J. R., L. Peterson, R. Hoffman and R. Arnoldy: J. Geophys. Res. **64**, 597 (1959).
[89] Meyer, P., and J. A. Simpson: Phys. Rev. **106**, 568 (1957).
[90] Freier, P. S., E. P. Ney and C. J. Waddington: Phys. Rev. **114**, 365 (1959).
[91] McDonald, F. B.: Phys. Rev. **116**, 462 (1959).
[92] McDonald, F. B., and W. R. Webber: Phys. Rev. **115**, 194 (1959).
[93] Cosmic-ray intensity during the IGY, Nos. 1, 2, and 3, March 1959, December 1959, and April 1960, National Committee for the IGY, Science Council of Japan, Veno Park, Tokyo, Japan.
[94] Lockwood, J. A.: J. Geophys. Res. **65**, 3859 (1960).
[95] Reid, G. C., and H. Leinbach: J. Geophys. Res. **64**, 1801 (1959).
[96] Webber, W. R.: Progress in cosmic ray physics, Vol. 6 (in press).
[97] Fan, C. Y., P. Meyer and J. A. Simpson: Phys. Rev. Letters **4**, 421 (1960).
[98] Fan, C. Y., P. Meyer and J. A. Simpson: Phys. Rev. Letters **5**, 269 (1960).

[99] COLEMAN jr., P. J., C. P. SONETT, D. L. JUDGE and E. J. SMITH: J. Geophys. Res. **65**, 1856 (1960).
[100] FORBUSH, S. E.: J. Geophys. Res. Letter **61**, 155 (1956).
[101] MEYER, P., E. N. PARKER and J. A. SIMPSON: Phys. Rev. **104**, 768 (1956).
[102] SIMPSON, J. A.: Symposium on Astronomical Aspects of Cosmic Rays, Suppl. Series. Astrophys. J., Suppl. No. 44, **44**, 369—422 (June 1960).
[103] DORMAN, L. I.: Cosmic ray variations, State Publishing House for Technical Literature, Moscow, U.S.S.R. 727 pp. (1957).
[104] COCCONI, G., K. GREISEN, P. MORRISON, T. GOLD e S. HAYAKAWA: Nuovo Cim. **10**, Suppl. 8, 161 (1958).
[105] CARMICHAEL, H., and J. F. STELJES: International Union of Geodesy and Geophysics, Symposium on the July 1959 Events and Associated Phenomena, Helsinki 1960, Monograph No. 7, p. 10 (November 1960).
[106] QUENBY, J. J., and W. R. WEBBER: Phil. Mag. **4**, No. 37, 90—113 (January 1939).
[107] International Union of Geodesy and Geophysics, Symposium on the July 1959 Events and Associated Phenomena, Helsinki, July 1960, Monograph No. 7 (November 1960).
[108] MCCRACKEN, K. G.: J. Phys. Soc. Japan **17**, 310, Suppl. A-II, Internat. Conference on Cosmic Rays and the Earth Storm, Part II (1962).
[108a] EHMERT, A., u. G. PFOTZER: Mitt. Max-Planck-Inst. Aeronomie, Nr. **8**, 54 (1962).
[109] BAILEY, D. K.: J. Geophys. Res. **62**, 431 (1957).
[110] LEINBACH, H., and G. C. REID: Phys. Rev. Letters **2**, 61 (1959).
[111] LITTLE, C. G., and H. LEINBACH: Proc. I.R.E. (Inst. Radio Engrs.) **46**, 334 (1959).
[112] BAILEY, D. K., and J. M. HARRINGTON: J. Phys. Soc. Japan **17**, 334, Suppl. A-II, Internat. Conference on Cosmic Rays and the Earth Storm, Part II (1962).
[113] NEY, E. P., J. R. WINCKLER and P. S. FREIER: Phys. Rev. Letters **3**, 183 (1959).
[114] REID, G. C., and H. LEINBACH: J. Geophys. Res. **64**, 1801 (1959).
[115] ANDERSON, K. A.: Phys. Rev. Letters **1**, 335 (1958).
[116] ROTHWELL, P., and C. E. MCILWAIN: Nature, Lond. **184**, 138 (1959).
[116a] PFOTZER, G., A. EHMERT and E. KEPPLER: Mitt. Max-Planck-Inst. Aeronomie, Nr. 9A and B (1962).
[117] FREIER, P. S., E. P. NEY and J. R. WINCKLER: J. Geophys. Res. **64**, 685 (1959).
[118] NEY, E. P.: Ann. Rev. Nuclear Sci. **10**, 461 (1960).
[119] THOMPSON, A. R., and A. MAXWELL: Nature, Lond. **185**, 89 (1960).
[120] KUNDU, M. R., and F. T. HADDOCK: Nature, Lond. **186**, 610 (1960).

Magnetic Properties of Rocks and Minerals.

By

TAKESI NAGATA.

With 46 Figures.

I. Magnetism of rock-forming minerals.

1. Magnetic properties of rocks as an assemblage of various kinds of minerals. Rocks are assemblages of a large number of minerals, and the chemical and crystallographic structures of those minerals are different from each other even in a group of minerals called with a single name. From the view point of magnetic properties, however, they can be classified into four groups, viz. *diamagnetic*, *paramagnetic*, *ferrimagnetic* and *antiferromagnetic* ones. Minerals composed of ions having no magnetic moment are *diamagnetic*, while those containing magnetic ions such as Fe^{2+}, Fe^{3+} and Mn^{2+} are either *ferrimagnetic* or *antiferromagnetic* when there is a certain interaction among those magnetic ions, and are *paramagnetic* when there is no interaction. Therefore only those minerals which contain magnetic ions as compositions are responsible for magnetic properties of rocks.

So long as magnetic properties are concerned, rocks may be schematically considered as such solid bodies that a large number of small grains of ferrimagnetic, antiferromagnetic and paramagnetic minerals are scattered among a non-magnetic (strictly speaking diamagnetic) medium. Provided that this magnetic model of rocks is accepted, the main problems concerning magnetic properties of rocks may be summarized as

(a) Intrinsic magnetic properties of magnetic minerals themselves,

(b) Effect of small grain size of magnetic minerals,

(c) Interaction among magnetic minerals.

Other problems which are particularly important in rock-magnetism may be the behaviour of change in magnetic properties and magnetization of rocks or magnetic minerals under the influence of particular conditions on the earth such, for example, as chemical changes of materials during extremely long geologic times, say, 10^4 to 10^9 years, and very large pressure or stress affecting rocks, say, 10^3 to 10^5 atmospheres. Another problem in relation with geophysics may be stability of magnetization, or in other words, the after-effect of magnetization during a very long time.

Besides changes in magnetization caused by chemical and crystallographic changes of substances, remanent magnetization of ferrimagnetic materials should be reduced owing to thermal agitation provided that the temperature is high enough. When the time-scale concerned is very long as is the case for a geologic period, this fact becomes primarily significant, even if its amount is negligibly small in case of laboratory time-scale experiments.

Not all of the above-mentioned problems have yet been completely solved, but some of them have been quantitatively clarified, as will be described in the following sections.

2. Paramagnetic minerals. Minerals which contain magnetic ions such as Fe^{3+}, Fe^{2+} or Mn^{2+} generally show either ferrimagnetic or paramagnetic properties. Natural rock-forming minerals having *paramagnetic* properties are, for example, *Olivine, Pyroxene, Monoclinic pyroxene, Garnet, Biotite, Amphibole, Cordierite,* and others. The chemical compositions of these minerals are generally expressed as follows:

olivines: $(Fe^{2+}, Mg)_2SiO_4$,

pyroxenes:

orthopyroxenes: $(Fe^{2+}, Mg)SiO_3$,

monoclinic pyroxenes:

$$\alpha\, Ca(Mg, Fe^{2+})Si_2O_6 + \beta\, (Mg, Fe^{2+})_2Si_2O_6 + \gamma\, Al_2O_3 + \delta\, Na(Al, Fe^{3+})Si_2O_6$$

garnets:

pyralspites: $(Mg, Fe^{2+}Mn)_3Al_2Si_3O_{12}$,

grandites: $Ca_3(Al, Fe^{3+})_2Si_3O_{12}$,

biotites:

$$\alpha\, K_2(Mg, Fe^{2+})_6(OH)_4Si_6Al_2O_{20} + \beta\, K_2(Mg, Fe^{2+})_5Al(OH)_4Si_5Al_3O_{20},$$

amphiboles:

alkali amphiboles:

$$\alpha\, Na_2Ca(Mg, Fe^{2+})_5Si_8O_{22}(OH)_2 + \beta\, Na_3(Mg, Fe^{2+})_4(Al, Fe^{3+})Si_8O_{22}(OH)_2 \\ + \gamma\, Na_2(Mg, Fe^{2+})_3(Al, Fe^{3+})_2Si_8O_{22}(OH)_2,$$

calciferous amphiboles:

$$\alpha\, Ca_2(Mg, Fe^{2+})_5Si_8O_{22}(OH)_2 + \beta\, NaCa_2(Mg, Fe^{2+})_5AlSi_7O_{22}(OH)_2 \\ + \gamma\, Ca_2(Mg, Fe^{2+})_3(Al, Fe^{3+})_2Si_6O_{22}(OH)_2$$

cordierites: $(Mg, Fe^{2+})_2Al_4Si_5O_{18} \cdot H_2O$.

Here, the sum of α, β, etc. in 1 Mol should be unity for each case.

In these minerals, metallic ions having a magnetic moment seem to behave as free magnetic dipoles in such a way as described in LANGEVIN's theory of paramagnetism. Experimental examinations of paramagnetic minerals, however, have been made quite recently, and the results of these studies show that the above-mentioned assumption holds as a good approximation.

Among metallic ions which occur usually in natural rocks, only Fe^{3+}, Fe^{2+} and Mn^{2+} have a magnetic moment, and other ions such as Si^{4+}, Al^{3+}, Mg^{2+}, Ca^{2+}, Na^+, K^+, Ti^{4+}, P^{5+}, etc. have no magnetic moment. According to STONER, the effective magnetic moment of the three magnetic ions are given as

$$M(Fe^{2+}) = (5.25 \text{ to } 5.53)\,\mu_B,$$
$$M(Fe^{3+}) = 5.58\,\mu_B,$$
$$M(Mn^{2+}) = 5.58\,\mu_B,$$

where $\mu_B = 0.9274 \times 10^{-20}$ emu denotes the Bohr magneton's moment and numerals show the number of effective Bohr magnetons, p_B, of the respective ions, (i.e. $M = p_B \cdot \mu_B$).

When the contents of Fe^{2+}, Fe^{3+}, and Mn^{2+} per gram of a mineral are x, y, and z grams, respectively and no effective interaction among those magnetic ions is assumed, the specific magnetic susceptibility χ of the mineral is given, according to LANGEVIN's theory, by

$$\left.\begin{aligned}\chi &= \frac{1}{3RT}\left[x M^2(Fe^{2+}) + y M^2(Fe^{3+}) + z M^2(Mn^{2+})\right]\\ &= \frac{\mu_B^2}{3RT}\left[x P_B^2(Fe^{2+}) + y P_B^2(Fe^{3+}) + z P_B^2(Mn^{2+})\right],\end{aligned}\right\} \quad (2.1)$$

where R and T respectively, denote, the gas constant and the temperature.

Experimental examinations of magnetic susceptibility of paramagnetic minerals were carried out with reference to this theoretical expectance, in 1957 by NAGATA *et al.* [*1*] for olivines, in 1958 by CHEVALLIER and MATHIEU [*2*] for monoclinic pyroxenes, by AKIMOTO *et al.* [*3*] for orthopyroxenes, and in 1960 by SYONO [*4*] for garnets, biotites, amphiboles and cordierite. The difficulty in

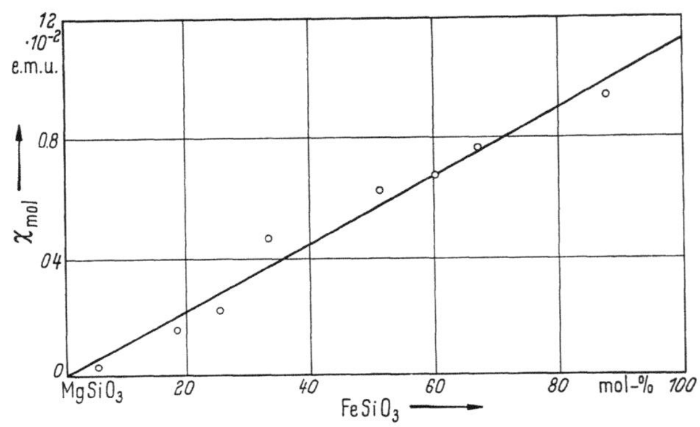

Fig. 1. Magnetic susceptibility of orthopyroxene at room temperature. (After AKIMOTO, HÔRAI and BOKU.)

this kind of examination lies mostly in the determination of the exact chemical composition of the samples.

Summarizing all results obtained, it can be concluded that the assumption of free ion behaviour of Fe^{2+}, Fe^{3+}, and Mn^{2+} well holds in those minerals, and therefore that Eq. (2.1) can be used for estimating the magnetic susceptibility of the paramagnetic minerals, when their chemical composition is given. Fig. 1 illustrates, for example, the relation between magnetic susceptibility χ and chemical composition of orthopyroxenes, which are the simplest case of paramagnetic minerals, y and z in (2.1) being zero.

In the diagram, a full line represents the theoretical value and hollow circles show the observed values of magnetic susceptibility and actually analyzed chemical compositions. Likewise, in cases of olivines, monoclinic pyroxene, and garnets, the agreement between observed and theoretical values is satisfactory.

It must be noted, however, that the observed values of magnetic susceptibility are always appreciably larger than theoretical values in minerals containing H_2O such as biotite, hornblende and cordierite. Although this discrepancy seems to be due to a certain effect of H_2O, its physical mechanism has not yet been studied.

3. Classification of ferrimagnetic minerals in rocks. Ferrimagnetic minerals in rocks can be classified into two groups, viz. metallic oxides and metallic sulphides. Ferrimagnetic metallic oxides, such as *magnetites*, *titanomagnetites*, *maghemites*, *titanomaghemites*, *haematites*, *haemo-ilmenites* and their mutual solid solutions are commonly contained in the majority of ordinary rocks. On the other hand, ferrimagnetic iron sulphides, which are known as *pyrrhotites*, are usually very localized as the minerals are produced under special conditions. It may be said therefore that metallic oxide minerals are exclusively general and important as origin of ferrimagnetic characteristics of general rocks.

Up to recent days, opaque minerals which can be attracted by a magnet had been called *"magnetites"*, as a whole, in general petrology. Actually, these *"magnetites"* are responsible for the most parts of ferrimagnetism of rocks, in which they are contained. However, the chemical compositions of the "magnetites" in natural rocks are very complex and varied, and their magnetic properties change with changes in their chemical composition.

According to the results of chemical analysis of a large number of ferrimagnetic oxide minerals in rocks, their main components are FeO, Fe_2O_3 and TiO_2, and their sub-components, very little in amount compared with the above mentioned three, are MnO, MgO, Al_2O_3, V_2O_3, Cr_2O_3, etc.

Magnetic and other physical properties of the $FeO—Fe_2O_3—TiO_2$ ternary system should, therefore, be the most basic knowledge for understanding the ferrimagnetic characteristics of general rocks.

4. Outline of magnetic properties of the ternary system $FeO—Fe_2O_3—TiO_2$. *α) The $FeO—Fe_2O_3—TiO_2$ ternary system.* Magneto-chemical studies on the $FeO—Fe_2O_3—TiO_2$ ternary system were made, first by POUILLARD on the $TiFe_2O_4—Fe_3O_4$ series and the $TiFeO_3—Fe_2O_3$ series, and by CHEVALLIER *et al.* on the $TiFeO_3—Fe_3O_4$ series, and then have been systematically and comprehensively extended by AKIMOTO, GORTER, NÉEL, KAWAI and others. From results of these studies, it has been found that there are three basic series of solid solution in the $FeO—Fe_2O_3—TiO_2$ ternary system. They are, as shown in Fig. 2,

I. the $Fe^{2+}Fe_2^{3+}O_4—Fe_2^{2+}Ti^{4+}O_4$ series having spinel structure,

II. the $Fe_2^{3+}O_3—Fe^{2+}Ti^{4+}O_3$ series having rhombohedral structure,

III. the $Fe_2^{3+}Ti^{4+}O_5—Fe^{2+}Ti_2^{4+}O_5$ series having orthorhombic structure.

In addition to these three basic solid solution series, a particular series of $Fe_3O_4—Fe_2O_3$, belonging also to the ternary system, exists commonly in rocks and can easily be produced in the laboratory. γ-Fe_2O_3 is a particular metastable crystal of Fe_2O_3 having cubic structure, and is called *maghemite* in comparison with the stable α-Fe_2O_3 *(haematite)* having rhombohedral structure. It has been found that maghemite has such a crystal structure as having a vacancy of a Fe ion from the inverse spinel type structure of Fe_3O_4. Generally, the $Fe_3O_4—TiFe_2O_4$ series (titanomagnetites) having inverse spinel type structure can get a vacancy of position of Fe ion without changing their crystal structure.

Production of a vacancy of Fe ion position in titanomagnetites may correspond to their oxidation, accompanied by $Fe^{2+} \rightarrow Fe^{3+}$. Thus, it has been shown that ferrimagnetic minerals of inverse spinel type having chemical composition in an area bounded by the $Fe_3O_4—TiFe_2O_4$ join and the $Fe_2O_3—TiFeO_3$ join in the diagram of $FeO—Fe_2O_3—TiO_2$ ternary system, shown in Fig. 2. This problem will be dealt with in more detail in Sect. 17.

β) $Fe_3O_4—TiFe_2O_4$ series (Titanomagnetites). The $x TiFe_2O_4 \cdot (1-x) Fe_3O_4$ series are solid solutions, for which magnetite (Fe_3O_4) and ulvöspinel ($TiFe_2O_4$) are the end members, being called *titanomagnetites*. Change in amount of x in this series may be considered to be due to substitution, $2Fe^{3+} \rightleftharpoons Fe^{2+} + Ti^{4+}$.

Since the first success of artificial synthesis of ulvöspinel, magneto-chemical studies on this series have been partly made by ERNST [5], POUILLARD [6], KAWAI, KUME, and SASAJIMA [7], while complete research on single phase synthetic samples of $x Fe_2TiO_4 \cdot (1-x) Fe_3O_4$ throughout the whole region of x

$(1 \geqq x \geqq 0)$ has been followed up by AKIMOTO, KATSURA and YOSHIDA [8], and NAGATA, AKIMOTO and YAMA-AI.

According to the latest results, the lattice parameter α of this series changes with x almost linearly, from 8.39 Å of Fe_3O_4 to 8.53 Å of Fe_2TiO_4, as illustrated in Fig. 3.

The Curie temperature Θ of this series decreases continuously with increase in x from 578° C of Fe_3O_4, as shown in Fig. 4. It is interesting from the geophysical viewpoint that titanomagnetites of $x > 0.8$ have their Curie point below the ordinary atmospheric temperature. Extrapolating the curve showing the

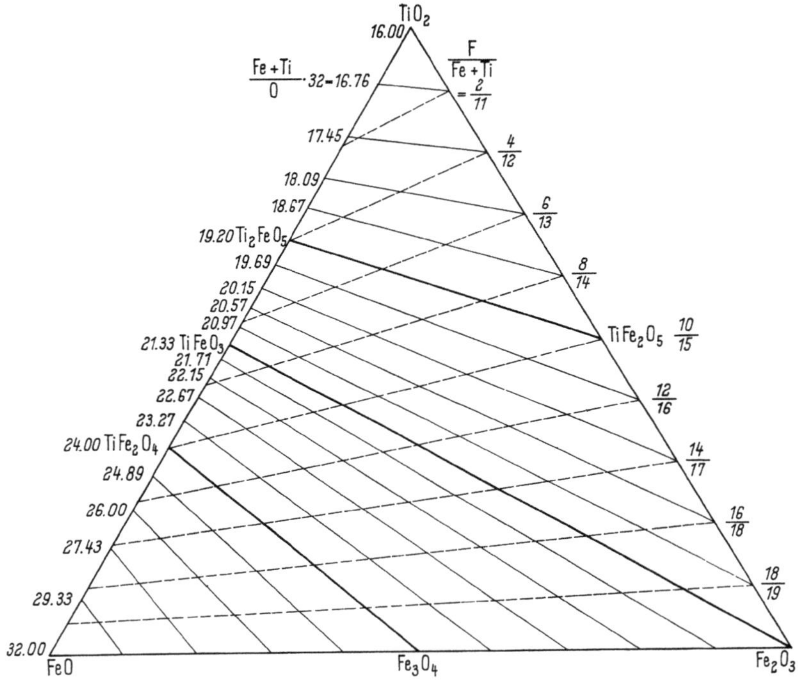

Fig. 2. Ternary system of $FeO-Fe_2O_3-TiO_2$. The oxidation lines are indicated by broken lines.

relation between Θ and x in Fig. 4 down to $x=1$, it may be found that Curie temperature of pure ulvöspinel is about 120° K. However, provided that ulvöspinel crystal keeps the regular inverse spinel structure, it should be antiferromagnetic, its spontaneous magnetization being zero.

The relation between intensity of saturation magnetization σ and x is given in Fig. 5, where σ decreases continuously with increase in x. As for the constitution of *spontaneous magnetization*[1] σ_0, two different interpretations have been proposed: In one of these interpretations, the configuration of metallic ions in the crystal of titanomagnetites is expressed by

$$Fe^{3+}_{1-x}Fe^{2+}_{x}(Fe^{2+}Fe^{3+}_{1-x}Ti^{4+}_{x})O^{2-}_{4}$$

[1] The intensity of *saturation magnetization* σ measured by an ideal method at 0° K should be the same as the intensity of *spontaneous magnetization* σ_0. In this article, however, spontaneous magnetisation will be denoted σ_0 in order to distinguish from the actually measured value σ.

throughout the range $0 \leqq x \leqq 1$, where the inside and the outside of the parenthesis show the ion configurations of octahedral and tetrahedral sites respectively of sublattices of the spinel structure. In this interpretation, σ_0 per molecule should be expressed as

$$\sigma_0 = \mu_B P_B(Fe^{2+})(1-x), \tag{4.1}$$

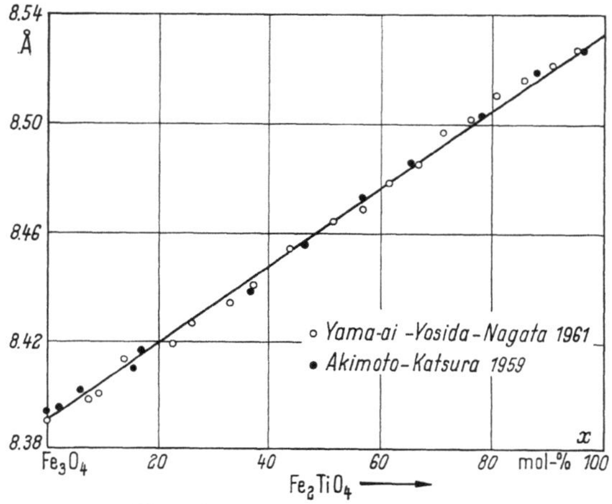

Fig. 3. Lattice constant of titanomagnetites.

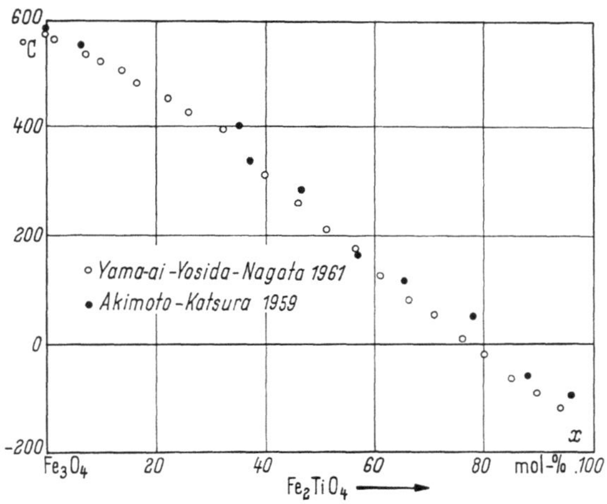

Fig. 4. Curie temperature of titanomagnetites.

showing that σ_0 decreases linearly with increase in x from Fe_3O_4 to Fe_2TiO_4, which has no spontaneous magnetisation, as illustrated by curve a in Fig. 5, In another interpretation, suggested by NÉEL, CHEVALLIER and others by taking into consideration VERWAY's empirical law of metallic ion distribution in spinel structure, it is assumed that first $2Fe^{3+}$ in octahedral site is substituted by $Fe^{2+}+Ti^{4+}$ until Fe^{3+} there becomes zero and then Fe^{3+} in tetrahedral site begins to be substituted. Then, the configuration of metallic ions can be expressed by

$$Fe^{3+}(Fe^{2+}_{1+x}Fe^{3+}_{1-2x}Ti^{4+}_x)O^{2-}_4 \quad \text{for} \quad 0<x<\tfrac{1}{2},$$

$$Fe^{3+}_{2-2x}Fe^{2+}_{2x-1}(Fe^{2+}_{2-x}Ti^{4+}_x)O^{2-}_4 \quad \text{for} \quad \tfrac{1}{2}<x<1.$$

In this case, σ_0 is given by

$$\left.\begin{aligned}\sigma_0 &= \mu_B\left[P_B(\mathrm{Fe}^{2+}) + x\{P_B(\mathrm{Fe}^{2+}) - 2P_B(\mathrm{Fe}^{3+})\}\right] && \text{for} \quad 0<x<\tfrac{1}{2}\\ \sigma_0 &= \mu_B(1-x)\{3P_B(\mathrm{Fe}^{2+}) - 2P_B(\mathrm{Fe}^{3+})\} && \text{for} \quad \tfrac{1}{2}<x<1,\end{aligned}\right\} \qquad (4.2)$$

and at $x=\frac{1}{2}$, σ_0 becomes $\frac{1}{2}[3P_B(\mathrm{Fe}^{2+})-2P_B(\mathrm{Fe}^{3+})]$. The relation of σ_0 to x given by the above is illustrated by curve *b* in Fig. 5. Comparing plots of actually measured values with Curves *a* and *b* in Fig. 5, it may be hardly possible to determine which interpretation is right. Furhter, it might be possible, as pointed

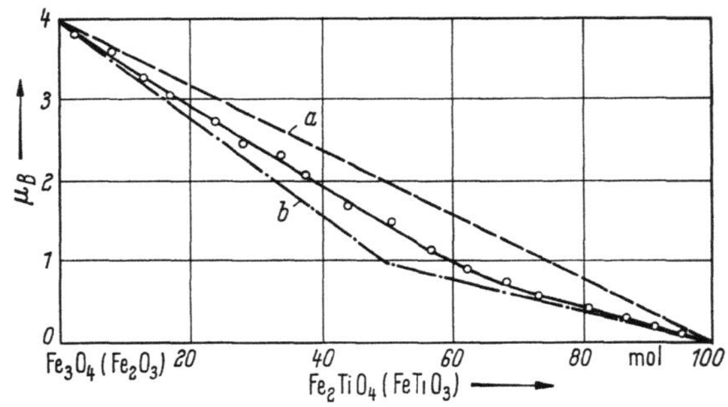

Fig. 5. Saturation magnetization of titanomagnetites in units of Bohr magnetons per mol. (After YAMA-AI, YOSHIDA and NAGATA.)

out by GORTER [9], that tetrahedral site positions also are occupied more or less by Ti ions in actual minerals. In such a case, the configuration of ions becomes

$$\mathrm{Fe}^{3+}_{1-a},\ \mathrm{Ti}^{4+}_{a}(\mathrm{Fe}^{2+}_{1+x}\mathrm{Fe}^{3+}_{1-2x+a}\mathrm{Ti}^{4+}_{x-a})\mathrm{O}^{2-}_{4},$$

or

$$\mathrm{Fe}^{3+}_{2-2x-a}\mathrm{Fe}^{2+}_{2x-1}\mathrm{Ti}^{4+}_{a}(\mathrm{Fe}^{2+}_{2-x}\mathrm{Fe}^{3+}_{a}\mathrm{Ti}^{4+}_{x-a})\mathrm{O}^{2-}_{4}.$$

These problems have not yet been completely solved.

γ) Oxidation of titanomagnetites (Titanomaghemites). Chemical compositions of ferrimagnetic minerals having inverse spinel structure (same as titanomagnetites) do not always lie on the straight line joining Fe_3O_4—Fe_2TiO_4 in Fig. 2, but are fairly scattered within a quadrangle Fe_2TiO_4—Fe_3O_4—Fe_2O_3—$FeTiO_3$ (see Fig. 8).

CHEVALLIER and GIRARD assumed that a part of those non-titanomagnetite cubic ferrimagnetic minerals are solid solutions between Fe_3O_4 and $FeTiO_3$, and succeeded to synthesize $FeTiO_3$—Fe_3O_4 solid solutions up to 37 mol % of $FeTiO_3$. On the other hand, it was proved by AKIMOTO *et al.* [8] that ferrimagnetic minerals of spinel structure with lattice defect of almost all chemical compositions within the quadrangle Fe_2TiO_4—Fe_3O_4—Fe_2O_3—$FeTiO_3$ can be produced by oxidation of Fe_3O_4—Fe_2TiO_4 series at adequate temperatures. Dependence of lattice parameter, Curie temperature, and intensity of saturation magnetisation at the atmospheric temperature of those specimens produced by the oxidation, upon their chemical composition are shown in Figs. 6 to 8.

Thus, it may be generally concluded that ferrimagnetic minerals of spinel structure having chemical compositions deviating from the Fe_3O_4—Fe_2TiO_4 join are titanomagnetites with vacancy of Fe^{2+} positions in the lattice. The titanomagnetites with the lattice defect may be called *titanomaghemites*, since their extreme case of Ti content $= 0$, that is, $Fe_3O_4—\gamma Fe_2O_3$ solid solutions, have been called maghemite. The process of change from titanomagnetites to titano-

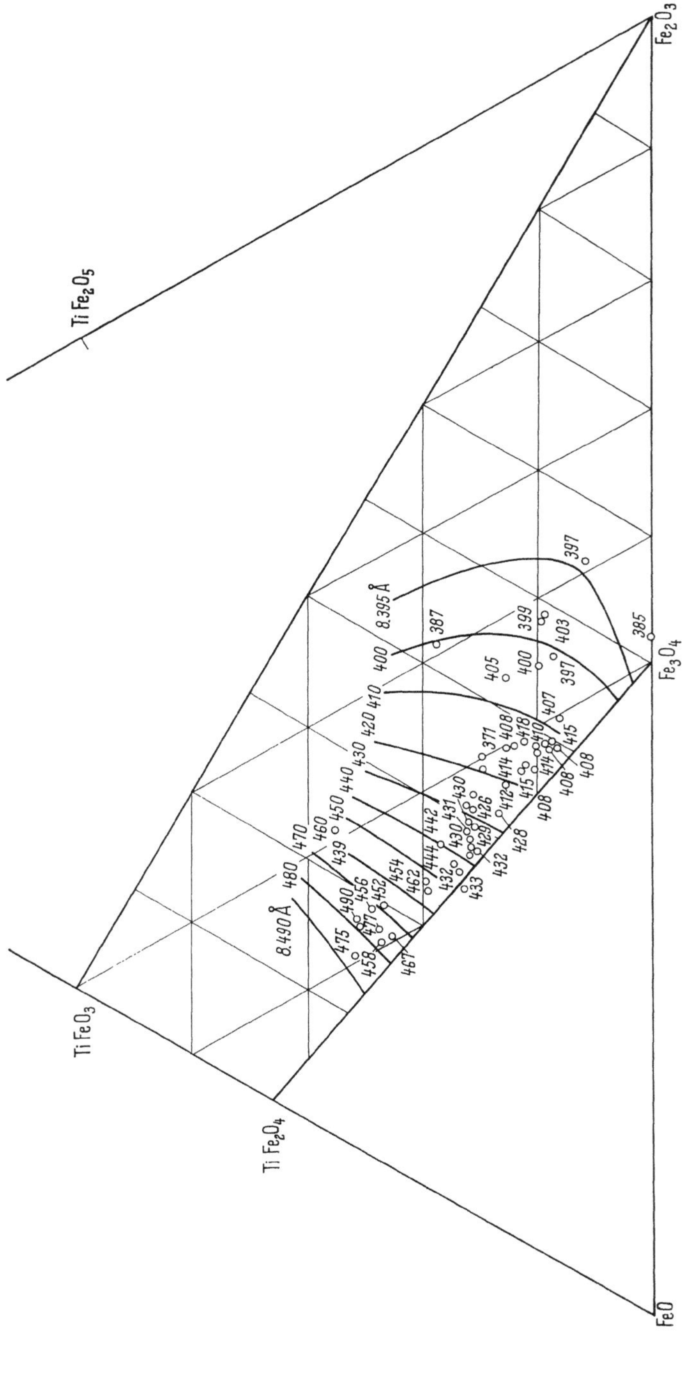

Fig. 6. Lattice constant and chemical composition of magnetic minerals of spinel phase. Circles: natural minerals. Contour lines: synthetic minerals. (After AKIMOTO and KATSURA.)

maghemites is equivalent to an oxidation process of $Fe^{2+} \rightarrow Fe^{3+}$. In Fig. 2, the oxidation-reduction line showing the process of oxidation of titanomagnetites and the degree of their oxidation are illustrated respectively by lines of

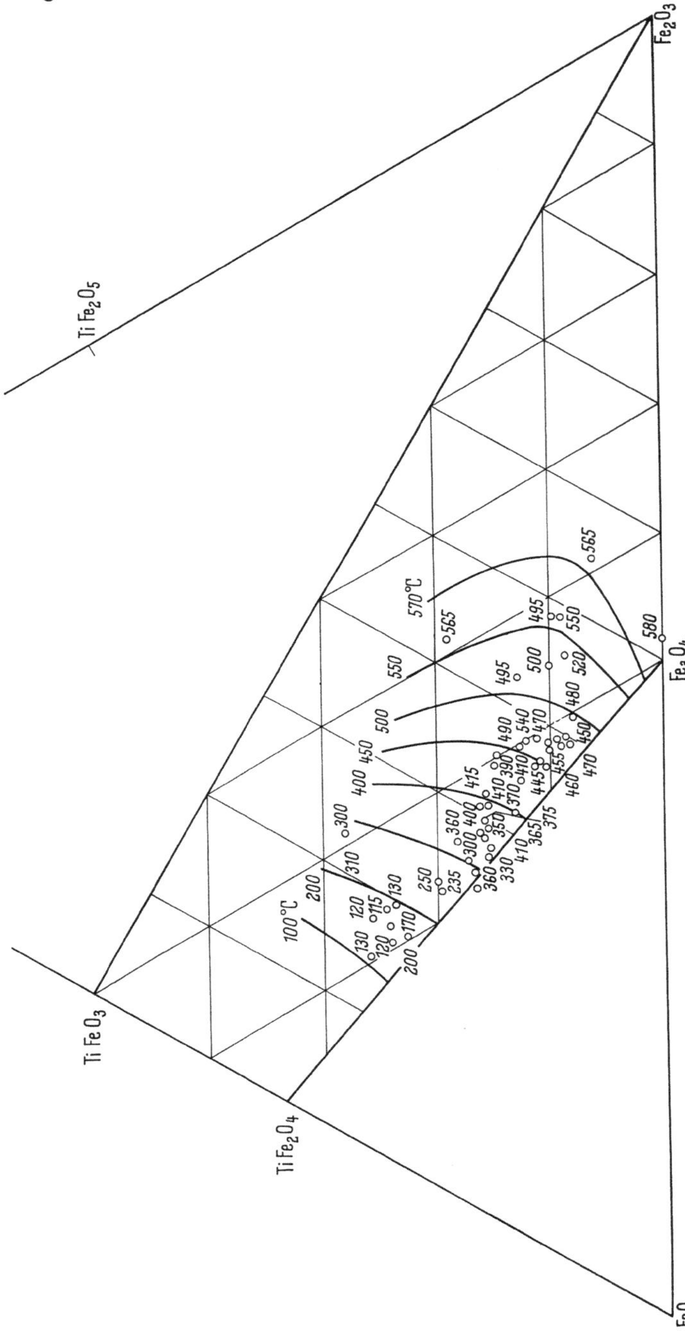

Fig. 7. Curie temperature and chemical composition of magnetic minerals of spinel phase. (After AKIMOTO and KATSURA.)

Fe/(Fe + Ti) = constant and lines of $\frac{\text{Fe}+\text{Ti}}{\text{O}} \times 32 = \text{const}$, which indicate the number of metallic ions in a unit cell of spinel. It might be considered that titanomaghemites in nature could be produced mostly by the oxidation of titanomagnetites, owing, for example, to hydrothermal intrusions or other mechanisms of rather rapid oxidation.

δ) Titanomagnetites and titanomaghemites in igneous rocks. Practical procedures of examining ferrimagnetic minerals in rocks are as follows: a rock sample to be examined is crushed and pulverized into small grains of suitable sizes, i.e. the same as or a little smaller than ferrimagnetic mineral's size which can be detected by microscope observation; then groups of ferrimagnetic minerals are separated out

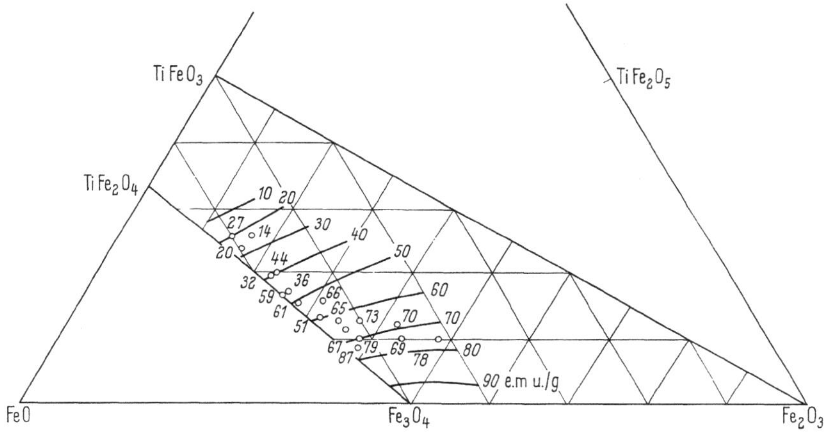

Fig. 8. Saturation magnetization (at the atmospheric temperature) and chemical composition of magnetic minerals of spinel phase. (After Akimoto and Katsura.)

by means of repeated use of a magnetic separator; sometimes a procedure of magnetic separation at various different temperatures [*10*] is necessary to obtain single phase ferrimagnetic minerals, because the compositions of ferrimagnetic minerals in even a piece of rock are occasionaly distributed over a fairly large range; composition and crystal structure of the ferrimagnetic minerals thus separated out are determined by chemical analysis and X-ray analysis respectively; their magnetic properties, such as intensity of saturation magnetization, Curie temperature, coercive force, initial susceptibility, etc, are determined with usual magnetometers of various adequate types with a little higher sensitivity than in the case of ferromagnetic metals.

Generally speaking, more than 90 percent of ferrimagnetic minerals contained in rocks are those of inverse spinel structure, namely, titanomagnetites and titanomaghemites. One may say therefore that ferrimagnetism of rocks is due in most cases either to titanomagnetites or titanomaghemites. Fig. 9 shows the chemical composition of a large number of spinel type ferrimagnetic minerals contained in various igneous rocks of Japan, which were ascertained by X-ray analyses to be in almost perfect single phase.

It is shown in this figure that the chemical composition of most spinel type ferrimagnetic minerals is situated between the Fe_3O_4—Fe_2TiO_4 join and the Fe_2O_3—$FeTiO_3$ join, but some strongly oxidized minerals have chemical composition even in the Fe_2O_3—$FeTiO_3$—TiO_2 region. The latters are the most typical titanomaghemites, and they are easily distinguished by their typical bluish grey colour under a reflection microscope.

The lattice parameters of these titanomagnetites and titanomaghemites in igneous rocks shown in Fig. 9 range from 8.35 to 8.49 Å, and their Curie temperatures and intensities of saturation magnetisation at the atmospheric temperatures are between 580 and 100° C and between 90 and 10 emu/g respectively, and dependences of these values of natural minerals on chemical compositions are well represented by Figs. 6 to 8, which are derived from data of artificially synthesized samples.

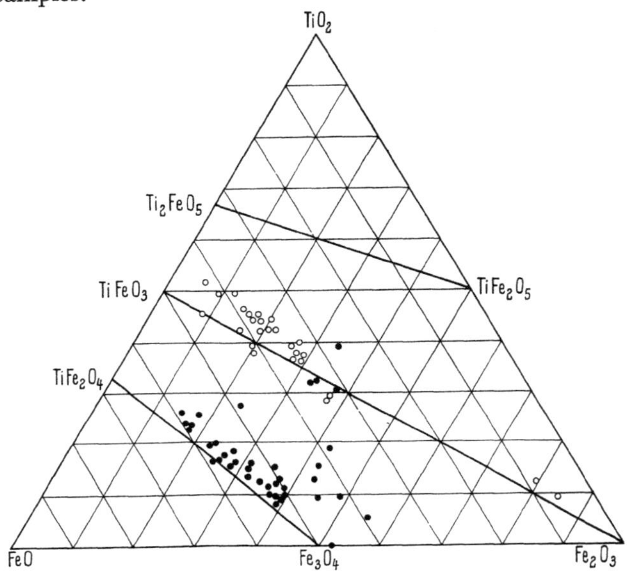

Fig. 9. Chemical composition of natural ferrimagnetic minerals on $FeO-Fe_2O_3-TiO_2$ diagram in mol percent. Full circles: inverse spinel crystal. Hollow circles: rhombohedral crystal. (After NAGATA and AKIMOTO.)

It may be worth while to note that the value of Fe/(Fe + Ti) of titanomagnetites and titanomaghemites in igneous rocks decreases with decrease in those rocks' acidity.

ε) $FeTiO_3$—αFe_2O_3 series (Hemo-ilmenite and titano-haematite). Since the first success of artificial synthesis of solid solution between ilmenite ($FeTiO_3$) and haematite (αFe_2O_3), made in 1934 by BARTH and POSNJAK [*11*], it had been believed that there would be complete solid solution throughout the whole range between $FeTiO_3$ and Fe_2O_3 at a certain high temperature. Recently, synthesis and crystallographic studies of a part of this series was carried out by POUILLARD [*6*], and then the synthesis of the whole range of this series and their crystallographic and magnetochemical studies were completed by NAGATA and AKIMOTO [*12*]. Further, particular magnetic properties of this series have been cleared in detail by ISHIKAWA and AKIMOTO [*13*], [*14*] and BOZORTH *et al.* [*15*], [*16*].

Crystal structure of $x FeTiO_3 \cdot (1-x) Fe_2O_3$ is always rhombohedral throughout the whole range of x, but their magnetic properties are separated into two groups, namely, those of antiferromagnetic behaviour (generally superposed by parasitic ferromagnetism caused by a small amount of deviation from antiferromagnetic configuration of spin) in the range $0.5 \geqq x \geqq 0$, and those of ferrimagnetic behaviour in the range $1 > x \geqq 0.45$, while pure ilmenite ($x = 1$) is antiferromagnetic.

Fig. 10 illustrates the relation between intensity of saturation magnetization (σ) and chemical composition (x). The magnetic differences between the two groups are due to difference in the ordering of Ti ions in sub-lattices. According to result of neutron diffraction test and X-ray analysis of single crystals, it has been found that, in ferrimagnetic state ($1 > x \geqq 0.45$), Ti ions are of order configuration, namely Ti ions have their positions selectively on every other sub-lattice planes perpendicular to the C-axis of crystal, while in antiferromagnetic state ($0.5 > x \geqq 0$), all positions on the sub-lattice planes are occupied equally by Fe ions and Ti ions, namely, configuration of Ti ions is in disorder. Then, configuration of two sets of sublattices, magnetically opposite to each other, in crystals is shown in the first case by

$$(Fe^{3+}_{1-x}Fe^{2+}_{x})\,(Fe^{3+}_{1-x}Ti^{4+}_{x})O^{2-}_{3},$$

its spontaneous magnetization being consequently given by

$$\sigma_0 = x\mu_B P_B\,(Fe^{2+}), \tag{4.3}$$

and in the second case by

$$(Fe^{3+}_{1-x}Fe^{2+}_{x/2}Ti^{4+}_{x/2})\,(Fe^{3+}_{1-x}Fe^{2+}_{x/2}Ti^{4+}_{x/2})O^{-2}_{3},$$

whence $\sigma_0 = 0$. Thus, we better ought to distinguish these two groups from each other and may call minerals of the first case *hemo-ilmenites* and those of the second case *titano-haematites*. A thick straight line in Fig. 10 represents the theoretical ideal value of σ_0 of the ordered state given by (4.3), for the whole range of $0 \leqq x \leqq 1$. It will be seen in the figure that the ordered configuration of Ti is attained approximately at $x = 0.8$ in actual minerals, and for $0.5 < x < 0.8$ the ordering is in transient state between ordered and disordered structures.

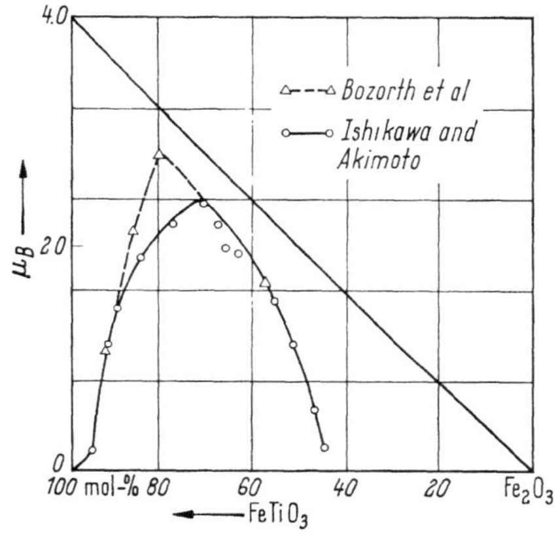

Fig. 10. Saturation magnetization of ilmenite and haematite series at 0° K.

As shown in Figs. 11a, b and 12, lattice parameters and Curie temperature of $FeTiO_3—\alpha Fe_2O_3$ series change continuously with x throughout the whole range of titano-hematites and hemo-ilmenites from Fe_2O_3 to $FeTiO_3$. Fig. 11a, b shows a linear relation between the volume of unit cell of the rhombohedral crystal and α_{rh} for the whole range of x. The Curie temperature curve in Fig. 12 is very smooth for change in x from 675° C of αFe_2O_3 to 55° K of $FeTiO_3$, Θ becoming 0° C approximately at $x = 0.8$. Therefore ferrimagnetism of hemo-ilmenites at the ordinary atmospheric temperature takes place only within the range $0.8 \geqq x \geqq 0.5$.

Magnetic properties of the $FeTiO_3—Fe_2O_3$ series have some particular characteristics, namely,

(a) the curve showing relation between saturation magnetization σ and temperature T of ferrimagnetic hemo-ilmenites is almost a straight line, (see an example given in Fig. 13), and

(b) a specimen having composition about $x = 0.5$ is extremely sensitive to heat-treatment, easily changing from the ferrimagnetic state to the antiferromagnetic state. The latter fact indicates an easy transition of configuration of Fe and Ti ions in the specimen of the critical composition from the ordered state to the disordered state. This order-disorder transformation is in close relation with production of self-reverse thermo-remanent magnetization in rocks, as will be dealt with in Sect. 15c. Fig. 14 shows a magneto-crystallographic phase diagram of this series, obtained by ISHIKAWA, where general characteristics of this series are well summarized.

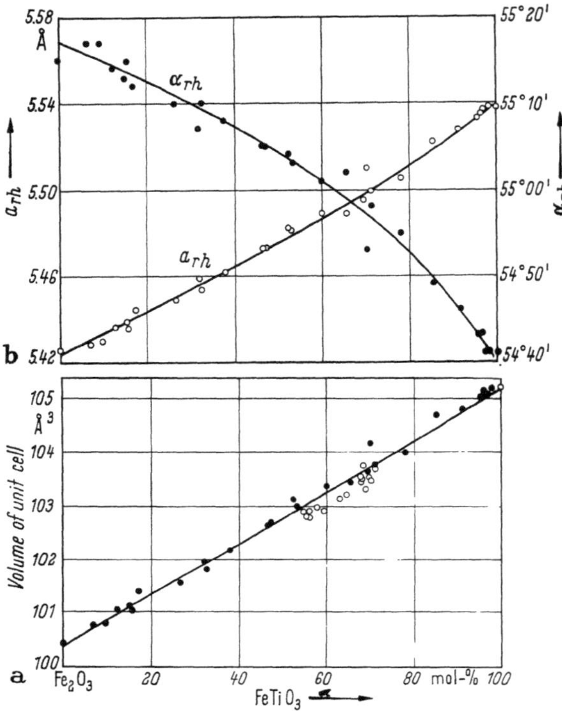

Fig. 11. a) Crystal parameters of ilmenite-haematite series. (After NAGATA and AKIMOTO.) b) Unit cell volume of ilmenite-haematite series. Full circles: synthetic sample. Hollow circles: natural sample. (After NAGATA and AKIMOTO.)

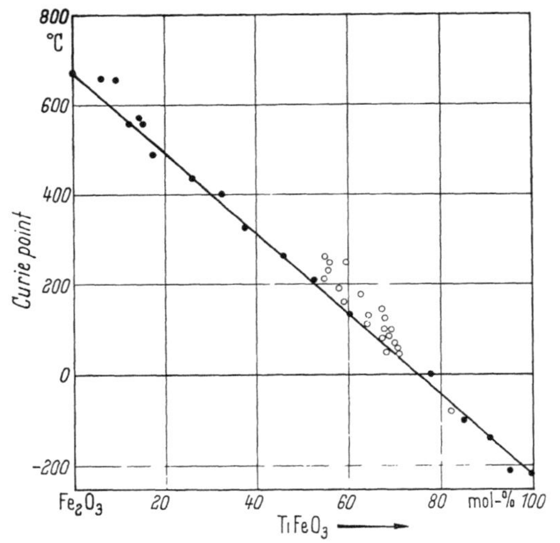

Fig. 12. Curie temperature of ilmenite-haematite series. (After NAGATA and AKIMOTO.)

ζ) Hemo-ilmenites and titanohaematites in rocks. It has been well known that most volcanic rocks contain ilmenites, but they are generally non-magnetic (namely, having composition near $FeTiO_3$). The content of $FeTiO_3—Fe_2O_3$ series

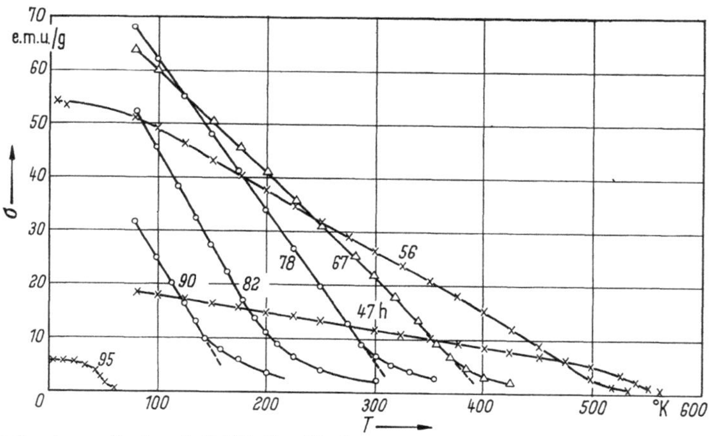

Fig. 13. Intensity of magnetization of $x FeTiO_3 \cdot (1-x) Fe_2O_3$ vs. temperature in $H = 8500$ Oe. Numerals indicate x in mol percent. (After UYEDA.)

mineral in volcanic rocks is definitely small compared with those of titanomagnetities and titanomaghaemites, the ratio of the former to the latter being less than 20%. Therefore, magnetic properties of volcanic rocks depend chiefly upon those of the spinel type ferrimagnetic minerals and little upon those of the $FeTiO_3—Fe_2O_3$ series, except some particular cases such as self-reversal of thermo-remanent magnetism.

On the contrary in plutonic rocks, the ratio of content of $FeTiO_3—Fe_2O_3$ series to that of titanomagnetites and titanomaghaemites is much larger than in the case of volcanic rocks, and consequently the contribution of hemo-ilmenites to magnetism of plutonic rocks is appreciable.

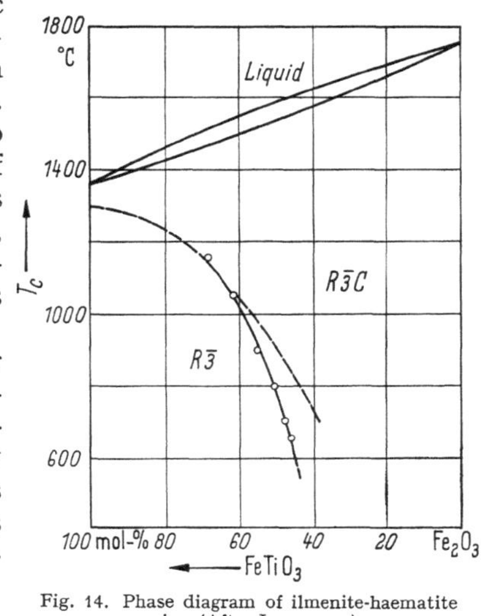

Fig. 14. Phase diagram of ilmenite-haematite series. (After ISHIKAWA.)

In metamorphic rocks, most parts of their natural remanent magnetization can be attributed to magnetization of the ilmenite-haematite series, as pointed out by BUDDINGTON and BALSLEY [*17*], [*18*], NAGATA and others [*19*], [*20*]. It seems therefore that this series is of special significance in magnetism of metamorphic rocks.

In some sedimentary rocks such, for example, as red sandstone, most parts of magnetic minerals contained in them are haematites or titanohaematites having little amount of Ti ions. As for magnetic properties of this kind of rock, only titanohaematites play an important role.

In Fig. 9, the chemical compositions of ferrimagnetic minerals of rhombohedral structure contained in volcanic rocks of Japan are plotted by hollow circles. These natural minerals are ascertained by means of X-ray analysis to be composed of rhombohedral structure phase alone. As will be seen in the figure, natural minerals of rhombohedral structure have a chemical composition on or quite near the $FeTiO_3-Fe_2O_3$ join. This is in remarkable contrast with the case of spinel type ferrimagnetic minerals.

η) $FeTi_2O_5-Fe_2TiO_5$ series. In addition to spinel type and rhombohedral type ferrimagnetics, there is a solid solution series of $Fe_2TiO_5-FeTi_2O_5$ in the $FeO-Fe_2O_3-TiO_2$ system (AKIMOTO, NAGATA and KATSURA [21]). The $xFe_2TiO_5\cdot(1-x)\,FeTi_2O_5$ series has an orthorhombic crystal structure and is paramagnetic

Table 1.

Composition $FeTi_2O_5$ mol-% %	Lattice parameters a Å	b Å	c Å	Volume of unit cell Å³	χ at 20° C emu/g
0	9.767	9.947	3.717	361.1	$43\cdot10^{-6}$
26	9.787	9.965	3.734	364.2	$55\cdot10^{-6}$
50	9.795	9.987	3.739	365.8	—
66	9.800	10.011	3.741	367.0	$58\cdot10^{-6}$
82	9.801	10.031	3.745	368.0	$48\cdot10^{-6}$
100	9.798	10.041	3.741	368.0	$53\cdot10^{-6}$

at the atmospheric temperature for the whole range of x, as shown in Table 1, but it is expected that this series becomes antiferromagnetic below a certain low temperature.

The crystal of Fe_2TiO_5 has been called pseudobrookite, and it has been pointed out (BASTA [22], CURNOW and PARRY [23], [24], UYEDA [25]) that this $Fe_2TiO_5-FeTi_2O_5$ series can be produced by oxidation of $FeTiO_3-Fe_2O_3$ series minerals or further of $Fe_2TiO_4-Fe_3O_4$ series minerals. In nature, however, minerals of this series were found occasionally only in strongly oxidized igneous or metamorphic rocks.

From the viewpoint of a general description of magnetic properties of the whole system of $FeO-Fe_2O_3-TiO_2$, it may be concluded that the $Fe_2TiO_5-FeTi_2O_5$ line composes, together with the $FeTiO_3-Fe_2O_3$ and $Fe_2TiO_4-Fe_3O_4$ lines, three solid solution lines in this system, but from the practical viewpoint, minerals of this series are very rare, and even if there are some of them in rocks, their magnetic effect is negligible.

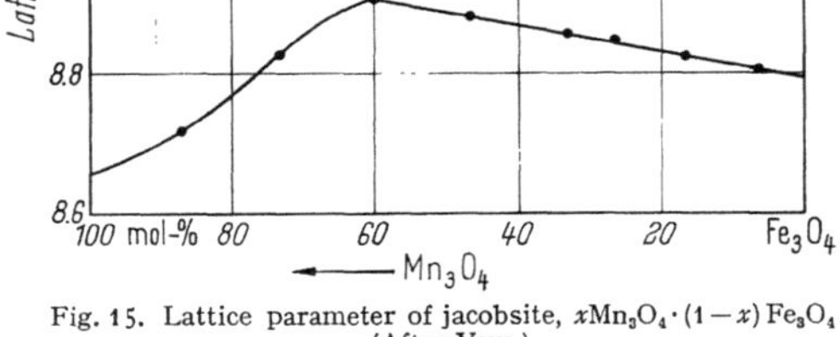

Fig. 15. Lattice parameter of jacobsite, $xMn_3O_4\cdot(1-x)\,Fe_3O_4$. (After YUN.)

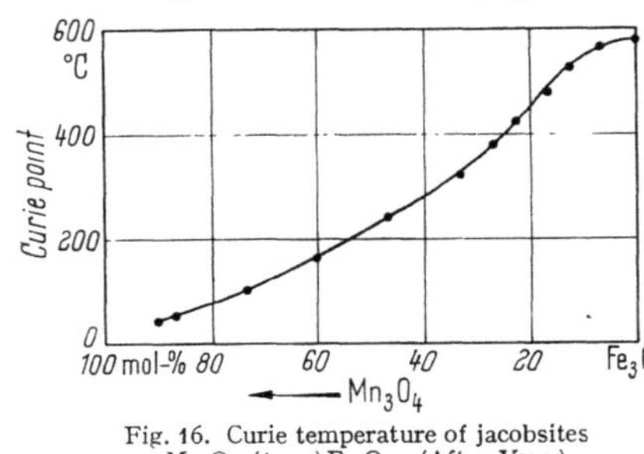

Fig. 16. Curie temperature of jacobsites $xMn_3O_4\cdot(1-x)\,Fe_3O_4$. (After YUN.)

5. $Fe_3O_4-Mn_3O_4$ series (Jacobsites). Solid solutions between magnetite (Fe_3O_4) and *hausmannite* (Mn_3O_4) have been called "*jacobsite*". Jacobsites are frequently found in manganic ores, and they have sometimes a very strong intensity of natural remanent magnetization. So far as magnetism of rocks and minerals are concerned, therefore, jacobsites are important metallic oxides, being next to those of the $FeO-Fe_2O_3-TiO_2$ system.

Crystallographic and magnetic properties of this series, obtained by YUN [*26*], are illustrated in Figs. 15 and 16. It is seen in Fig. 15 that solid solution $x Mn_3O_4 \cdot (1-x) Fe_3O_4$ of spinel structure can exist for the range $0 \leqq x < 0.6$, but for $x > 0.6$ its crystal structure becomes tetragonal. As shown in Fig. 16, the Curie temperature Θ decreases continuously with increasing x, but the intensity of saturation magnetisation σ shows a maximum at $x = 0.167$, being 4.5 μ_B per mol.

Table 2. *Divalent ion substituted magnetites*

x	Substituting ion							
	Co		Ni		Mg		Ca	
%	Θ	a Å	Θ	a Å	Θ	a Å	Θ	a Å
0	570	8.38	570	8.38	570	8.38	570	
10							570	
20					555			
25	555	8.375						
30							560	
40			575	8.36			550	
50	540	8.37						
60			580	8.35	530	8.37	420	
70							445	
75	530	8.36			490			
85							460	
100	520	8.355	590	8.34	315	8.37	480	

6. Other substituted magnetites. Ferrimagnetic minerals of metallic oxides found in nature contain sometimes Mg^{2+}, Al^{3+}, V^{3+} and other metallic ions in addition to Fe^{2+}, Fe^{3+}, Ti^{4+}, Mn^{2+} ions which have already been dealt with in the preceding sections. In most cases, Fe^{2+} and Fe^{3+} in the ferrimagnetic minerals are substituted by these divalent and trivalent ions respectively. Although the amount of these metallic ions in the ferrimagnetic minerals is, generally speaking, very little compared with that of Fe^{2+}, Fe^{3+} and Ti^{4+}, they still affect more or less magnetic and crystallographic characteristics of the minerals.

Table 3. *Trivalent ion substituted magnetite*

x	Substituting ion			
	Al		Cr	
%	Θ	a Å	Θ	a Å
0	575	8.41	575	8.41
9	560	8.39		
13.5			569	8.38
15	548	8.38		
21	545	8.37		
30	535	8.36	563	8.34
36			560	8.31

The basic forms of the ferrimagnetic minerals containing Mg, Al, etc, are such where Fe^{2+} or Fe^{3+} in magnetite is substituted by these ions, and they are called *substituted magnetites*. If any of these substituting metallic ions is denoted by M, then the compositions of the substituted magnetites are expressed by

$$(Fe_2^{3+} Fe_{1-x}^{2+} M_x^{2+}) O_4^{2-} \quad \text{(substituted by divalent ions)},$$

or

$$(Fe_{2-x}^{3+} Fe^{2+} M_x^{3+}) O_4^{2-} \quad \text{(substituted by trivalent ion)}.$$

According to result of researches by MICHEL, CHAUDRON and BÉNARD [*27*], and POUILLARD [*6*], the Curie temperature Θ and the lattice parameter a of substituted magnetites are given as in Tables 2 and 3.

Few studies have been made on substituted magnetites found in nature. However it may be presumed that there are substituted maghemites also in nature in addition to the substituted magnetites.

7. Pyrrhotites (FeS_{1+x}). Most of natural pyrrhotites have more or less spontaneous magnetisation, so that they have long been known as "a natural permanent magnet" together with magnetites. However, pyrrhotites exist in nature as ore minerals mostly together with *pyrite* (FeS_2), and very rarely take place in natural rocks. Only few examples, such as tuff in Tasmania, of rocks containing pyrrhotites have been reported so far.

The magnetic properties of pyrrhotite are so peculiar that a large number of scientists have engaged in experimental and theoretical researches of this particular mineral. The latest knowledge of its magnetic properties and their interpretations are as follows.

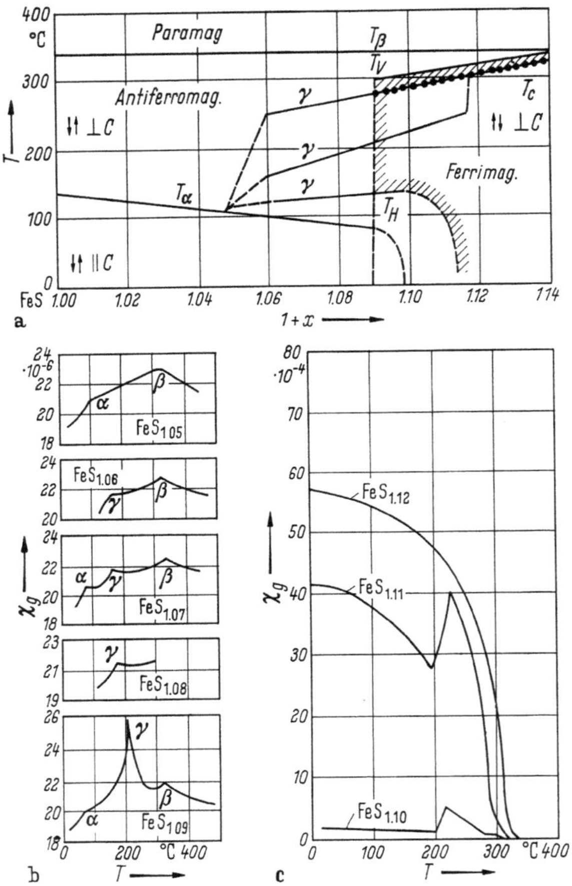

Fig. 17a—c. Magnetic phase diagram of pyrrhotite.

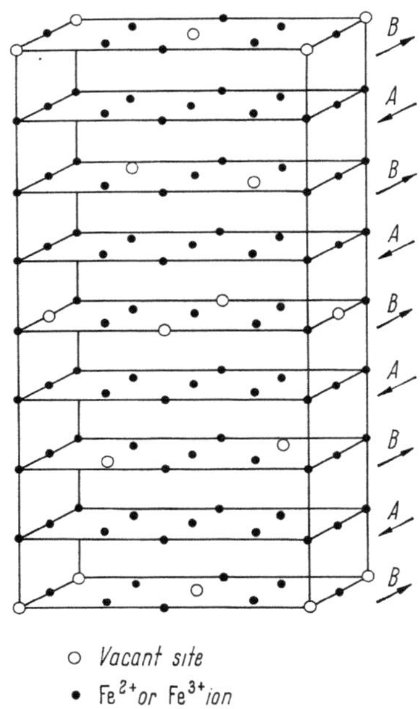

Fig. 18. Crystal structure of pyrrhotite. (After BERTAUT.)

Pyrrhotite, FeS_{1+x} ($0 \leqq x \leqq \frac{1}{7}$), has an excess amount x of sulphur ions compared with FeS, but when $x > \frac{1}{7}$, FeS_{1+x} is not of a single phase but is composed of two phases together with pyrite (FeS_2).

Pyrrhotites of $0 \leqq x \leqq 0.1$ are antiferromagnetic, while those of $0.1 < x < \frac{1}{7}$ are ferrimagnetic, their spontaneous magnetization being about $0.25\ \mu_B$/mol. Pyrrhotites of transient composition between the above-mentioned two, namely, those of $x \sim 0.1$, show ferrimagnetic characteristics for a certain temperature range.

The crystal structure of pyrrhotite is NiAs type hexagonal, having C 6 mc crystal symmetry, and the direction of its spontaneous magnetisation is parallel to the C-plane which is perpendicular to the C axis. Since positions of the cation in the NiAs type crystal become vacant, according as x increases, it may be reasonable to express composition of pyrrhotite by $Fe_{1-\delta}S$ ($\delta > 0$) rather than by FeS_{1+x}.

Fig. 17 illustrates the phase diagram and magnetic properties of pyrrhotites, mainly according to HARALDSEN [28]. As shown in Fig. 17a, FeS has two transitions, α and β, and according as x (or δ) increases, T_β is kept nearly constant at 320° C, but T_α decreases and looks like disappearing at $FeS_{1.09}$ (or $Fe_{0.917}S$). The α transition corresponds to a change of direction of the easiest magnetisation (of spin of Fe ions) from the C-axis to the C-plane, resulting in change in magnetic anisotropy, and this transition is accompanied by a change in electric conductivity. On the other hand, the β transition corresponds to an order-disorder transition in spin configuration, accompanied by an anomaly in specific heat; in fact, it is the NÉEL point of antiferromagnetics.

When x of FeS_{1+x} becomes larger than in $FeS_{1.06}$, a peak develops between the temperatures of α and β transitions on a curve showing the antiferromagnetic susceptibility of FeS_{1+x} as a function of temperature, this peak being called γ-anomaly (see Fig. 17a and b). The magnitude of magnetisation at the γ-anomaly peak increases with increasing x, and at about $x \simeq 0.10$ it attains to spontaneous magnetization of Λ shape with respect to temperature (Fig. 17c); then the Λ shape magnetisation is superposed by an ordinary Weiss type magnetisation for the range of $x = 0.10$ to 0.20; for further increase in x (i.e. $x > 0.20$), the Λ shape magnetisation disappears, the Weiss type magnetisation alone remaining.

As for the origin of ferrimagnetism of pyrrhotites, BERTAUT [29] has found regular configuration of vacancy of Fe ion position in a single crystal of Fe_7S_8, as shown in Fig. 18, at temperatures below the β-transition. As will been seen in the figure, the numbers of Fe ions in A- and B-layers are different from each other, resulting in appearance of ferrimagnetism when spin configuration becomes an antiferromagnetically ordered state.

Denoting vacancy by $\square$, $F_{1-\delta}S$ can be written as

$$Fe^{2+}_{1-3\delta}Fe^{3+}_{2\delta}\square_\delta S^{2-}.$$

Comparing this general expression with Fig. 18, it may be noticed that there is still freedom for distribution of Fe^{2+} and Fe^{3+} ions either in the A-layer or the B-layer which contains the vacancy. Final configuration, which has been obtained by taking into account magnetic properties, is given by

$$Fe_{1-\delta}S = \{Fe^{2+}_{\frac{1}{2}}\}\,[Fe^{2+}_{\frac{1}{2}-3\delta}Fe^{3+}_{2\delta}\square_\delta]S^{2-},$$

where { } means the A-layer, and [ ] the B-layer. Then the spontaneous magnetisation for Fe_7S_8 ought to be 2 μ_B/mol, which is in agreement with the observed value.

An order-disorder transition temperature (T_v) of the vacancy in FeS_{1+x} was found by LOTGERING [30], the result being shown in Fig. 17a also. As seen in Fig. 17a, $T_v < T_\beta$ for $x = 0.09$ to 0.14, and therefore FeS_{1+x} is antiferromagnetic between T_β and T_v, and is ferrimagnetic below T_v even in the range of $x > 0.10$.

II. Magnetic properties of assemblage of small grains of ferrimagnetic minerals.

8. Introduction. Concerning their magnetic properties, rocks can be considered such materials that a large number of ferrimagnetic minerals, as described before, of various different sizes and shapes are scattered over a practically non-magnetic medium composed of silicate minerals.

In *igneous* rocks, phenocrysts of the ferrimagnetic minerals are generally of size of order of $10^2\,\mu$, while the mean diameter of microcrystals of those minerals in groundmass ranges from 1 to 10 μ.

In *sediments*, the size of ferrimagnetic minerals ranges from a millimeter to $10^{-2}\,\mu$ according to condition of circumstances during the production of the sedi-

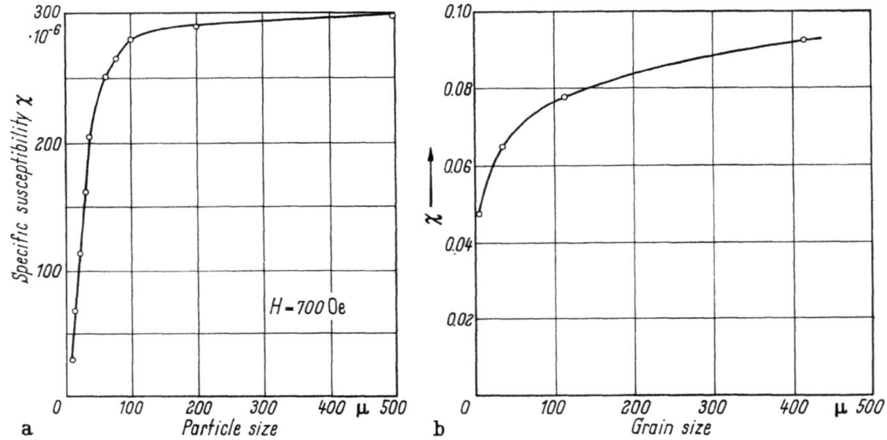

Fig. 19. a) Specific magnetic susceptibility of αFe_2O_3 as dependent on mean diameter of particle. (After CHEVALLIER and MATHIEU.) b) Specific magnetic susceptibility of Fe_3O_4 as dependent on mean diameter of particle. (After AKIMOTO.)

ments. As examples of sediments containing very fine grains of ferrimagnetic minerals (mean diameter $10^{-2}\,\mu$), those of varved clay in New England, U.S.A., and red sandstone in England may be pointed out.

Metamorphic rocks sometimes contain fairly large single crystals of magnetite. For example, single crystals of magnetite of octahedral shape of several mm to 1 cm in mean diameter can be frequently found in chlorite shists. Further, in some sedimentary and metamorphic rocks, an anisotropic distribution of particular minerals is remarkable, such, for example, as in the case of shistosity of gneiss. Magnetic properties of rock, as a whole, depend not only on those of the ferrimagnetic minerals it contains but also on shape, size and distribution of grains in the rock. Problems of magnetic properties of assemblage of such ferrimagnetic minerals will be dealt with in this chapter.

9. Dependence of magnetic susceptibility and coercive force on grain size. Magnetic susceptibility of ferromagnetic or ferrimagnetic grains becomes smaller according as their grain size becomes smaller. As shown in Fig. 19a and b, the susceptibility of α-Fe_2O_3 powders [*31*] and titanomagnetite grains [*32*] decreases remarkably in accordance with decrease in grain-size below 100 μ. On the other hand, the magnetic coercive force of those grains becomes larger according as the grain size becomes smaller, as shown in Fig. 20 [*33*] for example. A sharp increase in coercive force with decrease in grain size takes place for the same range as in case of the sharp change in susceptibility, namely, less than several ten μ in mean diameter. The relation between coercive force, H_c, and mean diameter of grain, d, can be approximately expressed by $H_c \sim d^{-1}$.

A qualitative interpretation of the dependency of susceptibility and coercive force upon grain size may be as follows: A large grain is composed of many magnetic domains, and the number of domains decreases as grain-size decreases, becoming a single domain for extremely small grain-size. In a grain of multidomain structure, magnetization in a weak field is mostly due to displacement of the boundary walls between domains, while in a single domain particle, magnetization can only be achieved by its magnetic rotation. Since energy for the domain rotation is larger than that of the wall displacement, the magnetization per unit volume of the smaller grains, in which the number of domains and consequently that of the walls are smaller, necessitates larger external magnetic field than in case of the larger grains. This condition may be the cause of the lower susceptibility of the smaller grains. In a quite similar way, the higher coercivity of the smaller grains might be interpreted.

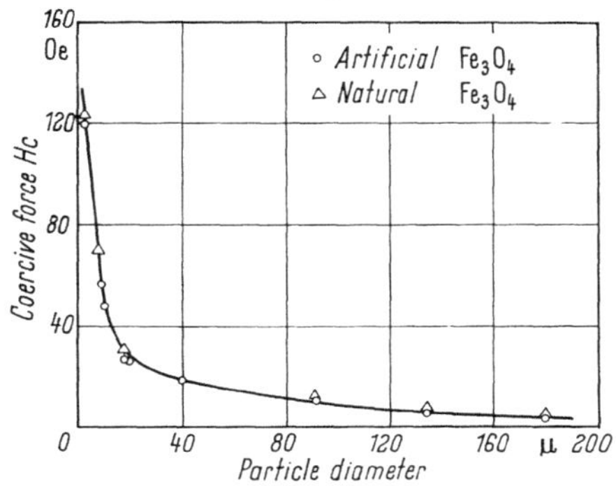

Fig. 20. Coercive force of magnetite as dependent on grain size. (After GOTTSCHALK.)

10. Initial susceptibility of an assemblage of large grains. The initial magnetic susceptibility $\varkappa_a$ of an assemblage of magnetic grains, mean initial susceptibility and mean demagnetizing factor of which are $\varkappa$ and $\bar{n}$ respectively, is given approximately by

$$\varkappa_a = \frac{p\varkappa}{1+\bar{n}\varkappa} \tag{10.1}$$

provided that these magnetic grains are scattered over a non-magnetic medium in such a way as in natural rocks and consequently the ratio of their total volume to the bulk volume, p, is sufficiently small compared with unity ([*34*] to [*37*]). Here, the expression „a large grain" means a grain composed of a large number of magnetic domains. When it is taken into consideration that the average shape of magnetic minerals can well be represented by a triaxial ellipsoid, demagnetizing coefficients of which are given by n_1, n_2 and n_3 for the three principal axis-directions, and further that a large number of triaxial ellipsoids are scattered with random orientation, the bulk susceptibility, $\varkappa_a$, is expressed by

$$\varkappa_a = \frac{p\varkappa}{3} \sum_{i=1}^{3} \frac{1}{1+n_i\varkappa}. \tag{10.2}$$

In most cases, where the average shape of magnetic minerals can be roughly represented by a spheroid, (10.2) becomes

$$\varkappa_a = \frac{p\varkappa}{3}\left(\frac{1}{1+n_1\varkappa} + \frac{2}{1+n_2\varkappa}\right), \tag{10.2a}$$

where n_1 denotes the demagnetizing factor for the direction along the revolution axis, and n_2 for the direction perpendicular to the former.

In some sedimentary and metamorphic rocks, the direction of the major axis of the ellipsoid is not at random but is statistically predominant into a particular direction. In some sediments, for example, a large number of magnetic grains have their major axis horizontally. In such a case the bulk susceptibility is of anisotropic character. If we assume, for simplicity, that the ellipsoid is a prolate spheroid, the demagnetizing factors of which are n_1 and n_2 respectively for major and minor axes, where $n_1 < 4\pi/3 < n_2$, and that the major axes of all grains are parallel to each other, then the bulk susceptibility into a direction making an angle θ with the direction of the major axis becomes

$$\varkappa_a = p\varkappa\left(\frac{\cos^2\theta}{1+\varkappa n_1} + \frac{\sin^2\theta}{1+\varkappa n_2}\right). \tag{10.3}$$

If the major axes of all the prolate spheroids are distributed at random but always horizontally, the bulk susceptibility in a direction making an angle θ with the horizontal plane is given by

$$\varkappa_a = p\varkappa\left[\frac{\sin^2\theta}{1+\varkappa n_2} + \frac{\cos^2\theta}{2}\left(\frac{1}{1+\varkappa n_1} + \frac{1}{1+\varkappa n_2}\right)\right]. \tag{10.4}$$

Feasibility of these expressions (10.1 to 3) has been proved [*38*] by examining the magnetic susceptibility of rocks and the rock-forming minerals as well as artificially made scattered assemblages of grains of natural magnetic minerals. According to the results, $\bar{n} = \left\{3\Big/\sum_i \frac{1}{1+n_i\varkappa} - 1\right\}\Big/\varkappa$ is smaller than the value of the demagnetizing factor of a sphere, namely $n = 4\pi/3$; $\bar{n}$ amounts to between 3.0 and 3.5 generally. This means that the ratio of the major diameter to the minor of a prolate spheroid representing the average shape of magnetic minerals is between 3 and 5, since the initial susceptibility of titanomagnetites $x\mathrm{TiFe_2O_4}(1-x)\,\mathrm{Fe_3O_4}$ of $x \ll 1$, which are the most common ferrimagnetic minerals, is about 0.4 e.m.u./cm³.

Anisotropy of magnetic susceptibility of sediments amounts sometimes to 20% or more in the ratio of $\varkappa_a$ into the direction of the easiest magnetization to into that of the most difficult magnetization. Such anisotropy could be kinematically interpreted by referring to (10.3) or (10.4) for individual cases.

Generally speaking, the magnetic susceptibility of most rocks can be interpreted as that of an assemblage of large magnetic grains, as discussed in this section.

11. Isothermal remanent magnetization of an assemblage of large grains. Remanent magnetization which is produced by applying a magnetic field upon a specimen throughout a constant temperature will be called *isothermal remanent magnetization* (IRM). The meaning of the isothermal remanent magnetization is identical to that of the term "remanent magnetization", ordinarily used in physics of ferro- and ferrimagnetic materials. However, this term has been especially used in the science of magnetism of rocks and assemblages of small grains of ferromagnetics and ferrimagnetics in order to distinguish it from thermo-remanent magnetization which is produced in these materials by cooling from a certain high temperature in a magnetic field (Sect. 15).

Results of actual measurements of isothermal remanent magnetization, J_R, and coercive force H_C at the atmospheric temperature indicate that the ratio of IRM to the saturation magnetization, J_S, i.e. J_R/J_S, is roughly proportional to H_C, as illustrated in Fig. 21 [*38*]. The observed IRM of rocks is of course the sum of the magnetizations (j_R) of all grains at $H=0$.

However, even where $H=0$, the effective magnetic field H_{eff} for each individual magnetic grain is not zero, but is given by $-\bar{n}j_R$ provided that $\bar{n}$ denotes the average demagnetizing factor of an individual grain [*39*]. Then, in a diagram showing the magnetization curve of an individual grain, as shown in Fig. 21, J_R is given by the intensity of magnetization j at the intersection of the magnetization curve with a straight line, $h=-\frac{1}{n}j$. When n is sufficiently large, the value of H corresponding to $H_{\text{eff}}=0$ is nearly the same as $-H_C$, and therefore

$$J_R/J_S \simeq H_C/n\,J_S. \tag{11.1}$$

(11.1) well represents the observed relation between J_R/J_S and H_C, and in the cases shown by Figs. 20 and 21, the coefficient $1/n\,J_S$ amounts to about $1/6\times10^2$ e.m.u. Since the average of the observed values of J_S of titanomagnetite grains having multidomain size is 230 e.m.u., n is estimated to be about 2.6, which is a reasonable value for the demagnetizing factor of magnetic grains in rocks. It may be concluded from the above result that the IRM of most rocks, in which the coercive force does not exceed several hundred Oersteds, can be interpreted as that of an assemblage of a large number of multidomain grains.

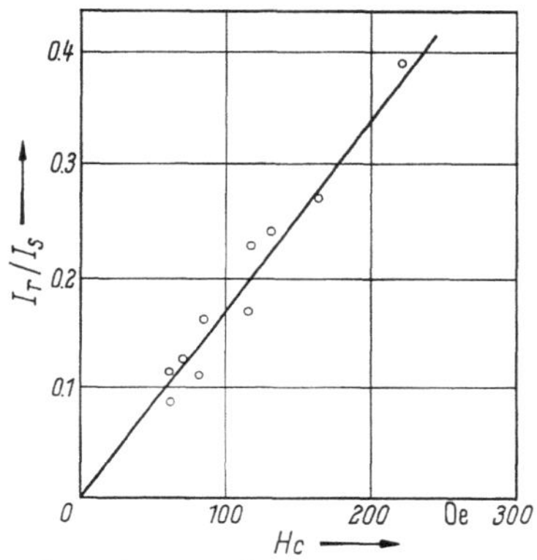

Fig. 21. The ratio J_r/J_s vs. coercive force H_c in igneous rocks. (After NAGATA.)

12. Magnetization of single-domain grains. In rocks, there are frequently a large number of very small grains of ferrimagnetic minerals present besides their large grains, and the former sometimes play a very important role in magnetic properties of rocks, especially in relation to remanent magnetization. A multidomain grain of ferrimagnetic minerals is divided by walls into individual domains. The wall thickness of ferromagnetic and ferrimagnetic materials is theoretically estimated to be $10^2\sim10^3$ Å, and the estimated values are in agreement with directly measured values. In case that the linear dimension of a magnetic grain becomes nearly equal to, or smaller than, this wall thickness, the grain itself can be nothing but a single domain. The critical linear size for a single domain grain of Fe_3O_4 is less than 0.1 μ, probably being 0.02 to 0.03 μ.

Magnetization within a single domain is uniform, its intensity being equal to that of *spontaneous magnetization*, J_S, and change in magnetization can be achieved only by its rotation. In this case, the internal magnetization energy of the single domain grain depends only on the angle of magnetization direction with a coordinate fixed to the grain. As a simple example, the magnetization energy E of a magnetically uniaxial single domain grain of volume v is given by

$$E=Kv\sin^2\theta, \tag{12.1}$$

where θ denotes the angle between magnetic direction and the axis and K is an anisotropy constant, which can be interpreted as due to the following three different origins.

(1) *Magnetocrystalline anisotropy*; this anisotropy in a small grain is the same as in a large crystal of the same material, being subject to orientation of electronic spins with respect to the crystal axes.

(2) *Shape anisotropy*; this depends on the shape of the single domain grain. For example, when the grain is a prolate spheroid, demagnetizing factors of which are N and M respectively along the major and minor axes, K is given by

$$K=\tfrac{1}{2}(M-N)\,J_S^2. \tag{12.2}$$

(3) *Anisotropy caused by mechanical stress*; K in this case is expressed by

$$K=\tfrac{3}{2}\lambda\sigma, \tag{12.3}$$

where σ is the tension parallel to the axis, and λ the longitudinal saturation magnetostriction.

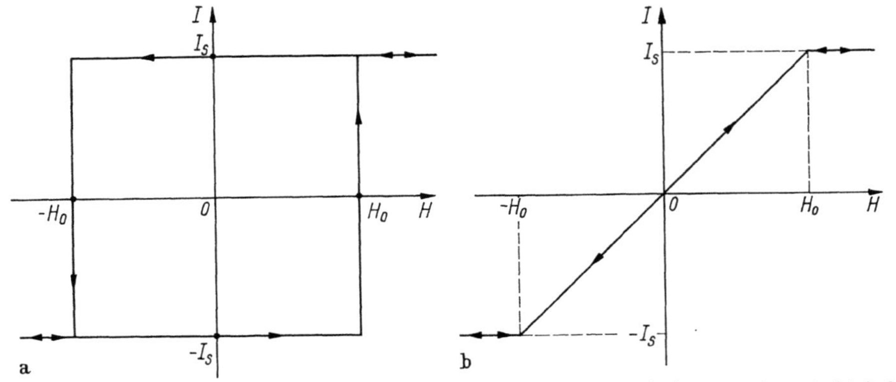

Fig. 22. a) Magnetization curve of an uniaxial single domain particle ($\theta=0$). b) Magnetization curve of an uniaxial single domain particle ($\theta=\pi/2$).

In actual cases, one of these anisotropies may be sometimes predominant in determining the K-value, but in some other cases all of these three may work together.

The curve of magnetization *hysteresis cycle* of the magnetically uniaxial specimen depends markedly upon the angle between the axis and the direction of the external magnetic field, H. If the major axis of the prolate spheroid is parallel to H, the hysteresis cycle has a rectangular form, as illustrated in Fig. 22a, J being either J_S or $-J_S$ between $H=-H_C$ and $H=H_C$, and J changes discontinuously at $H=-H_C$ and H_C where $H_C=2K/J_S$. On the other hand, when H is perpendicular to the axis, there exists no hysteresis phenomenon at all in the magnetization cycle as illustrated in Fig. 22b. The magnetization curve of an assemblage of such grains of random direction is shown in Fig. 23, where $J_R=\tfrac{1}{2}J_S$, $H_C=0.96\,K/J_S$, and the value of initial susceptibility is given by $J_S^2/3K$. It must be noted here that J_R in this case is independent of H_C, being $\tfrac{1}{2}J_S$. It seems very rare in natural rocks that all ferrimagnetic minerals are of *single domain*, having the average characteristics as mentioned here; it seems that most natural rocks contain a fair amount of ferrimagnetic minerals of the *large grain size*.

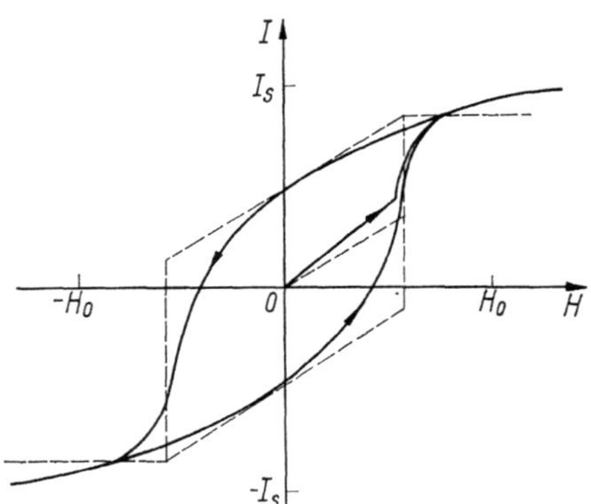

Fig. 23. Average magnetization curve of an assemblage of uniaxial single domain particles.

13. Magnetization of very fine grains (superparamagnetism). If the size of single domain grains becomes very small, the magnetic energy of a grain, Kv, (given by putting $\theta=\pi/2$ in (12.1)) becomes in order of magnitude comparable with the energy of thermal fluctuation of the grain's magnetic moment, kT. In such a case, an assemblage of the particles can achieve thermal equilibrium in a relatively short time. This behaviour is called "*superparamagnetism*".

The theory of this superparamagnetism as proposed by NÉEL [*40*], [*41*] may be outlined as follows: Consider an assemblage of uniaxial particles fully magnetized along the easy magnetization axis. When the magnetic field is removed, the thermal fluctuation energy kT makes the direction of each particle undergo a Brownian rotation against the energy barrier Kv. The resultant remanence J_r of the assemblage will vanish with time t after the removal of the field in the form of

$$J_r = J_0 \exp(-t/\tau_0), \tag{13.1}$$

where J_0 is the initial full magnetization and τ_0 is a quantity called relaxation time for the process. The relaxation time τ_0 depends upon the mechanism by which the thermal energy is transformed into the rotational energy of the magnetic moment as well as upon the energy barrier Kv and the thermal energy kT.

According to NÉEL, the relaxation time τ_0 in case of an assemblage of single domain particles is expressed by

$$1/\tau_0 = f_0 \exp(-Kv/kT), \tag{13.2}$$

with

$$f_0 = \frac{2eK}{mJs}(3G\lambda + DJs^2)\sqrt{\frac{2v}{\pi GkT}}, \tag{13.3}$$

where G is the shear modulus, λ the longitudinal saturation magnetostriction, D a constant depending on the demagnetization factor, and m and e the mass and charge of the electron (see more detail in Sect. 15b).

In the ordinary cases, f_0 is of the order of $10^9\,\text{sec}^{-1}$. Magnitudes of τ_0 thus calculated for various grain size of magnetite are given in Table 4. This shows that the remanent magnetization of magnetite particles of less than 300 Å average radius vanishes very rapidly. Actually CREER [*42*] has reported that remanent magnetization of Keupper Marl in England, which contain very fine grains of ferrimagnetic minerals, is extremely unstable and shows the behaviour of superparamagnetism.

Table 4. *Dependence of τ_0 on v/T and on the radius R of a particle at $T = 300°$ K*

τ_0	10^{-1}	10	10^3	10^5	10^7	10^9	10^{15}	(sec)
v/T	2.5	3.2	3.8	4.4	5.1	5.7	7.6	(cm³/degree)
R at $T = 300°$ K	260	280	300	320	330	340	380	(Å)

For the process of magnetization of the superparamagnetic substances in an external magnetic field, H_{ex}, the expression of τ_0 becomes much more complicated than (13.2) and (13.3), but the fundamental character of τ_0 in this case does not much differ from that in the case of demagnetization of remanent magnetization in $H_{ex} = 0$ regions (see more detail in Sect. 15b).

III. Remanent magnetism of rocks.

14. Introduction. It has long been known that some natural rocks, especially igneous and metamorphic rocks, have sometimes an extremely strong remanent magnetization. It was once believed that most of such strong remanent magnetization of the rocks is due to isothermal remanent magnetization produced by an instantaneous strong magnetic field such as caused by thunder lightnings, and in fact it was actually proved for several cases that strong remanent magnetization of rocks was indeed caused by lightnings.

However, several other processes resulting in fairly strong remanent magnetization of rocks under influence only of the earth's weak magnetic field have been discovered. Physical effects which work together with the weak magnetic field in the course of producing remanent magnetization are (a) thermal, (b) chemical and (c) dynamical. Therefore, remanent magnetizations thus produced are called respectively

(a) *Thermo-remanent magnetization* (Abbreviation: TRM);

(b) *Chemical remanent magnetization* (CRM);

(c) *Pressure remanent magnetization* (PRM).

In addition to the above-mentioned three different kinds of remanent magnetization, it has been found that rather weak remanent magnetization is produced by statistical alignment of a large number of small magnetized grains in fluid or in viscous substance under the effect of a magnetic field. This is called

(d) *Detrital remanent magnetization*; (DRM),

and this is the cause of remanent magnetization in most sedimentary rocks.

Rocks can also have the ordinary remanent magnetization, which is acquired by the ordinary process of making a hysteresis loop at a constant temperature. This is called

(e) *Isothermal remanent magnetization* (IRM).

IRM, acquired in the geomagnetic field, however, is generally very weak compared with the other four remanent magnetizations acquired in the same magnetic field. When an alternating magnetic field $\widetilde{H}$ is superposed upon a constant magnetic field h in the magnetizing procedure, an acquired remanent magnetization becomes much stronger and more stable than the ordinary IRM acquired in the constant field h alone. The remanent magnetization thus acquired is called

(f) *Anhysteretic remanent magnetization* (ARM).

In the above-mentioned six different processes of acquisition of remanent magnetization, the effect of *time* required for individual processes is not taken into account. In other words, each procedure is assumed to be completed within a sufficiently short time, the resultant remanent magnetization being apparently independent of the time required for the procedure.

All kinds of remanent magnetization, however, decrease more or less in their intensity with time, generally the more so the higher is the temperature T of the specimens. If further a specimen is kept in a magnetic field during an appreciably long time at $T > 0°$ K, the resultant remanent magnetization becomes larger than the ordinary IRM of the same specimen adopted in a short time. This effect of time caused by thermal agitation is called *magnetic viscosity*, and the additional remanent magnetization due to the magnetic viscosity is called

(g) *Viscous remanent magnetization* (VRM).

The seven different types of remanent magnetization mentioned above are now considered to be the basic types of the origin of natural remanent magnetization of rocks.

Some additional reasons for remanent magnetization of rocks may be considered, though they might possibly be classified into one of the seven basic types. One of them is the

(h) *Inverse thermo-remanent magnetization* (ITRM),

which is acquired by *heating* a specimen from a certain low temperature in a magnetic field.

Another is the

(i) *Thermo-chemical remanent magnetization* (TCRM),

which is a type of CRM, but the chemical change necessary for acquiring remanent magnetization takes place only at temperatures higher than a critical value.

Basing upon data of remanent magnetization of rocks, on the other hand, several applications have been proposed for examining the history of the behaviour of the earth's crust and the earth's interior in the past. Among them, "*palaeomagnetism*" which deals with secular variations in direction and intensity of the earth's magnetic field during the past geologic times has been developed remarkably in recent years. In this sort of palaeomagnetic research, the basic assumption should be that the remanent magnetization of rocks concerned has kept its direction as it was when it was produced, throughout a long geologic period up to the present. This basic assumption must be examined before dealing with details of palaeomagnetic results and their applications on physics of the earth. This is a problem of the *stability of magnetization of materials*. The term stability should cover a fairly wide range of meaning; that is stability of magnetization against thermal agitation, namely, magnetic viscosity, stability against chemical changes of minerals, and that against dynamical effect upon minerals and their mother rocks[1].

15. Thermo-remanent magnetization (TRM). α) *Physical characteristics of TRM*. Stable remanent magnetization of rocks, an assemblage of small grain of ferromagnetic or ferrimagnetic materials, or similar materials, which is acquired by cooling them down from a certain high temperature in a magnetic field, is called "*thermoremanent magnetization*", and may be abbreviated as TRM.

In nature, TRM of igneous or some metamorphic rocks is produced by cooling from a high temperature (above the Curie point) to the atmospheric temperature in the geomagnetic field.

Basic characteristics of TRM were investigated and cleared by KÖNIGSBERGER [*43*], THELLIER [*44*], and NAGATA [*45*], [*46*] and their physical interpretation was given by NÉEL [*42*], but there are still a number of problems to be solved about the exact physical mechanism of this phenomenon.

When a specimen is subjected to a magnetic field only from temperature T_i to T_j during the course of its cooling down to T_0, $(T_i > T_j \geqq T_0)$, so that $H=0$ for $T > T_i$ and for $T < T_j$, then the TRM whose direction is parallel to that of H at a temperature $T_0 \leqq T_j$ is a function of T_i, T_j, H and T_0, and may be expressed as $J_{T_i,H}^{T_j}(T_0)$. For the sake of practical convenience in geophysics, T_0 is usually taken to be the atmospheric temperature, but, in the sense of physics, T_0 can take any value between 0° K and Curie temperature. When T_i becomes higher, while T_j and T_0 are kept constant respectively, $J_{T_i,H}^{T_j}(T_0)$ increases, but when T_i becomes the Curie temperature Θ, J reaches a saturation value, and does not increase for $T_j > \Theta$. Therefore, when

$$T_i \geqq \Theta > T_j = T_0,$$

$J_{T_i,H}^{T_j}(T_0) = J_{\Theta,H}^{T_0}(T_0)$, and it is called *total thermo-remanent magnetization*, while when

$$\Theta > T_i > T_j > T_0,$$

$J_{T_i,H}^{T_j}(T_0)$ is called *partial thermo-remanent magnetization* for the interval T_i to T_j. Total thermo-remanent magnetism thus defined just corresponds to natural

[1] For palaeomagnetism cf. especially the contribution of S. K. RUNCORN to vol. 47 of this Encyclopedia (p. 470—497).

remanent magnetism of rocks *in situ*, which was acquired by cooling, in the geomagnetic field, from a temperature higher than their Curie temperature.

Fundamental characteristics of thermo-remanent magnetism may be summarized as follows.

αα) Addition law for temperature range. When $T_1 > T_2 > T_3 > T$, there is a relation that

$$J_{T_1 H}^{T_2}(T) + J_{T_2 H}^{T_3}(T) = J_{T_1 H}^{T_3}(T), \tag{15.1}$$

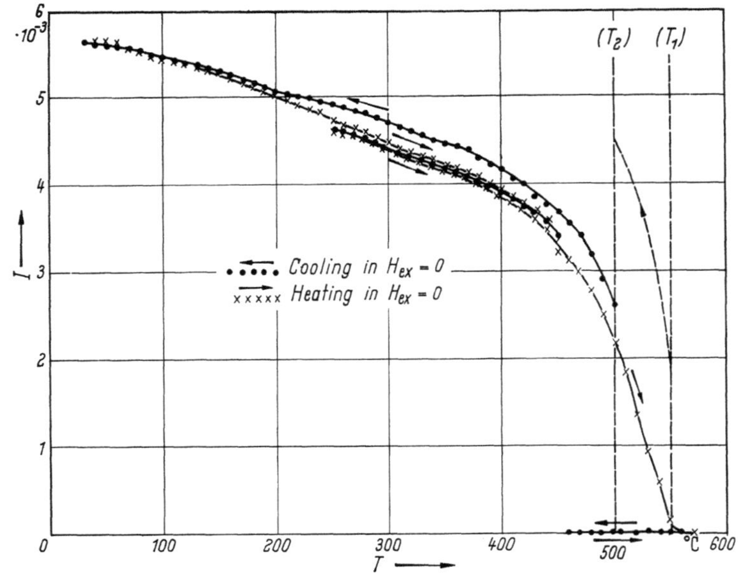

Fig. 24. Change in $J_{T_1, H}^{T_2}(T)$ for $T_1 = 509°$ C, $T_2 = 550°$ C and $H_{ex} = 0$. Sample: basalt. (After NAGATA.)

or more generally, when $T_1 > T_2 > \cdots > T_j > T$, then

$$J_{T_1 H}^{T_j}(T) = \sum_{i=1}^{j-1} J_{T_i H}^{T_{i+1}}(T). \tag{15.2}$$

This empirical law may indicate that partical TRMs for different temperature ranges are independent of each other. As a typical demonstration of this independence, $J_{T_i, H}^{T_j}(T)$ changes reversibly for change in temperature T provided that $T < T_j$ and no magnetic field exists for $T < T_j$, but when T exceeds T_j even in $H = 0$ regions, the magnetization decreases with increase in temperature, finally vanishing at $T = T_i$, as shown, for example, in Fig. 24. This may indicate that remanent magnetization acquired by cooling in a magnetic field from T_i to T_j is fixed at temperatures below T_j, only spontaneous magnetization of the fixed magnetic domains changing reversibly with temperature, while increase in temperature above T_j destroys the fixing of the magnetic domains.

*αβ) Dependence on applied magnetic field **H**.* The *direction* of TRM is parallel to that of the applied magnetic field ***H***, and the *intensity* of TRM is approximately expressed by

$$J_{T_i, H}^{T_j} = \frac{J_{T_i}^{T_j}}{K} \tanh K H, \tag{15.3}$$

where $J_{T_i}^{T_j}$ denotes the TRM for unit intensity of H, and K is a constant depending on material. When H is small, say a few Oersteds, (15.3) becomes

$$J_{T_i,H}^{T_j} \simeq H \cdot J_{T_i}^{T_j}. \tag{15.4}$$

In case of TRM of natural rocks *in situ*, (15.4) holds with a good accuracy.

αγ) Stability for A.C. field demagnetization. TRM is extremely stable against alternating field demagnetization. As illustrated in Fig. 25, for example, the total TRM of basalt composed of fine grains of titanomagnetites, acquired by cooling in a magnetic field of 1 Oe is reduced only by 20% by an A.C. demagnetization procedure, in which the maximum amplitude amounts to 400 Oe. On the other hand, the *isothermal* remanent magnetization of the same intensity of the same

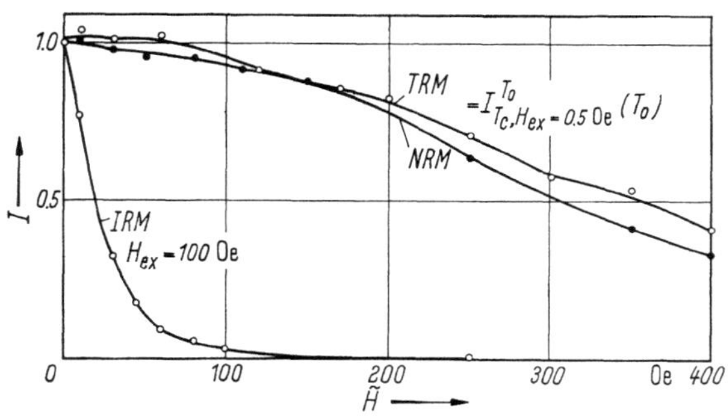

Fig. 25. AC demagnetization curves for NRM, TRM and IRM of basalt. (After AKIMOTO and NAGATA.)

sample at atmospheric temperature can be almost fully demagnetized by A.C. demagnetization of only 40 Oe in maximum amplitude. This striking contrast between TRM and IRM shows that fixing of magnetic domains in the case of TRM is extremely strong compared with that in the case of IRM.

αδ) Dependence on coercive force H_c. When the coercive force H_c is large, the TRM of the specimen is generally large. According to the results of experiments on a large number of igneous rocks, there holds a relation

$$J_{\Theta,H}^{T_0}/J_s \sim H_c. \tag{15.5}$$

αε) Ratio of TRM to induced magnetization. The ratio of the intensity of total TRM produced in H to that of the induced magnetization in the same magnetic field is usually denoted by Q; namely,

$$J_{\Theta,H}^{T_0}/\varkappa_a H \equiv Q, \tag{15.6}$$

where $\varkappa_a$ denotes the magnetic susceptibility. The Q-value has no particular physical meaning, but it can well represent the relative magnitude of TRM in case of geophysical discussions. That is because H is a magnetic field in which the sample concerned has made all its thermal history and $\varkappa_a H$ is the intensity of magnetization *now* induced by the magnetic field; in most cases of igneous and metamorphic rocks *in situ*, $Q=1$ to 10, and sometimes the Q value amounts to more than 100. This fact may indicate that TRM plays the most significant role in remanent magnetization of natural rocks.

β) Theoretical interpretation of TRM. The TRM of an assemblage of a large number of grains of single magnetic domain having uniaxial anisotropy has been well interpreted theoretically by NÉEL [42]. This theory is based on the dependence of relaxation of magnetization upon temperature. Denoting the degree of magnetic ordering just at the time when the external magnetic field $\boldsymbol{H}$ is removed by S_0, and that at time t after the initial state by S, there is generally a relation

$$S = S_0 \exp(-t/\tau_0), \tag{15.7}$$

where τ_0 is the relaxation time depending on temperature as well as material characteristics such as saturation magnetization, coercive force, grain size, etc. However, it is clear that $\tau_0 = \infty$ at 0° K and that it is approximately zero at Curie temperature Θ, as long as the very short relaxation time for response of paramagnetic behaviour to the external magnetic field is neglected.

The relaxation time τ_0 should vary from 0 to ∞ corresponding to decrease in temperature through a finite range, namely, from Θ to 0° K. This means that any long relaxation time τ^*, which is equal to the time t of cooling the concerned specimen in a magnetic field, does correspond to a certain temperature between 0° K and Θ, being denoted by T^* and called *relaxation temperature*, and that practically $\tau_0 \ll \tau^*$ for $T > T^*$ and $\tau_0 \gg \tau^*$ for $T < T^*$. Or, we may assume an extreme case of the above-mentioned condition as its fairly good approximation, in such a way that $\tau \simeq 0$ for $T > T^*$, $\tau = \tau^* = t$ for $T = T^*$, and $\tau = \infty$ for $T < T^*$. Consider now that a specimen is cooled in a magnetic field h from a temperature higher than T^*, and the field is taken away at $T = T^*$. At that instant, the average magnetization m of the specimen is given by a Boltzmann distribution so that

$$m = v\, J_s(T^*) \tanh\left(\frac{v\, J_s(T^*)\, h}{k\, T^*}\right), \tag{15.8}$$

where v denotes the average volume of each fine grain. In course of further cooling of the specimen with $h = 0$, the magnetization is expressed by

$$m(T) = v\, J_s(T) \tanh\left(\frac{v\, J_s(T^*)\, h}{k\, T^*}\right), \qquad T \leqq T^*, \tag{15.9}$$

because the term of tanh is invariant owing to the condition $\tau_0 = \infty$. Therefore reversible change in $m(T)$ for the $T \leqq T^*$ range is caused by the reversible change in spontaneous magnetization $J_s(T)$.

On the other hand, the assemblage of fine magnetic grains has a distribution with respect to T^*. Then TRM, $J_{T_i, H}^{T_j}$, should be a sum of magnetizations of grains having their T^* between T_i and T_j, and consequently the addition law (15.1) and (15.2) must naturally hold in the present case.

In case of an assemblage of single domain grains having uniaxial anisotropy, it is shown by NÉEL [42] that the relaxation times for inversion of grain magnetization from the axial direction parallel to the applied magnetic field to the reverse direction, $\tau(0, \pi)$, and that for the reverse process, $\tau(\pi, 0)$, are given as

$$\left.\begin{aligned}
\frac{1}{\tau(0,\pi)} &= C\left(1 + \frac{h}{H_c}\right)\left(1 - \frac{h^2}{H_c^2}\right)^{\frac{1}{2}} \exp\left\{-\frac{v\, J_s(H_c + h)^2}{2H_c k\, T}\right\},\\
\frac{1}{\tau(\pi,0)} &= C\left(1 - \frac{h}{H_c}\right)\left(1 - \frac{h^2}{H_c^2}\right)^{\frac{1}{2}} \exp\left\{-\frac{v\, J_s(H_c - h)^2}{2H_c k\, T}\right\},
\end{aligned}\right\} \tag{15.10}$$

$$C = \frac{e\, H}{2m} \left|3 G \lambda + D\, J_s^2\right| \left\{\frac{2v}{\pi\, G\, k\, T}\right\}^{\frac{1}{2}} \tag{15.11}$$

where e, m, and G denote respectively the mass and charge of electron and shear modulus of the grains, D is a numerical constant depending only on the shape of grain, and λ the longitudinal magnetostriction at saturation.

The above-mentioned theory looks like well explaining various characteristics of TRM, at least qualitatively. However, it is not true that all of ferrimagnetic minerals in rocks are small enough to have single domain structure. For instance, the uppermost critical diameter of single domain grain for magnetite should be, at the largest, less than 0.1 μ, while most ferrimagnetic grains in rocks, having capability of TRM, are ranged from 1 to $10^3\,\mu$, having multi-domain structure. Therefore TRM of an assemblage of multi-domain ferrimagnetic grains must be theoretically understood.

A kinematical theory of TRM of multi-domain grains has also been proposed by NÉEL [39]. Its outline is as follows:

The dependence of spontaneous magnetization J_s and of coercive force H_c of any multidomain substance of ferromagnetics or ferrimagnetics on temperature

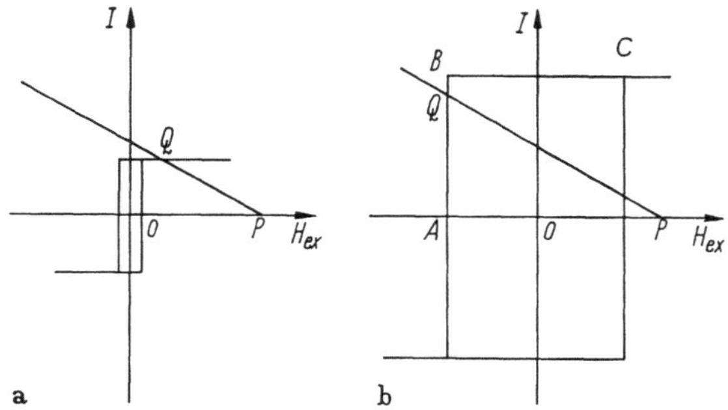

Fig. 26 a and b. NÉEL's model for a mechanism of blocking a domain wall by field cooling.

can be empirically represented, as an approximation, by

$$J_s \sim (\Theta - T)^{\frac{1}{2}}, \tag{15.12}$$

$$H_c \sim (\Theta - T). \tag{15.13}$$

Therefore there holds an approximate relation

$$\frac{H_c}{H_{c_0}} = \left(\frac{J_s}{J_{s_0}}\right)^2, \tag{15.14}$$

with H_{c_0} and J_{s_0} denoting respectively H_c and J_s at the atmospheric temperature T_0. Further, we may assume that the magnetic hysteresis cycle of the grain is approximated by a rectangular cycle, $2J_s$ in height along the J-axis and $2H_c$ in width along the H-axis, as illustrated in Fig. 26, where (a) represents the hysteresis cycle at a higher temperature T_1, while (b) that at a lower temperature T_2. With n denoting the demagnetizing factor of the grain, the intensity of magnetization corresponding to $h = \overline{OP}$ in Fig. 26 (a) is given by the height of the intersection Q of the hysteresis cycle with a straight line $\overline{PQ}$, the inclination of which with respect to the h-axis is $-1/n$. Let us now assume that the cooling of the sample in h is started at a temperature T_1. When the temperature decreases, the hysteresis cycle becomes enlarged according to (15.12) and (15.13), but the straight line $\overline{PQ}$ remains invariant, as shown in Fig. 26 (b). In the process of cooling, the point Q moves counter-clockwise along the upper branch of the hysteresis curve namely, $\overrightarrow{CB}$ and then $\overrightarrow{BQ}$ as long as the ratio $\overline{AQ}/\overline{AB} \equiv r$ continues to decrease, that is, $dr/dT > 0$. Here r is given by

$$r = \frac{h + H_c}{n\,J_s}. \tag{15.15}$$

The condition for stopping the downward movement of Q, i.e. $dr/dT=0$, is represented by

$$\frac{dH_c}{dT} = \frac{h+H_c}{J_s}\,\frac{dJ_s}{dT}. \tag{15.16}$$

At the temperature T' satisfying (15.16), Eq. (15.11) gives

$$h = H_c(T'), \quad \text{and} \quad (r)_{T=T'} = r' = \frac{2h}{n\,J_s}. \tag{15.17}$$

Under the given conditions and assumptions, further decrease in temperature must result in an increase of the r-value in the above-mentioned formulae. But physically this is not true, because the point Q cannot move upwards along the upper branch of the hysteresis curve, i.e., the $\overline{AB}$ line. Therefore, the magnetization $r' J_s$ should be fixed at $T=T'$ and for $T<T'$; in other words, magnetic domains contributing to the magnetization $r' J_s(T')$ should be fixed for further cooling. Then, remanent magnetization at $T=T_0<T'$, even if the external field h is taken away, will be given as

$$J_{T_1,h}^{T_0} = r' J_{s0}. \tag{15.18}$$

Or, by referring to (15.14) and (15.17), TRM is expressed as

$$J_{T_1,h}^{T_0} = \frac{2}{n}\,(h\cdot H_{c0})^{\frac{1}{2}}. \tag{15.19}$$

It must be noted here that T_1 can be replaced by Θ, provided that the point Q is on the line BC at the initial temperature T_1.

On the other hand, from (10.1), the magnetic susceptibility $\varkappa_a$ of most rocks is approximately given by p/n, because generally $n\,\varkappa \gg 1$. For an assemblage of magnetic grains, with p being their volume content, the Q-value defined by (15.6) is $Q \equiv p J_{H,h}^{T_0}/\varkappa_a h = J_{H,h}^{T_0}\Big/\frac{h}{n}$. Then, using (15.19) in the present case, we get

$$Q = 2\left(\frac{H_{c0}}{n}\right)^{\frac{1}{2}}. \tag{15.20}$$

Comparing these theoretical results, (15.19) and (15.20), with the observed characteristics of TRM, given in Sect. 14, we may find their qualitative agreement for the TRM's dependence on H_c, and on h for *large* values of h. But there is also a definite discrepancy for the TRM's dependence on h for *small* values of h.

VERHOOGEN [47] has recently pointed out that TRM may be retained in regions where mechanical stress is concentrated locally around dislocations. His idea is that small strained volumes around local nuclei of dislocation within magnetic substances can behave magnetically independently of the surrounding unstrained space, and consequently can behave just as the single domain particles with relaxation time proportional to v/T.

It seems that further quantitative theories of TRM are still desirable.

γ) Reverse thermo-remanent magnetism (RTRM). In rare cases, rocks are occurring in nature which acquire thermo-remanent magnetization whose direction is just opposite to that of the external magnetic field applied during their cooling process. It has been known that particular ferrimagnetic minerals contained in the rocks have the characteristics of reversed magnetization after a field cooling. This particular TRM may be called *reverse thermo-remanent magnetization.*

General characteristics of the reverse TRM are represented by an example shown in Fig. 27, where production of normal partial TRM in a magnetic field H, its changing into reverse magnetization and the further development of the reversed magnetization during continued cooling process without magnetic field are illustrated.

It has been made clear that the self-reversal of TRM is originated in peculiar physical properties of a certain particular ferrimagnetic grain itself. The physical mechanism of the origination of self-reversal of TRM has been investigated both experimentally and theoretically. Any interpretation of this phenomenon has reached a conclusion that two different constituents, A and B, in a grain are closely coupled with each other, and in process of magnetic field cooling, the A-phase which has a higher Curie-temperature Θ_A possesses magnetization parallel to the applied magnetic field $\boldsymbol{H}$, and then, when temperature becomes the (lower) Curie-temperature of the B-phase, Θ_B, and lower, the magnetization of the B-phase is forced to be anti-parallel to $\boldsymbol{H}$ under influence of a strong coupling between A and B; fixing of magnetic domains of both A- and B-phases taking place, as in the normal TRM phenomenon, during the cooling process. Then, provided that $\partial |J_B|/\partial T$ is sufficiently larger than $\partial |J_A|/\partial T$, the resultant TRM $J(T) = n_A J_A(T) - n_B J_B(T)$ can become zero, at a certain temperature T^+ and negative at $T > T^+$, where n_A and n_B denote volumes of the fixed domains of A- and B-phases, respectively. In some cases, however, $J(T) > 0$ though $J(T) < n_A J_A$. Only the case that n_B *and* $\partial J_B/\partial T$ are sufficiently large can result in the reverse TRM.

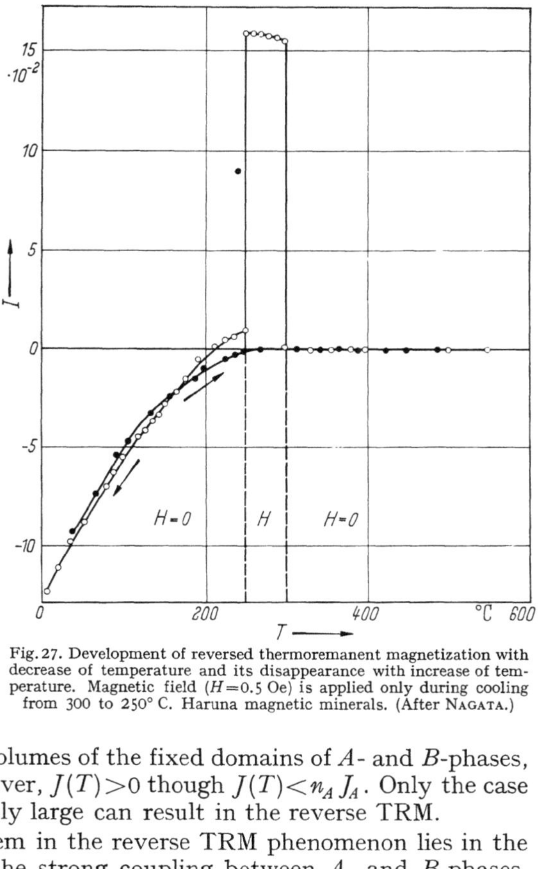

Fig. 27. Development of reversed thermoremanent magnetization with decrease of temperature and its disappearance with increase of temperature. Magnetic field ($H = 0.5$ Oe) is applied only during cooling from 300 to 250° C. Haruna magnetic minerals. (After NAGATA.)

The most fundamental problem in the reverse TRM phenomenon lies in the possible physical mechanism of the strong coupling between A- and B-phases. Three different ways of interpretation have been postulated for possible mechanisms of the coupling, namely,

1. magneto-static interaction between neighbouring lamellae of the different phases, by NÉEL [48] and NAGATA [49];

2. super-exchange interaction between two sub-lattices in a crystal lattice, by NÉEL [48];

3. exchange interaction through a wall between neighbouring domains of the different phases, by NAGATA and UYEDA [50].

The possibility of theory 1 has been proved mathematically and also experimentally. It seems likely that the rather weak tendency of a reverse TRM phenomenon of some specimens is due to this mechanism. But this is not the case for typical reverse TRM, discovered first from pumice of Mt. Haruna in Japan and then from some other rocks. In nature, the most typical and strong reverse TRM phenomenon appears in $x\,FeTiO_3 \cdot (1-x)\,Fe_2O_3$ of $0.45 < x < 0.60$, i.e. in the transition range between ferrimagnetic and antiferromagnetic states. Fig. 28 illustrates the dependence of total TRM of synthetic ilmenite-hematite $0.48\,FeTiO_3 \cdot 0.52\,Fe_2O_3$, showing the typical reverse TRM character, upon the applied magnetic field H_{ex}.

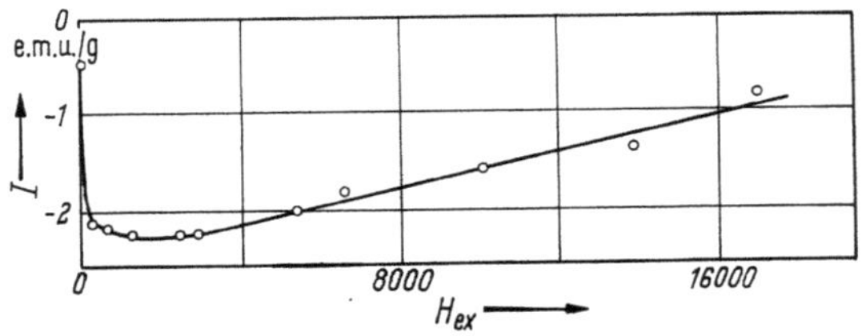

Fig. 28. Total TRM $J^{20}_{300, H_{ex}}(20°)$ of $0.48\,FeTiO_3 \cdot 0.52\,Fe_2O_3$ vs. H_{ex}. (After NAGATA and UYEDA.)

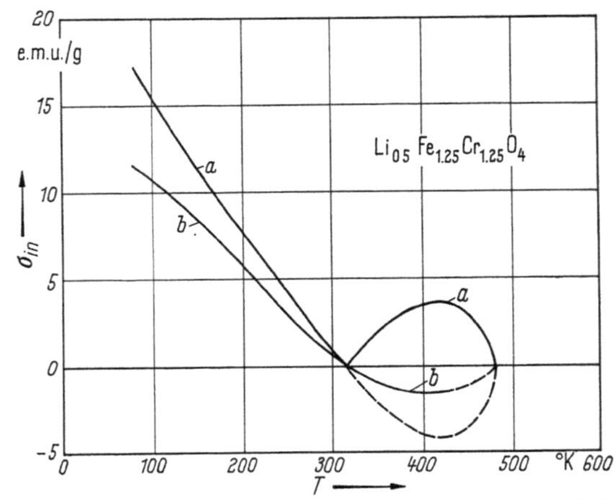

Fig. 29. N-type ferrimagnetism of $Li_{0.5}Fe_{1.25}Cr_{1.25}O_4$. Curve *a*. Spontaneous magnetization. Curve *b*. Remanent magnetization. (After GORTER and SCHULKES.)

In this figure, it is shown that the direction of the total TRM is opposite to that of H_{ex} even when H_{ex} exceeds 16000 Oe; that is, the strength of the coupling is equivalent to more than 16000 Oe in magnetic field intensity. Such a strong magnetostatic coupling cannot be expected by assuming any particular shape and configuration of the neighbouring two phases.

In theory 2, the spontaneous magnetization J_s itself of the material should be reversed at a certain transition temperature. That is to say, $J_s = |J_A - J_B|$ in such a material and $J_A = J_B$ at transition temperature T^+. Saturation magnetization in a sufficiently strong magnetic field of this material changes with temperature in such a way as shown by full line in Fig. 29. If the domain of this material having J_s in spontaneous magnetization is fixed at temperature between T^+ and Θ, magnetization changes with decrease in temperature below T^+ in a non-magnetic space or in a weak magnetic field in the way shown by the dotted line in Fig. 29, because the super-exchange interaction between A- and B-sublattices in a crystal is strong enough for maintaining their antiparallel magnetic coupling.

In ferrimagnetic crystals composed of two or more sub-lattices, there is the possibility of resulting in several different types of thermal change of spontaneous magnetization according to different modes and intensities of the super-exchange interaction among the sub-lattices. The type shown in Fig. 29 is one of them, and is called *N-type ferrimagnetism* by NÉEL [51]. Actually, $Li_{0.5}Cr_{1.25}Fe_{1.25}O_4$ was proved by GORTER and SCHULKES [52] to have the characteristics of the N-type ferrimagnetism. Several other metallic compounds having the N-type characteristics have recently been discovered.

If natural ferrimagnetic minerals, such as haemo-ilmenites, have the characteristics of N-type ferrimagnetism, they can acquire reverse TRM. However,

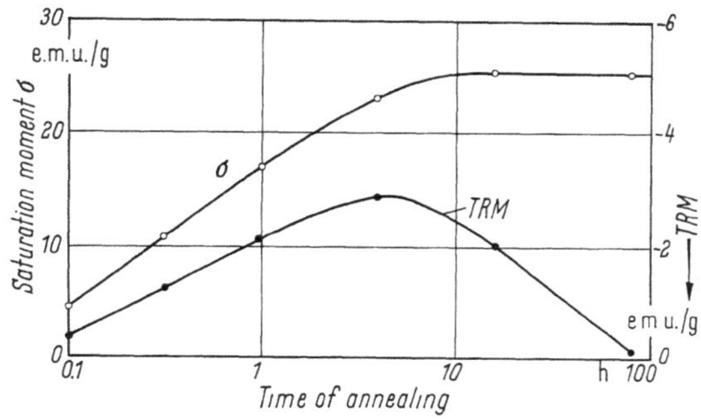

Fig. 30. Saturation magnetic moment (σ) and reverse TRM intensity at room temperature of $0.56\ FeTiO_3 \cdot 0.44\ Fe_2O_3$ as dependent on time of annealing at 700° C. (After ISHIKAWA and SYÔNO.)

no sample of $x\ FeTiO_3 \cdot (1-x)\ Fe_2O_3$ of $0.45<x<0.6$ shows any evidence of this type.

Then, excluding the possibility of the N-type ferrimagnetism, the only possible source of such an intense interaction seems to be the exchange interaction across the boundary wall between two neighbouring phases, i.e., the case of interpretation 3.

ISHIKAWA and SYONO [53] have shown that such a strong antiparallel super-exchange interaction takes place between the ordered ferrimagnetic phase (see Sect. 19) and a Fe-rich metastable phase which appears as a transient phase from the disordered antiferromagnetic phase to the ordered phase. Fig. 30 illustrates changes in TRM and saturation magnetic moment of a specimen of $0.56\ FeTiO_3 \cdot 0.44\ Fe_2O_3$, which is initially in the disordered state, with time of annealing at 700° C, where the saturation moment represents approximately the total volume of the ordered phase developed by the annealing. The reversed TRM in this diagram takes a maximum value at an appropriate time of the annealing, and then it decreases with longer times of the annealing, the characteristic reverse TRM disappearing beyond a critical length of the annealing time. This experimental result indicates that a metastable transient phase between the disordered and ordered phases plays an essential role for producing the reverse TRM. It was also proved by their experiments that this metastable phase has a Curie temperature higher than that of the ordered phase, owing to its richness in Fe ion content, and that a strong superexchange interaction exists between the metastable phase and the ordered phase (more than 10^4 Oe order of magnitude of the equivalent magnetic field).

The appearance of the metastable phase has been interpreted theoretically as follows. The diffusion of cations when changing from the disordered phase to the ordered phase should form local patches of the ordered phase surrounded by the Fe-rich transient phases and the Ti-rich ones. The Fe-rich parts have higher Curie temperatures than the ordered phase but the Ti-rich parts have lower ones, and consequently only the Fe-rich parts magnetically affect the magnetization of the ordered phases in course of acquisition of TRM. It seems therefore

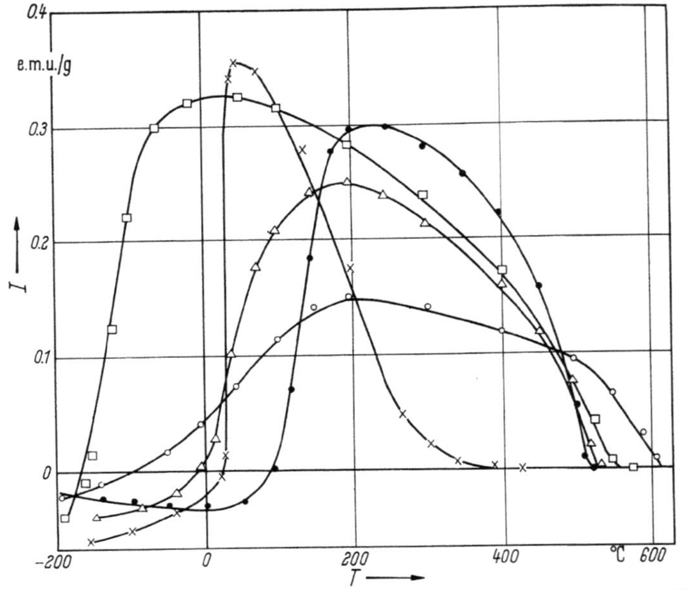

Fig. 31. Change of TRM $J^{T_c-100}_{T_c, H_{ex}=0.5\,\text{Oe}}(T)$ of the Allard Lake ilmenite-haematite with temperature. Heat treatment: —×— 1150° C for 8 days, —•— 900° C for 100 days (scale × $\frac{1}{3}$), —▵— 800° C for 100 days, —▫— 700° C for 100 days, —○— natural crystal (scale × $\frac{1}{5}$), (component along H_{ex}). (After CARMICHAEL.)

that the reverse TRM mechanism of 0.5 $FeTiO_3 \cdot 0.5\ Fe_2O_3$ has been well understood.

CARMICHAEL [*54*], on the other hand, has found that haematite and ilmenite series specimens of $x=7$ to 25% in the compositional range have also the characteristics of reserve TRM. He has shown that TRM of exsolved ilmenohematite of $x=7$ to 10% in form of fine lamellae in hemoilmenite of 70% ilmenite — 30% haematite in bulk composition is reversed at about −100° C, and that TRM of the same specimen heat-treated for 100 days at 900° C, the x-value of which becomes about 20%, is reversed at about +100° C (Fig. 31).

CARMICHAEL has suggested that the reverse TRM might be due to an ordering of ferric and ferrous ions, namely Fe^{2+} on A sites and Fe^{3+} on B sites of the sublattices. It seems likely, however, that the appearance of a transient metastable phase such as found in the case of $x \simeq 0.5$ ilmenohematite might be considered to play an essential role for the reverse TRM of this case, because a completely exsolved specimen has not the characteristics of the reverse TRM. This problem may deserve a further detailed study.

δ) Inverse type of thermo-remanent magnetization (ITRM). When titanomagnetites are heated in a magnetic field from temperatures below a certain critical value to the atmospheric temperature, they acquire a fairly stable remanent magnetization (NAGATA, OZIMA and YAMA-AI [*55*]). Since the procedure of changing the temperature in this case is opposite to that in the case of the ordinary TRM, the remanent magnetization acquired by the present procedure may be called the *inverse type of thermoremanent magnetization* (ITRM).

Fig. 32 illustrates the dependence of the partial ITRM of synthesized stoichiometric magnetite (an assembly of fine grains of Fe_3O_4) on temperature. The maximum production of the partial ITRM of magnetites takes place at about $-140°$ C, at which temperature the crystal anisotropy constant K_1 becomes zero accompanied by a maximum peak of magnetic susceptibility. As shown in Fig. 33, the partial ITRM produced in the temperature range near $-140°$ C is as stable as the ordinary TRM against the A.C.-demagnetization, while the partial ITRMs produced at higher temperature ranges are definitely more unstable, the stability decreasing with increase in temperature.

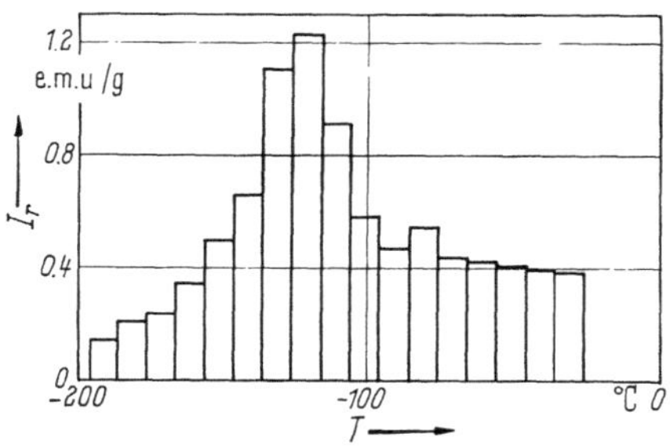

Fig. 32. Partial ITRM of magnetite acquired in heating. (After NAGATA, OZIMA and YAMA-AI.)

The mechanism of acquisition of ITRM may be interpreted as follows. As directly proved by a sharp increase in magnetic susceptibility, the energy of domain walls becomes extremely small at temperatures where K_1 is very small or zero resulting in the larger magnetization at those temperatures in a magnetic field. When the temperature of the specimen increases in the magnetic field, the rapid change of the crystal anisotropy energy may tend to constrain the domain

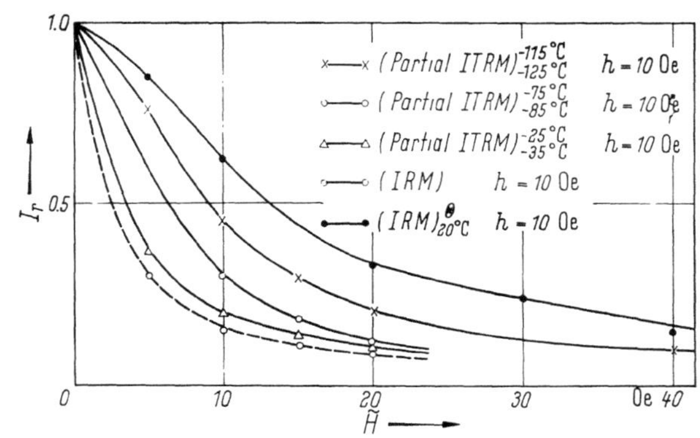

Fig. 33. AC demagnetization curves for partial ITRM, IRM and TRM of magnetite. (After NAGATA, OZIMA and YAMA-AI.)

walls to the displayed positions to keep the already acquired magnetization. When the magnetic field is removed at the atmospheric temperature the domain wall energy may have become large enough to keep a certain amount of stable remanent magnetization. In the above discussion, an effect of a random distribution of internal stress coupled with the crystal anisotropy is implicitly assumed. The mechanism of acquisition of ITRM seems thus to be made clear, at least qualitatively.

The present interpretation of the mechanism was justified by a similar experiment on cobalt chips which have the Curie temperature at 1115° C and in which crystal anisotropy vanishes ($K=0$) at 250° C. Moreover the argument that the energy concerning ITRM is mainly due to the domain wall energy, which is subjected to the crystal anisotropy constant, has been supported by the experimental fact that single crystals of magnetite have almost the same characteristics of ITRM into any direction with respect to the crystal axes as those of an assemblage of fine grains of magnetites.

16. Chemical remanent magnetization (CRM). *α) Characteristics of CRM.* HAIGH [*56*] and NAGATA and KOBAYASHI [*57*] have performed a laboratory demonstration to produce a remanent magnetization by a chemical change of $\alpha\, Fe_2O_3 \rightarrow Fe_3O_4$ (a reduction process) at a temperature much below the Curie temperature of the chemically produced ferrimagnetics (i.e. Fe_3O_4) in a magnetic field. The remanent magnetization J_r acquired by Fe_3O_4 in this procedure is in the direction of the magnetic field $\boldsymbol{H}$ which is applied during the process of chemical changes and is approximately proportional to H, provided that H is smaller than 50 Oe. Although the intensity of this remanent magnetization is appreciably smaller than that of total TRM of the same specimen produced in the same magnetic field, it is as stable as TRM against the A.C.-demagnetization and is definitely larger and more stable than IRM. Since this remanent magnetization is produced during the course of formation of ferro- (or ferri-) magnetic phases by a chemical change in a magnetic field, it has been called *chemical remanent magnetization* (CRM). Obviously, CRM is produced in a magnetic field when fine crystals of ferro- (or ferri-) magnetic phase are formed in a non-magnetic matrix, whence CRM is occasionally read as *crystallization remanent magnetization* [*43*]. Fig. 34 illustrates the dependence of intensity of CRM upon the magnetic field H applied during the $\alpha\, Fe_2O_3 \rightarrow Fe_3O_4$ process, together with the dependence on H of total TRM and IRM of Fe_3O_4 having the same grain size.

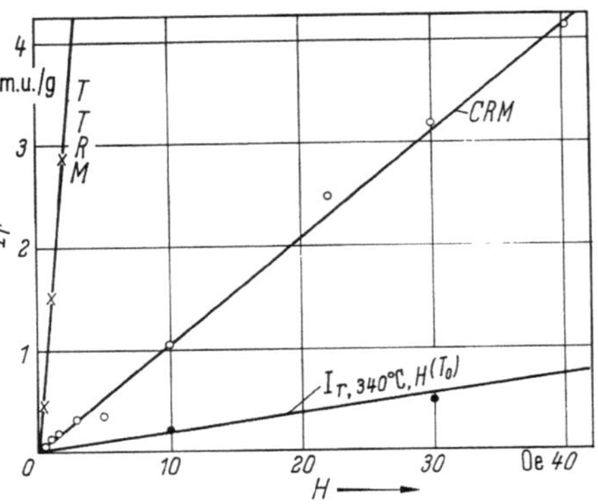

Fig. 34. CRM acquired by chemical change of $\alpha Fe_2O_3 \rightarrow Fe_3O_4$. IRM and total TRM of Fe_3O_4 as dependent on magnetic field. —o— $I_{cr,340°C,H}(T_0)$, chemical remanent magnetization generated at 340° C, —×— $I^{T_0}_{T_c,H}(T_0)$, total thermo-remanent magnetization, —•— $I_{r,340°C,H}(T_0)$. (After NAGATA and KOBAYASHI.)

Another example of CRM is a remanent magnetization produced by an oxidation process from titanomagnetite to titanomaghemite in a magnetic field.

Oxidation of titanomagnetites, $x\, Fe_2TiO_4 \cdot (1-x)\, Fe_3O_4$, results generally in an increase in Curie temperature and in a decrease in lattice parameter, as long as their crystal structures are kept in a spinel form.

Let a titanomagnetite specimen having Curie temperature Θ_a be oxidized at a constant temperature T, where $\Theta_a < T$. T can be so chosen that the Curie temperature Θ_b of the oxidized specimen, which is a solid solution of titanomagnetite and titanomaghemite, is higher than T; i.e., $\Theta_a < T < \Theta_b$.

The original titanomagnetite is ferrimagnetic at temperatures below Θ_a, but is paramagnetic at T. However, the resultant sample produced by the oxidation at T is ferrimagnetic at T. When the oxidation procedure is carried out in a magnetic field, the oxidized sample acquires a stable remanent magnetization. This remanent magnetization may be called *thermochemical remanent magnetization* (TCRM), because the oxidation temperature between Θ_a and Θ_b plays an essential role in the process of acquisition of this kind of CRM [*58*].

Fig. 35 illustrates an example of development of TCRM with the oxidation time, where the original titanomagnetite has a chemical composition of 40 Mol.% $Fe_2TiO_4 \cdot 60$ Mol.% Fe_3O_4, and Curie temperatures of the original and oxidized specimen and the oxidizing temperature are $\Theta_a = 365°$ C, $\Theta_b = 560°$C and $T = 470°$ C respectively. The TCRM thus acquired is as stable as the ordinary CRM.

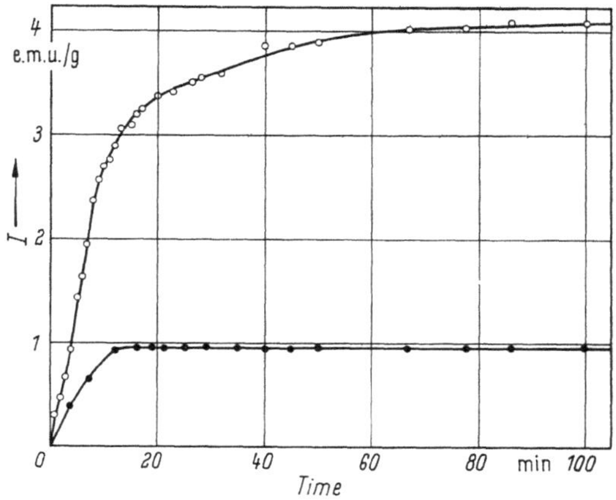

Fig. 35. Development of TCRM in 0.40 Fe_2TiO_4 0.60 Fe_3O_4 with time during its oxidation at 470° C. (After KOBAYASHI and NAGATA.)

β) Mechanism of acquisition of CRM and thermo-chemical remanent magnetization (TCRM). Mechanism of acquisition of the stable CRM has been interpreted as due to growth of volume of individual ferrimagnetic phases in a nonmagnetic matrix. In (15.7), the relaxation time τ_0 for magnetization of magnetic grains of single domain is expressed generally as

$$1/\tau_0 = f_0 \exp\left(-\frac{K v}{k T}\right), \qquad (16.1)$$

where f_0 and K denote respectively quantities depending on material constants and on the applied magnetic field. At the very beginning of precipitation of ferri- (or ferro-) magnetic phases, when v is very small, τ_0 too is very small and conse-

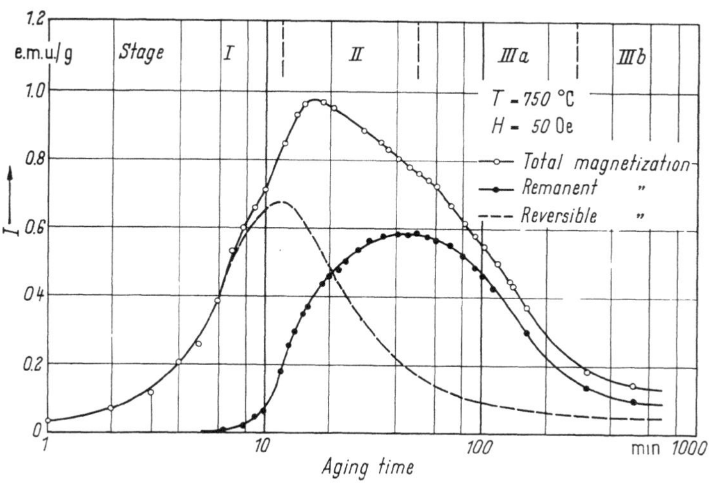

Fig. 36. Change in remanent magnetization, reversible magnetization and their sum of 2 weight% Co—Cu alloy as dependent on aging time at 750° C in $H_{ex} = 50$ Oe. (After KOBAYASHI.)

quently the specimen may have a superparamagnetic character, as mentioned in Sect. 13.

An increase in v caused by progress of chemical change results in an exponential increase in τ_0. So long as newly produced magnetic phases keep their single domain structure, τ_0 increases markedly with increase in v, the magnetization being consequently blocked along the direction of an applied magnetic field. If, however, a further increase in v results in a multi-domain structure of individual magnetic phases, reversely magnetized domains will be formed, so that the apparent magnetization will decrease according to the increase in v.

The above-mentioned three stages in magnetic characteristics of an assemblage of chemically produced magnetic phase will be summarized as follows;

(I) superparamagnetic behaviour when $v < v_c$;

(II) Magnetically stable single domain when $v_D > v > v_c$;

(III) multidomain structure when $v > v_D$;

here v_c and v_D are idealized critical values for grain size between superparamagnetics and single domain structure and between single domain and multidomain structures.

The above-mentioned theory of production of CRM was experimentally demonstrated by KOBAYASHI [57] using development of the grain size of ferromagnetic cobalt precipitates generated by an isothermal aging of 98% Cu to 2% Co alloy which is initially non-magnetic.

Fig. 36 illustrates the mode of change in remanent magnetization, reversible magnetization and their sum of the specimen during its aging in a magnetic field. Stages I, II and III in the diagram indicate respectively the superparamagnetic, single domain and multi-domain stages.

17. Detrital (depositional) remanent magnetization (DRM). *α) Characteristics of DRM.* The recent sedimentary deposits on the ocean bottom surface have natural remanent magnetization whose declination is in agreement with that of the geomagnetic field, whereas the inclination is generally a little less than the ambient geomagnetic inclination. The results of the laboratory deposition experiments of pulverized sedimentary rocks (containing ferrimagnetic mineral grains) in a magnetic field have shown that the horizontal direction of the remanent magnetization of the artificial deposits always coincides with that of the horizontal magnetic force within an error of several degrees, while the inclination of the magnetization is somewhat less than that of the applied magnetic field, the difference amounting to $10 \sim 20°$. The natural remanent magnetization of sedimentary deposits can therefore be considered in many cases to be caused by the statistical alignment of ferrimagnetic mineral grains along the ambient magnetic field during their sedimentation. This kind of remanent magnetization is called *detrital or depositional remanent magnetization* (DRM).

Basic experiments to demonstrate the formation of DRM of sedimentary rocks were performed by NAGATA et al. [59] to [61], JOHNSON et al. [62], GRAHAM [63], KING [64], GRIFFITH [67], WRIGHT [66], and GRIFFITH et al. [67]. Their experimental results will be summarized in the following:

(a) The horizontal direction of DRM agrees with that of the applied magnetic field.

(b) The inclination of DRM of sediments deposited on a horizontal plane is always less than that of the applied field, and the amount of deviation is dependent on the inclination of the magnetic field, as illustrated for example in Fig. 37. This deviation has often been called the *inclination error*.

(c) The intensity (J_n) of DRM increases with increase in depth of water for the deposition. J_n reaches its saturated value within several tens of centimetres in depth (Fig. 38).

(d) J_n of DRM increases with increase in the intensity (H) of the applied magnetic field as shown in Fig. 39. However, the effect of superposition of IRM on the real DRM becomes appreciably large when H is larger.

(e) J_n of DRM is approximately proportional to the content of ferrimagnetic minerals in the depositing sediments.

(f) When the bedding plane for deposits is inclined towards the horizontal plane, the inclination error becomes different.

(g) When the deposition experiments are carried out in running water, the direction of DRM becomes different from that in case of the deposition in still water. The apparent deviation of the direction of the DRM from the magnetic field direction may be resolved into two parts, namely (i) the inclination error (δ), which is represented by a rotation about a horizontal axis perpendicular to the field and (ii) a rotation angle (ϱ) about a horizontal axis perpendicular to the current direction. The inclination error δ seems to be reduced when the running water speed increases from zero. The rotation angle ϱ increases with increase in the water current velocity, but it reaches an asymptotic value of less than 20° at only several cm/sec in the current velocity.

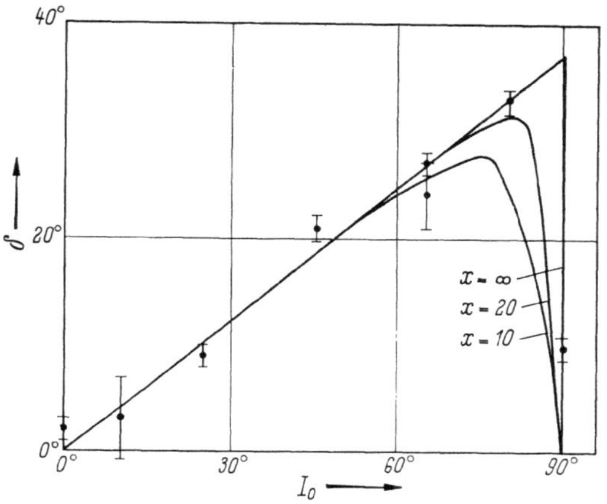

Fig. 37. The inclination error δ as a function of inclination I_0 of the affecting magnetic field. Full circles: empirical data (after KING). Full lines: theoretical values ($f = 0.58$).

Summarizing the above-mentioned various factors controlling characteristics of DRM, we must assume at the present stage of knowledge that the direction of DRM of sediments in nature is accompanied by errors of about 20° and even after adequate corrections for the probable errors, the resultant error may still amount to about 10°. It will be desirable therefore to study more exactly the connection between the direction of DRM of sediments and an applied magnetic field during their deposition.

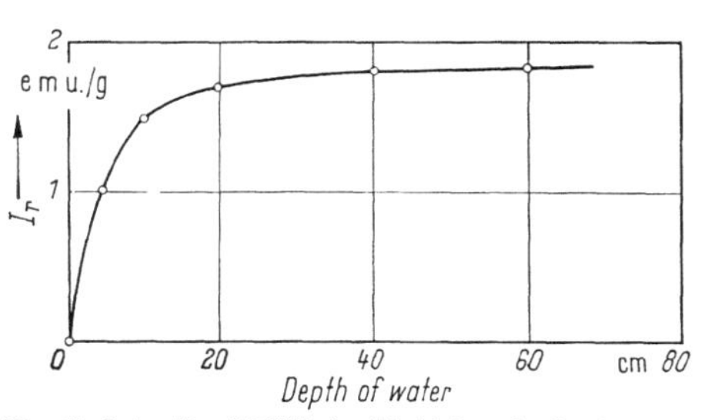

Fig. 38. Intensity of DRM of artificial deposit of red sandstone in H_{ex}=8.6 Oe as dependent on depth of still water. (After NAGATA and KOBAYASHI.)

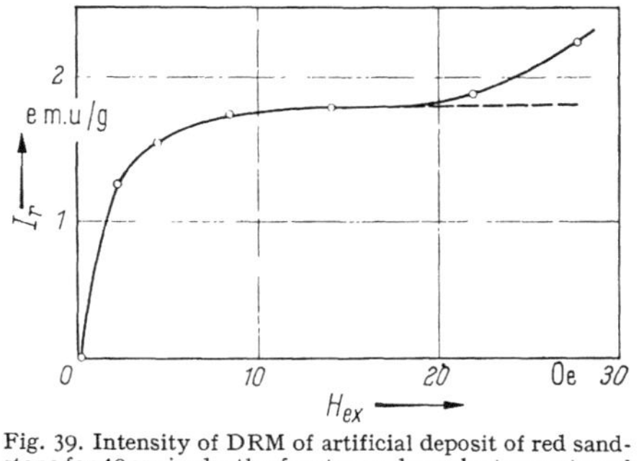

Fig. 39. Intensity of DRM of artificial deposit of red sandstone for 40 cm in depth of water as dependent on external magnetic field, H_{ex}. (After NAGATA and KOBAYASHI.)

β) Theoretical interpretation of some characteristics of DRM. According to ISING [68] and GRANAR [69], varved clays have a definite anisotropy of magnetic susceptibility, in which the direction of maximum susceptibility is within the bedding plane and the minimum direction is vertical. This fact seems to indicate that the average shape of ferrimagnetic minerals in the varves is an oblate whose major axis is within the horizontal plane. WRIGHT [66] has shown, on the other hand, that the sphericity of particle shape of the Swedish varves is represented by the dimension ratio (the major radius to the minor radius) being ranged between 1.1 and 1.3, not exceeding 1.5. The shape of ferrimagnetic particles in sediments may therefore be approximated by an oblate whose dimension ratio ranges from unity to 1.5. The oblate particles having the dimension ratio very near unity may be considered as a sphere, while those having the comparatively large dimension ratio may have to be considered as a real oblate.

Both types of particle will make a rotational motion during their deposition through still water in a magnetic field. A rotation (θ) of a magnetized particle of the magnetic moment σ in a magnetic field H, in still water is expressed by

$$\lambda \frac{d\theta}{dt} + \sigma H \sin\theta \simeq 0, \tag{17.1}$$

where θ denotes the angle between magnetic moment and the field direction, and λ is the viscosity coefficient for rotation of the particle. The solution of (17.1) is given as

$$\tan\frac{\theta}{2} = \tan\frac{\theta_0}{2}\exp\left(-\frac{\sigma H}{\lambda}t\right), \tag{17.2}$$

where θ_0 means the initial value of θ at $t=0$.

t indicates the time for deposition of a particle from the water surface to the bedding plane at the bottom, whence t is related to the depth (h) of still water as

$$h = \frac{m}{v} g\, t, \tag{17.2a}$$

where m, g, and v denote respectively the mass of a particle, the force of gravity and the viscosity coefficient of translation of a particle.

Since the direction of magnetization is assumed to be at random at $t=0$, the total resultant magnetic moment (J) of a deposit of a large number (N) of *spherical* particles at $t=t$ is derived [70] from (17.2) as

$$J = \frac{N\sigma}{2}\int_0^{\pi}\cos\theta\sin\theta_0\, d\theta_0 = N\sigma\{\coth x - x\,\mathrm{cosech}^2 x\} \tag{17.3}$$

where $x \equiv \frac{\sigma H}{\lambda} t$.

On the other hand, the *oblate* particles may be assumed to have their magnetization within the major axial plane, and the major axis plane may be set parallel to the bedding plane at their final setting position. Hence these particles are considered to have the two-dimensional freedom around their minor axis. Then the total resultant magnetic moment of a large number N of oblate particles at time t is given as

$$J' = \frac{N\sigma}{\pi}\int_0^{\pi}\cos\theta\, d\theta_0 = N\sigma\tanh\left(\frac{\sigma H'}{2\lambda}t\right), \tag{17.4}$$

where H' denotes the effective magnetic field for rotation around the minor axis, i.e. $H' = H\cos I$, I being the inclination of the $\boldsymbol{H}$ field. Then, (17.4) can be expressed as

$$J' = N\sigma\tanh\left(\frac{x}{2}\cos I\right). \tag{17.5}$$

An idealized model for DRM of sediments will be that magnetized particles in the sediments consist of a fraction f of spherical shape and the remaining $(1-f)$ portion of spheroidal shape. In the sediments deposited on a horizontal bedding plane, $\boldsymbol{J}$ has the direction of $\boldsymbol{H}$, while $\boldsymbol{J}'$ is horizontal and along the horizontal component of $\boldsymbol{H}$, whence the resultant total magnetic moment J_0 of the deposits becomes

$$\left.\begin{aligned} J_0 &= N\sigma f\left[D^2(x)\sin^2 I + \left\{D(x)\cos I + \frac{1-f}{f}\tanh\left(\frac{x\cos I}{2}\right)\right\}^2\right]^{\frac{1}{2}}, \\ D(x) &\equiv \coth x - x\,\mathrm{cosech}^2\, x. \end{aligned}\right\} \tag{17.6}$$

The results of laboratory deposition experiments on the $J_0 \sim h$ relation and the $J_0 \sim H$ relation have been well explained on the basis of (17.6).

Defining then the inclination error δ by

$$\delta = I - I_0, \tag{17.7}$$

where I_0 denotes the inclination of the total resultant magnetic moment $\boldsymbol{J}_0 = \boldsymbol{J} + \boldsymbol{J}'$, we get

$$J\sin\delta = J'\sin I_0. \tag{17.8}$$

Putting (17.3) and (17.5) into (17.8), we get

$$\tan I_0 = \frac{f\sin I}{f\cos I + (1-f)\tanh(x\cos I/2)/D(x)}. \tag{17.9}$$

The values of δ calculated from (17.7) and (17.9) for the case of $f=0.58$ for different values of I are shown by full lines in Fig. 37. The theoretical curve for $x=20$ is in good agreement with the experimental results.

In most cases of DRM of natural sediments, however, x is rather large, so that we may approximately put $x\to\infty$. Then, (17.6) and (17.9) are approximately represented by

$$\left.\begin{aligned} J_0 &\simeq N\sigma[2f(1-f)(\cos I - 1) + 1]^{\frac{1}{2}}, & |I| &< \frac{\pi}{2} \\ &= N\sigma f, & I &= \frac{\pi}{2} \end{aligned}\right\} \tag{17.6a}$$

and

$$\left.\begin{aligned} \tan I_0 &\simeq \frac{f\sin I}{f\cos I + (1-f)}, & |I| &< \frac{\pi}{2} \\ &\simeq 0, & I &= \frac{\pi}{2}. \end{aligned}\right\} \tag{17.9a}$$

A question may be raised about the behaviour of settling of deposited particles on the bedding plane. It is possible to assume that a particle deposited onto an undulating surface of the bed cannot maintain its orientation which has been subject to the applied magnetic field, but rather rolls down to a more stable position by its rotation. If a spherical particle having a magnetization $(J_x, J_y, J_z) = (J\cos I, 0, J\sin I)$, where x, y, z are rectangular coordinates, rotates by an angle φ around a horizontal axis making an angle λ with the x-axis, the resultant magnetization (J_x', J_y', J_z') becomes

$$\begin{aligned} J_x' &= J\cos I(\cos^2\lambda + \sin^2\lambda\cos\varphi) + J\sin I\sin\lambda\sin\varphi, \\ J_y' &= \tfrac{1}{2}J\cos I\sin^2\lambda(1-\cos\varphi) + J\sin I\cos\lambda\sin\varphi, \\ J_z' &= J\sin I\cos\varphi - J\cos I\sin\lambda\sin\varphi. \end{aligned}$$

If we further assume that the azimuth of the rotation axis is at random within the horizontal plane, the average magnetization components of a large number of such rotated particles are represented by

$$\left(\frac{J}{2}\cos I\,(1+\cos\varphi)\,,\,0,\,J\sin I\cos\varphi\right).$$

Therefore,

$$\tan I_0 \equiv \frac{\overline{J_{z'}}}{\overline{J_{x'}}} = \frac{2\cos\varphi}{1+\cos\varphi}\tan I\,. \qquad (17.10)$$

Since $2\cos\varphi\,(1+\cos\varphi) \leqq 1$, I_0 is always smaller than I. In other words, the rotation of magnetized particles around horizontal axes whose azimuths have a random distribution results in a reduction of the inclination of magnetization.

This effect may contribute to a certain extent to the observed inclination error of DRM. When the bedding plane is inclined, deposited particles tend to roll downwards appreciably more than upwards. This effect may modify the inclination error of DRM of sediments deposited onto an inclined bed.

The fact that the inclination of DRM of sediments is reduced by about 9° when the sample is fully compressed vertically [*71*] will also be interpreted as due to the effect of rotation of magnetized particle, because vacancies in each horizontal layer of the initial specimen of loose accumulation of particles should be filled up by particles rolling from the neighbouring layers when the specimen is made compact by vertical compression.

Thus, various characteristics of DRM of sediments described in Sect. 17a seem to be qualitatively interpreted. The actual shape of particles in natural sediments, however, is neither spherical nor spheroidal, and consequently their dynamic motion and geometrical configuration should be much more complicated. Only few experimental and theoretical examinations of DRM have been made hitherto.

18. Viscous remanent magnetization (VRM). *α) Characteristics of VRM.* It has long been known that all ferromagnetic substances change more or less their magnetization gradually with time. Such a slow change in magnetization with time is generally called the effect of *magnetic viscosity*. The magnetic viscosity affects the remanent magnetization of rocks in two different ways; one is an acquisition of a remanent magnetization which is called *viscous remanent magnetization* (VRM) when a rock specimen is kept in a magnetic field during a long time, and the other is a time-decay of the remanent magnetization such as TRM, CRM, or DRM.

Fig. 40 illustrates an example of acquisition of VRM of natural magnetites in a constant magnetic field at various temperatures. A number of experimental studies [*72*] to [*78*] on this problem have been reported. The results of these investigations show that the amount of change in magnetization ΔJ is expressed as a function of time such as

$$\Delta J = \text{constant} + S\log t\,, \qquad (18.1)$$

where the time t is reckoned from the moment of application of a magnetic field H. The magnetic viscosity coefficient S in (18.1) depends on H and the temperature T as well as on the material constants; S is proportional to H as far as the Rayleigh region of magnetization is concerned and is proportional to T, except at temperatures close to the Curie temperature or to the temperature of transformation of crystal structure (for example, $-160°$ C in case of magnetite).

Let $J_r(H, 0)$ and $J_r(H, t)$ denote respectively the IRM of a specimen in a magnetic field H and the remanent magnetization of the same specimen after keeping it in H during a time t. Then

$$J_r(H,t)=J_r(H,0)+(\text{viscous remanent magnetization for } t \text{ in } H). \quad (18.2)$$

In (18.2), $J_r(H,t)$ is considered as the remanent magnetization measured immediately after the removal of H.

When the remanent magnetization is measured at t' after the removal of H, the measured magnetization $J_r(H,t,t')$ depends upon t' also. The results of experiments show that

$$J_r(H,t,t')=J_r(H,0)+S'(\log t-\log t'), \quad (18.3)$$

where S' is a viscosity coefficient for remanent magnetization.

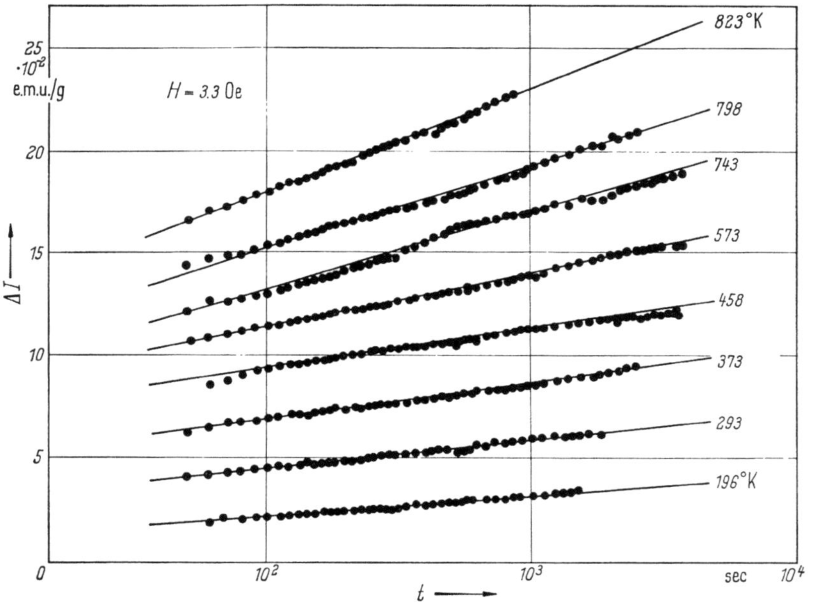

Fig. 40. Development of VRM of magnetite with time at various temperatures. (After SHIMIZU.)

On the other hand, the intensity of IRM decays with time when the specimen is kept in a non-magnetic space. Fig. 41 illustrates an example of such a decay of IRM with time t'.

The results of experiments to put a specimen in H during the time t and measure the remanent magnetization at t' after the removal of H are summarized as that the measured remanent magnetization $J_r(0,t')$ is given by

$$J_r(0,t')=J_r(H,0)-S'(\log t'-\log t). \quad (18.4)$$

S' in (18.4) is always the same as S' in (18.3) and they are about one half of S in (18.1) for the same specimen and for the same magnetic field H. For example, $S'=4.5\times 10^{-3}$ emu/g and $S=8.5\times 10^{-3}$ emu/g at the room temperature for the natural magnetite grains in $H=3.3$ Oe [75].

These observed facts may lead to the conclusion that VRM is generally expressed as $S'(\log t-\log t')=\frac{S}{2}(\log t-\log t')$.

VRM acquired during a longer time or in a larger magnetic field is more stable against the A.C.-demagnetization. The intensity of the critical alternating magnetic field ($\widetilde{H}_d$) for destroying VRM acquired during time t is roughly proportional to $\log t$, namely, $\widetilde{H}_d \propto \log t$. For example, IRM acquired in $H=10$ Oe is destroyed by $\widetilde{H}_d=10$ Oe, but the critical field $\widetilde{H}_d$ to destroy $J_r(H,t)=(\text{IRM})+(\text{VRM})$ for $H=10$ Oe and $t=$one month amounts to more than 200 Oe [77].

β) Theoretical interpretation of VRM. Theoretical interpretations of the magnetic viscosity as an effect of thermal agitation have been proposed by NÉEL [39], [42], [79]. According to the experimental results (SHIMIZU [75]), it seems likely that S or S' becomes zero at $T=0°$ K regardless of the magnitude of H. When T is appreciably higher than 0° K, irregular thermal fluctuations of forces acting on magnetic domains may enable the magnetization to get across the energy barriers which it otherwise could not do in a given magnetic field. It was proposed by NÉEL that the effect of thermal agitation on the domain wall can be replaced by a fluctuating magnetic field $H_f(t)$ in addition to the applied field H, where $H_f(t)$ is represented by

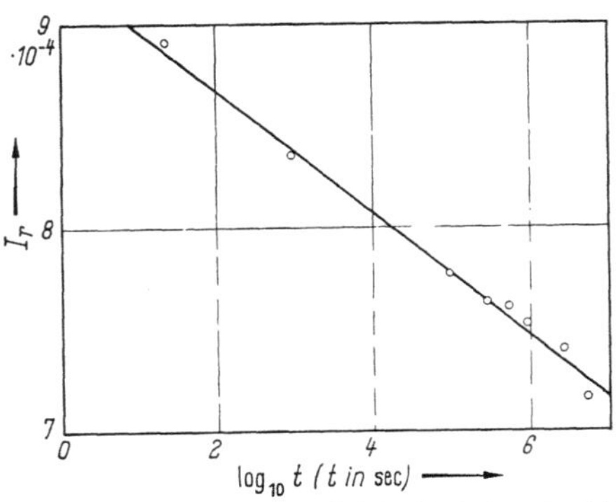

Fig. 41. Demagnetization of IRM of baked earth with time. (After THELLIER.)

$$H_f(t)=S_v(Q+\log t)\,, \tag{18.5}$$

where Q is a numerical constant of the order of 50 when time is measured in seconds, and S_v denotes a characteristic constant of a specimen depending on temperature T. Then, the intensity of magnetization at t after the application of H on a specimen is represented by

$$J(H,t)=J_0+\varkappa_i S_v(Q+\log t)\,, \tag{18.6}$$

where J_0 and $\varkappa_i$ denote respectively the magnetization at $t=0$ and the irreversible differential susceptibility at $H=H$. In the Rayleigh region of magnetization, we find $J_0=AH+BH^2$ and $\varkappa_i=2BH$ in (18.6). Then (18.6) becomes

$$J(H,t)=AH+BH^2+2BH\,S_v(Q+\log t)\,. \tag{18.7}$$

When H is removed, the remanent magnetization is represented as $BH^2/2$, and the irreversible susceptibility at the condition as $-BH$. Hence, the remanent magnetization at the time t' after the removal of H becomes

$$J_r(0,t')=\tfrac{1}{2}BH^2-BH\,S_v(\log t'-\log t)\,. \tag{18.8}$$

Comparing (18.7) and (18.8) with (18.1) and (18.4), respectively, we get that $S=2BH\,S_v$ and $S'=BH\,S_v$, and consequently $S'=S/2$.

NÉEL predicted that, if (18.5) holds as a sufficiently good representation for the thermal agitations, the alternating magnetic field ($\widetilde{H}_d$) to destroy (IRM)+(VRM) acquired in H for a time t is expressed as

$$\widetilde{H}_d=H+S_v\log t/\log\tau\,, \tag{18.9}$$

where τ denotes a relaxation time for the magnetization concerned. As discussed in Sect. 18a, (18.9) can well explain the observed characteristics of the A.C.-demagnetization of VRM.

Theoretical calculation of S_v has been performed by NÉEL and by STREET and WOOLY [80] based on different models. In NÉEL'S result, S_v is proportional to $T^{\frac{1}{2}}$, while S_v in STREET-WOOLY's calculation is proportional to T. It seems likely that more experimental and theoretical studies on magnetic viscosity of rock-forming magnetic minerals must be carried out in the future.

γ) VRM in rock magnetism. If some rock specimens are stored at fixed positions in a laboratory, the intensity and direction of their remanent magnetizations change with time. Fig. 42 illustrates an example of changes in the direction of the natural remanent magnetization (NRM) of rocks with time during their storage in laboratory. As seen in this diagram, the direction of NRM has shifted

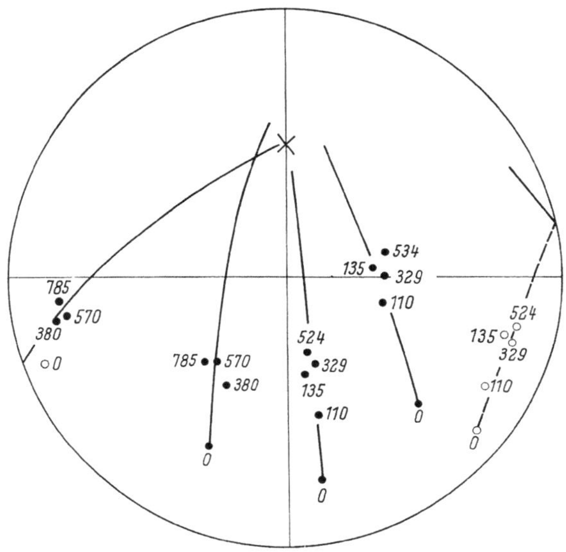

Fig. 42. Change in the direction of remanent magnetization during storage in the laboratory. Numerals denote time of storage in number of days. × The direction of the geomagnetic field. Full circles: lower hemisphere. Hollow circles: upper hemisphere. (After AKIMOTO.)

gradually with time along the great circles towards the direction of the geomagnetic field. From the result of examining the stability of remanent magnetization, the change of the direction of NRM is interpreted as consisting of three components; namely,

(a) a "hard component" ($\boldsymbol{J}_0$) which keeps the magnetization along the initial direction of NRM and whose intensity decreases extremely slowly with time,

(b) a "soft component" ($\boldsymbol{J}_f$) along the initial direction of NRM which decreases rapidly with time, and

(c) a "soft component of VRM" ($\boldsymbol{J}_v$) which is acquired along the geomagnetic field direction. The resultant vector ($\boldsymbol{R}$) of the remanent magnetization may therefore be expressed as

$$\boldsymbol{R} = (\boldsymbol{J}_0 + \boldsymbol{J}_f) + \boldsymbol{J}_v,$$

where $\boldsymbol{J}_0$ and $\boldsymbol{J}_f$ have the same direction, but $\boldsymbol{J}_v$ is parallel to the geomagnetic field $\boldsymbol{H}$. In some rocks, $J_0 \ll J_f$ and $J_0 \ll J_v$. These rocks cannot be used for the purpose of palaeomagnetic studies. In many rocks, however, $J_0 \gg J_f$ and $J_0 \gg J_v$. These rocks only can be the objects of the *palaeomagnetic studies. The stability test of rock magnetization is therefore an unavoidable process in the experimental studies of rocks for palaeomagnetic purposes.*

On the other hand, VRM takes place in any kind of ferromagnetic or ferrimagnetic substances as long as $T > 0°$ K. For an assemblage of magnetite grains, for example, the ratio of VRM to TRM in the geomagnetic field amounts to 0.019 and 0.024 for $t = 10^6$ years and $t = 10^9$ years, respectively, being the acquisition period of the VRM. In such a case, the effect of VRM may be ignored compared with the intensity of TRM.

19. Anhysteretic remanent magnetization (ARM). *α) Characteristics of ARM.* When a magnetic specimen is magnetized by applying a constant magnetic field h superposed by an alternating magnetic field $\tilde{H}$ whose amplitude is reduced from $\tilde{H}$ to zero, the remanent magnetization after removal of h is larger and more stable than the IRM of the same specimen acquired in h alone. The remanent magnetization acquired in the above-mentioned process is called the *anhysteretic remanent magnetization* (ARM). The process of acquisition of ARM does not seem to take place often in natural rocks *in situ*. However, ARM seems to be closely related to the internal structures of the materials which cause various other kinds of remanent magnetization.

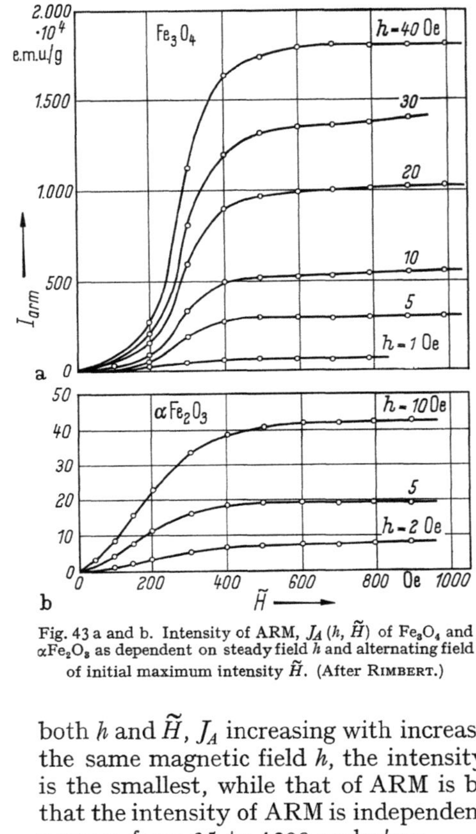

Fig. 43 a and b. Intensity of ARM, $J_A(h, \tilde{H})$ of Fe_3O_4 and αFe_2O_3 as dependent on steady field h and alternating field of initial maximum intensity $\tilde{H}$. (After RIMBERT.)

Basic characteristics of ARM have to a large extent been clarified by THELLIER and RIMBERT [*81*], PETROVA [*82*], PATTON and FINCH [*83*] and others. Their results may be briefly summarized as follows: The *direction* of ARM (J_A) is parallel to that of the applied constant field $\boldsymbol{h}$, being independent of that of the alternating field $\tilde{\boldsymbol{H}}$. The *intensity* J_A of ARM, however, depends on both h and $\tilde{H}$, J_A increasing with increase in h and $\tilde{H}$, as illustrated in Fig. 43. For the same magnetic field h, the intensity of TRM is the largest and that of IRM is the smallest, while that of ARM is between those of the former two. It seems that the intensity of ARM is independent of the frequency of alternating field over a range from 35 to 1200 cycles/sec.

One of the most important characteristics of ARM may be its addition law with regard to $\tilde{H}$, which is similar to the addition law of TRM.

General characteristics of the additivity of ARM observed by RIMBERT and PATTON and FINCH are as follows.

(a) ARM $J_A(h, \tilde{H})$ acquired in the alternating field $\tilde{H}$ together with the steady field h can be demagnetized by the ordinary A.C.-demagnetization process whose maximum alternating field is $\tilde{H}$, provided that $h \ll \tilde{H}$; when $\tilde{H}$ and h are of the same order of magnitude, the maximum intensity of alternating field

$\widetilde{H}_d$ to demagnetize $J_A(h, \widetilde{H})$ is expressed as

$$\widetilde{H} < \widetilde{H}_d < \widetilde{H} + h. \tag{19.1}$$

(b) During the ARM production process, the alternating field $\widetilde{H}$ will be changed only from $\widetilde{H}_{j+1}$ to $\widetilde{H}_j$ (where $\widetilde{H}_{j+1} > \widetilde{H}_j$), the field being taken away at $\widetilde{H} = \widetilde{H}_j$. The ARM thus acquired may be called partial ARM, being denoted by $\Delta J_A(h, \widetilde{H}_{j+1}, \widetilde{H}_j)$. If $h \ll \widetilde{H}$,

$$\Delta J_A(h, \widetilde{H}_{j+2}, \widetilde{H}_j) = \Delta J_A(h, \widetilde{H}_{j+1}, \widetilde{H}_j) + \Delta J_A(h, \widetilde{H}_{j+2}, \widetilde{H}_{j+1}). \tag{19.2}$$

These two characteristics of the additivity of ARM are nearly the same as those of the addition law of TRM. It must be noted, however, that the addition law of ARM can hold only if the *initial* non-magnetized state of a specimen is prepared by the A.C.-demagnetization procedure. The addition law does not hold when the non-magnetized state is produced by thermal demagnetization. In other words, the addition law of ARM can hold if the demagnetized state of a specimen consists of a uniaxial configuration of magnetic domains, namely, an equal distribution of domain orientation into plus and minus directions along a uniaxis.

When the procedure of production of ARM in h and $\widetilde{H}$ is applied on a specimen which has already TRM or IRM or another kind of remanent magnetization, the intensity of the resultant remanent magnetization is approximately expressed as

(A.C. demagnetization of the already acquired remanent magnetization up to $\widetilde{H}$) + + (ARM in $\widetilde{H}$ and h).

When the complete A.C.-demagnetization of a specimen is required, therefore, the steady field h must be made exactly zero. This is an important requirement in laboratory procedure which has sometimes been neglected.

β) *Theoretical interpretation of ARM.* Theoretical interpretation of ARM has been developed mostly by NÉEL [39], [84] and LLIBOUTRY [85]. An outline of NÉEL's theory of ARM may be summarized as follows:

Let the coercive forces of a rectangular hysteresis cycle of magnetization of a fine magnetic element on plus and minus sides of the magnetic field be denoted by H_c and $-H_c'$ respectively. As shown in Fig. 44, take H_c as x-variable and $-H_c'$ as y-variable in an xy coordinate plane. Then, any combination of H_c and $-H_c'$ can be plotted as a point in the xy plane. A positive field h causes all the negatively magnetized elements having $H_c \leqq h$ to reverse in sign, and a negative field $-h$ causes all the positively magnetized elements having $H_c' \leqq h$ to reverse in sign.

When h becomes equal to the saturation field strength, h_s, all the elements become positively directed as illustrated in Fig. 44 (a). When a negative field $-h_1$ where $h_1 < h_s$ is applied as the next procedure, all elements having $H_c' \leqq h_1$ are reversed, those only having $h_1 < H_c' \leqq h_s$ remaining positive, as shown in Fig. 44 (b). Further application of an $h_2 < h_1$ field on the above-mentioned state causes the configuration of magnetization illustrated in Fig. 44 (c). Thus, successive applications of plus and minus magnetic fields from h_1 to $h_n = 0$ where $h_1 > h_2 > h_3 > \cdots > 0$ results in a distribution of magnetization as given by Fig. 44 (e). If $\delta h = h_i - h_{i+1}$ is made sufficiently small, the resultant distribution of magnetization may be given by Fig. 44 (f). Obviously, the state represented by Fig. 44 (f) is the A.C.-demagnetized state having no remanent magnetization, positive and negative magnetizations cancelling each other.

When a steady field $+h$ is applied into the direction of the axis of alignment of magnetization of elements of this A.C.-demagnetized specimen, the apparent magnetization is represented by a triangular area ABO in Fig. 45 (a), which remains as it is when h is taken away if the hysteresis cycle has a rectangular form. This may be the case of IRM.

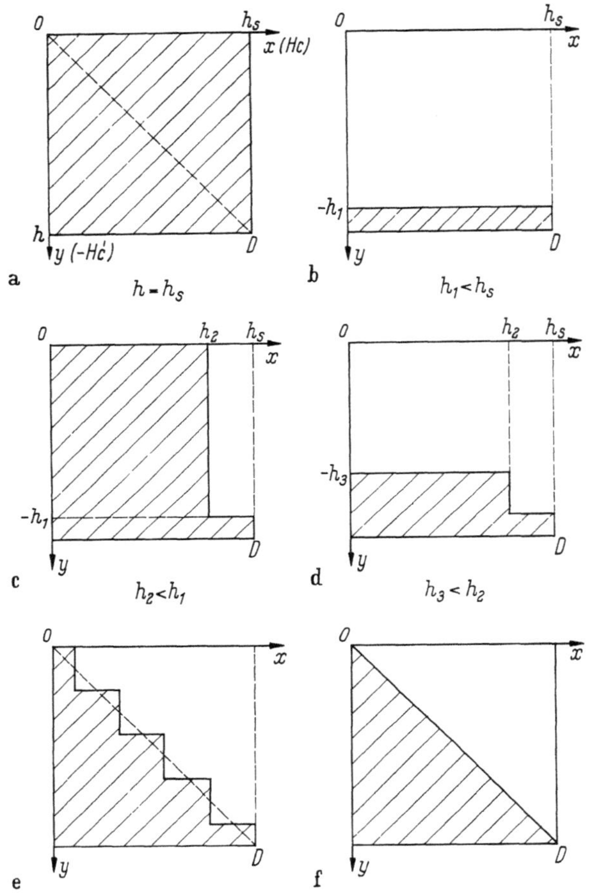

Fig. 44a—f. Schematic representation of processes of acquiring remanent magnetization and AC demagnetization

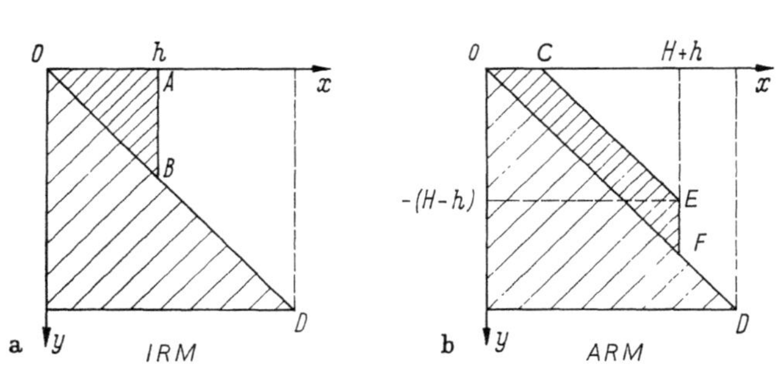

Fig. 45a and b. Schematic representation of IRM and ARM.

When a steady field h and an alternating field $\tilde{H}$ are applied together and the intensity of $\tilde{H}$ is gradually decreased to zero, the resultant remanent magnetization is represented by an area $OCEF$ in Fig. 45 (b). This may be the case of ARM. The area OAB in Fig. 45 (a) corresponding to IRM is $\frac{1}{2}h^2$, while the area $OCEF$ in Fig. 45 (b) for ARM is $\frac{1}{2}\{(\tilde{H}+h)^2-(\tilde{H}-h)^2\}=2\tilde{H}h$. Therefore, if $\tilde{H}\gg h$ as demonstrated in the experiments, ARM is much larger than IRM.

In order to estimate the intensity of ARM, an appropriate distribution of (H_c, H_c') on the xy-diagram must be assumed. If H_c' is not so much different from H_c for individual elements, or in other words, if H_c and H_c' are distributed mostly around the OD-line, the addition law of ARM can easily be explained.

20. Pressure remanent magnetization (PRM). A geometrical deformation of a magnetized specimen caused by mechanical stress can obviously result in deviation of the direction of remanent magnetization. An example of such a case will be the effect of uniaxial compression of a loose-packed sample of sediment, as already discussed in Sect. 17β. Another example was demonstrated by an uniaxially compressed magnetized Cu—Co alloy of circular cylindrical shape. As already mentioned (Sect. 16), an appropriately heat-treated Cu—Co alloy contains a large number of fine precipitations of the ferromagnetic Co phase, which resembles scattering of fine grains of ferrimagnetic minerals in rocks. Now consider a circular cylinder of such an alloy having IRM along a direction within the circular plane being compressed into a direction making 45° of angle with the direction of IRM. Then, the cylinder is plastically deformed, the circular cross section obtaining an elliptic form. A line mark put on a circular surface parallel to the direction of IRM changes accordingly as the cylinder is deformed, and at the same time the direction of IRM changes also. The angles between the axis of compression on the one side, and the direction of the line mark and that of IRM on the other side are given in Table 5.

Table 5.

Profile (Initial)	Angle between the line mark and the axis of compression	Angle between the direction of IRM and the axis of compression
Circle $a/b=1.00$	45° ± 5°	45° ± 5°
Ellipse $a/b=1.26$	43° ± 5°	39° ± 5°
Ellipse $a/b=1.52$	37° ± 5°	35° ± 5°
Ellipse $a/b=2.05$	29° ± 5°	29° ± 5°

As will be seen from this result, the direction of IRM changes almost together with a rotation of a straight line mark of the body caused by the plastic deformation. If individual magnetic elements can be assumed to be represented by small dipole arrows having the direction of magnetization, the above-mentioned deviation of IRM caused by the deformation can be explained by a simple theory of plastic deformation. However, the feasibility of the representation has not yet been exactly proved.

Although the important effect of geometrical deformation of magnetized specimens which may frequently take place in nature has not yet been fully understood, this effect may not be a problem of *magnetic* properties of materials. If there exists a process of producing or modifying remanent magnetization under the direct effect of pressure or stress in a magnetic field, it would be called *pressure remanent magnetization* (PRM). A possibility of existence of PRM has been frequently suggested on basis of observations of distribution of remanent magnetization in various geological structures [86] to [88] and laboratory experiments [89], [90]. It does not seem however that extensive fundamental studies on PRM have been carried out. A summary of several reports of experimental studies on this problem may be given in the following paragraphs.

DOMEN [89] has stated that the direction of a remanent magnetization of an assemblage of powdered magnetites compressed vertically in the geomagnetic field is horizontal within the geomagnetic meridian plane and the intensity of the remanent magnetization increases with increase in pressure, reaching a saturation value at a pressure of about 6×10^3 kg/cm². He has called this particular remanent magnetization *piezo-remanent magnetization* (PRM). The specific intensity of the saturated PRM amounts to about 0.8 emu/g, which is about one hundredth of the spontaneous magnetization of magnetite. It seems that this result is quite different from the packing effect of already magnetized particles mentioned before.

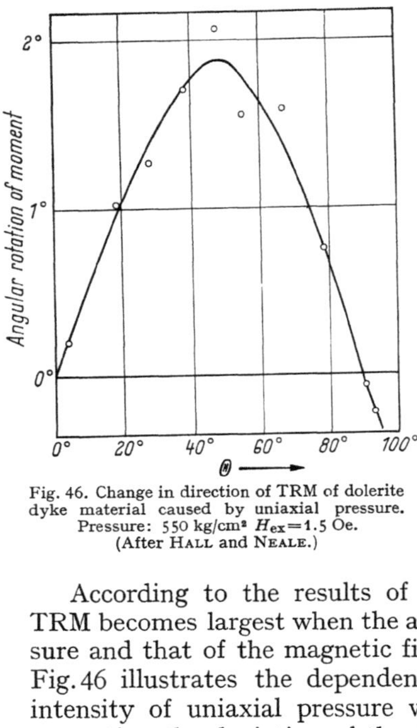

Fig. 46. Change in direction of TRM of dolerite dyke material caused by uniaxial pressure. Pressure: 550 kg/cm² $H_{ex}=1.5$ Oe. (After HALL and NEALE.)

It has also been found in preliminary studies that the *direction* of TRM and CRM deviates systematically from the direction of the applied field towards a direction perpendicular to that of pressure, when the sample is compressed by pressures of several thousand kg/cm². HALL and NEALE [90] have shown that the application of uniaxial pressure on magnetites separated from dolerite during the acquisition of TRM gives rise to a systematic rotation of remanent magnetization towards the plane perpendicular to the vertical stress axis, together with a reduction in the intensity of magnetization.

According to the results of their experiments, the amount of rotation of TRM becomes largest when the angle θ between the direction of the uniaxial pressure and that of the magnetic field is $\pi/4$, and becomes zero at $\theta=0$ and $\theta=\pi/2$. Fig. 46 illustrates the dependence of the amount of rotation of TRM on the intensity of uniaxial pressure when $\theta=\pi/4$. The remanent magnetization representing the deviation of the direction of TRM from that of the applied field may be called the *pressure effect* on TRM, or PRM *in the course of acquisition of TRM*. The intensity of PRM thus acquired becomes larger if the grain size is smaller.

STACEY [91] and NÉEL [92] have examined theoretically the mechanism of PRM. A certain change of reversible magnetization under both effects of magnetic field and pressure can be expected as an inverse effect of *magnetostriction*. However PRM is the irreversible magnetization which remains after removal of the pressure. In order to clarify the mechanism of PRM, therefore, experimental evidence to show a certain modification of blocking the remanent magnetization owing to stress applied during the course of magnetization will be required.

In recent years, characteristics of piezo-remanent magnetization have been clarified in some detail by NAGATA and KINOSHITA [94], [95]. It has been known [96], [97] that the magnetic susceptibility $\varkappa_a$ of rocks or of an assemblage of magnetic minerals decreases with increasing uniaxial compression, P, when P is applied along the direction of the magnetic field $\boldsymbol{H}$. The relation between $\varkappa_a$ and P

has been empirically expressed as

$$\varkappa_a = \varkappa_0/(1+\beta P),$$

where $\varkappa_0$ denotes the susceptibility for $P=0$ and β is a constant. The magnitude of β is about 1.0×10^{-4} cm²/kg for most rocks and 1.5×10^{-3} cm²/kg for titanomagnetites. KERN [*98*] and STACEY [*99*] have suggested some complicated theoretical explanations of this dependence of $\varkappa_a$ upon P, and NAGATA [*100*] has shown, in a simple way, that $\varkappa_a$ for magnetic minerals having spontaneous magnetization, J_s, anisotropy energy constant, K, and averaged magnetostriction coefficient λ_s is given by

$$\varkappa_a = \varkappa_0\Big/\left(1+\frac{9\lambda_s}{2K}P\right) = \varkappa_0\Big/\left(1+\frac{3\lambda_s\varkappa_0}{J_s^2}P\right),$$

and $\varkappa_a$ of rocks by

$$\varkappa_a = \varkappa_0\Big/\left(1+\frac{3\lambda_s P}{\frac{2}{3}K+N J_s^2}\right),$$

with N denoting the average demagnetizing factors of magnetic mineral particles in rocks. Since $K\approx1.4\times10^5$ c.g.s., $\lambda_s\approx39\times10^{-6}$ c.g.s. (SYONO [*101*]) and $J_s\approx 5\times10^2$ c.g.s. for magnetite-rich titanomagnetite, and since the average value of N for most rocks is about 3.4, the theoretical value of β can be evaluated to be 1.3×10^{-3} cm²/kg for magnetite-rich titanomagnetites and 1.25×10^{-4} cm²/kg for most rocks. It seems that the agreement between observed and theoretical values of $\varkappa_a$ is satisfactory.

On the other hand, NAGATA and KINOSHITA have shown experimentally that a removal of uniaxial compression P in the presence of a magnetic field H only results in an addition of remanent magnetization which has been defined as PRM, while a removal of H in the presence of P results, on the contrary, in a reduction of the remanent magnetization. This phenomenon has been theoretically interpreted by NAGATA [*102*], using the model of magnetization described in Sect. 12 of this chapter, in the following way. The uniaxial compression by P along the direction of the magnetic field H modifies the coercive force H_c in such a way, that the apparent coercive force h_c for $\theta=\pi/2$ in Fig. 22b becomes $h_c=H_c+3\lambda_s P/J_s$ and that for $\theta=0$ in Fig. 22a it becomes $h_c=H_c-3\lambda_s P/J_s$ for $P<H_c J_s/3\lambda_s$; if $P>H_c J_s/3\lambda_s$, in the case $\theta=0$ we get a reversible magnetization curve as illustrated in Fig. 22b instead of the hysteresis curve of Fig. 22a, the apparent coercive force h_c then becoming $h_c=\frac{3\lambda_s P}{J_s}-H_c$. It will be obvious from the above-mentioned results, that the magnetization in the vicinity of a direction along the $\theta=\pi/2$ axis is always reversible with respect to H and also to P, and that the susceptibility expressed by $\varkappa_0\Big/\left(1+\frac{3\lambda_s P}{2K}\right)$ for this magnetization always decreases with increasing values of P. These components of magnetization nearly along the $\theta=\pi/2$ axis play the major role for the dependence of magnetic susceptibility upon uniaxial compression.

On the other hand, the uniaxial compression P causes a decrease of apparent coercive force h_c and, consequently, those magnetic grains which have

$$h_c = H_c - 3\lambda_s P/J_s < H$$

may be magnetized in the presence of a magnetic field H, even if $H<H_c$. When the magnetic field is reduced to zero in the presence of P, the magnetization of the grains with $H_c<3\lambda_s P/J_s$ is reversibly reduced to zero because of their reversible magnetization characteristic, on the given condition.

On the contrary, a removal of P in the presence of H may result in the irreversible remanent magnetization J_R which has already been acquired by H and P, because the change of magnetization curve is reversible with respect to P and, consequently, the additional magnetization caused by a decrease in h_c with P remains as an additional remanence when h_c increases from $H_c - 3\lambda_s P/J_s$ to H_c. Thus, the additional remanent magnetization nearly along the $\theta=0$ axis can be identified with the observed PRM. The above-mentioned qualitative interpretation of the PRM phenomenon has been more quantitatively discussed, with satisfactory success, in agreement with experimental results.

References.

[1] NAGATA, T., T. YUKUTAKE and S. UYEDA: J. Geomag. Geoelectr., Kyoto **9**, 51 (1957).
[2] CHEVALLIER, R., et S. MATHIEU: Bull. Soc. chim. France pp. 726—729 (1958); ibid. pp. 9—10 (1959).
[3] AKIMOTO, S., K. HÔRAI and T. BOKU: J. Geomag. Geoelectr., Kyoto **10**, 7 (1958).
[4] SYONO, Y.: J. Geomag. Geoelectr., Kyoto **11**, 85 (1960).
[5] ERNST, T.: Z. angew. Min. **4**, 349 (1943).
[6] POUILLARD, E.: Ann. Chim. **5**, 164 (1950).
[7] KAWAI, N., S. KUME and S. SASAJIMA: Proc. Japan Acad. **30**, 588 (1954).
[8] AKIMOTO, S., T. KATSURA and M. YOSHIDA: J. Geomag. Geoelectr., Kyoto **9**, 165 (1957).
[9] GORTER, E. W.: Adv. Physics **6**, 336 (1957).
[10] NAGATA, T., S. AKIMOTO and S. UYEDA: J. Geomag. Geoelectr., Kyoto **5**, 168 (1953).
[11] BARTH, T. F. W., u. E. POSNJAK: Z. Kristallogr. **82**, 325 (1932).
[12] NAGATA, T., e S. AKIMOTO: Geofis. pura e appl., Milano **34**, 36 (1956).
[13] ISHIKAWA, Y., and S. AKIMOTO: J. Phys. Soc. Japan **12**, 1083 (1957).
[14] ISHIKAWA, Y., and S. AKIMOTO: J. Phys. Soc. Japan **13**, 1110, 1298 (1958).
[15] BOZORTH, R. M., D. E. WALSH and A. J. WILLIAMS: Phys. Rev. **108**, 157 (1957).
[16] BOZORTH, R. M., and V. KRAMER: J. Phys. Radium **20**, 393 (1959).
[17] BALSLEY, J. R., and A. F. BUDDINGTON: J. Geomag. Geoelectr., Kyoto **6**, 176 (1954).
[18] BALSLEY, J. R., and A. F. BUDDINGTON: Econ. Geol. **53**, 777 (1958).
[19] NAGATA, T., and Y. SHIMIZU: Antarctic Record, Japan, No. 10, 661 (1960).
[20] NAGATA, T., and Y. SHIMIZU: Nature, Lond. **184**, 1472 (1959).
[21] AKIMOTO, S., T. NAGATA and T. KATSURA: Nature, Lond. **179**, 37 (1957).
[22] BASTA, E. Z.: Thesis, Univ. of Bristol, 1953.
[23] CURNOW, C. E., and L. G. PARRY: Nature, Lond. **174**, 1101 (1954).
[24] CURNOW, C. E., and L. G. PARRY: J. Proc. Soc. New South Wales **89**, 64 (1956).
[25] UYEDA, S.: J. Geomag. Geoelectr., Kyoto **8**, 39 (1956).
[26] YUN, I.: Mem. Coll. Sci. Univ. Kyoto B **25**, No. 2, 125 (1958).
[27] MICHEL, A., G. CHAUDRON et J. BÉNARD: J. Phys. Radium **12**, 189 (1951).
[28] HARALDSEN, H.: Z. anorg. allg. Chem. **231**, 78 (1937); **246**, 169, 195 (1941).
[29] BERTAUT, F.: C. R. Acad. Sci., Paris **234**, 1298 (1952). — J. Phys. Radium **13**, 372 (1952). — Acta Cryst. **6**, 557 (1953).
[30] LOTGERING, F. K.: Thesis, Univ. of Utrecht, 1956. — Philips Res. Rep. **11**, 189 (1956).
[31] CHEVALLIER, R. C., et S. MATHIEU: C. R. Acad. Sci., Paris **204**, 854 (1957).
[32] AKIMOTO, S.: J. Geomag. Geoelectr., Kyoto **3**, 47 (1951).
[33] GOTTSCHALK, V. H.: Physics **6**, 127 (1935).
[34] SLICHTER, L. B.: Amer. Inst. M. E. E. Tech. Publ. 120 (1928).
[35] SLICHTER, L. B.: Adv. Physics **6**, 333 (1957).
[36] NAGATA, T.: Bull. Earthq. Res. Inst. **18**, 102 (1941).
[37] NAGATA, T.: Rock Magnetism, Rev. Ed., pp. 129—135 (1961).
[38] NAGATA, T.: Bull. Earthq. Res. Inst. **21**, 1 (1943).
[39] NÉEL, L.: Adv. Physics **4**, 191 (1955).
[40] NÉEL, L.: C. R. Acad. Sci., Paris **228**, 664 (1949).
[41] NÉEL, L.: Ann. Geophys. **5**, 99 (1949).
[42] CREER, K. M.: Trans. Amer. Geophys. Univ. **41**, 613 (1960).
[43] KÖNIGSBERGER, J. G.: Terr. Magn. **43**, 119, 299 (1938).
[44] THELLIER, E.: Ann. Inst. Phys. Globe, Paris **16**, 157 (1938).
[45] NAGATA, T.: Bull. Earthq. Res. Inst. **18**, 281 (1940).
[46] NAGATA, T.: Bull. Earthq. Res. Inst. **21**, 1 (1943).
[47] VERHOOGEN, J.: J. Geophys. Res. **64**, 2441 (1959).

[48] Néel, L.: Ann. Geophys. 7, 90 (1951).
[49] Nagata, T.: Nature, Lond. 169, 704 (1951); 172, 85 (1953).
[50] Nagata, T., and S. Uyeda: Nature, Lond. 184, 890 (1959).
[51] Néel, L.: Ann. Phys., Paris 3, 137 (1948).
[52] Gorter, E. W., and J. A. Schulkes: Phys. Rev. 90, 487 (1953).
[53] Ishikawa, Y., and Y. Syono: J. Phys. Soc. Japan 17, Suppl. B-1, 714 (1962).
[54] Carmichael, C. M.: Proc. Roy. Soc. Lond., Ser. A 243, 508 (1961). — Proc. Phys. Soc. Japan 17, Suppl. B-1, 711 (1962).
[55] Nagata, T., M. Ozima and M. Yama-Ai: Nature, Lond. 197, 444 (1963).
[56] Haigh, G.: Phil. Mag. 3, 267 (1958).
[57] Nagata, T., and K. Kobayashi: Proc. Japan Acad. 34, 269 (1958).
[58] Nagata, T., and K. Kobayashi: Nature, Lond. 197, 476 (1963).
[59] Kobayashi, K.: J. Phys. Soc. Japan 15, 1352 (1960).
[60] Nagata, T., K. Akasi and T. Rikitake: Bull. Earthq. Res. Inst. 21, 276 (1943).
[61] Nagata, T., Y. Harada and K. Hirao: Bull. Earthq. Res. Inst. 23, 79 (1945).
[62] Nagata, T., K. Hirao and H. Yoshikawa: J. Geomag. Geoelectr. 1, 52 (1949).
[63] Johnson, E. A., T. Murphy and O. W. Torreson: Terr. Magn. 53, 349 (1948).
[64] Graham, J. W.: J. Geophys. Res. 54, 131 (1949).
[65] King, R. F.: Monthly Not. Roy. Astron. Soc. Geophys. Suppl. 7, 115 (1955).
[66] Griffiths, D. H.: Monthly Not. Roy. Astron. Soc. Geophys. Suppl. 7, 103 (1955).
[67] Wright, A. E.: J. Sediment. Petrol. 27, 306 (1957).
[68] Griffiths, D. H., R. F. King, A. I. Rees and A. E. Wright: Proc. Roy. Soc. Lond. Ser. A, 256, 359 (1960).
[69] Ising, G.: Ark. Geofys. 29, Ser. A, No. 5 (1943).
[70] Granar, A.: Ark. Geofys. 3, No. 1 (1958).
[71] Nagata, T.: J. Geomag. Geoelectr., Kyoto 14, 99 (1962).
[72] Vlassov, A-Ya, i G. Kovalenko: Izv. Acad. Sci. USSR. Geophys. Ser., No. 5, 639 (1962).
[73] Thellier, E.: Ann. Inst. Phys. Globe 16, 157 (1938).
[74] Brynjolfsen, A.: Adv. Physics 6, 247 (1957).
[75] Creer, K. M.: Phil. Trans. Roy. Soc. Lond., Ser. A, 250, 130 (1957).
[76] Shimizu, Y.: J. Geomag. Geoelectr., Kyoto 11, 125 (1960).
[77] Roquet, J.: Ann. Géophys. 10, 226, 282 (1954).
[78] Rimbert, F.: Thesis, Univ. of Paris, 1958.
[79] Yanovsky, B. M., L. E. Sholpo i E. S. Gorshkov: Izv. Acad. Sci. USSR, Geophys. Ser., No. 6, 719 (1962).
[80] Néel, L.: J. Phys. Radium 11, 49 (1950).
[81] Street, R., and J. C. Wooley: Proc. Phys. Soc. Lond., A 62, 562 (1949); B 63, 509 (1950); B 65, 461, 647 (1952).
[82] Thellier, E., et F. Rimbert: C. R. Acad. Sci., Paris 239, 1399 (1954).
[83] Petrova, G. N.: Ann. Géophys. 15, 60 (1950).
[84] Patton, B. J., and J. L. Finch: J. Geophys. Res. 67, 307 (1962).
[85] Néel, L.: Cahiers Phys. No. 12, 2 (1942); No. 13, 18 (1943).
[86] Lliboutry, L.: Ann. Phys., Paris 6, 731 (1951).
[87] Graham, J. W.: J. Geophys. Res. 61, 735 (1956).
[88] Graham, J. W., A. F. Buddington and J. R. Balsley: J. Geophys. Res. 62, 465 (1957).
[89] Doell, D. W.: Nature, Lond. 187, 225 (1960).
[90] Domen, H.: Bull. Fac. Education, Yamaguchi Univ. Japan 7, 41 (1957).
[91] Hall, J. M., and R. N. Neale: Nature, Lond. 188, 805 (1960).
[92] Stacey, F. D.: J. Geophys. Res. 63, 361 (1958).
[93] Néel, L.: Conférence à Newcastle, 4. Mai 1959.
[94] Nagata, T., and H. Kinoshita: Nature, Lond. 204, 1183 (1964).
[95] Nagata, T., and H. Kinoshita: J. Geomag. Geoelectr. 17, 121 (1965).
[96] Kalashinkov, A. G., i S. P. Kapitsa: Akad. Nauk USSR. 86 (3), 521 (1952).
[97] Grabovsky, M. A., i E. I. Parkhomenko: Izv. Akad. Nauk USSR., Geophys. Ser 5, 405 (1953).
[98] Kern, J. W.: J. Geophys. Res. 66, 3807 (1961).
[99] Stacey, F. D.: Phil. Mag. 7, 551 (1962).
[100] Nagata, T.: J. Geomag. Geoelectr. 18 (1966) (in press).
[101] Syono, Y.: Japanese J. Geophys. 4, 71 (1965).
[102] Nagata, T.: J. Geomag. Geoelectr. 18 (1966) (in press).

Sachverzeichnis.

(Deutsch-Englisch.)

Bei gleicher Schreibweise in beiden Sprachen sind die Worte jeweils einfach aufgeführt.

Subject Index.

(English-German.)

Where English and German spelling of a word is identical the German version is omitted.

MIX
Papier aus verantwortungsvollen Quellen
Paper from responsible sources
FSC® C105338

FSC
www.fsc.org

If you have any concerns about our products,
you can contact us on
ProductSafety@springernature.com

In case Publisher is established outside the EU,
the EU authorized representative is:
Springer Nature Customer Service Center GmbH
Europaplatz 3, 69115 Heidelberg, Germany

Printed by Libri Plureos GmbH
in Hamburg, Germany